OCEANOGRAPHY
and
MARINE BIOLOGY

AN ANNUAL REVIEW

Volume 48

T0199570

OCEANOGRAPHY and MARINE BIOLOGY

AN ANNUAL REVIEW

Volume 48

Editors

R.N. Gibson

Scottish Association for Marine Science
Scottish Marine Institute
Oban, Argyll, Scotland
robin.gibson@sams.ac.uk

R.J.A. Atkinson

University Marine Biological Station Millport
University of London
Isle of Cumbrae, Scotland
r.j.a.atkinson@millport.gla.ac.uk

J.D.M. Gordon

Scottish Association for Marine Science
Scottish Marine Institute
Oban, Argyll, Scotland
john.gordon@sams.ac.uk

Founded by Harold Barnes

CRC Press
Taylor & Francis Group
Boca Raton London New York

CRC Press is an imprint of the
Taylor & Francis Group, an **informa** business

International Standard Serial Number: 0078-3218

CRC Press
Taylor & Francis Group
6000 Broken Sound Parkway NW, Suite 300
Boca Raton, FL 33487-2742

First issued in paperback 2019

© 2010 by Taylor & Francis Group, LLC
CRC Press is an imprint of Taylor & Francis Group, an Informa business

No claim to original U.S. Government works

ISBN-13: 978-1-4398-2116-9 (hbk)
ISBN-13: 978-0-367-38410-4 (pbk)

Visit the Taylor & Francis Web site at
http://www.taylorandfrancis.com

and the CRC Press Web site at
http://www.crcpress.com

Contents

Preface

The 48th volume of this series contains five reviews written by an international array of authors. As usual, these reviews range widely in subject, taxonomic and geographical coverage. The editors welcome suggestions from potential authors for topics they consider could form the basis of appropriate future contributions. Because the annual publication schedule places constraints on the timetable for submission, evaluation and acceptance of manuscripts, potential contributors are advised to make contact with the editors at an early stage of manuscript preparation. Contact details are listed on the title page of this volume.

The editors gratefully acknowledge the willingness and speed with which authors complied with the editors' suggestions, requests and questions and the efficiency of CRC Press, especially Marsha Hecht, in ensuring the timely appearance of this volume.

It is with great regret that we report the death of Margaret Barnes in October 2009. Margaret was associated with *Oceanography and Marine Biology: An Annual Review* for 40 years and was editor from 1978 to 2002. An appreciation of her life and work is included in this volume.

Margaret Barnes DSc FRSE FIBiol
1919–2009

Managing Editor Oceanography and Marine Biology: An Annual Review 1978–1994
Editor 1995–2002

Margaret Barnes began her scientific career in 1939 soon after receiving her BSc. Her further education was interrupted by the outbreak of WWII, and she went to work in industry and spent the following 6 years using her training as a chemist to investigate colloidal graphite lubricants. During this period, she continued her education in her spare time and at the end of the war in 1945 was awarded an MSc. She had met her future husband, Harold, while at college, and they married in 1945. Harold was also a chemist but in 1943 had been seconded to the Marine Station of the Scottish Marine Biological Association (SMBA) at Millport in the Firth of Clyde, Scotland. There he was involved in the development of antifouling paints. After their marriage, Margaret joined him in Millport, and it was there that their lifelong partnership in science began. His early work was varied but he had developed an interest in barnacles during his antifouling work and began publishing on the group in the early 1950s. Margaret acted as his assistant (officially designated by the Marine Station in the restrictive practices of the SMBA of the time as an 'unpaid permanent visiting worker'), and their first joint article appeared in 1953, albeit on *Calanus finmarchicus*. Subsequently, their barnacle articles came on stream covering a wide range of topics, including general biology, morphology, distribution, reproduction and development, settlement, biochemistry, physiology and metabolism. In 1967 the SMBA opened its new laboratory in Oban and Harold and Margaret moved there from Millport to continue their barnacle studies.

Before moving, however, in 1963 Harold had started the review series *Oceanography and Marine Biology: An Annual Review*. The husband and wife team, now becoming recognised as world authorities in barnacle biology, continued their partnership in editing 'The Review', as they called it. Not content with starting one journal, and with Margaret's support, Harold followed *Oceanography and Marine Biology* 4 years later in 1967 with the *Journal of Experimental Marine Biology and Ecology* (JEMBE). The first issue of JEMBE was published in September, and it is significant that the first article in that issue was coauthored by Harold and Margaret. Margaret was an integral, experienced and tireless other half of the editorial team on both periodicals so that on his sudden and untimely death in early 1978, it was natural for her to assume the editorship of both publications and so ensure their smooth continuation. The year following Harold's death was a difficult one for Margaret but she showed little outward signs of her grief and buried herself in finishing the writing of manuscripts that had been unfinished and in the considerable amount of editorial work the two periodicals entailed. At that time *Oceanography and Marine Biology* had reached its 15th volume and Margaret's immediate task was to ensure that the manuscripts for Volume 16 were prepared to meet the deadline for publication by Aberdeen University Press (AUP) in the summer. She also had to be involved in the painful task of discussing with the publishers her future role. Fortunately, AUP was aware of her contribution to the regular appearance of past volumes and was content to allow her to continue as editor. The transition for JEMBE was not as smooth and Elsevier insisted that others join her on the editorial team. Although Margaret was not initially happy with this arrangement, she realised it was for the best because one person could not have managed the burden of editing both journals single-handed. In the late 1980s she invited colleagues to become assistant editors on *Oceanography and Marine Biology* to share the load. In 1998, and approaching her 80th birthday, she decided it was time to take a back seat in the editorial team, and Alan Ansell took over the reins as managing editor. Prior to this, however, AUP had collapsed as a result of what was known at the time as the 'Maxwell affair', and the rights were bought by University College London Press. Another change of publisher took place in 1998 (to Taylor & Francis). Margaret dealt calmly with all these changes and continued as editor until Volume 40 was published in 2002, when she decided to stand down, having retired from JEMBE in 1999, thus ending a 57-year contribution to marine science.

She was a meticulous editor with a fine eye for detail who insisted on high standards of English and spent many hours improving the texts both of authors whose first language was not English and of many whose it was. She dealt diplomatically but firmly with tardy or recalcitrant authors, and I well remember her patience when meticulously compiling the indexes for early volumes of *Oceanography and Marine Biology* from entries on scraps of paper, which were then sorted and typed by hand, a task now done in a fraction of the time by computer. She brought to both publications standards that few others could match.

Margaret travelled extensively in the course of her barnacle studies and was a founder member of the European Marine Biology Symposium (EMBS), acted as minutes secretary for the organisation for a while and in 1988 was elected for a term as president. She was intimately involved with the two symposia that were held in her hometown of Oban in 1974 and 1989 and was instigator, organiser and senior editor of the proceedings of the latter meeting. In later years when she no longer felt able to attend the symposia, I was frequently asked "How's Margaret?" and to pass on regards. At the EMBS and during her visits to numerous laboratories throughout Europe and the United States she made contact with many people the world over, and many of these contacts developed into lasting friendships. Always encouraging to young scientists, especially young women, she was an independent and determined woman largely overshadowed by her husband and her true scientific and editorial abilities only became apparent after his death. She was also a gentle, modest, courteous and charming person, a good listener, and she had a terrific sense of humour. In her younger days she was very active as a keen cross-country skier, mountaineer and long-term member of the Austrian Alpine Club. She remained sprightly until her death, working in her garden throughout the year, and we had numerous conversations about hill walking and the state of her crops. However, I suspect that many will particularly remember her for her coffee mornings and dinner parties. They were deservedly famous for their wide-ranging and relaxed conversation and their cuisine, and it gave her great pleasure to entertain students and visiting scientists of all ages and nationalities at her home overlooking the sea.

Margaret died peacefully on 30 October 2009 after an accident while working in her garden. She will be greatly missed by all who were privileged to call her friend or colleague.

Robin N. Gibson

Oceanography and Marine Biology: An Annual Review, 2010, **48**, 1-42
© R. N. Gibson, R. J. A. Atkinson, and J. D. M. Gordon, Editors
Taylor & Francis

TOWARD ECOSYSTEM-BASED MANAGEMENT OF MARINE MACROALGAE—THE BULL KELP, *NEREOCYSTIS LUETKEANA*

YURI P. SPRINGER[1], CYNTHIA G. HAYS[2], MARK H. CARR[1,3,4] & MEGAN R. MACKEY[3]

[1]*Department of Ecology and Evolutionary Biology, University of California Santa Cruz, Long Marine Laboratory, 100 Shaffer Road, Santa Cruz, CA 95060, USA*
E-mail: yurispringer@gmail.com
[2]*Bodega Marine Laboratory, University of California Davis, P.O. Box 247, Bodega Bay, CA 94923, USA*
E-mail: cghays@ucdavis.edu
[3]*Pacific Marine Conservation Council, 4189 SE Division Street, Portland, OR 97202, USA*
E-mail: carr@biology.ucsc.edu, megan_mackey@speakeasy.net
[4]*Corresponding author*

Abstract Ecosystem-based management is predicated on the multifaceted and interconnected nature of biological communities and of human impacts on them. Species targeted by humans for extraction can have multiple ecological functions and provide societies with a variety of services, and management practices must recognize, accommodate, and balance these diverse values. Similarly, multiple human activities can affect biological resources, and the separate and interactive effects of these activities must be understood to develop effective management plans. Species of large brown algae in the order Laminariales (kelps) are prominent members of shallow subtidal marine communities associated with temperate coastlines worldwide. They provide a diversity of ecosystem services, perhaps most notably the fuelling of primary production and detritus-based food webs and the creation of biogenic habitat that increases local species diversity and abundance. Species of kelp have also been collected for a variety of purposes throughout the history of human habitation of these coastlines. The bull kelp, *Nereocystis luetkeana*, provides a clear example of how the development of sustainable harvest policies depends critically on an understanding of the morphological, physiological, life-history, demographic, and ecological traits of a species. However, for *Nereocystis* as well as many other marine species, critical biological data are lacking. This review summarizes current knowledge of bull kelp biology, ecological functions and services, and past and ongoing management practices and concludes by recommending research directions for moving toward an ecosystem-based approach to managing this and similarly important kelps in shallow temperate rocky reef ecosystems.

Introduction

Why the interest in ecology and ecosystem-based management of Nereocystis?

Among the many tenets of ecosystem-based management (EBM) of marine resources, two are central to the goal of a more comprehensive approach to resource management. First, EBM recognizes that species targeted for extraction can have multiple ecological functions and provide society with a variety of ecosystem services. Management practices therefore should strive to accommodate these diverse values (Field et al. 2006, Francis et al. 2007, Marasco et al. 2007). Second, EBM recognizes that multiple and diverse human activities, from local fisheries to global climate change, affect the state and sustainability of marine resources and the ecosystems that support them, and that a thorough understanding of both the independent and interactive effects of these activities must underpin management plans for these to be effective (Leslie & McLeod 2007, Levin & Lubchenco 2008, McLeod & Leslie 2009). As management goals move from maximizing the sustainable use of marine resources along a single axis (e.g., single species-based sustainable fishery yields) to a more comprehensive balancing of *multiple* services with each other in a manner that ensures the sustainability of those services and their associated ecosystems, knowledge of the ecological functions and services of species and of how human activities influence them will be critical. Models for both EBM and strategies to move toward EBM must recognize species that provide multiple, well-characterized ecological functions and services and that are known to be influenced by a variety of human activities.

Species of large brown macroalgae of the order Laminariales, commonly referred to as kelps, are a conspicuous component of coastal rocky reef habitats in temperate oceans throughout the world. Kelps have been harvested throughout the history of human habitation of temperate coastlines for a variety of purposes, including human consumption, the production of pharmaceuticals, and as food for commercial mariculture. However, kelps also provide a diversity of ecosystem services to the biological communities of which they are part. As such, the consequences of human impacts on kelps are not limited to the direct effects on kelp populations themselves, but also influence indirectly the many species that depend on or benefit from the presence of these macroalgae in nearshore habitats.

Along the western coast of North America, two genera, the giant kelp *Macrocystis* spp. (hereafter *Macrocystis*), and the bull kelp *Nereocystis luetkeana* (hereafter *Nereocystis*), form extensive forests in shallow (<30-m depth) rocky habitats. Because of their fast growth rate and large stature, these algae are thought to contribute markedly both to the productivity of shallow coastal marine ecosystems and as habitat for a diversity of fishes and invertebrates (Foster & Schiel 1985, Graham 2004, Graham et al. 2008). Both of these fundamental ecosystem functions of kelps are realized not only by those species that reside in kelp forests throughout their lives (i.e., kelp forest residents) but also by species that use these habitats as foraging grounds (e.g., shorebirds, sea otters) and nurseries (particularly fishes) because of the enhanced growth and survival provided to them by the productivity and structural refuge created by kelp (see review by Carr & Syms 2006). Many of the species that utilize kelp habitat have been strongly affected by overfishing and are themselves the focus of conservation efforts (e.g., abalone, rockfishes, sea otters). In addition to these effects on primary and secondary productivity in nearshore habitats, the physical barrier created by kelp forests along the shoreline dampens ocean waves, thereby reducing coastal erosion (Lovas & Torum 2001, Ronnback et al. 2007). Kelps also represent important biological links between marine ecosystems. The biomass and nutrients they produce, in the forms of detritus or entire detached plants, are exported by storms to sandy beaches and submarine canyons, where they fuel food webs in the absence of other sources of primary production (Kim 1992, Vetter 1995, Harrold et al. 1998). Floating kelp rafts may also serve as habitat for larval and juvenile fishes and invertebrates, effectively transporting them among spatially isolated local populations of adults (Kingsford 1992, Kokita & Omori 1998, Hobday 2000, Thiel & Gutow 2005). Furthermore, kelps are of great social, cultural, and economic importance because of the many human activities they foster (e.g., recreational fishing, scuba diving, bird watching, kayaking); tourism and recreation are included in one of the fastest-growing sectors of California's economy today (Kildow & Colgan 2005). Separately and in combination, the direct and indirect benefits that kelp forests provide can translate into socioeconomic values of extreme importance to local coastal communities.

Due to their close proximity to shore, kelp forests are subject to deleterious anthropogenic impacts that can impair the functions and services they provide. In addition to direct extraction, kelps can be exposed to coastal pollution in the form of nutrient discharge from urban and agricultural sources and thermal pollution associated with cooling water outflow from coastal power plants. Increases in turbidity and rates of sedimentation associated with all of these activities impair photosynthesis (i.e., growth and survival of adult plants) and smother reproductive stages and spores, preventing reproduction and germination. Beyond these localized and regional threats, kelp forests are vulnerable to environmental modification caused by global climate change. The existence and tremendous productivity of these forests rely on the upwelling of deep offshore nutrient-rich waters. This upwelling process is driven by coastal winds that move surface waters offshore, driving their replacement by the deeper nutrient-rich waters. As atmospheric conditions fluctuate in response to large-scale climate trends, changes in the timing, location, and intensity of coastal winds alter the distribution and magnitude of upwelling, thereby changing the environmental conditions required to sustain kelp forests. Large storms associated with El Niño are major causes of mortality and the loss of entire kelp forests (Tegner & Dayton 1987), and increases in the frequency, duration, and strength of El Niño in recent years may be a direct consequence of concurrent regional climate changes (Trenberth & Hoar 1996).

The direct and indirect impacts of kelp extraction depend very much on the species and means by which it is removed. Historically, extraction has been focused on the giant kelp *Macrocystis*, primarily by the pharmaceutical industry. Specially designed harvesting vessels were used to remove large swathes across forests from the upper 2 m of the canopy. The direct impact on the forests is considered minimal because the canopy is often replaced rapidly by the growth of fronds from the base of the plants. Moreover, the alga is perennial, and the reproductive tissues are located at the base of the plant and remain intact during and subsequent to harvesting. Thus, the algae are able to reproduce, and associated forests to persist, in the face of large-scale mechanical extraction. However, the indirect effects on the fishes and invertebrates that use the forest canopy as nursery habitat, and on the many species that require the flux of kelp blades from the canopy to the reef habitat below to fuel a detritus-based food web (akin to litter fall in terrestrial forests), have not been rigorously investigated.

The extraction of *Nereocystis* is a more recent development, fuelled by the demands of abalone mariculture and human consumption. Although relatively smaller in volume and geographic extent, the harvest of *Nereocystis* is problematic. Extraction is primarily by hand from a boat and, like giant kelp, limited to the upper 2 m of the forest canopy. However, the source of buoyancy that keeps *Nereocystis* plants upright, along with the alga's reproductive organs, is located at the top of the plant and is often removed during harvest. In the absence of this source of buoyancy and associated photosynthetic tissue, individual plants may sink to the bottom and die. Furthermore, because *Nereocystis* is an annual species, removal of the upper portion of plants prior to reproduction can potentially preclude the production of subsequent generations. The spores of *Nereocystis* are thought to move very short distances (tens of metres) on average; thus, local impairment of reproduction might eventually result in the disappearance of a forest, although local recruitment could be subsidized by input of spores from other populations delivered by either drifting reproductive sporophytes or abscised sori. In addition, the presence of dormant spores produced by previous generations of *Nereocystis* could potentially reseed local populations that have been depleted by harvesting. However, because there are few data on the dispersal potential and dormancy durations of spores, these mechanisms of local 'rescue' cannot at present be incorporated into management plans in a quantitative manner.

Approach, scope of synthesis, and products

The EBM of coastal marine resources is based, in part, on scientific understanding of the broad (i.e., ecosystemwide) consequences of human uses of the coastal environment, including resource extraction and degradation of habitats. To effectively manage these resources, a clear understanding of the potential threats and consequences of human activities to the resource and the ecosystem is essential. To contribute to this understanding, this report synthesizes the state of knowledge of (1) the ecology of *Nereocystis* and its role in coastal ecosystems, (2) the past and present human uses of and threats to this species and, by extension, the coastal ecosystem, and (3) the past and present approaches to managing this resource. This synthesis identifies gaps in current knowledge of *Nereocystis* biology and ecology and recommends priority research needs to inform management of the human activities that impinge on this species and its ecosystem functions and services. The scope of this review spans studies and management programs from Alaska to central California and includes data from both peer-reviewed scientific journals and non-peer-reviewed sources (e.g., reports produced by governmental agencies and non-governmental organizations [NGOs]).

Review and synthesis of the ecology of *Nereocystis luetkeana*

Species description and geographic distribution

Nereocystis is a conspicuous brown macroalga in nearshore environments along the Pacific Coast of North America (Figure 1). The blades of the alga (30–60 on an adult sporophyte, each up to 4 m long) are held near the surface of the water by a gas-filled, spherical pneumatocyst at the end of a long, slim stipe (~1/3 inch in diameter), attached to the substratum with a hapterous holdfast (Figure 2). Up to one-third of the upper portion of the stipe is hollow, and it is extremely elastic; when exposed to wave force it can stretch more than 38% (Koehl & Wainwright 1977). Because all of an individual's blades are at or near the water surface, the canopy provides virtually all substrata for photosynthesis and nutrient uptake, and photosynthate is subsequently translocated throughout the rest of the thallus via sieve elements in the medulla (Nicholson & Briggs 1972, Schmitz & Srivastava 1976).

Figure 1   A stand of *Nereocystis* on a shallow rocky reef off the coast of central California. Schooling surf perch (Embiotocidae) are visible at the bottom right. (Photograph courtesy of Steve Clabuesch.)

Figure 2 Morphology of *Nereocystis* plants. *Bulb* refers to the gas-filled pneumatocyst. (Diagram from G.M. Smith, *Marine Algae of the Monterey Peninsula*, copyright © 1944 by the Board of Trustees of the Leland Stanford Jr. University, renewed 1972. Photograph of young plants emerging from a sparse cover of the understory kelp *Pterygophora californica* courtesy of Steve Clabuesch.)

Nereocystis forms extensive beds from Point Conception, California, to Unimak Island, Alaska (Figure 3; Druehl 1970, Abbott & Hollenberg 1976, Miller & Estes 1989) on bedrock reefs and boulder fields 3 to 20 m deep (Nicholson 1970, Vadas 1972). Across its geographic range, the relative functional importance of *Nereocystis* as a source of surface canopy varies with the occurrence of other species of canopy-forming kelps. In some regions of its range, it is the sole or predominant canopy-forming kelp, while in others it co-occurs with either dragon kelp *Eualaria fistulosa* (formerly *Alaria*) or species of giant kelp *Macrocystis pyrifera* or *M. integrifolia*. The relative abundance of these species varies with respect to both latitude and exposure to ocean swells (Figure 3). In the more protected southern portion of the range, south of Año Nuevo Island (Santa Cruz County, California), *Nereocystis* occurs together with the predominant *Macrocystis*, sometimes forming mixed beds (Foster 1982, Dayton et al. 1984, Dayton 1985, Foster & Schiel 1985, Harrold et al. 1998). From Año Nuevo Island to Alaska, *Nereocystis* is often the sole or predominant canopy-forming kelp on both exposed and protected shores (e.g., Strait of Georgia and Puget Sound, Washington). *Nereocystis* and *Macrocystis* form mixed stands in British Columbia (e.g., western and northern Vancouver Island). *Nereocystis* is the predominant canopy-forming species in south-eastern Alaska, although *Macrocystis* is predominant in some locations along the outer coast (S. Lindstrom personal communication). At the northern end of its range, from north-western Prince of Wales Island to Unimak Island, *Nereocystis* and *Eualaria fistulosa* co-occur regionally, and local beds sometimes alternate between these species through time (B. Konar and S. Lindstrom personal communication). All three kelps co-occur in a few small regions: north-western Prince of Wales Island and Kodiak Islands (M. Norris personal communication). Unattached adult plants (i.e., their holdfasts dislodged from the substratum) have also been found rafting in waters farther south in California (Bushing 1994) and in the Commander Islands in Russia, the westernmost extension of the Aleutian Islands (Selivanova & Zhigadlova 1997).

Evolutionary history

Seaweeds are a polyphyletic group of organisms with varied evolutionary histories. *Nereocystis* is a brown alga (division Heterokontophyta) in the order Laminariales (the true kelps). There are at least 100 species of kelps worldwide (Guiry & Guiry 2010), and this group includes other common

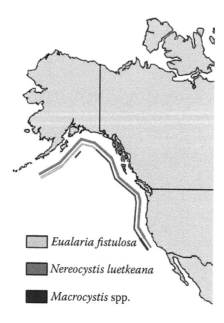

Eualaria fistulosa

Nereocystis luetkeana

Macrocystis spp.

Figure 3 (See also Colour Figure 3 in the insert following page 212.) Geographic distribution of *Nereocystis luetkeana* indicating areas of co-occurrence with two other surface canopy-forming kelps: giant kelp *Macrocystis* spp. and *Eualaria* (formally *Alaria*). Distributional patterns based on personal communications with M. Foster, M. Graham, B. Konar, and S. Lindstrom. Line width proportional to levels of relative abundance across the range of the species.

species, such as *Macrocystis* and *Postelsia* (sea palm). *Nereocystis* is a monotypic genus; traditional taxonomy, largely based on sporophyte morphology, places it within the family Lessoniaceae (Setchell & Gardner 1925). With the advent and increasing accessibility of molecular techniques, the evolutionary relationships among kelp taxa, particularly among the three 'derived' families (Alariaceae, Lessoniaceae, and Laminariaceae) have been the topic of increased scrutiny and debate (Saunders & Druehl 1991, 1993, Coyer et al. 2001). The most comprehensive genetic data to date suggest that *Nereocystis* should be grouped (along with *Macrocystis, Postelsia,* and *Pelagophycus*) in a revised Laminariaceae Postels et Ruprecht (Lane et al. 2006). Based on the results of crossing experiments (Lewis & Neushul 1995) and genetic analyses (Lane et al. 2006), *Nereocystis* is thought to be most closely related to *Postelsia*.

There has been some suggestion that *Nereocystis* will hybridize in the laboratory with *Macrocystis* (Lewis & Neushul 1995) in spite of differences in chromosome number (Sanbonsuga & Neushul 1978). However, this is likely to be an artifact of the laboratory and reflective of parthenogenesis or male apogamy rather than actual hybridization (Druehl et al. 2005). No hybrids between *Nereocystis* and *Macrocystis* have ever been found in the field.

Life history

Like all kelp species, *Nereocystis* exhibits alternation of generations between a large, diploid sporophyte stage and a microscopic haploid gametophyte stage (Figure 4). Young sporophytes typically appear in the early spring and grow to canopy height (10 to 17 m) by midsummer. Individuals grow to roughly match the depth at which they settle (i.e., until the pneumatocyst reaches the water surface); this appears to be regulated by a phytochrome-mediated response, such that stipe elongation is inhibited by red wavelengths of light (Duncan & Foreman 1980). *Nereocystis* sporophytes can grow at extremely high rates, up to 6 cm day^{-1} (Scagel 1947). Maximum photosynthesis occurs in

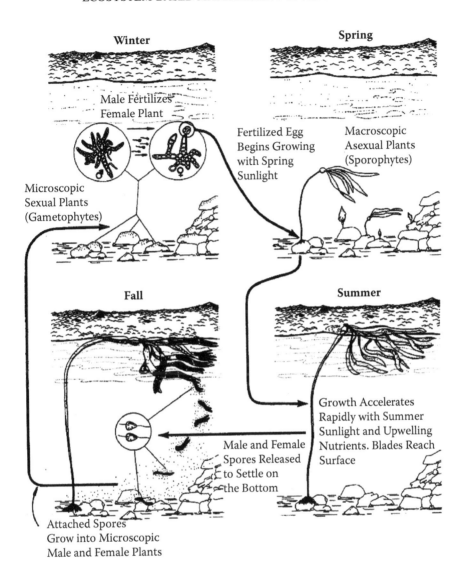

Figure 4 Diagram of the life cycle of *Nereocystis*. (Reproduced from a 1982 report by permission from TERA Corporation, now Tenera Environmental Inc., San Louis Obispo, California.)

summer and early fall, and mortality of *Nereocystis* sporophytes reaches a maximum in the winter, primarily due to dislodgement by winter storms. Lower kelp densities after a storm can also cause surviving individuals to experience increased grazing pressure from sea urchins (Dayton et al. 1992). Each sporophyte produces a single stipe in its lifetime and cannot regrow from its holdfast once the upper stipe is destroyed (Nicholson 1970). Thus, *Nereocystis* is essentially an annual species, although in some populations individuals that are produced late in the season may successfully overwinter and survive a second year (Chenelot et al. 2001). This biennial life history appears to be more common in shallow water populations or protected locations where wave stress is not as great as on the open coast.

Nereocystis sporophytes produce biflagellate haploid spores through reduction division on fertile patches of blades called sori. Sori may be more than 30 cm long and are produced near the proximal end of the blade (Scagel 1947). The maturity of sori therefore increases with increasing distance toward the distal edge of the blade (Nicholson 1970, Walker & Bisalputra 1975, Walker 1980a).

Nereocystis possesses a mechanism for spore dispersal that is rare among kelps: Sori that are releasing (or are about to release) spores abscise from the blade and are released into the water column. Abscission of sori results from a chain of cellular events causing structural weakening (e.g., necrosis of specific tissue layers and dissolution of the cuticle covering the sporangia) in conjunction with the physical force of water motion (Walker 1980b). Within 1 to 4 h of abscission, virtually all spores are released from the sorus (Nicholson 1970, Walker 1980b, Amsler & Neushul 1989).

Spores that successfully settle germinate into microscopic sessile gametophytes, which are uniserate branched filaments. Compared with the conspicuous sporophyte stage, little is known about the ecology of kelp gametophytes. For example, it is unclear how long *Nereocystis* gametophytes persist in the field. There is a distinct seasonality to the reappearance of sporophytes, so it is likely that the production of gametes requires an environmental cue. After 2 to 3 months and exposure to suitable light and nutrients, gametophytes produce oogamous gametes. Vadas (1972) showed that under limited light conditions in the laboratory gametophytes may survive and grow vegetatively for over a year before a change in conditions allows the production of gametes or the growth of very young sporophytes. Evidence of these light-dependent processes suggests that *Nereocystis* gametophytes may act in a manner analogous to a terrestrial seed bank (Santelices 1990, Edwards 2000). Alternatively, seasonality may be imposed by larger-scale phenomena such as strong winter storms and the abiotic environmental changes that accompany them.

Kelp eggs release sexual pheromones that attract sperm (Maier et al. 1987), but the spatial scale over which this mechanism promotes successful syngamy is very low. The density of settling spores, and resulting proximity of male and female gametophytes, is thus critical to fertilization and recruitment success. In giant kelp, spore density must exceed 1–10 spores mm^{-2} for successful recruitment to occur (Reed 1990, Reed et al. 1991). Critical spore density for *Nereocystis* recruitment is not known but is likely to be similar in scale. The recent development of a species-specific method based on polymerase chain reaction (PCR) for detecting microscopic stages (zoospores, gametophyes, and microscopic sporophytes) of *Nereocystis* holds great promise for revealing patterns of spatial dispersion and mortality associated with these phases of the life cycle of the kelp (Fox & Swanson 2007).

Population ecology

Dispersal and population genetic structure

Dispersal of kelp gametes is thought to be negligible. Extruded eggs typically remain attached to the ruptured oogonium on the female gametophyte, and the pheromones that kelp eggs produce (which induce gamete release from male gametophytes and attract sperm to the egg) are only effective when gametes are within about 1 mm of each other (Muller 1981, Maier & Muller 1986). Thus, there are three possible points in the life history of *Nereocystis* when dispersal may occur: as spores, as intact sporophytes, and as detached sori. Detached sori and intact dislodged sporophytes have the potential for long-distance dispersal and gene flow in this species. However, to our knowledge, the relative frequency and scale of dispersal by this mechanism has not been measured.

Nereocystis sporophytes produce an enormous quantity of spores; an average of 2.3×10^5 spores cm^{-2} of sori min^{-1} during initial release has been estimated (Amsler & Neushul 1989), and release may be up to six times faster than those associated with *Macrocystis* (Collins et al. 2000a). Individual plants produce sori on different blades at the same time, but sori mature and are released somewhat synchronously, in pulses that occur every 4–6 days (Amsler & Neushul 1989). Spore production and release occur with a monthly and daily periodicity that varies with geographic location. In British Columbia, *Nereocystis* are thought to release sori only at the beginning of spring tides (Walker 1980b), but near the southern range limit in central California this monthly pattern appears weak or non-existent (Amsler & Neushul 1989). Sori abscission does have a distinct diel pattern in central California. Most abscission occurs in the hours immediately before and after dawn (Amsler & Neushul 1989). Like other kelps (e.g., *Macrocystis*, *Laminaria farlowii*), *Nereocystis* spores are capable of photosynthesis, and although net photosynthesis is low (Watson & Casper 1984), spores should be able to contribute to their own carbon needs. Dawn release may thus reflect an adaptation to maximize photosynthetic opportunity (e.g., to increase viability in the plankton or maximize energy reserves for early germination and growth).

If spores are released from the intact blade or from detached sori drifting through the water column, this mechanism should result in broader dispersal of spores and an increase in the total area over which siblings are distributed (Strathmann 1974, Amsler & Neushul 1989). However, many (or most) spores are likely still retained in the sorus when it arrives at the substratum, which would both concentrate a large portion of siblings spatially and may ensure that most progeny remain near the parent plant (Amsler & Neushul 1989). Kelp spores (e.g., *Macrocystis*) can remain viable in the water column for several days (Reed et al. 1992, Brzezinski et al. 1993) and may be dispersed over long distances by ocean currents (Reed et al. 1988, Norton 1992). In Kachemak Bay, Alaska, *Nereocystis* is only found in the outer bay, so that sporophyte distribution is thought to be driven by estuarine current flow, which acts to prevent dispersal of spores into the inner bay (Schoch & Chenelot 2004).

A population genetic approach is necessary to resolve the spatial scale of population connectivity and would also provide insight into the relative importance of the three potential mechanisms of dispersal. Currently, no published studies of population genetic structure in *Nereocystis* are available.

Spatial and temporal variation in population dynamics

Nereocystis shows high spatial and temporal variability in distribution and abundance patterns, consistent with its annual life history and tendency to colonize recently disturbed areas. For example, in a study of the effects of harvest on *Nereocystis* dynamics, Foreman (1984) found greater interannual variability in abundance in 1-hectare control plots than in plots that had been harvested (see section on historical and current stock assessments beginning on p. 19 for descriptions of available data on spatial and temporal variation in *Nereocystis* cover/productivity across the range of the species).

The reproductive phenology of *Nereocystis* also varies spatially, and it seems that sporophyte recruitment and spore production occur earlier in more northern populations. Burge and Schultz (1973) studied *Nereocystis* in Diablo Cove, California, and documented initiation of new sporophytes from late March through August. Sori were present on blades before they reached the water surface, and complete abscission of sori occurred over a long time: as early as June and as late as March of the following year. More than 1600 km to the north, in Tacoma Narrows, Washington, *Nereocystis* appears to be a strict annual. Sporophytes recruit slightly earlier and more synchronously (early March through June), with peak spore release occurring in August (Maxell & Miller 1996). In the westernmost population in the current distribution of the species (Umnak Island, Alaska), Miller and Estes (1989) observed that sporophytes in July showed characteristics (i.e., size, maturity, and epiphyte cover) that typically reflect individual condition in fall and winter. There was no evidence of a second cohort of smaller individuals, so it seems unlikely that all individuals were second-year plants that had successfully overwintered; earlier recruitment or faster growth of sporophytes provides a more plausible explanation.

Leaman (1980) quantified seasonal variation in sporophyte fertility (number of fertile blades, average sori number and area) in Barkley Sound, British Columbia, from June through October and found that peak fertility occurred in early July, with a smaller peak in September and October. No comparable data on seasonal variation in spore production are available for California populations (according to Collins et al. 2000a).

Abiotic and biotic factors limiting distribution and abundance

Physical factors known to influence the distribution and abundance of subtidal kelp species include irradiance, substratum, sedimentation, nutrient levels, temperature, water motion, and salinity. As pointed out by Dayton (1985), these effects are often difficult to characterize because they seldom act in isolation (e.g., increased water motion may act to increase water turbidity, decreasing irradiance). Moreover, the interactive effects of these factors (or their interaction with biotic ones) may be complex and non-intuitive.

Light   Studies of *Nereocystis* in culture suggested that the total quantity of light (photoperiod × intensity) is the single most important factor in the development of both gametophytes and young sporophytes (Vadas 1972). Furthermore, the range of conditions under which vegetative growth is maintained is broader than the conditions necessary for reproduction. In laboratory cultures, gametophytes did not reach sexual maturity under light levels <15 foot candles. Given that light availability is typically well below this threshold in mature kelp forests, it is likely that *Nereocystis* recruitment is light limited in established kelp stands (Vadas 1972).

Temperature   Upper thermal limits are often a phylogenetically conserved trait, and thermal tolerance is thought to constrain the southern range limit of many algal species, including *Nereocystis* (Luning & Freshwater 1988). The decline of *Nereocystis* near warm water discharge from the Diablo Canyon power plant (Pacific Gas and Electric Company 1987) supports this idea. Culture studies with *Nereocystis* showed that the thermal conditions that allow sporophyte and gametophyte reproduction range from 3°C to 17°C (Vadas 1972). Much of the Aleutian Islands chain is influenced by the Kuroshio Current, so it seems unlikely that thermal constraints alone could be responsible for the sharp northern/western boundary observed at Umnak Island. Alternatively, light limitation driven by the high fog cover characteristic of the western islands, especially in the summer, may act to prevent spread (Miller & Estes 1989).

Nutrient levels   Both spatial and temporal variation in nutrient availability can strongly influence kelp productivity (Dawson 1966, Rosell & Srivastava 1984). The seasonal growth pattern of *Nereocystis* is such that initial growth occurs in late winter and early spring, when organic and inorganic nitrogen levels are relatively high. During the summer months, C:N ratios in *Nereocystis* peak, generally as a result of reductions in the availability and assimilation of nitrogen (Rosell & Srivastava 1985). Like other kelps, *Nereocystis* displays simultaneous uptake of both nitrate and ammonium but shows a preference for nitrate. Ahn et al. (1998) found that nitrate uptake by *Nereocystis* increased linearly with nitrate availability, up to the highest concentration tested (30 µM). In contrast, ammonium uptake rates reached a plateau at availabilities >10 µM. In addition to macronutrients and micronutrients known to influence algal productivity in general (e.g., phosphate, potassium, calcium, magnesium), *Nereocystis* has the capacity to take up other metallic and non-metallic compounds from seawater (Whyte & Englar 1980a,b). What role they may play in *Nereocystis* growth is unknown.

Wave action   There is a complex relationship between any benthic alga and the hydrodynamics of its environment. Hydrodynamics can directly affect individual fitness through multiple avenues, such as nutrient uptake rates and gas exchange, direct effects on reproduction and recruitment, as well as flow-induced mortality via dislodgement (e.g., wave action during winter storms is thought to be the main source of mortality for sporophytes, but see Duggins et. al. 2001). *Nereocystis* is relatively resistant to dislodgement compared with other large kelps and is typically found in nearshore habitats characterized by high wave action. This distributional characteristic is especially evident in the southern portion of its geographic range, where it frequently co-occurs with giant kelp. In the northern part of its range, *Nereocystis* survival and distribution show a non-linear relationship with flow, driven by an interaction with herbivory (Duggins et al. 2001). Herbivore abundance typically shows an inverse relationship with wave exposure, and damage by herbivores can compromise the structural integrity of the *Nereocystis* stipe and holdfast. This interaction between physical and biotic stresses is thought to be the reason why northern *Nereocystis* populations are seldom found in habitats with intermediate flow energy; that is, the combination of both high grazing pressure and periodic high drag forces exerted on herbivore-damaged kelp may result in a sharp increase in sporophyte mortality rates.

 Nereocystis is a striking exception to the general rule that wave-swept organisms tend to be smaller than sister taxa that occur in calmer waters. This intriguing observation has motivated empirical investigations, beginning in the mid 1970s, that produced a series of highly technical studies of the biomechanics of *Nereocystis* morphology. For example, see Koehl & Wainwright (1977) and Johnson & Koehl (1994) for a consideration of unidirectional flow and Denny et al. (1997) for an analysis of dynamic flow effects. *Nereocystis* also shows dramatic phenotypic plasticity in frond morphology in response to flow. At relatively calm sites, *Nereocystis* produce blades that are wide and undulate with wavy margins, whereas in more exposed habitats blades are narrow, flat, and strap-like. Work involving both laboratory and field transplant experiments demonstrated that this morphological variation is caused by water flow and associated hydrodynamic drag (Koehl et al. 2008). Physiologically, the ruffled blade morph is produced when longitudinal growth along the edge of the blade exceeds the rate of longitudinal growth along the blade midline. The two morphs appear to arise from a trade-off between dislodgement risk and photosynthetic efficiency. The fluttering of ruffled blades may reduce self-shading and enhance interception of light (by orienting perpendicular to current flow) (Koehl & Alberte 1988, Hurd et al. 1997), and water turbulence generated at the blade surface may act to enhance nutrient uptake (Hurd & Stevens 1997), but greater drag will increase the risk of breakage under high flow stress.

Grazers   Major grazers of *Nereocystis* include red and purple sea urchins (*Strongylocentrotus franciscanus* and *S. purpuratus,* respectively) and red abalone (*Haliotis rufescens*), as well as limpets (e.g., *Collisella pelta*), snails (e.g., *Tegula* spp., *Callistoma* spp.), and various crustaceans (Cox 1962, Nicholson & Briggs 1972, Burge & Schultz 1973). Sea urchin grazing in particular is well known to exert a powerful influence on kelp forest dynamics, and many studies have documented this effect (e.g., Paine & Vadas 1969, Duggins 1980, Pace 1981). When sea urchins are removed from the system, the presence and density of *Nereocystis* sporophytes can increase dramatically. Breen et al. (1976) found that the density and area of *Nereocystis* beds increased following removal of red sea urchins. In a study by Pace (1981) performed in Barkley Sound, British Columbia, *Nereocystis* density increased from 4.6 plants m^{-2} to 13.9 plants m^{-2} in a single year following experimental removal of red sea urchins. Work by Duggins (1980) showed that in the year following sea urchin removal in Torch Bay, Alaska, kelp biomass increased from zero standing crop to roughly 60 kg wet mass m^{-2}, most of which was *Nereocystis*. Increases in the size and density of *Nereocystis* beds near Fort Bragg, California, between 1985 and 1988 were correlated with the commercial harvest of roughly 32,500 t of red sea urchins from areas off the coast of Mendocino and Sonoma Counties (Kalvass et al. 2001). Several studies have also demonstrated that the seaward limit of *Nereocystis* beds may be set by sea urchin grazing (Breen et al. 1976, Pearse & Hines 1979). The capacity of the species for rapid growth under high light conditions permits fast recovery by *Nereocystis* sporophytes when the canopy opens up due to grazing or other disturbance. For example, Foreman (1977a) showed that *Nereocystis* underwent the largest variation in biomass of any algal species over the course of recovery from grazing by green urchins in the Strait of Georgia, British Columbia, and dominated the algal community for a period of 4 yr before declining toward predisturbance levels. A study by Chenelot & Konar (2007) that examined the effects of grazing by the mollusc *Lacuna vincta* on different age classes of *Nereocystis* in Kachemak Bay, Alaska, found that the snail fed significantly more on tissue of juvenile than adult plants, and that snail densities in nature can exceed 1500 m^{-1} on juvenile blades. This apparent preference for young plants, coupled with observation of high but spatially patchy snail densities in the field, led the authors to conclude that grazing by *L. vincta* has the potential to strongly influence the dynamics of local *Nereocystis* populations.

In addition to direct negative effects of grazing, the presence of grazers can have important interactive effects with other biotic and abiotic factors. For example, damage by grazers can weaken the structural integrity of the *Nereocystis* stipe and holdfast and increase an individual plant's vulnerability to wave action. Koehl & Wainwright (1977) reported that 90% of detached single individuals had broken at a flaw in the stipe. While this damage appeared to be caused by herbivore grazing, no conclusive evidence supporting this anecdotal connection could be found. Herbivory can also alter the competitive hierarchy among kelps and other macroalgae (Paine 2002), and the presence of herbivores may positively affect *Nereocystis* by decreasing competition with other algal species. In the absence of herbivory, species of understory and turf algae such as foliose reds (*Botryoglossum farlowianum, Polyneura latissima*) and midwater canopy species (*Laminaria* spp., *Pterygophora californica, Eisenia arborea*) can reach high levels of abundance and prevent the recruitment of *Nereocystis* through competition for primary space and overshadowing (discussed in Collins et al. 2000b). Such effects have been observed in association with a number of different mechanisms, such as after mass disease-related mortality of sea urchins in Carmel, California (Pearse & Hines 1979), the introduction of sea otters (predators of urchins and abalone) in Torch Bay and Surge Bay, Alaska, and Diablo Cove, California (Duggins 1980, Gotshall et al. 1984, Estes & Duggins 1995), and the commercial harvest of red sea urchins near Fort Bragg, California (Collins et al. 2000b). The beneficial effects of sea urchin grazing for *Nereocystis* may be particularly important in areas of heavy scour, and unstable substrata where the rapidly colonizing red algae that potentially outcompete *Nereocystis* are often the predominant component of stands of macroalgae (Duggins 1980). Thus, the net effects of herbivory on *Nereocystis* beds will be driven by both the abundance and feeding preferences of grazers and the nature of competitive interactions between *Nereocystis* and other species of algae with which it co-occurs at a given location. Furthermore, although grazing is clearly an important driver of *Nereocystis* population dynamics, the effects of different grazer species on per capita rates of *Nereocystis* growth, survival, and reproduction are largely unknown. Because of their size, kelp gametophytes may be vulnerable to mortality from grazers, but this interaction has not been examined quantitatively.

Competition As alluded to in this discussion, competition is another major driver of *Nereocystis* distribution, both within and across sites. *Nereocystis* is generally thought to be an opportunistic kelp that can rapidly colonize disturbed sites but is usually outcompeted by competitively perennial species in the absence of disturbance (Dayton et al. 1984, Dayton 1985). Where bull and giant kelp co-occur, *Nereocystis* is typically only found in more exposed areas where *Macrocystis* abundance is low and understory kelps are sparse (Figure 2). *Nereocystis* also displays temporal dynamics that are consistent with an r-selected species (e.g., rapid population growth in response to disturbance and increased light availability, eventual replacement by other species; Foreman 1977b).

Epiphytes A wide variety of different epiphytic algae and invertebrates colonize *Nereocystis*; over 50 species of epiphytic algae have been documented on *Nereocystis* blades and stipes, often showing distinct patterns of vertical distribution (Markham 1969). Common algal epiphytes include filamentous species of *Ulva, Enteromorpha,* and *Antithamnion* and the foliose red alga *Porphyra nereocystis*. As the species epithet implies, *P. nereocystis* is a common epiphyte on the stipe of *Nereocystis* (and occasionally other laminarian kelps) and displays a life history that synchronizes reproduction and recruitment with its host (Dickson & Waaland 1984, 1985). Epiphyte cover on *Nereocystis* sporophytes increases over the summer and through fall and winter and can cause strong reduction in photosynthesis through direct shading of blades. At high levels of epiphyte cover, this added weight may overcome the buoyancy of the pneumatocyst and cause the entire alga to sink to depths where light intensity is lower and blades are more likely to be in direct contact with grazers (Collins et al. 2000a). Epiphyte load leads to increased tattering of blades and may increase the likelihood of complete detachment during high wave forces due to increased drag (Foreman 1970). Some blade tissue may also be inadvertently lost to fish feeding on epiphytic plants or animals (e.g., Hobson & Chess 1988). No estimates of either sporophyte mortality or reduction in photosynthesis and productivity due to the direct or interactive effects of epiphytes on *Nereocystis* are currently available.

Disease   The only known parasitic algae that commonly infects *Nereocystis* is *Streblonema* sp., a brown alga that apparently causes distortions of the stipe ranging from galls to extended rugose areas. These deformations can weaken the stipe and could result in breakage during exposure to strong surge or storm conditions. *Nereocystis* does not appear to be susceptible to black rot disease or stipe blotch disease, conditions that affect other brown alga and can result in substantial loss of biomass through degradation and abscission of stipes and blades (Collins et al. 2000a).

Community ecology and its role in coastal marine ecosystems

Direct and indirect interactions with other species

Macroalgae can interact directly with other species by competing for limited resources (e.g., light, space, nutrients), providing food for herbivorous grazers and detritivores, and providing habitat for other algae, invertebrates, and fishes. Macroalgae can also indirectly influence other species through mechanisms that include modification of water flow and the delivery of larvae and other plankton, harboring of prey and predators of other species in a community, and trophic cascades (i.e., fueling grazer or detritus-based trophic pathways). Whereas such interactions have been the focus of numerous studies of *Macrocystis*, similar studies involving *Nereocystis* are few. Nonetheless, the diverse ecological functions that have been attributed to *Nereocystis* can have direct implications for the variety of ecosystem services that *Nereocystis* provides to human societies (Table 1).

Macroalgae   As described in the section on biotic factors that limit its distribution and abundance, *Nereocystis* appears to be competitively inferior to many other algae (Foster & Schiel 1985 and others cited previously). This conclusion is based in part on the ephemeral occurrence of individual plants and whole forests and the small holdfast and narrow morphology that constrain its usurpation of space on a reef and attenuation of light, respectively. As such, the impact of *Nereocystis* on other macroalgae is thought to be limited, although more research on this topic is warranted. One exception to this general conclusion is the facilitative effect *Nereocystis* has for epiphytes (see section on epiphytes). Whether modification of water flow by *Nereocystis* on reef habitats (diminishing current speed and turbulence) also facilitates or impairs the growth, survival, and replenishment of other macroalgae remains unclear.

Invertebrates   A variety of small invertebrates uses the stipe and canopy of *Nereocystis* for food and habitat (e.g., sessile invertebrates such as bryozoans, especially *Membranipora membranacea*, hydroids, and barnacles, and small mobile grazers such as isopods, caprellid amphipods, and snails; McLean 1962, Burge & Schultz 1973, Foster et al. 1979, Foster 1982, Gotshall et al. 1984, 1986). The benthic invertebrate assemblage associated with *Nereocystis* is similar to that associated with other annual kelp. Gotshall et al. (1984) documented lower invertebrate abundances around *Nereocystis* than *Macrocystis*, with the notable exception of red and purple sea urchins, which were more than twice as dense under *Nereocystis* beds. Also, like giant kelp, the holdfast of *Nereocystis* provides habitat for a large number and diversity of small invertebrates, including brittle stars,

Table 1 Summary of ecosystem functions and services provided by bull kelp *Nereocystis luetkeana*

Ecosystem function	Ecosystem service
Trophic functions	
Primary production	Source of carbon sequestering
Fuels secondary production: grazers (crustaceans, gastropods, echinoderms)	Production of culturally and recreationally important species (abalone), minor harvest for recreational and commercial consumption by humans
Fuels secondary production: detritivores (crustaceans, gastropods, echinoderms)	Production of commercially fished species (abalone, sea urchins), harvested for commercial mariculture of abalone
Fuels tertiary production: invertivores	Production of commercially fished species (crabs, fishes)
Structural functions	
Biogenic 3-dimensional habitat	Provides structural framework for nearshore ecosystems
Source of habitat for epiphytes	Increased local species diversity
Source of recruitment and nursery habitat for juvenile invertebrates and fishes	Production of recreationally and commercially fished species (rockfishes, salmon)
Physical structure dampens inshore swell and turbulence	Reduces swell and coastal erosion
Ecosystem connectivity	
Export of primary production to coastal marine ecosystems (sandy beaches, rocky intertidal, offshore soft-bottom and submarine canyons)	Fuels secondary production of detritivores in other coastal ecosystems

Note: Supporting documentation is provided in the text.

crabs, and small abalone (Andrews 1925), and may serve an important nursery function for juvenile invertebrates (sensu Beck et al. 2001), although this possibility has not been rigorously tested. Calvert (2005) and Siddon et al. (2008) conducted the only large-scale (1500-m^2) manipulations of the presence of *Nereocystis* canopies to examine the effect on the abundance of invertebrate species. They found no effects of canopy removal on invertebrates distributed in either the surface or bottom portion of the water column. However, their sampling was limited to collectors (light traps and standardized monitoring unit for recruitment of fishes [SMURFs]), not visual surveys.

Fishes Because of the commercial and recreational value of fishes that inhabit shallow rocky reef habitats throughout the western coast of North America, a great deal of research has been done on the relationships between macroalgae and fishes. Again, much of this research has focused on interactions between fishes and the giant kelp *Macrocystis* spp., and far less attention has been given to *Nereocystis*. Nonetheless, a few studies have described the relationship between fishes and *Nereocystis* throughout its range. Like other taxa, the relationships between fishes and *Nereocystis* can be divided into trophic and structural interactions and between the juvenile and adult stages.

The strongest relationships between macroalgae and fishes reflect the importance of habitat structure created by macroalgae for the juvenile stages. Although a number of studies have described the importance of algal structure as habitat for larval settlement and refuge from predators (see reviews by Carr & Syms 2006 and Steele & Anderson 2006), almost all of this work has focused on *Macrocystis*. Our understanding of the importance of *Nereocystis* for the recruitment of juveniles to populations of adult reef fishes suffers from a lack of studies targeting this relationship throughout the range of *Nereocystis*. In the few places and cases it has been examined, recruitment of several species of fishes, most notably the rockfishes (genus *Sebastes*), appears to increase in, or is associated with, the presence of *Nereocystis*.

Five examples of observational studies of the association of juvenile fishes with *Nereocystis* are particularly noteworthy. One includes the occurrence of recently settled copper rockfish *Sebastes caurinus* in the canopy formed by forests of *Nereocystis* in the Strait of Georgia, between Vancouver

Island and mainland Canada (Haldorson & Richards 1987). Haldorson and Richards concluded that *Nereocystis* forests were "especially important habitat" for very young copper rockfish that had recently settled into shallow reef habitats. These young fish eventually migrated down plants to the reef habitat. Carr surveyed fish assemblages associated with *Nereocystis* forests along the central coast of Oregon. Very high numbers of juvenile rockfishes, including copper (and perhaps quillback *Sebastes maliger*), and fewer juvenile black (*S. melanops*) rockfish were observed both in the canopy and on the bottom at multiple forests (M. Carr, unpublished data). Similarly, Bodkin (1986) observed aggregations of juvenile rockfishes (various species combined) at mid-depth and on the bottom of a *Nereocystis* forest in central California. In that study, it is unknown whether the fishes use the canopy habitat specifically because that portion of the water column was not sampled. Leaman (1980) mentioned that juvenile stripe surfperch *Embiotoca lateralis* were more abundant within the *Nereocystis* forest than in habitat adjacent to the forest. Comparison of densities of juvenile and adult fishes (primarily rockfishes and Pacific cod *Gadus macrocephalus*) among shallow rocky reefs that varied in the occurrence and density of *Nereocystis* and understory kelps along the coast of south-central Alaska revealed higher fish densities in the presence of *Nereocystis* (Hamilton & Konar 2007).

Central to determining whether *Nereocystis* forests are of particular importance to the growth and survival of juvenile fishes is determining whether the forest habitats contribute disproportionately to the number of juveniles that survive to become adults (i.e., 'nursery habitat' *sensu* Beck et al. 2001). In addition, the most direct evidence of the effect of kelp forests on the local recruitment of reef fishes is from experimental manipulations of the presence of giant kelp (e.g., Carr 1989, 1991, 1994). To date, only three studies have manipulated *Nereocystis* in attempts to assess its effect on recruitment of juvenile fishes (Leaman 1980, Calvert 2005, Siddon et al. 2008). All three studies identified effects on adult fishes, especially small cryptic species, but none detected strong effects on the density of young recruits as described by the observational studies mentioned. In addition to the juveniles of these rocky reef-associated fishes, juveniles of various species of salmon are also frequently observed schooling through, and associated with, stands of macroalgae. One example includes frequent encounters with juvenile pink (*Oncorhynchus gorbuscha*), coho (*O. kisutch*), and chum salmon (*O. keta*) associated with stands of *Laminaria saccharina* in south-eastern Alaska (Johnson et al. 2003), and another is the significantly higher density of juvenile coho and Chinook salmon (*O. tshawytscha*) associated with *Nereocystis* beds along the Washington coast of the central and western Strait of Juan de Fuca (Shaffer 2002).

Information on the association of adult fishes with *Nereocystis* forests is based largely on four observational studies broadly distributed across the geographic range of the alga. Bodkin (1986), Leaman (1980), Dean et al. (2000), and Calvert (2005) described the fish assemblages associated with *Nereocystis* forests in central California, British Columbia, south-eastern Alaska, and Prince William Sound, Alaska, respectively. Because of the close association of kelps with rocky habitat and because the presence of *Nereocystis* forests is highly seasonal, the extent to which the structure of fish assemblages (i.e., diversity and relative abundance of species) is related to *Nereocystis* or the rocky reef habitat is unclear. Bodkin (1986) compared fish assemblages between *Nereocystis* and *Macrocystis* forests but did not compare reefs with and without *Nereocystis*. Leaman (1980) compared reefs with and without *Nereocystis* and noted that three benthic species were particularly associated with the *Nereocystis* plants: the sculpin *Synchirus gilli*, the snailfish *Liparis* sp., and the blenny *Phytichthys chirus*. All three are small (<10 cm length) cryptic species that sit directly on the blades and stipes of the alga. In addition, the tubesnout *Aulorhynchus flavidus* was thought to be influenced by the presence of *Nereocystis* because it deposits its eggs directly on the pneumatocysts. Dean et al. (2000) compared fish assemblages among nearshore environments, including a variety of algal habitat. They found distinct fish assemblages associated with habitats of different vegetation and exposure; the most abundant benthic fishes within eelgrass beds were juvenile Pacific cod (*Gadus macrocephalus*), greenlings (Hexagrammidae), and gunnels (Pholidae), whereas pricklebacks

(Stichaeidae) and sculpins (Cottidae) dominated in *Agarum-Laminaria* headland and bay habitats, and kelp greenling and sculpins were numerically dominant in *Nereocystis* beds. Gunnels were less abundant in *Nereocystis* beds than elsewhere. Thus again, the addition of *Nereocystis* to the mix of vegetation habitats in a region was correlated with greater regional fish diversity.

Long-term manipulative experiments that create reefs with and without kelp forests are the only definitive way to determine the extent to which a fish assemblage is influenced by the presence or abundance of kelp (e.g., Carr 1989). Leaman (1980) conducted short-term manipulations of *Nereocystis* and found that effects of removal of the canopy varied between midwater and benthic fishes and whether the removal was conducted at the edge or middle of the forest. Removal of the canopy near the edge of the bed had little effect on the abundance of benthic species but decreased the abundance and diversity of neritic species. In contrast, removal of the canopy in the middle of the bed increased the density of neritic species and decreased both the abundance and the number of benthic species. Because neritic species feed on plankton transported across reefs, removal of the canopy from inner portions of the bed may have increased the transport and delivery of food to these species. Thus, this study indicated that both benthic and neritic fish assemblages responded to the removal of a *Nereocystis* canopy, and that differences in the response of the two assemblages to canopy removal depended on the location of plant removal within a forest.

Calvert (2005) and Siddon et al. (2008) also conducted two large-scale manipulations of the presence of *Nereocystis* canopy and the subcanopy formed by lower-growing algae (*Laminaria*). They found that fish abundance was greatest in plots with both canopy and subcanopy present, and that the removal of the canopy decreased the local abundance of fishes. However, as Leaman (1980) found, this effect varied between the benthic and neritic fish assemblages. On manipulation of the kelp canopy, significantly greater abundance and biomass of benthic fishes occurred at sites with *Nereocystis* than sites without. Juvenile benthic fishes from the families Pholidae, Cyclopteridae, and Hemitripteridae at sites with *Nereocystis* canopy were twice as abundant and had an estimated biomass more than four times that of fishes observed at canopy removal sites (Siddon et al. 2008). In contrast, a direct negative effect of *Nereocystis* was observed for schooling fishes; six times more schooling fishes (juvenile Pacific cod *Gadus macrocephalus* and walleye pollock *Theragra chalcogramma*) were observed at sites without a canopy kelp. These effects varied seasonally, with the influence on neritic fishes limited to the summer.

In general, where *Nereocystis* and *Macrocystis* forests co-occur, *Nereocystis* forests appear to support lower densities of reef fishes than those associated with *Macrocystis*, but the fish assemblages associated with the two forest types differ somewhat, suggesting that regional fish diversity is increased in nearshore waters where both *Nereocystis* and *Macrocystis* forests co-occur. Both Leaman (1980) and Bodkin (1986) compared fish assemblages between nearby *Nereocystis* and *Macrocystis* forests. Bodkin found that the overall composition of the fish assemblages was generally similar between the forest types; however, the density of many species was generally greater in the *Macrocystis* forest. Leaman (1980) also found greater fish densities in *Macrocystis* forests, but noted that the benthic fish assemblages associated with the two forest types were "decidedly different," with sculpins disproportionately abundant in the *Nereocystis* forest. Both Leaman (1980) and Bodkin (1986) noted greater abundance and representation of neritic species in *Macrocystis* than *Nereocystis* forests. Although based on limited observations, these results suggest that fish diversity is increased in nearshore waters where both *Nereocystis* and *Macrocystis* forests co-occur.

Fishes may benefit from trophic interactions associated with kelp forests as well, by feeding on (1) increased numbers of prey that graze directly on the *Nereocystis* plants (e.g., snails and amphipods), (2) prey that feed on detritus produced by the canopy that detaches and falls to the floor of the forest, or (3) prey associated with the algae (e.g., juvenile rockfishes as described). However, no studies have examined this situation specifically with respect to the presence or abundance of *Nereocystis*.

Interactions with other ecosystems

The extent of exchange of resource subsidies between biological communities is an area of great current interest in ecosystem ecology. Given the massive productivity of *Nereocystis* sporophytes, bull kelp populations are likely to have considerable impacts on adjacent habitats and ecosystems through the allochthonous export of biomass (Colombini & Chelazzi 2003). *Nereocystis* production can be exported to other marine ecosystems (e.g., marine canyons, sandy beaches, rocky intertidal areas) as detritus or when blades or entire thalli are dislodged or broken from their holdfast. Allochthonous input from detached subtidal algae is known to be particularly important in ecosystems with limited primary production (Kim 1992, Vetter 1995, Harrold et al. 1998) and can influence community dynamics by changing the bacterial community (e.g., Tenore et al. 1984) and providing refugia (e.g., Norkko et al. 2000) and a food source for invertebrates (e.g., Pennings et al. 2000). The probability that drifting *Nereocystis* blades or thalli are retained in a habitat varies spatially, likely with local oceanographic features and with substratum characteristics (e.g., Orr et al. 2005). Commensurate with the rapid growth of *Nereocystis*, decomposition of lamina tissue is relatively rapid (Smith & Foreman 1984), although the impact of *Nereocystis* in detrital pathways has not yet been quantified. A study by Mews et al. (2006) of beach wrack decomposition found that *Nereocystis* decay rates were considerably higher than those of other common wrack species (*Macrocystis integrifolia*, *Fucus* spp., *Ulva* spp., and *Phyllospadix* spp.) in Barkley Sound, British Columbia. These results suggest that energetic resources exported from *Nereocystis* beds in the form of drift kelp may be quickly broken down and assimilated on arrival in other marine ecosystems.

Human activities and management

Harvest

Nereocystis has been harvested for human consumption, agricultural purposes, and for use as mariculture feed. The thick central stalk is pickled and marketed as a speciality food product, and the dried parts are used for arts and crafts (Kalvass et al. 2001). *Nereocystis* is very similar to wakame (*Undaria pinnatifida*) used in traditional Asian cooking and may have potential as a culinary substitute (Malloch 2000). It is thought to be a blood-cleansing product by Koreans, and new mothers traditionally eat it every week for a year after giving birth (Malloch 2000). Bull kelp tissue has been harvested for use in the production of liquid fertilizer and as feed in abalone mariculture (Kalvass et al. 2001). Unlike *Macrocystis*, female herring will not deposit their eggs on fronds of *Nereocystis*, and as such it is not used in the spawn-on-kelp (SOK) industry. SOK is a speciality seafood product that consists of kelp fronds covered in herring roe and then stored/preserved in brine. Popular in Japan, where it is known as komochi (or kazunoko) konbu, SOK is produced commercially in enclosed bays or inlets called ponds. Harvest kelp blades are strung across the pond on lines, and herring are then introduced to or given access to the ponds for egg deposition.

In contrast to *Macrocystis*, for which harvesting involves removal of tissue from the upper 4 ft of canopy and leaves the rest of the plant essentially intact and capable of continued vegetative growth and reproduction, harvesting of *Nereocystis* often involves the removal of the pneumatocyst and associated blades (Figure 2). By removing most of the photosynthetic and meristematic tissue of an individual plant, this method of collection eliminates the potential for further vegetative growth and eventually kills the plant by removing its source of buoyancy, causing the stipe to sink to the substratum (Mackey 2006). As a result, if collection occurs prior to the release of reproductive spores, plants harvested in this manner do not contribute reproductively to the maintenance of the populations of which they are a part. Collection involving pneumatocyst removal can thus have immediate, dramatic, and long-lasting effects on the extent of *Nereocystis* canopy cover in harvested beds. To avoid these outcomes, it has been recommended that harvesting (1) involve only the removal of distal portions of the fronds to allow for vegetative regrowth of adult plants and (2) be timed according to the reproductive schedules of the plants such that it does not occur before the production and release of reproductive spores (Wheeler 1990). Admittedly, more data on these schedules will need to be collected to determine the timing and predictability of reproductive events associated with local *Nereocystis* populations.

In further contrast to giant kelp, for which numerous harvesting-related studies have been performed, the effects of harvesting on *Nereocystis* remain largely unexplored. The provincial government of British Columbia funded two studies of the effects of *Nereocystis* harvesting on kelp forest ecology (Wheeler 1990). In an earlier study performed by Foreman (1984), no significant harvesting effects could be detected on recruitment and regrowth of the beds in subsequent years. The impact of harvesting was simulated by removing all *Nereocystis* sporophytes from within 100-m^2 experimental plots at Malcolm Island and comparing abundance in those plots in subsequent years with control plots where no harvesting occurred. Whereas the results suggested that harvesting had no measurable effect on temporal patterns of plant density or mean plant biomass, the limited replication and short duration of the experiment (two to three plots, 6 yr total with 2 yr of postmanipulation monitoring) severely limit the spatiotemporal scale of inference. Foreman pointed out that harvesting by hand could allow for selective removal of stipes from plants only after their sori had been released. It is now known, however, that *Nereocystis* blades will continuously produce sori until the plant dies. Given this fact, data on rates of sori production during the course of the growing season would be needed to identify the optimal timing for lamina harvest that minimizes impacts on lifetime sori production of harvested plants. In a study of harvest involving partial removal of fronds rather than complete removal of the pneumatocyst, Roland (1985) explored the influence of timing of harvest and extent of tissue removal on sporophyte growth and reproduction. *Nereocystis* sporophytes growing in a bed near Victoria, British Columbia, had laminae cut 30 cm above the pneumatocyst once, every 30 days, or not at all (unharvested control). Harvest via partial removal of laminae did not significantly increase mortality of plants relative to unharvested controls, but postharvest lamina production and the proportion of blades bearing reproductive sori were significantly reduced in both harvest treatments (Roland 1985). These effects could influence both the amount of canopy biomass available as habitat and as detritus for associated fish and invertebrate communities and the number of reproductive propagules contributing to recruitment to the bed in subsequent years.

The history of harvest, current and historical stock assessments, and current harvest and management of bull kelp are described next, organized by the region under which management occurs (i.e., British Columbia and in the United States by state).

California

History of harvest   Kelp has been harvested commercially off the California coast since the early 1900s. The vast majority of this collection involved *Macrocystis*, extracts from which were used in the manufacturing of explosives, livestock and mariculture feed, and algin. ISP Alginates (formerly Kelco), the largest commercial kelp-harvesting operation in California, accounts for at least 95% of the annual harvest in the state. The company has been in operation since 1929 and by 2002 had acquired lease rights to 15 beds (~28 miles^2) from Monterey Bay to Imperial Beach. Approximately 22 other harvesters held licenses to collect kelp in 2002 (Little 2002). Like ISP Alginates, nearly all of these firms also targeted giant kelp.

In contrast to *Macrocystis*, there was essentially no targeted harvesting of *Nereocystis* in California until the 1980s. Prior to that time, small amounts of incidental harvesting of *Nereocystis* likely occurred during the harvest of giant kelp in mixed-stand beds, but amounts were never quantified. Abalone International, a Crescent City mariculture company, began collecting *Nereocystis* from a region between Point St. George and the Crescent City Harbor in 1988 and received exclusive lease privileges from the California Department of Fish and Game (CDFG) for collection in bed 312 in 1997 (Kalvass et al. 2001). Based on estimates of local *Nereocystis* abundance in this region, their harvest limit was set at 821 t yr^{-1} (Kalvass et al. 2001). Peak harvest by this firm was only 149 t in 1999, and collection dropped substantially thereafter, with only 11 and 44 t landed in 2000 and 2001, respectively (harvesting statistics are given in a table in Collins et al. 2000a). This decline has been attributed to decreasing demand rather than reduced availability of the resource (Kalvass et al. 2001). As of 2002, only 3 of the state's 13 *Nereocystis*-dominated beds were open to harvest, and only 1 is currently leased to a commercial harvesting operation (bed 312, to Abalone International).

Historical and current stock assessment   The first survey of kelp abundance in California that recognized *Nereocystis* was part of a larger mapping effort spanning the Gulf of Alaska to Cedros Island (Baja California) between 1911 and 1913. This work, overseen by Dr. Frank Cameron, was undertaken by the U.S. Bureau of Soils to investigate the potential use of kelps as a source of potash fertilizer (Cameron 1912). Historical records of *Nereocystis* abundance are limited because subsequent surveys often did not differentiate between *Macrocystis* and *Nereocystis*. In addition, because most of these subsequent surveys were motivated by a desire to map the distribution of the more economically valuable *Macrocystis*, few were conducted in northern areas of the state where *Nereocystis* predominates. Current estimates of the sizes of *Nereocystis* populations in northern California are based largely on surveys performed in 1989 and 1999 and on information from the Crescent City area (Del Norte County) provided by Abalone International (Kalvass et al. 2001). While the 1912 and 1989 surveys estimated roughly 6.5 miles^2 of *Nereocystis* canopy north of Point Montara, the 1999 survey indicated a decline of approximately 42% in kelp coverage in the area between Point Montara (San Mateo County) and Shelter Cove (Humboldt County) (Kalvass et al. 2001). This apparent decline, which runs counter to observations of extensive beds in this region in late 1999, may be attributable in part to (1) the timing of the 1999 survey, which occurred after a major storm; (2) improved interpretation methods for aerial photos; or (3) natural fluctuations in kelp bed coverage and density (Kalvass et al. 2001). On a more local level, the Crescent City harvesting operation conducted a 1996 survey of *Nereocystis* abundance in bed 312 as part of their harvesting lease agreement with CDFG. This yielded an estimate of 5475 t of *Nereocystis* in the 205 acres of bed 312 between Point St. George and Whaler Island.

No recent *Nereocystis* surveys have been done in central California. Results of the 1912 survey suggested that 32% of the 17.55 mi^2 kelp canopy in this region was associated with *Nereocystis* (Kalvass et al. 2001). In central California, *Nereocystis* seems to be outcompeted by *Macrocystis* and is generally restricted to areas (1) on the outer fringes of giant kelp beds, (2) within the surge zone, or (3) from which giant kelp has been temporarily removed by disturbance associated with winter storms or strong waves. Evidence of the temporally dynamic nature of *Nereocystis* abundance comes from Diablo cove, where density levels declined from 200 t acre^{-1} in 1975 to 4.8 t acre^{-1} in 1982 (Kalvass et al. 2001).

Management and recent harvest   California's kelp bed management, a responsibility of the CDFG, has focused mostly on *Macrocystis*. Collection of kelp for commercial purposes requires a 1-yr, $100 license, and harvesters are required to keep collection records (discussed in the next paragraph). Kelp beds may be leased from the State Land Commission for up to 20 years with a deposit of no less than $40 miles^{-2}. Leased areas may not exceed 25 miles^2 or 50% of the total kelp resource, whichever is greater (Mackey 2006). There is a royalty for edible seaweeds of $24 wet t^{-1} harvested from waters other than San Francisco Bay and Tomales Bay (Mackey 2006). No collection is allowed in marine life refuges or specially designated aquatic parks. If *Nereocystis* is collected for human consumption, the harvest limit is 2 t each year, and the entire plant must be harvested. Collection is to be performed by cutting, and harvesting must be at a depth of less than 4 ft below the water surface (Hillmann 2005). Collection for personal and scientific use requires a permit and is limited to 10 pounds wet weight per permit (Mackey 2006). Personal, non-commercial harvest is prohibited in marine life refuges, marine reserves, ecological reserves, national parks, or state underwater parks (Mackey 2006).

All commercial harvesters are required to keep records of the weight, species, collector, and location of harvest and report these figures to CDFG on a monthly basis (Kalvass et al. 2001, Mackey 2006). Although these harvest summary data have been collected regularly since 1915, routine and formal stock assessments of the state's kelp resources have never been performed. CDFG conducts aerial surveys of California kelp beds only periodically, and while many commercial harvesters (e.g., ISP Alginates) conduct additional aerial surveys of their own (probably with greater frequency and precision), the resulting data are often proprietary and not available to the public or management agencies (Little 2002). As such, although the Fish and Game Code (§6654) gives the CDFG the authority to close a kelp bed to harvest for up to 1 yr if it is determined that the bed is being damaged by collection, the information necessary to detect detrimental impacts of harvest on kelp resources is largely unavailable (Kalvass et al. 2001).

Given these management resources, the CDFG commission took the following formal precautionary steps to protect kelp beds in northern California (especially *Nereocystis*) in 1996. First, the kelp bed-numbering system initiated in 1915 for beds in southern and central California was extended by adding a 300-series designation for kelp beds north of San Francisco (Kalvass et al. 2001, Little 2002). These beds are composed primarily of *Nereocystis*. Prior to this action, because of the lack of formal CDFG recognition, any northern bed could have been harvested for commercial purposes. Second, beds 303–307 were closed to future commercial harvest. Finally, collection in the remaining beds in the 300 series was limited to a maximum harvest of 15% of the biomass as determined by a CDFG-approved annual survey conducted by the lessee. In 2001, the commission added the following additional restrictions (Kalvass et al. 2001). First, beds 301, 302, 310, and 311 were closed. Harvesting of *Nereocystis* was restricted north of Point Arguello by California code of regulations title 14, section 165(c)(4) (Monterey Bay National Marine Sanctuary 2000). Second, harvest was restricted from April 1 through July 31 within the Monterey Bay National Marine Sanctuary. Third, harvesters were required to have a commission-approved harvest plan prior to taking kelp with a mechanical harvester in open beds north of Santa Rosa Creek (San Luis Obispo County). Finally, the commission assumed the authority to designate open beds, or portions thereof, as harvest control areas where harvest is limited for a specific period of time.

As of 2006, there were five active commercial permits in California for *Macrocystis* harvest and none for commercial harvest of *Nereocystis* (Mackey 2006). A 3-yr experimental kelp-harvesting permit has been granted to The Nature Conservancy to study the effects of giant kelp harvesting on associated fish assemblages (Mackey 2006). Very little is known about the nature and magnitude of recreational harvest (Little 2002).

Oregon

History of harvest The only recent documented commercial harvest of *Nereocystis* in Oregon occurred from 1988 through 1992, when a company collected approximately 70 t of tissue from kelp beds associated with Orford Reef in southern Oregon (Kalvass et al. 2001, Mackey 2006). A 5-yr experimental lease granted to a different commercial entity by the Oregon Department of State Lands (DSL) in 1996 expired in 2000 with no harvest ever having occurred.

Historical and current stock assessment In 1954, Waldron conducted perhaps the first formal survey of *Nereocystis* distribution and abundance focused specifically on the coast of Oregon (Waldron 1955). Using aerial surveys of the coastlines of Lincoln, Coos, and Curry Counties, he noted that beds were typically located within 1 mi of the shore in less than 60 ft of water and tended to be concentrated in areas protected from prevailing winds. Beds were small or absent on the seaward sides of reefs and along exposed sections of headlands. Across the three survey areas, he estimated

3704 total acres of kelp beds, of which 1766 acres were classified as being of moderate or high density and accessible for harvest. Over 70% of the total acreage was located in Curry County, with the bed at Orford Reef accounting for over 20% (791 acres) of this total. A team from the Oregon Department of Fish and Wildlife surveyed *Nereocystis* beds on five subtidal reefs in Oregon (Orford, Blanco, Redfish Rocks, Humbug Mountain, and Rogue Reefs) in 1996, 1997, and 1998 (Fox et al. 1998). They compared kelp biomass estimates derived from three measures: weights of individual plants, plant density derived from canopy percentage cover estimates, and total canopy area at the ocean surface. The Orford Reef bed was consistently the largest (estimated biomass between 6454 and 3442 t), but interannual variation in bed size was high, and no consistent temporal trends were apparent. The results suggested that estimates of the surface area of kelp beds might not be an accurate proxy for annual biomass. Across all five reefs, total surface area ranged from 179 to 371 ha, and estimated biomass fluctuated between 8137 and 16,583 t.

Management and recent harvest   The DSL has jurisdiction over submerged subtidal lands (any land "lying below the line of ordinary low water of all navigable waters within the boundaries of this state") and can issue permits for commercial leasing of state-owned portions of these lands (Mackey 2006). Prior to October 2008, up to 40 miles of submerged land could be leased to a single individual for a period of no more than 50 yr with no stated restrictions of the amount of kelp that may be harvested during the lease period. However, on 14 October 2008, the State Land Board approved a rule change that effectively prohibits the harvest of kelp and other seaweed for commercial purposes from state-owned submerged lands [OAR 141-125-0120(13)]. Persons collecting <2000 pounds of kelp yr^{-1} from these lands for the purpose of personal consumption do not require a lease. No commercial collection is permitted within the Oregon Shore Recreation Area, and kelp harvesting is prohibited in 12 specially managed marine areas: Haystack Rock Marine Garden (Cannon Beach), Cape Kiwanda Marine Garden (Pacific City), Boiler Bay Research Reserve (Depoe Bay), Pirate Cove Research Reserve (near Depoe Bay), Whale Cove Habitat Refuge (near Depoe Bay), Otter Rock Marine Garden (Devil's Punchbowl), Yaquina Head Marine Garden (north of Newport), Yachats Marine Garden (south of Yachats), Neptune State Park Research Reserve (north of Florence), Gregory Point Research Reserve (Charleston/Coos Bay), Harris Beach Marine Garden (Brookings), and Brookings Research Reserve (Brookings) (Mackey 2006). The Oregon Parks and Recreation Department (ORPD) requires a scientific research permit for all activities that involve specimen collection or fieldwork or that have the potential to damage the natural resources on lands owned and managed by the DSL (Mackey 2006).

As of April 2006, there were no current or pending commercial leases through the DSL for harvest of kelp in submerged lands in Oregon. There was a single active lease for kelp harvest in the intertidal zone of southern Oregon, but this permit was set to expire in mid-2006, and associated harvest should not have involved *Nereocystis*. The levels of personal harvest of kelp in both intertidal and subtidal regions in Oregon have not been quantified and are believed to be low (Mackey 2006).

Washington

History of harvest   No evidence of attempts to commercially harvest *Nereocystis* in Washington State could be located.

Historical and current stock assessment   Members of the Nearshore Habitat Program of the Washington Department of Natural Resources (DNR) have used aerial photographs to monitor kelp beds fringing the Olympic Peninsula since 1989. The annual surveys are designed to track changes in the size and shape of beds as well as the relative abundance of the two dominant canopy-forming species, *Macrocystis* and *Nereocystis*. The kelp canopy-monitoring study area includes the mainland coastline along the Strait of Juan de Fuca as well as the outer coast of Washington from Port

Figure 5 (See also Colour Figure 5 in the insert.) (A) Study area for long-term monitoring of canopy-forming kelp in Washington State conducted by the Department of Natural Resources. (B) Colour-infrared imagery collected by areal surveys. Floating kelp canopies appear as red areas on the dark water surface. Photo interpretation is used to classify red floating kelp as canopy area. Bed area is delineated by grouping classified kelp canopies with a distance threshold of 25 m. (Courtesy of Helen D. Berry, Nearshore Habitat Program, Washington State Department of Natural Resources.)

Townsend to the Columbia River (~360 km of total shoreline; Figure 5A) (Berry et al. 2001). Data are collected according to the following protocol: First, color-infrared photographs of the survey areas, taken at a scale of 1 in:2500 ft, are collected from a fixed-wing aircraft using a 70-mm camera (Figure 5B). The annual inventory is completed in late summer to coincide with the maximum kelp canopy (most often in September). Target conditions for photographic survey days are tidal levels less than +1.0 MLLW (mean low low water), surface winds less than 10 knots, sea/swell less than 5 ft, sun angle greater than 30° from vertical, cloud and fog-free skies. Work evaluating this photo-based assessment technique, conducted in *Nereocystis* beds adjacent to San Juan Island, Washington, demonstrated the potential for tidal height and currents to significantly affect estimates of bed size based on canopy area (Britton-Simmons et al. 2008). Beds appear relatively smaller as current velocity and tidal height increases because subducted plants are more difficult or impossible to detect by aerial photos and near-infrared (NIR) imaging.

Analyses of data collected between 1989 and 2000 revealed pronounced interannual variability in total aerial extent of kelp beds with no consistent long-term trend. Relative to *Macrocystis*, *Nereocystis* beds almost always covered a larger area, were less dense, and exhibited greater interannual variability in their extent. *Nereocystis* also may be more sensitive to climatic anomalies; during the 1997 El Niño, *Nereocystis* populations along the outer coast experienced a 75% reduction in size compared with an 8% reduction for *Macrocystis* (Berry et al. 2001). It has been proposed that reductions in sea urchin abundance and associated kelp grazing, due both to increases in sea otter abundance and direct harvesting of grazer species by humans, may contribute to spatiotemporal variability in the size of kelp beds in Washington. Rigorous quantitative tests of these conjectures have not been performed.

Management and recent harvest   Commercial harvest of seaweed, including collection on privately owned tidelands (60% of Washington's intertidal zones), is prohibited except with the approval of both the Washington DNR and the Department of Fish and Wildlife. In 1993, the Washington legislature identified marine aquatic plants as a source of 'essential habitat' in light of their biological importance and economic value and urged the implementation of stricter harvesting regulations (Mackey 2006). At present, seaweeds are only harvested for recreational purposes in Washington (Mackey 2006). Harvesters must be over 15 years of age and can collect no more than 10 pounds of algae (wet weight) per person per day. For *Nereocystis*, fronds are to be cut no closer than 24 in. above the pneumatocyst using a knife or similar instrument (Hillmann 2005). There are three types of non-scientific collection permits in Washington: (1) annual combination permits allow for harvest of seaweed, shellfish, and both fresh and saltwater fishes; (2) annual shellfish and seaweed permits allow for harvest of seaweed and shellfish; and (3) 1- to 5-day combination permits allow the same harvest as the annual combination permits but are valid for no more than 5 days. As of November 2005, the numbers of active permits of these types were 165,983, 161,550, and 196,280, respectively. All but three state parks are closed to seaweed harvesting, and scientific permits, granted only when the proposed collecting has a demonstrable scientific purpose, are required in these parks (Mackey 2006).

In an attempt to conserve nearshore subtidal ecosystems, the Washington DNR has introduced legislation that would authorize the leasing of "submerged lands" for restoration and conservation purposes. Leasing would effectively place these lands, which could include kelp beds, under the stewardship of conservation-oriented individuals or agencies, further protecting coastal environments from commercial harvesting (Mackey 2006).

British Columbia

History of harvest   The first attempt at commercial harvesting of marine plants in British Columbia was undertaken by Canada Kelp Company Limited in 1949. Financial complications led to the failure of this endeavour, and no further harvesting operations were initiated until 1967, when nearly the entire coastline of British Columbia was subdivided into 44 harvesting licenses collectively granted to six companies. Two of these never initiated development of their operations, and the remaining four (Sidney Seaweed Products, North Pacific Marine Products [bought out by Kelpac Industries], Pacific Kelp Co., and Intertidal Industries) either failed to reach the harvesting phase or experienced financial difficulties and were operational only briefly. The one exception was Sidney Seaweed Products, a manufacturer of algae-based agricultural products that experienced small-scale economic success from 1965 to 1974. In 1981, the provincial government, through solicitation of harvesting proposals by the Marine Resources Branch of the Ministry of the Environment, adopted a more active approach to establishing a commercial kelp-harvesting industry in British Columbia. Of the applicants, Enmar Resources Corporation was selected and awarded a 5-yr license to operate off the coast of Porcher Island. Despite the support of provincial authorities, the company was ultimately unwilling to initiate development because of a refusal by the federal government of Canada to approve the project. Since that time, harvesting has been confined to small-scale operations collecting a total of less than 100 t yr^{-1} (from Wheeler 1990, Malloch 2000).

Historical and current stock assessment   To gather baseline information on spatiotemporal variability in marine plant populations, the Ministry of Fisheries initiated a kelp inventory program in 1975. Surveys are based on the Kelp Inventory Method (KIM-1) developed by Foreman (1975) that uses aerial photographs to estimate the area, density, and species composition of kelp beds. These data are combined with field-collected density and plant weight information to derive biomass estimates for 1-km wide sections of surveyed coastline. By 2000, there were 12 surveys completed, covering the majority of kelp beds that could support large-scale harvesting. Approximately 94% of the standing stock in these beds consisted of *Nereocystis*. As part of a study at Malcom Island, Foreman (1984) concluded that KIM-1 estimates were generally proportional to standing crop values but tended to overestimate these values by approximately 30%. More detailed descriptions of the survey methodology can be found in Foreman (1975) and Wheeler (1990).

Management and Recent Harvest   While the responsibility to manage marine plants is assigned to the federal government of Canada by sections 44–47 of the Federal Fisheries Act, a 1976 agreement between national and provincial governments transferred authority to adopt and enforce management regulations to the Ministry of Agriculture, Food, and Fisheries (MAFF) in British Columbia. Licensing applications for commercial harvesting of kelp must still be reviewed by the national-level Department of Fisheries and Oceans (Malloch 2000). The minister has the authority to decline to issue a license if proposed harvesting (1) tends to impair or destroy any bed or part of a bed on which kelp or other aquatic plants grow, (2) tends to impair or destroy the supply of any food for fish, or (3) is detrimental to fish life. Section 35(1) of the Federal Fisheries Act states that no person shall carry on any work or undertaking that results in the harmful alteration, disruption, or destruction of fish habitat. For a permit request to be granted, the applicant must present evidence that the overall operation is economically feasible and that the raw material requirement is low in absolute terms or compared with the estimated standing crop in the desired area or both. Licensing is to be preceded by a stock assessment regardless of the harvest quota requested. If data are not available, a license may be issued with the view to gathering management-related data concurrently with the commercial operation (Wheeler 1990). Licenses are granted annually and issued on a first-come, first-served basis. In an attempt to promote sustainable use of the resource, exclusive access to defined geographic areas is awarded, and harvesters are given the right of first refusal for their assigned localities during licensing renewal reviews (Malloch 2000).

A license costing $110 annually is required only for commercial harvest of kelp. Only Canadian citizens, members of the Canadian armed forces, and persons who are legal permanent residents of Canada are eligible to apply for a license. No more than 20% of the total biomass of a marine plant bed may be harvested, and a royalty of between $10 and $100 per wet ton of tissue is to be paid to the federal government (amounts vary by species). For *Nereocystis*, blades may be cut no closer than 20 cm from the pneumatocyst, and no harvest of the bulb or stipe is permitted. There are no permits required for personal, non-commercial harvest, and collection is prohibited in specially managed areas such as ecological and marine reserves and provincial and federal parks (Hillmann 2005).

Between 1992 and 2000, the number of companies or individuals licensed to commercially harvest marine plants in British Columbia never exceeded 15 (excluding licenses for *Macrocystis* harvesting as part of the herring SOK industry). Non-commercial harvesting is unregulated, and this poses a problem for enforcement of management regulations because the intended use of harvested materials is not always clear. The government of British Columbia has taken steps to incorporate use of marine plants by native/aboriginal groups (First Nations) into the evaluation process for commercial harvesting licenses (Malloch 2000).

Alaska

History of harvest   No information on historical harvesting of *Nereocystis* in Alaska could be located.

Historical and current stock assessment   The only comprehensive assessment of canopy-forming kelp in Alaska was the 'potash from kelp' survey carried out by the U.S. Department of Agriculture (USDA) in 1913. In south-eastern Alaska, 1133 beds, with an estimated area of 18,300 ha and biomass of 7.15×10^6 metric tonnes (mt), were counted. Of these beds, 87% consisted principally of *Nereocystis*, 6% of *Macrocystis*, and 7% of *Eualaria fistulosa*. In the northern Gulf of Alaska, 358 beds, representing an estimated 4610 ha and 3.26×10^6 mt, were recorded. Here, *Nereocystis* made up approximately 55% of the beds, with the remaining 45% being *Eualaria fistulosa* (Frye 1915, Rigg 1915). Based on the results of more recent small-scale surveys, it has been suggested that these values overestimated actual abundance by approximately 10% (Frye 1915, Rigg 1915). The Alaska Department of Fish and Game (ADFG) carries out kelp surveys in conjunction with the commercial herring harvest and the SOK industry, but these rarely involve *Nereocystis* because herring will not spawn on *Nereocystis* (M. Stekoll personal communication). In addition, Alaska

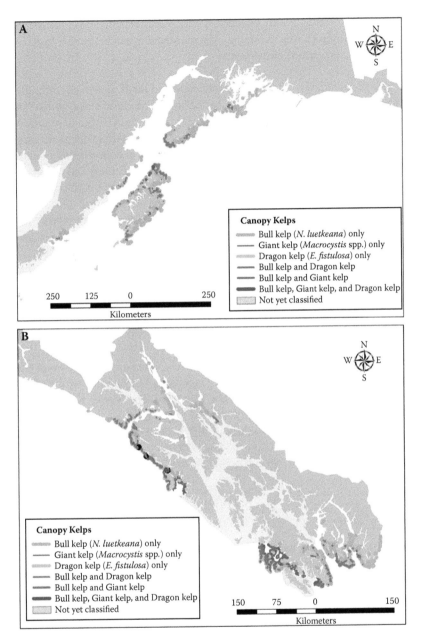

Figure 6 (See also Colour Figure 6 in the insert.) Canopy kelp distribution of *Nereocystis luetkeana*, *Macrocystis* spp., and *Eualaria fistulosa* in (A) Gulf of Alaska and (B) south-east Alaska. (Courtesy Alaska ShoreZone. Program materials available at http://alaskafisheries.noaa.gov/habitat/shorezone/szintro.htm)

ShoreZone, a multiagency coastal mapping program, collects a variety of high-resolution biophysical data from aerial imagery, including the geographic distribution of each of canopy-forming kelp species (Figure 6). Imagery from over 44,000 km of coastline in central and south-eastern Alaska has been recorded, and data from the majority of this region have been mapped and are available through an interactive Web site (http://alaskafisheries.noaa.gov/habitat/shorezone/szintro.htm). The development of an aerial digital multispectral camera (DMSC) imaging system to more accurately and precisely estimate the area and biomass of *Nereocystis* beds in south-eastern Alaska is currently being investigated (Stekoll et al. 2006). Compared with traditional methods of assessment based on aerial photos and NIR imagery, this technology has the advantage of being able to detect submerged plants up to 3 m below the surface of the water.

Management and recent harvest   Intertidal and submerged lands in Alaska, from the mean high-tide line out to 3 geographic miles, are owned by the state, and enforcement of harvest regulations is the responsibility of ADFG. Commercial permits are issued by ADFG and required for all commercial harvest. Local ADFG offices decide on the harvest guidelines for their area (M. Stekoll personal communication). Harvesters must report daily records of collection amounts and locations to ADFG once a year. Harvest must be by hand or mechanical cutting and cannot be performed using diving equipment. There are no fees associated with the permit. Collection of *Macrocystis* for herring SOK is subject to different regulations (Hansen & Mumford 1995, Hillmann 2005). A sportfishing license is required for personal collection ($15 annually for Alaska residents, $100 annually for non-residents, no charge for collectors under 16 or over 60 years of age) (Hillmann 2005), but there are apparently no restrictions on take with the exception of the SOK industry (Hansen & Mumford 1995). Scientific permits are available at no cost and require the submission of an annual report of take (number of each species collected, date and location of collection, location of specimen deposition) and of scientific findings associated with the collection (Hansen & Mumford 1995, Hillmann 2005).

Simple Pleasures of Alaska, a small commercial operation out of Sitka, processes *Nereocystis* for making pickles and relish. They harvest approximately 1 t yr^{-1} (B. Pierce personal communication). The Alaska Kelp Company (formerly Pacific Mariculture Company Inc.) of Point Baker, Alaska, was issued a *Nereocystis* harvest permit from the Petersburg ADFG office for 200,000 pounds yr^{-1}. This amount was reduced to 51,000 pounds a few years ago. The Alaska Kelp Company has made a plant fertilizer enhancer from the *Nereocystis,* and it is sold under the names Opticrop, Garden Grog, and Alaska Kelp. One year, they sold about 10,000 t to a company trying to make potting soil from sawdust, fish wastes, and kelp, but it is unclear whether commercial production of the agricultural product was ever initiated (M. Stekoll personal communication).

Pollution

Thermal pollution

Increases in ambient water temperature associated with anthropogenic point-source discharge can cause adverse effects on both gametophytes and young sporophytes of *Nereocystis*. As part of mediation associated with the Diablo Canyon power plant, TERA Corporation (now Tenera Environmental Inc., San Louis Obispo, CA) conducted temperature sensitivity experiments using *Nereocystis* in 1982 (TERA Corporation 1982). Under laboratory conditions, juvenile sporophytes were exposed to water temperatures ranging from 10°C to 20°C for 44 days. The results indicated that prolonged exposure to water temperatures above 18°C is lethal. Furthermore, 25% of the plants held at 15.9°C died after 36 days. A primary cause of mortality appeared to be a reduction in the healing ability of damaged tissue. In the field, Pacific Gas and Electric (PG&E), which operates the plant, noted that in 1985 and 1986 *Nereocystis* plants that came in contact with the thermal discharge plume of the power plant experienced premature blade loss, and *Macrocystis*, a more heat-tolerant species, eventually colonized those sites. *Nereocystis* beds persisted in areas where the thermal plume was deflected (e.g., Diablo Rock) or where cold water conditions were more common due to prevailing currents (discussed in Collins et al. 2000a). These observations were supported by a comparative study performed by Schiel et al. (2004), who used data from an 18-year intertidal and subtidal monitoring program and before-after control-impact (BACI) analyses to demonstrate quantitatively that *Nereocystis* density and abundance were significantly reduced by a 3.5°C rise in water temperature associated with thermal discharge from the Diablo Canyon plant.

Sediment and nutrient run-off (sewage, agriculture, development, dredging, freshwater intrusion)

For *Nereocystis*, the availability of light is perhaps the factor most critical for the growth and sexual maturation of gametophytes and the growth of sporophytes (see population ecology discussion here and discussion in Collins et al. 2000a). Reductions in light penetration could result from a number of processes that increase water turbidity. Sewage discharge and nutrient run-off associated with agriculture could trigger phytoplankton blooms that significantly reduce water clarity. Particulate run-off from the terrestrial environment or the suspension of benthic sediments by dredging activity or storm-associated surge could similarly reduce light penetration. Finally, growth of other algal species near the substratum could overshadow and thereby reduce the germination and growth of gametophytes and young sporophytes. Studies of the effects of sedimentation in nearshore waters have documented reduced *Nereocystis* density in areas associated with landslides (Shaffer & Parks 1994, Konar & Roberts 1996). Burge & Schultz (1973) observed an increase in water turbidity in Diablo Cove, California, following exceptionally heavy rains and associated run-off during the winter of 1968–1969. *Nereocystis* sporophytes were not seen in the area again until mid-July 1969, and the re-emerging bed was reported to be one-quarter the size of the bed in 1968. This reduction in *Nereocystis* abundance was attributed to changes in nearshore light levels (discussed in Collins et al. 2000a), but the large pulse of freshwater run-off associated with this event may have also contributed to the *Nereocystis* dieback. Brown (1915) found that exposure to freshwater for periods of up to a week could cause tissue deterioration. Additional work by Hurd (1916) substantiated this finding, showing that *Nereocystis* sporophytes develop blisters and wilt when subjected to rapid reductions in environmental salinity.

Toxic chemicals

Very little is known about the effects of toxic chemicals on *Nereocystis*. James et al. (1987) showed that, of 10 species of brown algae examined, gametophytes of *Nereocystis* were most sensitive to hydrazine, a chemical used to decrease corrosion in high-pressure boilers. At levels of 0.025 ppm, gametophyte development was inhibited, and sporophyte production would not occur (discussed in Collins et al. 2000a). Antrim et al. (1995), in tests of the effects of diesel fuel, intermediate fuel oil (IFO), and crude oil on *Nereocystis* plants, verified that exposure to petroleum products has a negative effect on *Nereocystis*. Severe tissue necrosis occurred at meristematic tissue between the stipe and bulb. In contrast to these results, comparisons of *Nereocystis* biomass and percentage cover between oiled and control sites in Prince William Sound, Alaska, following the EXXON VALDEZ oil spill, did not indicate any effects of petroleum exposure. *Nereocystis* individuals at oiled sites tended to be smaller, but it was not clear whether this was due to chemical toxicity or preexisting differences arising from natural factors such as recent recruitment or slow growth (discussed in Collins et al. 2000a).

Human modification of species interactions

Human introduction of non-native species into kelp forest ecosystems has the potential to modify species interactions in ways that affect the distribution and abundance of *Nereocystis*. The invasive macroalga *Sargassum muticum* is an example. Introduced to Puget Sound from Japan in the 1940s (Giver 1999), this species occupies space in shallow areas of *Nereocystis* beds and has been shown to competitively exclude *Nereocystis* from these locations under some circumstances (Thom & Hallum 1990). *Sargassum muticum* distribution and abundance is limited to areas associated with lower wave energy, so this type of competitive exclusion is likely to be less common at exposed sites (O'Clair & Lindstron 2000).

More important to *Nereocystis* distribution and abundance than introduced species are natural or human-induced changes in the abundance of species native to kelp forest ecosystems. Central among these are red and purple sea urchins (*Strongylocentrotrus franciscanus* and *S. purpuratus*, respectively) and red abalone (*Haliotis rufescens*; see population ecology section), all species that are currently harvested and fluctuate in density in response to sea otter population dynamics (Vanblaricom & Estes 1988). Native or naturalized epiphytes growing on *Nereocystis* can have a negative impact on kelp growth and survival (see population ecology section and references therein); such epiphyte effects could be particularly pronounced in areas where elevated nutrient concentrations promote growth of epiphytic algae.

Climate change

Episodic El Niños: temperature changes and storms

The reduction in upwelling and increased frequency of severe storms and strong wave action associated with El Niños could all have negative impacts on the distribution and abundance of *Nereocystis*. Suppression of upwelling reduces the amount of cold, nutrient-rich water brought into shallow subtidal areas from depth (Collins et al. 2000a) and can lead to warming of surface waters by up to 4°C for extended periods (McPhaden 1999). As mentioned, *Nereocystis* is sensitive to increases in water temperature, and the availability of nutrients, particularly nitrogen in the form of nitrate, is perhaps second in importance only to light availability as a necessary condition for the growth and reproductive maturation of *Nereocystis*. Perturbations in the physical environment associated with El Niño events thus have the potential to drive reductions in *Nereocystis* abundance and recruitment on interannual timescales. Delayed recruitment and reduced growth of *Nereocystis* in the beds near Fort Bragg, California, in 1992 may have been associated with the El Niño of that year (Kalvass et al. 2001). During the 1997 El Niño, total kelp canopy in Washington decreased by 32%. *Nereocystis* populations along the outer coast were reduced by 75% (compared with only 8% reductions for *Macrocystis*).

While mortality associated with strong storm events and wave action has the potential to reduce the size of *Nereocystis* beds, the ability of *Nereocystis* to rapidly recolonize disturbed areas following the removal of more competitively dominant species of algae such as *Macrocystis* may often accelerate postdisturbance recovery. In 1998, following the reductions in *Nereocystis* abundance in Washington mentioned above, *Nereocystis* populations rebounded dramatically, increasing by 423% (Berry et al. 2001). This dramatic population growth may be evidence of a positive effect of storms on *Nereocystis* abundance arising from temporary release from competition with other algal species for light, nutrients, or primary space. The timing and intensity of storms and the identity and abundance of competing species of sympatric algae are probably important in determining the nature of storm and wave disturbance on *Nereocystis*.

Long-term global warming impacts

Because of its sensitivity to water temperature and preference for cold conditions, *Nereocystis* would likely be adversely affected by increases in sea surface temperature associated with global warming. No data to quantitatively substantiate this speculation are available. Changes in the carbon dioxide concentrations of nearshore waters could offset this impact to some extent. Because global warming is driven to a large extent by increases in atmospheric concentrations of carbon dioxide, it is thought that increased sequestration of carbon dioxide in nearshore marine environments could influence growth rates of photosynthetic organisms. One short-term study showed that doubling ambient carbon dioxide concentrations for 2 h increased the net apparent photosynthetic rate of *Nereocystis* by a factor of between 2.2 and 2.8 (Thom 1996). Temperature-related increases in *Nereocystis* mortality could thus be offset to some extent by enhanced growth under conditions of greater carbon dioxide availability. Swanson & Fox (2007) used a laboratory experiment to test the effects of predicted increases in carbon dioxide and ultraviolet B (UVB) levels on *Nereocystis* development and biochemistry and on grazing by herbivores. Carbon dioxide enrichment increased kelp growth by over threefold, an effect that ameliorated the negative influence of elevated UVB on the same parameter. While tissue phlorotannin (phenolic) concentrations increased as a result of experimental treatments, rates of herbivory by the gastropod *Tegula funebralis* did not differ significantly between experimental and control groups. Results showing slower rates of tissue decay under high carbon dioxide conditions suggest that predicted global climate change could slow rates of nutrient recycling and the export of kelp-associated resources into detritus-based ecosystems.

Incidental damage

Commercial fishing activities can cause direct physical damage to kelp via propeller cuts to blades and stipes. This occurs as boats travel through kelp beds and during the process of 'backing down' when engines are run in reverse to dislodge propellers fouled by kelp fronds and stipes (Collins et al. 2000b). In addition, the deployment and retrieval of fishing gear, particularly crab, lobster, and live fish traps, can cause breakage of fronds and stipes and have the potential to dislodge kelp plant holdfasts from the substratum. Similar effects can be produced during the retrieval of anchors. While deleterious effects on kelp arising from these activities can be appreciable in locations where commercial fishing activity is high or chronic, the extent of kelp damage due to boats and fishing gear is thought to be minimal (Collins et al. 2000b).

Recommendations for management and research to inform management decisions

Overview and challenges for managing the harvest of nereocystis

In contrast to *Macrocystis*, the relatively limited commercial utility and financial value of *Nereocystis* tissue has resulted in minimal attention being paid to the development and implementation of stock assessment programs, harvest record databases, and management guidelines for *Nereocystis*. To some extent, this neglect is understandable because harvest pressure on *Nereocystis* to date appears to be negligible. Across the biogeographic range of *Nereocystis*, commercial harvesting has been confined to a small number of short-lived operations that collected relatively little kelp when compared with the harvesting of *Macrocystis*. There appear to be fewer than 20 active permits for the collection of *Nereocystis* across range of the entire species. No quantitative records of personal (non-commercial) take could be found, but given the relative inaccessibility of subtidal algae this amount is probably minimal.

Management regulations vary widely and appear to be consistent only in the fact that they are based on little if any scientific data on either natural fluctuations in the abundance of *Nereocystis* or effects of harvesting on the demography of *Nereocystis* populations. Aside from recent but spatio-temporally localized surveys conducted by a handful of harvesters and periodic aerial inventories taken by state management agencies, current estimates of *Nereocystis* abundance in many areas of its range may still be based in large part on the results of one or a few comprehensive surveys conducted almost 100 years ago. In contrast to the growing body of literature focused on harvesting impacts on giant kelp populations, only three studies that directly examined the effects of harvesting on *Nereocystis* growth, reproduction, or population dynamics were found. Two of these studies involved such limited replication that the results cannot reasonably be used to inform sound management policies. In addition, because of the fundamental life-history differences between *Macrocystis* and *Nereocystis*, it would be expected that the demographic impacts of harvesting on these two species are fundamentally different, and the extent to which our understanding of *Macrocystis* can be used to create sound policies for the harvest of *Nereocystis* needs to be questioned.

The exceedingly superficial understanding of *Nereocystis* demography and the effects of harvest may be a principle cause of the dramatic variation in *Nereocystis* management regulations in the different political provinces where the species is found. In northern California, commercial harvesting in the 300-series beds that consisted mostly of *Nereocystis* is forbidden or severely restricted. North of the border in Oregon, commercial harvest of kelp and other seaweed has recently been administratively prohibited, although personal collection of up to 2000 pounds of tissue does not require a permit and is largely unmonitored. Further north, Washington State has arguably the most conservative and scientifically sound management policies (Berry et al. 2001, 2005). Surveys of kelp abundance along the shores of the Olympic peninsula, conducted annually since 1989 by members of the Nearshore Habitat Program, provide the only recent, broad-scale, high-resolution, quantitative characterization of the population dynamics of *Nereocystis*. In spite of these relatively detailed stock assessments, the state of Washington prohibits the commercial harvest of *Nereocystis* and limits personal take to 10 pounds permit^{-1} day^{-1}. Unlike all other states and provinces where *Nereocystis* is found, the commercial leasing of subtidal lands and associated kelp beds that often serves as the basis for establishing the boundaries of kelp-harvesting operations is not permitted in Washington. In British Columbia, approximately a dozen stock assessment surveys have been conducted since the 1970s, but most of these were limited in their geographic breadth and did not involve the resampling that is necessary to estimate long-term patterns of kelp abundance at particular localities. The leasing of kelp beds is permitted, but leases must be reapproved each year, and take is limited to less than 20% of the total bed biomass. While coarsely defined permitting regulations are in place in Alaska, we found it exceedingly difficult to collect any information on the number of permits issued and estimates of algal biomass collected by harvesters.

The variability in management regulations is probably evidence of two factors that must be addressed for scientifically sound management policies to be enacted. First, more basic research needs to be done to characterize natural demographic dynamics of *Nereocystis* beds and quantify the effects of harvesting on these dynamics. Data generated by more regular and comprehensive stock assessment surveys can be used by governmental agencies to produce an allocation plan based on accurate knowledge of the kelp resource base and to identify harvestable areas and associated quotas. Commercial operations can also use these data to select locations and capacity requirement of their facilities. Studies of the impacts of harvesting on *Nereocystis* physiology, growth, and demography are critical for the development of management policies that will sustain both the profitability of commercial harvesting operations and the fundamental ecological patterns and process associated with *Nereocystis* beds. Second, greater methodological and legislative consistency among management entities is essential for the type of broad-scale, ecosystem-based approach needed to manage highly interconnected marine populations. Because replenishment of local populations of marine organisms is often a function of both local reproductive output and input from more distant sites, policy makers must look beyond political boundaries to develop biologically comprehensive management strategies. Stock assessment would be greatly facilitated if regulatory agencies from different states could collectively design a survey approach and agree to implement it on a more regular basis. Exchange of data generated by these surveys could be used to develop harvest regulations in a similar consensus-based manner. The use of a common approach to regulate and monitor the impacts of harvesting will facilitate (1) comparison of data from different geographic regions since information is collected using the same methods, (2) maintenance of more comprehensive and intelligible databases on stock assessments and harvest levels, and (3) enforcement of harvest regulations since broadly adopted policies reduce uncertainty about local regulations for both collectors and enforcement agents.

The importance of the lack of biological, economic, and management information for *Nereocystis*, as summarized, is underscored by the strong management implications of many life-history, morphological, physiological, and ecological characteristics of this species (Table 2). It is clear that any ecosystem-based approach to management practices for this and other macroalgae in coastal marine ecosystems will benefit from a strong understanding of the diversity of traits that define a species.

Research recommendations

The synthesis of the literature presented generated a series of recommendations for future research that could substantially improve our ability to manage human impacts on *Nereocystis*. These recommendations include studies designed to better our knowledge of (1) the status, dynamics, and use of *Nereocystis* populations; (2) impacts of harvest on the sustainability (resilience and replenishment) of *Nereocystis* populations; (3) impacts of harvest of *Nereocystis* on shallow reef ecosystems; and (4) impacts of harvest on other coastal marine ecosystems. These recommendations are listed in the order that we believe they should be prioritized.

Stock and resource assessment methodologies

Central to any resource management programme is knowledge of the status, dynamics, and use of the resource. Like *Macrocystis* stock assessments, aerial digital image-based measures of canopy cover appear to be the most cost-effective method for assessing kelp abundance and distribution, although multispectral imaging techniques show promise (Stekoll et al. 2006). However, the accuracy and precision of these traditional methods are questionable (e.g., Schiel et al. 2004), and variation in potentially confounding factors (e.g., tides, currents, sea conditions, atmospheric conditions, timing) must be more closely examined. Surveys to test the functional relationship between diver-based estimates of plant density and biomass and estimates of abundance from aerial surveys or images would provide ground-truthing of the aerial-based estimates and perhaps allow translation of canopy cover to biomass estimates. Thus, a well designed study of the use of aerial digital imagery is recommended as a foundation for a more comprehensive stock assessment. Like other recommendations that follow, and as mentioned in the preceding section, such efforts should be coordinated across states to ensure consistency in stock assessments over the range of the species.

Potential impacts of harvest on the resources

Relative impacts of different harvest methods The commercial and recreational harvest of *Nereocystis* includes several variables, the relative ecological impacts of which have not been tested. Such variables include hand versus mechanical harvest; whole-plant versus partial-blade removal; the relative extent (percentage of stand) and location (outer, middle, or inner) of harvest of a bed; and timing of harvest relative to plant phenology. For example, is there a threshold percentage of a forest that should not be harvested to ensure local re-establishment? Can the impact of harvest on growth and reproduction of *Nereocystis* be minimized by timing harvest according to plant phenology? Does phenology vary geographically and, if so, in what ways? Because some literature hints at the possibility of spores overwintering (a biennial rather than annual reproductive cycle), further exploration of this possibility should be conducted to determine if generations can overlap. Equally important will be the identification of one or more quantifiable metrics to be used to assess the impacts of different harvesting strategies. These metrics (e.g., abundance of adult plants in the year following harvest) would ideally be (1) directly linked to demographic patterns or rates associated with the *Nereocystis* populations being harvested and (2) easy to measure. Finally, indirect effects of harvesting on the distribution or abundance of adults, manifested via impacts on gametophytes, must be investigated. Because interannual changes in sporophyte abundance are inexorably linked by a gametophyte stage, more information is needed on gametophyte abundance, distribution, and longevity and the

Table 2 Implications of species traits and ecological patterns of bull kelp *Nereocystis luetkeana* for harvest management

Biological trait	Ecological pattern	Implications for harvest management/ ecosystem management
Morphological and physiological traits		
Non-buoyant stipe with pneumatocyst at the surface as sole source of buoyancy.	Loss of pneumatocyst leads to mortality, loss of entire alga.	Unlike giant kelp, surface harvest will cause bull kelp to sink and die; may be vulnerable to epiphyte smothering in nutrient-rich waters.
All reproductive and majority of photosynthetic tissue attached to pneumatocyst at the surface.	Individuals and populations persistent in areas of high turbidity.	Plants highly vulnerable to destructive activities at the surface: harvest, damage from boat props, retrieval of fishing gear and anchors.
High morphological plasticity of blade shape (narrower in high-surge areas).	Reduced productivity in areas of high surge.	Harvest rate may require spatial (habitat-based) regulation to accommodate variation in productivity.
Sensitive to water temperature.	Geographic and latitudinal changes in distribution reported in literature.	Potentially vulnerable to ocean temperature shifts, including frequency and intensity of El Niño; may result in shift to deeper water and constrain and eliminate populations simultaneously limited by light attenuation.
Early life stages (i.e., gametes, spores) sensitive to light availability.	Depending on water clarity, sedimentation, and abundance of competitors (other algae), sporadic reproductive success, high interannual variability.	Germination, growth, and survival, especially of gametophytes and young sporophytes, may be strongly influenced by runoff, sedimentation, eutrophication.
Very high rate of individual growth.	Rapid replenishment of harvested populations.	Proper management practice could maintain high production.
Sperm/egg fertilization pheromone driven.	High population density (close proximity of individuals) critical to success and rate of reproduction; combined with limited propagule dispersal, highly vulnerable to Allee effect and local extinction.	Local harvest rates need to ensure sufficient density and propagule production.
Life-history traits		
Annual life history.	Population dynamics highly variable in space and time; highly seasonal variation in abundance.	Highly sensitive to timing and magnitude of harvest. Vulnerable to local extinction if harvested before reproductive maturity and spore production.
Spore dispersal probably highly limited.	Interannual rates of recolonization highly dependent on rate of local spore production.	Highly sensitive to timing and magnitude of harvest. Vulnerable to local extinction if harvested before reproductive maturity and spore production.
Ecological traits		
Local populations form beds of high plant density.	Dominant or only bed-forming kelp across much of its range.	Provides much of the physical, 3-dimensional structure of shallow rocky reef ecosystem, increasing diversity of physical habitats and potentially altering water flow, which can influence delivery of larvae and nutrients.

Table 2 (continued)   Implications of species traits and ecological patterns of bull kelp
Nereocystis luetkeana for harvest management

Biological trait	Ecological pattern	Implications for harvest management/ ecosystem management
Forms surface canopy.	Predominant or only surface canopy-forming algae throughout much of its geographic range.	Removal of canopy may reduce rates of recruitment and survival of juvenile fishes and invertebrates.
Poor competitor, especially as gametophyte and juvenile sporophyte.	If removal of bull kelp enhances development of understory algae, can reduce regeneration rate.	Slower recovery in response to overharvest.
Strongly impacted by herbivore grazing.	Single stipe that cannot regenerate lends entire alga vulnerable to damage by grazers; grazers can cause both direct damage and indirectly influence growth of gametophytes and young sporophytes by grazing on benthic algae, which compete for light and space.	Management of impacts on herbivore grazers and their predators can have strong direct and indirect effects on bull kelp.
Distribution is largely limited by competition to high-surge environments.	Association with sites of high wave exposure may limit accessibility for harvest.	Harvest may be inaccessible over large portions of the population and disproportionately high at sites with greater accessibility.

sensitivity of these factors to environmental changes that may arise from harvesting of sporophytes (e.g., sedimentation, light and nutrient availability, competition for space with understory algae and invertebrates). The recent development of a molecular technique to detect the microscopic life stages may greatly facilitate the collection of relevant information (Fox & Swanson 2007).

Spatial components of replenishment of harvested populations

Virtually nothing is known about spatial patterns of population connectivity (i.e., the transport of spores from one population to another) in *Nereocystis*. Such information is key to determining the distances over which dispersal from neighbouring kelp forests can be expected to help replenish harvested forests. What is the dispersal kernel for *Nereocystis* spores, and does this vary regionally? This information will help determine the relative vulnerability of forests based in part on their size and isolation. There is evidence that *Macrocystis* spores must settle at a threshold density to ensure successful subsequent fertilization between the male and female gametophytes. At lower densities, gametophytes are separated from one another by too great a distance for successful encounters of eggs and sperm to occur with appreciable frequency. Whether this Allee effect holds true for *Nereocystis* is unknown. If present, it could have profound implications for the level of harvest and remaining density of reproductive plants necessary to ensure replenishment of a forest. In addition to field-based studies of dispersal, analysis of population genetic structure would be instrumental in assessing population connectivity and would be useful in identifying regions with unique genetic composition with an eye toward preserving the genetic diversity of the species at the biogeographic scale.

Potential impact of harvest on kelp forest ecosystems

This review of the literature revealed a startling paucity of research on the role of *Nereocystis* in shallow rocky reef ecosystems. With the exception of a few widely separated studies, few rigorous assessments of the presence or density of *Nereocystis* on the structure and functions of associated algal, invertebrate, or fish assemblages have been made. Because of its ephemeral occurrence and lack of structural complexity, *Nereocystis* may not be nearly as influential as *Macrocystis* on kelp forest communities. However, because it is often the only source of habitat that extends to the water surface in many localities, and juvenile fishes have been observed to strongly associate with the kelp canopy, it may in fact be a particularly important source of habitat. Moreover, because of its rapid growth rate and production of detritus, *Nereocystis* may be an important source of nutrient resources for key species in shallow reef ecosystems (e.g., sea urchins, abalone). Some important questions that remain unanswered include the following: Are there invertebrate or fish species that are strongly influenced by the presence of *Nereocystis* forests, such that harvesting indirectly effects their local and regional distribution and abundance, and how does this vary geographically? What is the competitive relationship between *Nereocystis* and other species such that removal of a forest at a particular time allows competitors to increase and preempt regeneration (as has been observed between *Macrocystis* and *Sargassum* in some regions)? What is the role of *Nereocystis* in the detrital pathway on shallow reefs and the sustainability of other resources (e.g., sea urchins, abalone)? Answers to these questions will require both surveys to examine the generality of relationships and experiments to definitively test for causality in these relationships.

Potential impacts of harvest on other coastal marine ecosystems

Like *Macrocystis*, the great amount of biomass produced by and lost from *Nereocystis* forests each year can be transported to adjacent ecosystems on shore (e.g., sandy beaches, rocky intertidal areas) and off shore (e.g., deep rocky reefs, submarine canyons), where it fuels detritus-based trophic pathways and creates temporary habitat structure. The magnitude and consequence of this connectivity among ecosystems is poorly understood. What is the role of *Nereocystis* in maintaining connectivity between the nearshore kelp forests and other marine ecosystems, and how will reduction of canopy export to these ecosystems influence their structure and functions? Studies designed to survey and manipulate this influx would provide useful insight into the importance of this process.

Acknowledgments

We thank Margaret Bowman and Charlotte Hudson of the Lenfest Ocean Program at the Pew Charitable Trusts for supporting the development of this review and for their incredible patience during its writing. Jennifer Bloeser and Caroline Gibson at the Pacific Marine Conservation Council also provided valuable assistance in coordinating the completion of the review. Dan Malone contributed valuable comments and assisted in creating some of the figures. The manuscript benefited greatly from thoughtful reviews by Drs Michael Foster and Michael Graham. We are most grateful to the many individuals associated with the harvest or management of *Nereocystis luetkeana* through non-profit organizations, government agencies, and private business, who responded to our queries and shared their knowledge of bull kelp with us. Some of the publication costs were paid by the Partnership for Interdisciplinary Studies of Coastal Oceans (PISCO). PDF versions of much of the literature cited in this report are available by request from the corresponding author.

References

Abbott, I.A. & Hollenberg, G.J. 1976. *Marine Algae of California.* Stanford, California: Stanford University Press.

Ahn, O., Petrell, R.J. & Harrison, P.J. 1998. Ammonium and nitrate uptake by *Laminaria saccharina* and *Nereocystis luetkeana* originating from a salmon sea cage farm. *Journal of Applied Phycology* **10**, 333–340.

Amsler, C.D. & Neushul, M. 1989. Diel periodicity of spore release from the kelp *Nereocystis luetkeana* (Mertens) Postels et Ruprecht. *Journal of Experimental Marine Biology and Ecology* **134**, 117–127.

Andrews, H.L. 1925. Animals living on kelp. *Puget Sound Marine Station Publications* **5**, 25–27.

Antrim, L.D., Thom, R.M., Gardiner, W.W., Cullinan, V.I., Shreffler, D.K. & Bienert, R.W. 1995. Effects of petroleum products on bull kelp (*Nereocystis luetkeana*). *Marine Biology* **122**, 23–31.

Beck, M.W., Heck, K.L., Able, K.W., Childers, D.L., Eggleston, D.B., Gillanders, B.M., Halpern, B., Hays, C.G., Hoshino, K., Minello, T.J., Orth, R.J., Sheridan, P.F. & Weinstein, M.R. 2001. The identification, conservation, and management of estuarine and marine nurseries for fish and invertebrates. *Bioscience* **51**, 633–641.

Berry, H.D., Sewell, A., & Wagenen, B.V. 2001. Temporal trends in the areal extent of canopy forming kelp beds along the Strait of Juan de Fuca and Washington's outer coast. In *Proceedings of 2001 Puget Sound Research Conference.* Droscher, T.W (ed.). Puget Sound Action Team, Olympia, Washington.

Berry, H.D., Mumford, T.F., & Dowty, P. 2005. Using historical data to estimate changes in floating kelp (*Nereocystis luetkeana* and *Macrocystis integrifolia*) in Puget Sound, Washington. In *Proceedings of the 2005 Puget Sound Georgia Basin Research Conference.* T.W. Droscher and D.A Fraser (eds). Puget Sound Action Team, Olympia, Washington.

Bodkin, J.L. 1986. Fish assemblages in *Macrocystis* and *Nereocystis* kelp forests off central California. *Fishery Bulletin* **84**, 799–808.

Breen, P.A., Miller, D.C. & Adkins, B.E. 1976. An examination of harvested sea urchin populations in the Tofino area. Manuscript Report Series 1401, Fisheries Research Board of Canada, Pacific Biological Station, Nanaimo, British Columbia.

Britton-Simmons, K., Eckman, J.E. & Duggins, D.O. 2008. Effect of tidal currents and tidal stage on estimates of bed size in the kelp *Nereocystis luetkeana*. *Marine Ecology Progress Series* **355**, 95–105.

Brown, L.B. 1915. Experiments with marine algae in freshwater. *Puget Sound Biological Station Publication* **1**, 31–34.

Brzezinski, M.A., Reed, D.C. & Amsler, C.D. 1993. Neutral lipids as major storage products in zoospores of the giant kelp *Macrocystis pyrifera* (Phaeophyceae). *Journal of Phycology* **29**, 16–23.

Burge, R.T. & Schultz, S.A. 1973. The marine environment in the vicinity of Diablo Cove with special reference to abalone and bony fishes. Marine Research Technical Report 19, California Department of Fish and Game, Long Beach, California, 1–429.

Bushing, W.W. 1994. Biogeographical and ecological implications of kelp rafting as a dispersal vector for marine invertebrates. In *Proceedings of the Fourth California Islands Symposium: Update on the Status of Resources, March 22–25, 1994*, W. Halvorson & G. Maender (eds). Santa Barbara Museum of Natural History, Santa Barbara, California, 103–110.

Calvert, E.L. 2005. *Kelp beds as fish and invertebrate habitat in southeastern Alaska.* MS thesis, University of Alaska, Fairbanks.

Cameron, F.K. 1912. A preliminary report on the fertilizer resources of the United States. Senate Document 190, pp. 1–290, U.S. 62nd Congress, 2nd Session, U.S. Department of Agriculture, Washington, D.C.

Carr, M.H. 1989. Effects of macroalgal assemblages on the recruitment of temperate zone reef fishes. *Journal of Experimental Marine Biology and Ecology* **126**, 59–76.

Carr, M.H. 1991. Habitat selection and recruitment of an assemblage of temperate zone reef fishes. *Journal of Experimental Marine Biology and Ecology* **146**, 113–137.

Carr, M.H. 1994. Effects of macroalgal dynamics on recruitment of a temperate reef fish. *Ecology* **75**, 1320–1333.

Carr, M.H. & Syms, C. 2006. Recruitment. In *The Ecology of Marine Fishes: California and Adjacent Waters*, L.G. Allen et al. (eds). Berkeley, California: University of California Press, 411–427.

Chenelot, H. & Konar, B. 2007. *Lacuna vincta* (Mollusca, Neotaenioglossa) herbivory on juvenile and adult *Nereocystis luetkeana* (Heterokontophyta, Laminariales). *Hydrobiologia* **583**, 107–118.

Chenelot, H., Matweyou, J. & Konar, B. Investigation of the overwintering of the annual macroalga *Nereocystis luetkeana* in Kachemak Bay, Alaska. 2001. In *Cold Water Diving for Science. Proceedings of the 21st Annual Scientific Diving Symposium*, SC Jewett (ed.). American Academy of Underwater Sciences. University of Alaska Sea Grant, AK-SG-01-06, Fairbanks.

Collins, R., Wendell, F., Kalvass, P., Ota, B., Kashiwada, J., Tanaguchi, I., King, A., Larson, M., Gross, J., Wright, N., O'Brien, J., Bedford, D. & Veisze, P. 2000a. Environmental settings. In *2000 Final Environmental Document—Giant and Bull Kelp Commercial and Sport Fishing Regulations*, State Clearinghouse Number 2000012089, Chapter 3. Sacramento, California: California Department of Fish and Game, 155. Available HTTP: http://www.dfg.ca.gov/marine/kelp_ceqa/ (accessed 12 March 2009).

Collins, R., Wendell, F., Kalvass, P., Ota, B., Kashiwada, J., Tanaguchi, I., King, A., Larson, M., Gross, J., Wright, N., O'Brien, J., Bedford, D. & Veisze, P. 2000b. Environmental impacts. In *2000 Final Environmental Document—Giant and Bull Kelp Commercial and Sport Fishing Regulations*, State Clearinghouse Number 2000012089, Chapter 4. Sacramento, California: California Department of Fish and Game, 155. Available HTTP: http://www.dfg.ca.gov/marine/kelp_ceqa/ (accessed 28 Feb 2009).

Colombini, I. & Chelazzi, L. 2003. Influence of marine allochthonous input on sandy beach communities. *Oceanography and Marine Biology An Annual Review* **41**, 115–159.

Cox, K.W. 1962. California abalone, family Haliotidae. Fishery Bulletin 118, California Department of Fish and Game, Sacramento, California, 1–133.

Coyer, J.A., Smith, G.J. & Andersen, R.A. 2001. Evolution of *Macrocystis* spp. (Phaeophyceae) as determined by ITS1 and ITS2 sequences. *Journal of Phycology* **37**, 574–585.

Dawson, E.Y. 1966. *Marine Botany, an Introduction.* New York: Holt, Rinehart and Winston.

Dayton, P.K. 1985. Ecology of kelp communities. *Annual Review of Ecology and Systematics* **16**, 215–246.

Dayton, P.K., Currie, V., Gerrodette, T., Keller, B.D., Rosenthal, R. & Ventresca, D. 1984. Patch dynamics and stability of some California kelp communities. *Ecological Monographs* **54**, 253–289.

Dayton, P.K., Tegner, M.J., Parnell, P.E. & Edwards, P.B. 1992. Spatial and temporal patterns of disturbance and recovery in a kelp forest community. *Ecological Monographs* **62**, 421–445.

Dean, T.A., Haldorson, L., Laur, D.R., Jewett, S.C. & Blanchard, A. 2000. The distribution of nearshore fishes in kelp and eelgrass communities in Prince William Sound, Alaska: associations with vegetation and physical habitat characteristics. *Environmental Biology of Fishes* **57**, 271–287.

Denny, M.W., Gaylord, B.P. & Cowen, E.A. 1997. Flow and flexibility—II. The roles of size and shape in determining wave forces on the bull kelp *Nereocystis luetkeana*. *Journal of Experimental Biology* **200**, 3165–3183.

Dickson, L.G. & Waaland, J.R. 1984. Conchocelis growth sporulation and early blade development in *Porphyra nereocystis*. *Journal of Phycology* **20**.

Dickson, L.G. & Waaland, J.R. 1985. *Porphyra nereocystis*—a dual daylength seaweed. *Planta* **165**, 548–553.

Druehl, L.D. 1970. The pattern of Laminariales distribution in the northeast Pacific. *Phycologia* **9**, 237–247.

Druehl, L.D., Collins, J.D., Lane, C.E. & Saunders, G.W. 2005. An evaluation of methods used to assess intergeneric hybridization in kelp using Pacific Laminariales (Phaeophyceae). *Journal of Phycology* **41**, 250–262.

Duggins, D., Eckman, J.E., Siddon, C.E. & Klinger, T. 2001. Interactive roles of mesograzers and current flow in survival of kelps. *Marine Ecology Progress Series* **223**, 143–155.

Duggins, D.O. 1980. Kelp beds and sea otters—an experimental approach. *Ecology* **61**, 447–453.

Duncan, M.J. & Foreman, R.E. 1980. Phytochrome-mediated stipe elongation in the kelp *Nereocystis* (Phaeophyceae). *Journal of Phycology* **16**, 138–142.

Edwards, M.S. 2000. The role of alternate life-history stages of a marine macroalga: a seed bank analogue? *Ecology* **81**, 2404–2415.

Estes, J.A. & Duggins, D.O. 1995. Sea otters and kelp forests in Alaska: Generality and variation in a community ecological paradigm. *Ecological Monographs* **65**, 75–100.

Field, J.C., Francis, R.C. & Aydin, K. 2006. Top-down modeling and bottom-up dynamics: linking a fisheries-based ecosystem model with climate. *Progress in Oceanography* **68**, 238–270.

Foreman, R.E. 1970. *Physiology, ecology, and development of the brown alga* Nereocystis luetkeana *(Mertens) P. & R.* PhD thesis, University of California Berkeley.

Foreman, R.E. 1975. KIM-1: a method for inventory of floating kelps and its application to selected areas of Kelp License Area 12. Benthic Ecological Research Program Report 75-1, Federal Fisheries and Marine Service and Provincial Marine Resources Branch, Victoria, British Columbia, 1–81.

Foreman, R.E. 1977a. Benthic community modification and recovery following intensive grazing by *Strongylocentrotus droebachiensis*. *Helgoländer Wissenschäftliche Meeresuntersuchungen* **30**, 468–484.

Foreman, R.E. 1977b. Ecological studies of *Nereocystis luetkeana*. 1. Population-dynamics and life-cycle strategy in different environments. *Journal of Phycology* **13**, 78–78.

Foreman, R.E. 1984. Studies on *Nereocystis* growth in British Columbia, Canada. *Hydrobiologia* **116**, 325–332.

Foster, M.S. 1982. The regulation of macro algal associations in kelp forests. In *Synthetic and Degradative Processes in Marine Macrophytes*, L.M. Srivastava (ed.). Bamfield, British Columbia, Canada: Walter de Gruyter, 185–206.

Foster, M.S. & Schiel, D.R. 1985. The ecology of giant kelp forests in California: A community profile. *U.S. Fish and Wildlife Service Biological Report* **85**,1-152.

Foster, M.S., Agegian, C.R., Cowen, R.K., Van Wagenan, R.F., Rose, D.K. & Hurley, A.C. (1979). Toward an understanding of the effects of sea otter foraging on kelp forest communities in central California. Final report to the U.S. Marine Mammal Commission, Contract No. MM7AC023; National Technical Information Service, Springfield, Virginia. (Publ. No. PB293891)

Fox, C.H. & Swanson, A.K. 2007. Nested PCR detection of microscopic life-stages of laminarian macroalgae and comparison with adult forms along intertidal height gradients. *Marine Ecology Progress Series* **332**, 1–10.

Fox, D., Amend, M., Merems, A., Miller B. & J. Golden. 1998. 1998 Nearshore Rocky Reef Assessment. Coastal Zone Management Section 309 Final Report, Contract No. 99-020. Newport, Oregon: Oregon Department of Fish and Wildlife, Marine Resources Program, 53 pp.

Francis, R.C., Hixon, M.A., Clarke, M.E., Murawski, S.A. & Ralston, S. 2007. Fisheries management—ten commandments for ecosystem-based fisheries scientists. *Fisheries* **32**, 217–233.

Frye, T.C. 1915. The kelp beds of southeast Alaska. In *Potash From Kelp*, F.K. Cameron (ed.). Washington, DC: United States Department of Agriculture, report 100, part IV, 60–104.

Giver, K. 1999. *Effects of the invasive seaweed* Sargassum muticum *on marine communities in northern Puget Sound, Washington*. MS thesis, Western Washington University, Bellingham.

Gotshall, D.W., Raymond Ally, J.R., Vaughan, D.L., Hatfield, B.B. & Law, P. 1986. Pre-operational baseline studies of selected nearshore marine biota at the Diablo Canyon power plant site: 1979-1982., Long Beach, California. California Department of Fish and Game, (Marine Resources Technical Report, 50)

Gotshall, D.W., Laurent, L.L., Owen, S.L., Grant, J.J. & Law, P. 1984. A quantitative ecological study of selected nearshore marine plants and animals at the Diablo Canyon power plant site: a pre-operational baseline, 1973-1978. Long Beach, California. California Department of Fish and Game, (Marine Resources Technical Report, 48)

Graham, M.H. 2004. Effects of local deforestation on the diversity and structure of southern California giant kelp forest food webs. *Ecosystems* **7**, 341-357.

Graham, M.H., Halpern, B.S. & Carr, M.H. 2008. Diversity and dynamics of California subtidal kelp forests. In *Marine Sublittoral Food Webs*, T.R. McClanahan & G.M. Branch (eds). Oxford: Oxford University Press, 103-134.

Guiry, M.D. & Guiry, G.M. 2010. AlgaeBase. World-wide electronic publication, National University of Ireland, Galway. Available HTTP: http://www.algaebase.org (accessed 27 March 2009).

Haldorson, L. & L.J. Richards. 1987. Post-larval copper rockfish in the Strait of Georgia: habitat use, feeding, and growth in the first year. In *Proceedings of the International Rockfish Symposium*, B.R. Melteff (ed.). Anchorage, Alaska: Alaska Sea Grant Report 87-2. p. 129–142.

Hamilton, J. & Konar, B. 2007. Implications of substrate complexity and kelp variability for south-central Alaskan nearshore fish communities. *Fishery Bulletin* **105**, 189–196.

Hansen, G.I. & Mumford, T.F. 1995. 1994/1995 Regulations for seaweed harvesting on the west coast of North America. Conference handout distributed at 1994 Western Society of Naturalists Meeting in Monterey, California, 1–9. Available HTTP: http://www.oregonstate.edu/~hanseng/Regulations%20Paper.pdf (accessed 4 Feb 2009).

Harrold, C., Light, K. & Lisin, S. 1998. Organic enrichment of submarine-canyon and continental-shelf benthic communities by macroalgal drift imported from nearshore kelp forests. *Limnology and Oceanography* **43**, 669–678.

Hillmann, L.G. 2005. Summary of regulations for seaweed harvesting along the west coast of North America. Oregon Parks and Recreation Department, Salem, Oregon, 1–12. Available HTTP: http://www.egov. oregon.gov/OPRD/NATRES/docs/AlgaeRegulationsSummary.pdf (accessed 21 Feb 2009).

Hobday, A.J. 2000. Persistence and transport of fauna on drifting kelp (*Macrocystis pyrifera* L.C. Agardh) rafts in the Southern California Bight. *Journal of Experimental Marine Biology and Ecology* **253**, 75–96.

Hobson, E.S. & Chess, J.R. 1988. Trophic relations of the blue rockfish, *Sebastes mystinus*, in a coastal upwelling system off northern California. *Fishery Bulletin* **86**, 715–743.

Hurd, A.M. 1916. Factors influencing the growth and distribution of *Nereocystis luetkeana*. *Puget Sound Marine Station Publications* **1**, 185–197.

Hurd, C.L. & Stevens, C.L. 1997. Flow visualization around single- and multiple-bladed seaweeds with various morphologies. *Journal of Phycology* **33**, 360–367.

Hurd, C.L., Stevens, C.L., Laval, B.E., Lawrence, G.A. & Harrison, P.J. 1997. Visualization of seawater flow around morphologically distinct forms of the giant kelp *Macrocystis integrifolia* from wave-sheltered and exposed sites. *Limnology and Oceanography* **42**, 156–163.

James, D.E., Manley, S.L., Carter, M.C. & North, W.J. 1987. Effects of PCBs and hydrazine on life processes in microscopic stages of selected brown seaweeds. *Hydrobiologia* **151**, 411–415.

Johnson, A.S. & Koehl, M.A.R. 1994. Maintenance of dynamic strain similarity and environmental stress factor in different flow habitats— thallus allometry and material properties of a giant kelp. *Journal of Experimental Biology* **195**, 381–410.

Johnson, S.W., Murphy, M.L., Csepp, D.J., Harris, P.M. & Thedinga, J.F. 2003. A survey of fish assemblages in eelgrass and kelp habitats of southeastern Alaska. U.S Department of Commerce, NOAA Technical Memorandum NMFS-AFSC-139, 39 p, Alaska Fisheries Science Center, Seattle, Washington.

Kalvass, P., Larson, M. & O'Brien, J. 2001. Bull kelp. In *California's Living Marine Resources: A Statue Report*, W.S. Leet et al. (eds). California Department of Fish and Game, Sacramento California, 282–284. Available HTTP: http://www.dfg.ca.gov/marine/status/status2001.asp (accessed 1 March 2009).

Kildow, J. & Colgan, C.S. 2005. California's ocean economy. Prepared for the California Resources Agency by the National Ocean Economics Program, Moss Landing, California. Available HTTP: http://www.noep.mbari.org/Download/ (accessed 9 March 2009).

Kim, S.L. 1992. The role of drift kelp in the population ecology of a *Diopatra ornata* Moore (Polychaeta, Onuphidae) ecotone. *Journal of Experimental Marine Biology and Ecology* **156**, 253–272.

Kingsford, M.J. 1992. Drift algae and small fish in coastal waters of northeastern New Zealand. *Marine Ecology Progress Series* **80**, 41–55.

Koehl, M.A.R. & Alberte, R.S. 1988. Flow, flapping, and photosynthesis of *Nereocystis luetkeana*—a functional comparison of undulate and flat blade morphologies. *Marine Biology* **99**, 435–444.

Koehl, M.A.R., Silk, W.K., Liang, H. & Mahadevan, L. 2008. How kelp produce blade shapes suited to different flow regimes: a new wrinkle. *Integrative and Comparative Biology* **48**, 834–851.

Koehl, M.A.R. & Wainwright, S.A. 1977. Mechanical adaptations of a giant kelp. *Limnology and Oceanography* **22**, 1067–1071.

Kokita, T. & Omori, M. 1998. Early life history traits of the gold-eye rockfish, *Sebastes thompsoni*, in relation to successful utilization of drifting seaweed. *Marine Biology* **132**, 579–589.

Konar, B. & Roberts, C. 1996. Large scale landslide effects on two exposed rocky subtidal areas in California. *Botanica Marina* **39**, 517–524.

Lane, C.E., Mayes, C., Druehl, L.D. & Saunders, G.W. 2006. A multi-gene molecular investigation of the kelp (Laminariales, Phaeophyceae) supports substantial taxonomic re-organization. *Journal of Phycology* **42**, 493–512.

Leaman, B.M. 1980. The ecology of fishes in British Columbia kelp beds. I. Barkley Sound *Nereocystis* beds. Fisheries Development Report 22, British Columbia Ministry of the Environment, Nanaimo, British Columbia.

Leslie, H.M. & McLeod, K.L. 2007. Confronting the challenges of implementing marine ecosystem-based management. *Frontiers in Ecology and the Environment* **5**, 540–548.

Levin, S.A. & Lubchenco, J. 2008. Resilience, robustness, and marine ecosystem-based management. *Bioscience* **58**, 27–32.

Lewis, R.J. & Neushul, M. 1995. Intergeneric hybridization among five genera of the family lessoniaceae (Phaeophyceae) and evidence for polyploidy in a fertile *Pelagophycus* × *Macrocystis* hybrid. *Journal of Phycology* **31**, 1012–1017.

Little, A.D. 2002. Commercial and recreational fishing/kelp harvesting. Final Environmental Impact Report for the Tranquillon Ridge Oil and Gas Development Project, LOGP Produced Water Treatment System Project, Sisquoc Pipeline Bi-Directional Flow Project. Clearinghouse Number 2000071130, County of Santa Barbara Planning and Development Department, Santa Barbara, California.

Lovas, S.M. & Torum, A. 2001. Effect of the kelp *Laminaria hyperborea* upon sand dune erosion and water particle velocities. *Coastal Engineering* **44**, 37–63.

Luning, K. & Freshwater, W. 1988. Temperature tolerance of northeast Pacific marine algae. *Journal of Phycology* **24**, 310–315.

Mackey, M. 2006. *Protecting Oregon's bull kelp*. Astoria, Oregon: Pacific Marine Conservation Council, 1–12.

Maier, I. & Muller, D.G. 1986. Sexual pheromones in algae. *Biological Bulletin (Woods Hole)* **170**, 145–175.

Maier, I., Muller, D.G., Gassmann, G., Boland, W. & Jaenicke, L. 1987. Sexual pheromones and related egg secretions in Laminariales (Phaeophyta). *Zeitschrift für Naturforschung Section C Biosciences* **42**, 948–954.

Malloch, S. 2000. Marine plant management and opportunities in British Columbia. Prepared for BC Fisheries—Sustainable Economic Development Branch, British Columbia.

Marasco, R.J., Goodman, D., Grimes, C.B., Lawson, P.W., Punt, A.E. & Quinn, T.J. 2007. Ecosystem-based fisheries management: some practical suggestions. *Canadian Journal of Fisheries and Aquatic Sciences* **64**, 928–939.

Markham, J.W. 1969. Vertical distribution of epiphytes on the stipe of *Nereocystis luetkeana*. *Syesis* **2**, 227–240.

Maxell, B.A. & Miller, K.A. 1996. Demographic studies of the annual kelps *Nereocystis luetkeana* and *Costaria costata* (Laminariales, Phaeophyta) in Puget Sound, Washington. *Botanica Marina* **39**, 479–489.

McLean, J.H. 1962. Sublittoral ecology of kelp beds of open coast area near Carmel, California. *Biological Bulletin (Woods Hole)* **122**, 95–114.

McLeod, K.L. & Leslie, H.M. 2009. *Ecosystem-Based Management for the Oceans*. Chicago: Island Press.

McPhaden, M.J. 1999. Genesis and evolution of the 1997–98 El Niño. *Science* **283**, 950–954.

Mews, M., Zimmer, M. & Jelinski, D.E. 2006. Species-specific decomposition rates of beach-cast wrack in Barkley Sound, British Columbia, Canada. *Marine Ecology Progress Series* **328**, 155–160.

Miller, K.A. & Estes, J.A. 1989. Western range extension for *Nereocystis luetkeana* in the North Pacific Ocean. *Botanica Marina* **32**, 535–538.

Monterey Bay National Marine Sanctuary. 2000. Monterey Bay National Marine Sanctuary final kelp management report—background, environmental setting, and recommendations. Monterey Bay National Marine Sanctuary, Monterey, California, 1–54. Available HTTP: http://www.montereybay.noaa.gov/research/kelpreport/kelpreportfinal.pdf (accessed 28 March 2009).

Muller, D.G. 1981. Sexuality and sex attraction. In *The Biology of Seaweeds*, C.S. Lobban & M.J. Wynne (eds). Berkeley, California: University of California Press, 661–674.

Nicholson, N.L. 1970. Field studies on the giant kelp *Nereocystis*. *Journal of Phycology* **6**, 177–182.

Nicholson, N.L. & Briggs, W.R. 1972. Translocation of photosynthate in the brown alga *Nereocystis*. *American Journal of Botany* **59**, 97–106.

Norkko, J., Bonsdorff, E. & Norkko, A. 2000. Drifting algal mats as an alternative habitat for benthic invertebrates: species specific responses to a transient resource. *Journal of Experimental Marine Biology and Ecology* **248**, 79–104.

Norton, T.A. 1992. Dispersal by macroalgae. *British Phycological Journal* **27**, 293–301.

O'Clair, R.M. & Lindstron, S.C. 2000. *North Pacific Seaweeds*. Auk Bay, Alaska: Plant Press.

Orr, M., Zimmer, M., Jelinski, D.E. & Mews, M. 2005. Wrack deposition on different beach types: spatial and temporal variation in the pattern of subsidy. *Ecology* **86**, 1496–1507.

Pace, D. 1981. Kelp community development in Barkley Sound, British Columbia following sea urchin removal. In *Proceedings of the Eighth International Seaweed Symposium, August 18–23, 1974, North Wales*, G.E. Fogg & W.E. Jones (eds). Menai Bridge, Wales: Marine Science Laboratories, 457–463.

Pacific Gas and Electric Company. 1987. Thermal effects monitoring program. 1986 annual report. Diablo Canyon Power Plant. Submitted to the Central Coast Regional Water Quality Control Board, San Luis Obispo, California. Prepared by Tenera Environmental Inc. (formerly TERA Corp.) for Pacific Gas and Electric Company, San Francisco, California, DCL-87-087.

Paine, R.T. 2002. Trophic control of production in a rocky intertidal community. *Science* **296**, 736–739.

Paine, R.T. & Vadas, R.L. 1969. The effects of grazing by sea urchins, *Strongylocentrotus* spp. on benthic algal populations. *Limnology and Oceanography* **14**, 710–719.

Pearse, J.S. & Hines, A.H. 1979. Expansion of a central California kelp forest following the mass mortality of sea urchins. *Marine Biology* **51**, 83–91.

Pennings, S.C., Carefoot, T.H., Zimmer, M., Danko, J.P. & Ziegler, A. 2000. Feeding preferences of supralittoral isopods and amphipods. *Canadian Journal of Zoology* **78**, 1918–1929.

Reed, D.C. 1990. The effects of variable settlement and early competition on patterns of kelp recruitment. *Ecology* **71**, 776–787.

Reed, D.C., Amsler, C.D. & Ebeling, A.W. 1992. Dispersal in kelps—factors affecting spore swimming and competence. *Ecology* **73**, 1577–1585.

Reed, D.C., Laur, D.R. & Ebeling, A.W. 1988. Variation in algal dispersal and recruitment—the importance of episodic events. *Ecological Monographs* **58**, 321–335.

Reed, D.C., Neushul, M. & Ebeling, A.W. 1991. Role of settlement density on gametophyte growth and reproduction in the kelps *Pterygophora californica* and *Macrocystis pyrifera* Phaeophyceae. *Journal of Phycology* **27**, 361–366.

Rigg, G.B. 1915. The kelp beds of western Alaska. In *Potash From Kelp*, F.K. Cameron (ed.). Washington, DC: United States Department of Agriculture, report 100, part V, 105–122.

Roland, W.G. 1985. Effects of lamina harvest on the bull kelp, *Nereocystis luetkeana*. *Canadian Journal of Botany* **63**, 333–336.

Ronnback, P., Kautsky, N., Pihl, L., Troell, M., Soerqvist, T. & Wennhage, H. 2007. Ecosystem goods and services from Swedish coastal habitats: identification, valuation, and implications of ecosystem shifts. *Ambio* **36**, 534–544.

Rosell, K.G. & Srivastava, L.M. 1984. Seasonal variation in the chemical constituents of the brown algae *Macrocystis integrifolia* and *Nereocystis luetkeana*. *Canadian Journal of Botany* **62**, 2229–2236.

Rosell, K.G. & Srivastava, L.M. 1985. Seasonal variations in total nitrogen, carbon and amino acids in *Macrocystis integrifolia* and *Nereocystis luetkeana* (Phaeophyta). *Journal of Phycology* **21**, 304–309.

Sanbonsuga, Y. & Neushul, M. 1978. Hybridization of *Macrocystis* (Phaeophyta) with other float-bearing kelps. *Journal of Phycology* **14**, 214–224.

Santelices, B. 1990. Patterns of reproduction, dispersal and recruitment in seaweeds. *Oceanography and Marine Biology An Annual Review* **28**, 177–276.

Saunders, G.W. & Druehl, L.D. 1991. Restriction enzyme mapping of the nuclear ribosomal cistron in selected Laminariales (Phaeophyta)—a phylogenetic assessment. *Canadian Journal of Botany* **69**, 2647–2654.

Saunders, G.W. & Druehl, L.D. 1993. Revision of the kelp family Alariaceae and the taxonomic affinities of *Lessoniopsis reinke* (Laminariales, Phaeophyta). *Hydrobiologia* **261**, 689–697.

Scagel, R.F. 1947. An investigation on marine plants near Hardy Bay, B.C. Provincial Department of Fisheries, Victoria, British Columbia, Canada.

Schiel, D.R., Steinbeck, J.R. & Foster, M.S. 2004. Ten years of induced ocean warming causes comprehensive changes in marine benthic communities. *Ecology* **85**, 1833–1839.

Schmitz, K. & Srivastava, L.M. 1976. Fine structure of sieve elements of *Nereocystis luetkeana*. *American Journal of Botany* **63**, 679–693.

Schoch, G.C. & Chenelot, H. 2004. The role of estuarine hydrodynamics in the distribution of kelp forests in Kachemak Bay, Alaska. *Journal of Coastal Research* Special issue 45, Fall 2004, 179–194.

Selivanova, O.N. & Zhigadlova, G.G. 1997. Marine algae of the Commander Islands preliminary remarks on the revision of the flora .2. (Phaeophyta). *Botanica Marina* **40**, 9–13.

Setchell, W.A. & Gardner, N.L. 1925. *The Marine Algae of the Pacific Coast of North America*. Berkeley, CA: University of California Press.

Shaffer, J.A. 2002. Nearshore habitat mapping of the central and western strait of Juan de Fuca II. Preferential use of nearshore kelp habitats by juvenile salmon and forage fish. NOAA Technical Report to State of Washington Department of Fish and Wildlife G0100155, Washington Department of Fish and Wildlife, Port Angeles, Washington.

Shaffer, J.A. & Parks, D.S. 1994. Seasonal variations in and observations of landslide impacts on the algal composition of a Puget Sound nearshore kelp forest. *Botanica Marina* **37**, 315–323.

Siddon, E.C., Siddon, C.E. & Stekoll, M.S. 2008. Community level effects of *Nereocystis luetkeana* in southeastern Alaska. *Journal of Experimental Marine Biology and Ecology* **361**, 8–15.

Smith, B.D. & Foreman, R.E. 1984. An assessment of seaweed decomposition within a southern Strait of Georgia seaweed community. *Marine Biology* **84**, 197–205.

Steele, M.A. & Anderson, T.W. 2006. Predation. In *The Ecology of California Marine Fishes*, L.G. Allen et al. (eds). Berkeley, California: University of California Press, 428–448.

Stekoll, M.S., Deysher, L.E. & Hess, M. 2006. A remote sensing approach to estimating harvestable kelp biomass. *Journal of Applied Phycology* **18**, 323–334.

Strathmann, R. 1974. The spread of sibling larvae of sedentary marine invertebrates. *American Naturalist* **108**, 29–44.

Swanson, A.K. & Fox, C.H. 2007. Altered kelp (Laminariales) phlorotannins and growth under elevated carbon dioxide and ultraviolet-B treatments can influence associated intertidal food webs. *Global Change Biology* **13**, 1696–1709.

Tegner, M.J. & Dayton, P.K. 1987. El Niño effects on southern California kelp bed communities. *Advances in Ecological Research* **17**, 243–279.

Tenore, K.R., Hanson, R.B., McClain, J., Maccubbin, A.E. & Hodson, R.E. 1984. Changes in composition and nutritional value to a benthic deposit feeder of decomposing detritus pools. *Bulletin of Marine Science* **35**, 299–311.

TERA Corporation. 1982. Diablo Canyon power plant thermal discharge assessment. Pacific Gas and Electric Company, San Francisco, California (now Tenera Environmental Inc., San Louis Obispo, California).

Thiel, M. & Gutow, L. 2005. The ecology of rafting in the marine environment. II. The rafting organisms and community. In *Oceanography and Marine Biology An Annual Review* **43**, 279–418.

Thom, R.M. 1996. CO_2 enrichment effects on eelgrass (*Zostera marina* L) and bull kelp (*Nereocystis luetkeana* (Mert) P & R). *Water Air and Soil Pollution* **88**, 383–391.

Thom, R.M. & Hallum, L. 1990. Long-term changes in the areal extent of tidal marshes, eelgrass meadows and kelp forests of Puget Sound. Final Report to Office of Puget Sound, Region 10, US Environmental Protection Agency No. EPA 910/9-91-005 (FRI-UW-9008). School of Fisheries, University of Washington, Seattle, Washington.

Trenberth, K.E. & Hoar, T.J. 1996. The 1990–1995 El Niño Southern Oscillation event: Longest on record. *Geophysical Research Letters* **23**, 57–60.

Vadas, R.L. 1972. Ecological implications of culture studies on *Nereocystis luetkeana. Journal of Phycology* **8**, 196–203.

Vanblaricom, G.R. & Estes, J.A. 1988. *Community Ecology of Sea Otters.* Heidelberg: Springer-Verlag.

Vetter, E.W. 1995. Detritus-based patches of high secondary production in the nearshore benthos. *Marine Ecology Progress Series* **120**, 251–262.

Waldron, K.D. 1955. A survey of bull whip kelp resources off the Oregon coast in 1954. *Oregon Fish Commission Research Briefs* **6** (2), 15–20.

Walker, D.C. 1980a. A new interpretation of sorus inception and development in *Nereocystis luetkeana. Journal of Phycology* **16**, special issue 2, Abstracts from the annual meeting of the Phycological Society of America, Vancouver, British Columbia, Canada, 12–16 July 1980, 45.

Walker, D.C. 1980b. *Sorus abscission from laminae of* Nereocystis luetkeana *(Mert.) Post. and Rupr.* PhD thesis, University of British Columbia, Vancouver, British Columbia.

Walker, D.C. & Bisalputra, T. 1975. Fine structural changes at sorus margin during sorus release in *Nereocystis luetkeana. Journal of Phycology* **11**, 13–14.

Watson, M.A. & Casper, B.B. 1984. Morphological constraints on patterns of carbon distribution in plants. *Annual Review of Ecology and Systematics* **15**, 233–258.

Wheeler, W.N. 1990. Kelp forests of British Columbia: a unique resource. Fisheries Development Report 37, Province of British Columbia, Ministry of Agriculture and Fisheries, Aquaculture of Commercial Fisheries Branch, Victoria, British Columbia.

Whyte, J.N.C. & Englar, J.R. 1980a. Seasonal variation in the inorganic constituents of the marine alga *Nereocystis luetkeana.* 1. Metallic elements. *Botanica Marina* **23**, 13–17.

Whyte, J.N.C. & Englar, J.R. 1980b. Seasonal variation in the inorganic constituents of the marine alga *Nereocystis luetkeana.* 2. Non-metallic elements. *Botanica Marina* **23**, 19–24.

Oceanography and Marine Biology: An Annual Review, 2010, **48**, 43-160
© R. N. Gibson, R. J. A. Atkinson, and J. D. M. Gordon, Editors
Taylor & Francis

THE ECOLOGY AND MANAGEMENT
OF TEMPERATE MANGROVES

DONALD J. MORRISEY[1], ANDREW SWALES[2], SABINE DITTMANN[3],
MARK A. MORRISON[4], CATHERINE E. LOVELOCK[5] & CATHERINE M. BEARD[6]

[1]*National Institute of Water and Atmospheric Research Ltd.,*
P.O. Box 893, Nelson, New Zealand
E-mail: d.morrisey@niwa.co.nz
[2]*National Institute of Water and Atmospheric Research Ltd.,*
P.O. Box 11-115, Hamilton, New Zealand
E-mail: a.swales@niwa.co.nz
[3]*School of Biological Sciences, Flinders University,*
GPO Box 2100, Adelaide, South Australia 5001, Australia
E-mail: sabine.dittmann@flinders.edu.au
[4]*National Institute of Water and Atmospheric Research Ltd.,*
Private Bay 99940, Auckland, New Zealand
E-mail: m.morrison@niwa.co.nz
[5]*Centre for Marine Studies and School of Biological Sciences,*
University of Queensland, St. Lucia, Queensland 4072, Australia
E-mail: c.lovelock@cms.uq.edu.au
[6]*Environment Waikato, P.O. Box 4010, Hamilton East, New Zealand*
E-mail: catherine.beard@ew.govt.nz

Abstract Previous reviews of mangrove biology focused on the more extensive and diverse tropical examples, with those of temperate regions generally relegated to a footnote. Temperate mangroves are distinctive in several ways, most obviously by the lower diversity of tree species. Their occurrence in relatively developed countries has created different issues for mangrove management from those in the tropics. Mangroves in several temperate areas are currently expanding, due to changes in river catchments, in contrast to their worldwide decline. Information derived from the greater body of research from tropical regions has sometimes been applied uncritically to the management of temperate mangroves. The growing body of information on the ecology of temperate mangroves is reviewed, with emphasis on productivity, response to anthropogenically enhanced rates of sediment accumulation, and potential effects of climate change. There is no unique marine or estuarine fauna in temperate mangroves, but the poorly known terrestrial fauna includes mangrove-dependent species. Although productivity generally declines with increasing latitude, there is overlap in the range of reported values between temperate and tropical regions and considerable within-region variation. This, and variation in other ecologically important factors, makes it advisable to consider management of temperate mangroves on a case-by-case basis, for example, when responding to expansion of mangroves at a particular location.

Introduction

Why a review of information on temperate mangroves?

In the preface to his book *The Energetics of Mangrove Forests*, Alongi (2009) asked "Why another mangrove book?" and a similar question could be asked of the present review. While there has been no previous detailed review of the ecological energetics of mangrove forests, what possible need could there be for another general review of mangrove biology and ecology given the numerous previous reviews of this topic (e.g. Macnae 1968, Teas 1983, Hutchings & Saenger 1986, Tomlinson 1986, Kathiresan & Bingham 2001, Saenger 2002, Hogarth 2007)? Our justification is that all of these have focused primarily on mangroves in tropical and subtropical areas, and temperate mangroves have been relatively neglected. Macnae's (1968) review of Indo-Pacific mangroves, for example, devoted only 3 pages of 168 to "extratropical extensions". This inequality is not surprising given the larger areas and greater biological diversity of mangroves in warmer regions (Twilley et al. 1992) but has sometimes led to uncritical application of information about tropical mangroves to those in temperate regions or, conversely, to the assumption that, since temperate mangroves are less diverse and slower growing than those in the tropics, they are of relatively little ecological value.

The uncertainty created by the relative lack of information on temperate mangroves provided the impetus for this review in the form of a request by a statutory environmental management agency for a technical review of information on mangroves in New Zealand. The review was intended to inform a proposed change to the conservation status of mangroves in their jurisdiction. That proposal was, in turn, a response to requests from coastal property owners and others for permission to remove mangroves in areas where they are spreading in order to protect access to the coast and maintain open waterways. Low diversity and productivity of these mangrove assemblages, relative to their tropical counterparts and to estuarine habitats that they displace, are sometimes cited as mitigating factors in these proposed removals. Opponents of mangrove removal, on the other hand, may emphasise their potential ecological significance. Both claims have tended to lack supporting evidence or to use information selectively.

These issues identified several characteristics of mangrove assemblages in temperate regions that distinguish them from their tropical or subtropical equivalents. First, the number of mangrove species declines with increasing latitude (Ellison 2002), and temperate regions generally contain between one and three species, as discussed in this review. Whether their productivity and associated biological diversity are comparably lower is less clear. Second, temperate mangroves often occur in relatively developed countries, such as Australia, New Zealand, South Africa, and the United States, where issues relating to their management may be very different from those in tropical regions. For example, statutory protection of mangroves may be stricter, and mangrove conservation groups may have relatively greater influence. The areal extents of mangroves in some temperate areas, including New Zealand and southern Australia, are currently increasing (Saintilan & Williams 1999, Swales et al. 2007b), in contrast to the ongoing decline in mangrove areas worldwide, particularly the tropics (Duke et al. 2007). These changes are often in response to historic changes in terrestrial vegetation cover upstream following European colonisation and create specific management issues of their own.

Scope of the review

The present review expands the original New Zealand study to include mangroves throughout their temperate range. Temperate mangroves are considered in their own right and are not specifically compared with their tropical counterparts. Nevertheless, if comparisons are relevant to management of temperate mangroves (e.g., testing assumed similarities or differences that have been presented as arguments for or against removal of spreading mangroves), these comparisons are discussed.

The review starts with a working definition of *temperate mangroves* and their global distribution, followed by a discussion of their taxonomy and diversity. This discussion leads into a consideration of the physiological tolerances of temperate mangroves and their influence on distribution along latitudinal gradients and among habitats within latitudes.

Biomass and productivity of temperate mangroves are reviewed, with reference to equivalent information on tropical mangroves to address the question of whether they have similar ecological importance in both regions. This discussion also includes the role of other primary producers in mangrove forests, such as microalgae, macroalgae, ferns and vascular plants.

A review of mangrove-associated faunas is followed by discussion of primary consumption of mangrove material. Primary consumption includes direct consumption by herbivores, indirect consumption by detritivores, the uptake of mangrove-derived material within the forest and its export to adjacent habitats, and trophic pathways within mangrove systems.

Discussion of the role of mangroves in trapping sediment and mitigating coastal erosion and their associated role in the natural ageing of estuaries leads into a consideration of patterns and causes of changes in mangrove distribution. Changes include losses due to human activities such as infilling and clearance of coastal areas, effects of increased sediment loads from coastal catchments, and inputs of other anthropogenic contaminants. In some areas, in contrast, anthropogenically enhanced rates of sedimentation have resulted in the spread of mangroves, and the extent of their spread, the influencing factors and the consequences for estuarine habitat diversity are reviewed. Future changes in the distribution of mangroves in temperate areas are also considered, including effects of climate change and sea-level rise (SLR). Management of these changes, both to protect mangrove forests and to protect other coastal habitats, is discussed in relation to legislation and management initiatives in various temperate-zone countries, with New Zealand as a case study of managing conflicts resulting from the spread of mangroves. Management initiatives considered include restoration in addition to removal and the broader ecological effects of each. The review finishes with discussion of directions for future research.

Definition of temperate mangroves

The development of an appropriate working definition of *temperate mangroves* is important to ensure that all relevant areas of mangrove distribution are considered in the study. It is equally important, however, to ensure that the geographical and climatic ranges covered are not so broad that they defeat the object of the exercise, namely, to provide a review of information on this relatively distinct subset of high-latitude, low-diversity mangrove assemblages, and risk straying into subject areas that have been reviewed previously.

There are several relevant factors to consider in deriving the definition, the broadest of which (and the most clearly defined—at least at first glance) is latitude. Climate correlates with latitude in a general sense but with considerable smaller-scale variation at any given latitude. The third major factor for present purposes is the diversity of the mangrove assemblages present.

Latitude and climate

The Tropics of Cancer and Capricorn lie at latitudes of 23°30′N and S 23°30′S, respectively, and, according to classical geographical zonation, define the boundary between the tropical and the temperate zones. The separation of subtropical and temperate regions, however, is much more equivocal. In South Africa, for example, Whitfield (1994) categorised the eastern coast from the border with Mozambique (26°52′S) south to the Mbashe River estuary (32°15′S) as "subtropical", and the remaining stretch to the southern limit of the continent (34°52′S) as "warm temperate", based largely on water temperatures. This side of the Cape is subject to the warm, southward-flowing

Table 1 Latitudinal limits of mangroves in the main regions of their range

Region	Northern limit	Southern limit
America, western coast	30°15′N	5°32′S
America, eastern coast/Bermuda	30°02′/32°20′N[a]	28°56′S
Africa, western coast	19°50′N	12°20′S
Africa, eastern coast and Red Sea	28°24′N	33°04′S[b]
Australia, western coast	na	33°16′S
Australia, eastern coast	na	38°45′S
New Zealand	na	38°05′S[c]
Pacific continental Asia	31°22′N[a]	na

Source: Information from Saenger 1998 except as indicated.

Note: na = not applicable because mangroves are present to the north-ernmost (in the case of northern limits) or the southernmost (in the case of southern limits) limit of the region.

[a] From Spalding et al. 1997.

[b] From Ward and Steinke (1982).

[c] From de Lange and de Lange (1994).

Agulhas Current, whereas the western side is exposed to the cooling influence of the Benguela Current. Consequently, the western coast is generally classified as 'cool temperate', even though it lies within the same latitudes as the warm-temperate region on the eastern side.

Mangroves straddle the tropical-temperate boundary but are generally restricted to latitudes between 30°N and 30°S. Exceptions occur in Bermuda (32°20′N), Japan (31°22′N), Australia (38°45′S), New Zealand (38°05′S) and South Africa (33°04′S) (Table 1; see Ward & Steinke 1982, Hughes & Hughes 1992, de Lange & de Lange 1994, Spalding et al. 1997). They also occur along the 30° parallel in the south-eastern United States (Spalding et al. 1997).

As discussed later (p. 51 et seq.), the main latitude-related factor limiting mangrove distribution appears to be the occurrence of low temperatures (sea surface and air) and, in particular, extremes of temperatures. Patterns of rainfall are also important (Spalding et al. 1997, Saenger 1998). The modifying role of factors other than latitude is demonstrated by the notable absence of mangroves on the western coasts of Africa and America south of 12°20′ and 5°32′, respectively. The most obvi-ous of these modifying factors are local climatic features, including warming or cooling effects of coastal currents and patterns of rainfall.

The limits to the distribution of mangroves on the western coasts of Africa and South America (Table 1) correspond with the limits of arid regions, defined as summer rainfall and winter drought, >30 mm rainfall in any month of the year, and a precipitation to a potential evaporation ratio of <0.03 (Saenger 1998). Aridity is also likely to restrict mangrove distribution in western Asia and the Middle East.

The Köppen-Geiger climate classification scheme, based in part on the distribution of types of vegetation (Peel et al. 2007), has been widely used since its development in the early twentieth century. Despite some criticisms, it is still commonly used and has been modified and updated over time. Recent updates include that of Peel et al. (2007), on which the present discussion is based.

The Köppen-Geiger system divides climates into five main groups (tropical, arid, temperate, cold and polar), each containing several types and subtypes (see, e.g., Peel et al. 2007). The temper-ate group (group C), defined as regions with the temperature of the hottest month >10°C and tem-perature of the coldest month >0°C but <18°C, is clearly the most relevant for present purposes. It is subdivided into three types: Cs, dry summer (precipitation in the driest summer month <40 mm and less than a third of the precipitation in the wettest winter month); Cw, dry winter (precipitation in the driest winter month less than a tenth of the precipitation in the wettest summer month); and Cf, without dry season (fitting neither of the previous criteria). Each of these three types (Cs, Cw and Cf) is further divided according to summer temperature: a, hot summer (temperature of the hottest month ≥22°C); b, warm summer [not (a) and the number of months when the temperature is above 10°C ≥ 4]; c, cold summer [not (a) or (b) and the number of months when the temperature is above 10°C ≥ 1 but <4].

Comparison of the Peel et al. (2007) climate map with maps of the global distribution of mangroves (e.g., those in Spalding et al. 1997) suggests that the Cs and Cf climate types are most relevant for the present study (particularly the a and b subtypes of both). In some parts of the world, these climate types extend into latitudes below the tropics, for example, in south-eastern China and eastern Australia.

Mangrove species diversity

The decreasing number of mangrove species along gradients of increasing latitude is well documented (Ellison 2002), and areas at the extremes of mangrove distribution, such as Bermuda, southern Kyushu in Japan, southern South Africa, southern Australia and northern New Zealand, have only one to three species.

Species found at these latitudinal extremes are not cold specialists but tend to be those with wide latitudinal distribution. Among the eastern, or old-world, mangroves, they include *Aegiceras corniculatum, Avicennia marina, Bruguiera gymnorrhiza, Kandelia candel* and *Rhizophora mucronata*, and among the new-world species are *Avicennia germinans* and *Rhizophora mangle*.

Working definition

The primary interest in undertaking this review was to summarise information on the species-poor mangrove assemblages of temperate regions because of the potential ecological differences between them and the better-known mangroves of tropical and subtropical regions. These differences may arise from being located at the limits of mangrove distribution, with consequent potential effects on factors such as rates of growth and primary production, and from the restricted diversity of the mangrove species present, which may influence, for example, the range of habitats or shore heights that mangroves may occupy when constrained by the tolerances of only one or two species.

The focus of this review, therefore, is populations of mangroves occurring within climate zones Cs and Cf, most of which lie at latitudes higher than 29–30°, and containing limited numbers of mangrove species. Consequently, those parts of climate types Cs and Cf where mangrove communities are relatively diverse (generally more than three species occurring in the same location), such as south-eastern China (seven mangrove species at 27°N; Li & Lee 1997*), northern New South Wales and southern Queensland (north of, say, 30°S), Brazil north of 24°S, and the Florida peninsula are not considered. Also excluded are those parts of the world where mangrove species diversity is restricted by aridity, such as the northern part of eastern Africa and the Red Sea and Baja California and the Sea of Cortez.

The geographical areas included in the present review, with the types of mangroves that occur in them, are summarised in Table 2 and Figure 1.

Global distribution of temperate mangroves

According to estimates by Twilley et al. (1992, their Table 1), of about 24×10^6 ha of mangroves worldwide, 13.28×10^6 ha (55.3%) occur between the latitudes of 10°N and 10°S, 7.25×10^6 ha

---

* *Kandelia candel* was transplanted to Zhejian province (27–31°N) in the 1950s (Li & Lee, 1997), but most have been destroyed by human disturbance, and only 8 ha remained at the time of Li & Lee's review.

Table 2 Geographical areas included in the present review showing the climate type according to the Köppen-Geiger classification, the species of mangrove present in each area, and their latitudinal limits (distributional information as indicated in table footnotes)

Region	Climate type	Species	Latitudinal limit	Total mangrove species at this latitude
Louisiana, USA	Cfa	*Avicennia germinans*	30°02′N[a]	1
Bermuda	Cfa	*Avicennia germinans*	32°20′N[a]	3
	Cfa	*Conocarpus erectus*	32°20′N[a]	3
	Cfa	*Rhizophora mangle*	32°20′N[a]	3
Southern Kyushu, Tanegashima, Yakushima, Japan	Cfa	*Kandelia candel*	31°22′N[a]	1
North Island, New Zealand	Cfb	*Avicennia marina*	38°05′S[b]	1
Southern Brazil	Cfa	*Acrostichum aureum*	28°30′S[c]	3
	Cfa	*Avicennia germinans*	28°30′S[c]	3
	Cfa	*Laguncularia racemosa*	28°30′S[c]	3
	Cfa	*Rhizophora mangle*	27°53′S[c]	4
Eastern South Africa	Cfb	*Avicennia marina*	32°59′S[d,e]	1
	Cfb	*Bruguiera gymnorrhiza*	32°14′S[d]	2
	Cfa	*Rhizophora mucronata*	31°42′S[d]	4
Southern Australia	Cfb	*Avicennia marina*	38°45′S[f]	1
	Cfb	*Aegiceras corniculatum*	36°53′S[f]	2
	Cfa	*Exoecaria agallocha*	31°52′S[g]	3
	Cfa	*Rhizophora stylosa*	30°03′S[f]	4
	Cfa	*Bruguiera gymnorrhiza*	29°25′S[f]	5
	Cfa	*Lumnitzera racemosa*	27°30′S[f]	6
	Cfa	*Acrostichum aureum*	26°05′S[h]	7
Southwestern Australia	Csa	*Avicennia marina*	33°16′S[f]	1

[a] Nakasuga et al. 1974, Spalding et al. 1997.

[b] de Lange and de Lange 1994.

[c] Schaeffer-Novelli et al. 1990.

[d] Hughes and Hughes 1992.

[e] The stand of *A. marina* near the mouth of the Nahoon River (32°59′S) arose from transplanted material, but a stand further upstream may have derived from natural dispersal of propagules (Steinke 1995). The next highest latitude where this species occurs in South Africa is the Gqunube River (32°56′S).

[f] A.G. Wells 1983.

[g] West et al. 1985.

[h] Saenger 1998.

(30.2%) between latitudes 10° and 20°, 3.14 × 10⁶ ha (13.1%) between 20° and 30°, and only 0.33 × 10⁶ ha (1.4%) at latitudes greater than 30°. An estimate based on the data for individual temperate regions listed in Table 3, however, suggests that the total area of mangroves at latitudes >30° is in the range 0.05–0.06 × 10⁶ ha. Of this last range, 48–55% are in southern Australia, 41–47% in New Zealand, 2.9–3.9% in Louisiana and 0.5–0.6% in South Africa.

Mangroves at latitudes >30° in southern Australia contain *Avicennia marina*, *Aegiceras corniculatum* and *Exoecaria agallocha*, while *Bruguiera gymnorrhiza* and *Rhizophora stylosa* also occur between 29° and 30°S (West et al. 1985). New Zealand contains only *Avicennia marina* and Louisiana only *A. germinans*. Bermuda contains three species (*A. germinans*, *Conocarpus erectus* and *Rhizophora mangle*), while in South Africa *Avicennia marina*, *Bruguiera gymnorrhiza* and *Rhizophora mucronata* occur south of latitude 30°. Only *Kandelia candel* occurs at Kiire, southern Kyushu, Japan, and on the neighbouring islands of Tanegashima and Yakushima (all north

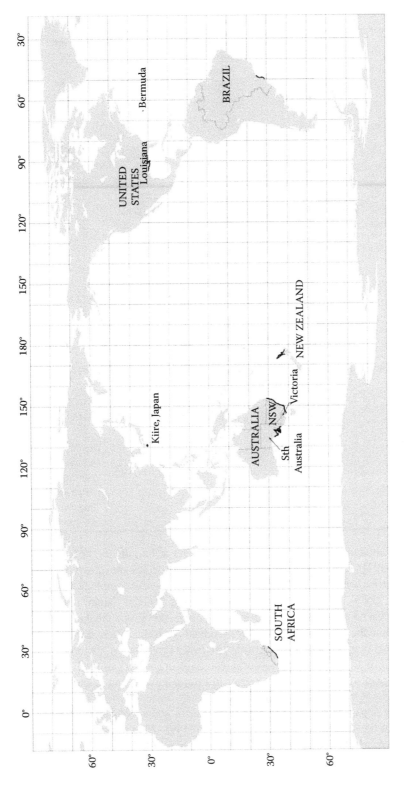

Figure 1 Global distribution of temperate mangroves as defined for the purposes of this review.

Table 3 Global distribution of temperate mangroves

Region	Area of mangroves (ha)	Notes	Reference
Louisiana	1560–2360	Value for late 1970s; since then contraction, due to freezes in the 1980s, and subsequent expansion have occurred.	Montz 1980 (cited in Lester et al. 2005)
Bermuda	10 16 17.5		Spalding et al. 1997 Wilkie & Fortuna 2003 Thomas 1993
Kyushu, Tanegashima, Yakushima and Amami Oshima, Japan	7.2 (>28°S) 7.1 (>30°S)	Total mangrove area for Japan from Nakasuga et al. (1974) is 517 ha. Wilkie & Fortuna (2003) gave a value of 400 ha for all Japan, including Kyushu (Kiire) and Okinawan Islands.	Nakasuga et al. 1974
New Zealand	28,700 22,200		Spalding et al. 1997 Wilkie & Fortuna 2003
South Africa	355 (>29°S) 295 (>30°S)	Adams et al. (2004) updated Ward & Steinke's estimates for the Transkei region (272 ha) to 270.6 ha based on data from 1999.	Ward & Steinke 1982
Southern Australia	31,074 (>29°S) 29,500 (>30°S)	NSW south of 29°S 11,674 ha; Victoria 3800 ha; South Australia 15,600 ha.	NSW, West et al. 1985; Victoria and South Australia, Duke 2006
Southwest Australia Total >29° Total >30°	6.4 54,860–62,520 53,580–60,890		http://dbforms.ga.gov.au/pls/ www/npm.ozest.show_ mm?pBlobno=9483 (cited on Wikipedia http://en.wikipedia.org/wiki/ Leschenault_Estuary#cite_ref-2)

of latitude 30°), although *Bruguiera gymnorrhiza* occurs at latitude 28°14′N on Amami Oshima (Nakasuga et al. 1974). The latitudinal limits of individual species are shown in Table 2.

Avicennia marina sensu lato (grey mangrove) has the greatest geographical range of all mangrove species, with its global limits occurring around latitudes 25°N in Japan and 38°S in Australia. However, *A. marina* subsp. *australasica* grows only in northern New Zealand, Lord Howe Island, New Caledonia, and the south-eastern coast of mainland Australia, where it forms the southernmost natural populations of mangrove at Corner Inlet, Wilson's Promontory, Victoria, Australia (latitude 38°45′S) (Table 2). With an ability to grow and reproduce in a variety of tidal, climatic and edaphic conditions, this species occupies a diverse range of littoral habitats and displays great variability of growth form (Duke et al. 1998, Maguire et al. 2002).

Links among populations of mangroves that occur in temperate regions have been most thoroughly studied in the species *A. marina* (Duke et al. 1998, Maguire et al. 2002, Arnaud-Haond et al. 2006) and *A. germinans* (Dodd et al. 2002). Populations of *A. marina* from temperate and often isolated locations have lower genetic diversity and higher levels of inbreeding than do core tropical populations, suggesting that populations are isolated with little gene flow among peripheral and core populations (Arnaud-Haond et al. 2006). Genetic differentiation is proposed to have occurred through bottlenecks and founder effects as well as through strong selection by temperate environments (Arnaud-Haond et al. 2006). For *A. germinans*, similar differentiation of peripheral populations has also been observed, although there is evidence for long-distance dispersal (Dodd et al. 2002).

Diversity of temperate mangroves

Mangroves are a taxonomically diverse group of halophytic (salt-tolerant) plants that, worldwide, comprise approximately 70 species within some 19 families. They are typically woody trees or shrubs taller than 0.5 m and inhabit the intertidal margins of low-energy coastal and estuarine environments over a wide range of latitude (Tomlinson 1986, Duke 1991). They normally occupy the zone between mean sea level (MSL) and high tide, growing on a variety of substrata, including volcanic rock, coral, fine sands and muddy sediments.

Although many species of mangrove are taxonomically unrelated, they all share a number of important traits that allow them to live successfully under environmental conditions that exclude many other plant species. Morphological, physiological and reproductive specialisations, such as aerial roots, support structures (buttresses or above-ground roots), and salt tolerance (Tomlinson 1986). Other traits, like vivipary (seeds that germinate while still on the adult tree) and positively buoyant propagules are also common in mangrove lineages (Rabinowitz 1978, Tomlinson 1986, Farnsworth & Farrant 1998).

Mangroves are most commonly associated with tropical and subtropical coastlines, and only a few species extend their range into the cooler warm-temperate climates typical of parts of New Zealand, Australia, Japan, South America and South Africa (Macnae 1966, Chapman 1977). A latitudinal pattern of species richness is evident, with diversity and extent both greatest at the equator and diminished towards the north and south (Ellison 2002). Mangrove communities near their northern global limits may include up to six mangrove species, whereas those at the southern limits are species poor; supporting between one and three species (Table 2). The most common species of mangroves that persist within temperate regions belong to the genus *Avicennia*.

Following Tomlinson's (1986) classification, *Avicennia* are true mangroves in that their habitat is defined solely by the intertidal zone, and they also possess specialized physiological and reproductive adaptations that allow them to grow there. Taxonomic treatments place the genus *Avicennia* either within the family Verbenaceae Jaume Saint-Hilaire (Green 1994) or as the sole genus within family *Avicenniaceae* Endlicher. However, more recent molecular evidence indicated that it may have closer affinities to the Acanthaceae sensu lato (Schwarzbach & McDade 2002).

Physiology of temperate mangroves

Key drivers of mangrove distribution: latitudinal limits

The global distribution of mangroves is approximately restricted to tropical climates where mean air temperatures of the coldest months are warmer than 20°C and where the seasonal range does not exceed 10°C (Chapman 1976, 1977). The geographic limits of mangrove growth are also coincident with ground frost occurrence and are closely linked with the 20°C winter isotherm for seawater. However, the occurrence of mangroves in New Zealand, parts of Australia, and eastern South America are notable exceptions to this pattern. Duke et al. (1998) suggested that these outlying distributions either coincide with extensions of irregular warm oceanic currents or are relict populations established during periods of warmer climate and greater poleward distributions.

Low temperatures limit the distribution of mangroves through their effects on a range of processes. One of the main hypotheses that has been proposed to account for the southern and northern latitudinal boundaries of mangroves is the lethal effects of extreme low winter temperatures (i.e., frosts) that kill trees (Chapman & Ronaldson 1958, Sakai & Wardle 1978, Sakai et al. 1981, Kangas & Lugo 1990, Saintilan et al. 2009). However, other factors may also limit the latitudinal distribution of mangroves. For example, work from New Zealand suggests that the distribution of *Avicennia* may also be constrained by its physiological limitations under low temperate (non-freezing) conditions (Walbert 2002, Beard 2006). Declining net primary production with increasing latitude (Saenger & Snedaker 1993) also suggests that photosynthetic carbon gain diminishes relative to respiratory demands. Thus, mangroves may have a more precarious carbon balance at high latitudes that may make them more vulnerable to abiotic stressors (e.g., tissue damage from freezing), competition and predation.

Limited productivity of mangroves at high latitudes may be associated with adaptations that improve resistance to freezing. Periodic freezing temperatures place a strong selective pressure for small xylem vessels that are the conduits for water transport between roots and roots. Small xylem vessels (with narrow diameters) reduce the probability of embolism of the xylem during freezing (Stuart et al. 2007), but the trade-off is that they place constraints on rate of water transport within stems (hydraulic conductivity), which in turn limits photosynthesis and carbon gain (Stuart et al. 2007). Thus, hydraulic characteristics that are required for safety during freezing temperatures come at the cost of lower carbon gain, which reduces growth rates.

Slower annual growth increments of trees at higher latitudes may also contribute to setting the latitudinal limits of mangroves by reducing the competitive ability of mangroves with co-occurring saltmarsh plants. Competitive interactions with saltmarsh plants are proposed to restrict mangrove forest development at their latitudinal limits (Saintilan et al. 2009). In addition, where herbivory of mangrove propagules is high, plant–animal interactions may also set latitudinal limits, especially when herbivore damage is combined with unfavourable abiotic conditions for growth (Patterson et al. 1997).

The timing and success of reproduction of *A. marina* vary predictably with latitude (Duke 1990). In a detailed study of its phenology over a wide range of latitudes, Duke (1990) found that the timing of reproductive events (e.g., initiation of flowering and maturation of propagules) was highly dependent on temperature. The success of flowers declined with decreasing temperature such that, at a mean annual temperature of 18°C, flowers did not develop into fruit. Thus, low temperatures at high latitudes may directly limit metabolic processes associated with reproduction, which would limit population growth and dispersal at latitudinal limits.

Finally, in addition to low temperatures, the availability of suitable habitat for mangrove growth and suitable conditions for propagule dispersal have been proposed to limit the distribution of mangroves (de Lange & de Lange 1994). Increases in suitable habitat for mangroves, for example through enhanced sedimentation (Burns 1982, Lovelock et al. 2007b, Swales & Bentley 2008) or through changes in groundwater availability (McTainsh et al. 1986, Rogers & Saintilan 2008), may increase the abundance and extend distributions of mangroves, particularly if these factors are combined with warming of air or sea temperatures or with other factors that enhance growth rates or resistance to stressors (e.g., nutrient enrichment) (Martin 2007). Thus, latitudinal limits are likely to be set by plant metabolic responses to low temperatures but are moderated by a complex suite of interacting biotic and abiotic factors.

Unique features of high-latitude mangroves

Temperate mangrove forests have many features in common with lower-latitude forests. They share common tree species and similar requirements for establishment and growth (Krauss et al. 2008).

Here, the focus is on assessing how the characteristics of temperate mangroves differ from forests in subtropical and tropical regions. This theme is continued in the following sections on biomass and productivity of temperate mangroves.

Reproductive characteristics

Reproductive traits differ among temperate and tropical mangroves. The duration required for fruit maturation increases from 200 days to 550 days from 15–30° latitude (Duke 1990). Propagule mass also correlates positively with latitude (Figure 2), suggesting that cool winter temperatures have resulted in selection for propagules that take longer to develop and that are provisioned with larger maternal reserves. The trend of increasing propagule weight with latitude is particularly evident for *A. marina* and less so for *A. germinans*, but the data for *A. germinans* are fewer and cover a more restricted latitudinal range than those for *A. marina*. In addition, the time during which newly established seedlings of *Avicennia* are dependent on maternal reserves appears to be shorter in the tropics (4 months; Smith 1987) compared with the temperate zone (12 months; Osunkoya & Creese 1997). Longer propagule development times, larger propagules and longer periods of dependency on maternal reserves within the propagule may result in differences in seedling ecology between temperate and tropical regions. Temperate seedlings may be more attractive to herbivores and predators, although declines in herbivores and predators with increasing latitude may offset increases in palatability of propagules at high latitude (Clarke & Myerscough 1993).

Prolonged periods of propagule development and dependency on cotyledonary reserves may also result in greater susceptibility of seedlings to stochastic processes such as storms and freezing. Successful establishment may become highly variable among years (e.g., Swales et al. 2007b), dependent on prolonged periods of suitable weather. It may therefore be expected that mangrove expansion at high latitudes will be pulsed and highly sensitive to climate change, with rapid establishment occurring with amelioration of key climate drivers that limit propagule development and establishment (low temperatures and storms).

Genetic differences among temperate and tropical mangroves may underlie the differences observed in some traits. Genetic differences among mangrove populations is high (Maguire et al. 2000). Introductions of new genotypes across long distances (oceanic basins) have been observed,

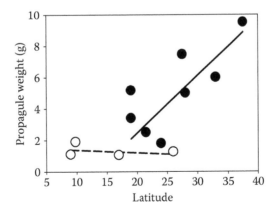

Figure 2 Variation in propagule weight of *Avicennia marina* (filled circles) and *A. germinans* (open circles) with latitude. There is a significant increase in propagule weight with latitude in *A. marina*, described by the line $y = 0.030x - 0.214$, $R^2 = 0.56$, $p = 0.013$. (Data are from Rabinowitz 1978, Downton 1982, Smith 1987, McKee 1995a,b, Clarke et al. 2001, Delgado et al 2001, Bhat et al. 2004, Ye et al. 2005, He et al. 2007, Yan et al. 2007.)

but distant populations are isolated and may have different responses to environmental change (Dodd & Afzal Rafii 2002). Few experiments have been done to test for differences in phenotypes of distant populations under common conditions. However, Duke (1990) reported that *A. marina* seedlings from a range of latitudes grown together at 19°S did not show significant differences in growth. Probing differences among temperate and tropical populations may improve predictions of the effects of climate change on temperate mangroves.

Tolerance of key environmental drivers

The constraints on the carbon balance of mangroves imposed by low temperatures and adaptations to low temperatures, as discussed, may make trees at high latitudes more sensitive to environmental drivers than those at lower latitudes. Salinity of soils is one of the most important factors that limit the growth of individual mangrove trees and of mangrove forests, with high salinity resulting in reduced growth rates, tree height and productivity (Ball 1998). Assessment of seedling growth studies conducted over a range of latitudes that examined salinity tolerance of *Avicennia marina* did not reveal any differences in tolerance of high-salinity conditions among plants from differing latitudes. When normalised for maximum growth rate (usually at ~25% seawater), growth declined similarly with increasing salinity over a range of latitudes (Figure 3).

The seaward limits of mangrove growth also appear to be higher in temperate compared with tropical latitudes (Clarke & Myerscough 1993). In Tauranga Harbour, New Zealand (37°40′S), undisturbed populations of *A. marina* occupied habitat from approximately 0.23 m above MSL (Park 2004), whereas in the southern Firth of Thames (37°13′S) the lower elevation limit (LEL) of mangroves was ≥0.5 m above MSL. In Florida, mangroves occurred higher in the intertidal zone than salt marshes. Restriction of mangroves to above the MSL have been attributed to the inhibitory effects of waves on propagule establishment (exposure) or a lag between shore accretion and mangrove colonisation (Clarke & Myerscough 1993). In addition, tolerance of inundation may differ among latitudes, although there are insufficient data to test this hypothesis.

Nutrient availability limits mangrove growth in both temperate and tropical locations (Feller 1995, Lovelock et al. 2007b, Martin 2007). In many temperate sites, nitrogen is the key limiting

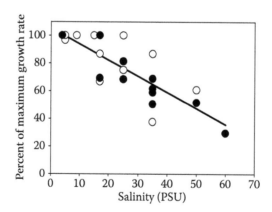

Figure 3 Decline in growth rate of *Avicennia marina* seedlings from temperate (filled circles, >28° latitude) and tropical (open symbols, <27° latitude) locations, grown over a range of salinities. There is no difference in the decrease in growth rates with increasing salinity between temperate and tropical species ($p > 0.05$). The line of best fit is described by $y = -1.16x + 105.4$, $R^2 = 0.72$, $p < 0.0001$. (Data are from Clarke & Hannon 1970, Downton 1982, Burchett et al. 1984, Clough 1984, Ball 1988, Yates et al. 2004, Ye et al. 2005, Yan et al. 2007.)

nutrient for plant growth (Lovelock et al. 2007a) and reproduction (C. Lovelock unpublished data), while both nitrogen and phosphorus may limit growth in tropical mangroves (Lovelock et al. 2007a). Differences in limiting nutrients between temperate and tropical locations are likely due to underlying differences in geochemistry over latitudes, in which tropical sites are often more highly weathered and deficient in phosphorus compared with temperate sites (Lovelock et al. 2007a).

Despite the positive effects nitrogen enrichment has on growth, nitrogen may increase susceptibility of trees to additional environmental stressors. For example, frost (Martin 2007) and drought (Lovelock et al. 2009) lead to tissue losses and increased mortality in nitrogen-fertilised trees compared with trees not enriched with nitrogen. Thus, nitrogen enrichment of estuaries may contribute to loss of resilience of mangrove ecosystems, which may be particularly important in temperate mangroves where climate conditions are variable, encompassing extreme low temperatures and extended droughts.

Productivity and biomass of temperate mangroves

Productivity of temperate mangroves

One of the strongest patterns observed over variation in latitude is that net primary production and maximum height of mangrove trees decline with increasing latitude (Twilley et al. 1992, Saenger & Snedaker 1993, Bouillon et al. 2008, Mendez-Alonso et al. 2008, Alongi 2009). Based on a review of 178 published measurements of litterfall, Bouillon et al. (2008) found significantly higher production at latitudes between 0 and 10° (average 10.4 ± 4.6 t ha^{-1} yr^{-1}, $n = 53$) than at higher latitudes (10–20°, 20–30° and >30°) and significantly lower production at latitudes >30° (4.7 ± 2.1 t ha^{-1} yr^{-1}, $n = 16$) than at lower latitudes. Average values for 10–20° and 20–30° were 9.1 (± 3.4, $n = 47$) and 8.8 (± 4.2, $n = 62$) t ha^{-1} yr^{-1}, respectively, and were not significantly different. There was no significant correlation between above-ground production of wood by mangroves and latitude (none of the available data were from latitudes above 30°).

Productivity is commonly measured as litter production, principally leaves, twigs and woody debris. Although this does not represent net primary production completely (because it does not include net increase in plant biomass), it represents an important component of it. It also reflects changing events in the life cycle of the mangrove (such as reproduction and senescence) and responses to environmental events (storms, variation in rainfall) and provides a measurement of inputs of organic matter and nutrients by mangroves to the estuarine system. Estimates of litter production may also include propagules, which are potentially viable material and not, strictly speaking, detritus.

Published estimates of litter production in temperate mangroves (Table 4) are generally consistent with the average value given by Bouillon et al. (2008) for latitudes >30°. Although recorded rates of litterfall for *A. marina* from New Zealand are below the maximum values reported from other parts of its distribution (e.g., those from tropical Australia), they are comparable with values from subtropical and temperate Australia (Table 4). Bunt (1995) gave an average value of 6.2 t ha^{-1} yr^{-1} from measurements taken throughout the range of *A. marina* in Australia (from Victoria to tropical Queensland, Northern Territory and Western Australia) but with a range from 1.10 to 15.98 t ha^{-1} yr^{-1}, indicative of considerable variation within and among locations.

Bouillon et al. (2008) also noted the relatively large variation about the average values for latitudinal ranges (0–10°, 10–20°, 20–30° and >30°) and pointed out that this was to be expected given the variation in primary production with numerous environmental factors that, themselves, vary at a range of spatial scales smaller than the latitudinal ranges concerned. There was variation among measured values from temperate regions, even within the same geographic area (Table 4). For example, in New Zealand the rates of litter production reported for adult trees (i.e., not saplings)

Table 4 Summary of published information on litterfall and above-ground biomass (AGB) of temperate mangroves

Country	Location	Latitude	Species	Height of trees (m)	Litter production (t ha^{-1} yr^{-1} DW)	AGB (t ha^{-1})	Notes	Reference
Australia	Northwestern Australia and Gulf of Carpentaria	10°50′–17°57′S	A.m.		2.34–4.30 (SD 2.11)		Hot, dry winter (climate zones Aw and Awi)	Bunt 1995
Australia	Darwin Harbour	12°26′S	A.m.	13	12.51 (SD 1.793)			Woodroffe et al. 1988
Australia	Darwin Harbour	12°26′S	A.m.	10.8	0.68 (SD 0.43)			Woodroffe et al. 1988
Australia	Northern Queensland	15°28′–18°21′S	A.m.		10.49 (SD 4.57)		Hot, generally short dry season (climate zone Am)	Bunt 1995
Australia	Central western coast	20°18′–24°53′S	A.m.		8.79 (SD 1.68)		Hot, arid (climate zone BWh)	Bunt 1995
Australia	Mary River, Queensland	25°26′S	A.m. A.c.			200–250 35	Upstream reaches of estuary	Saintilan 1997b
Australia	Mary River, Queensland	25°26′S	A.m. A.c.			120 50	Seaward edge of intertidal flats	Saintilan 1997b
Australia	Mary River, Queensland	25°26′S	A.m.			10	Landward edge of intertidal flats	Saintilan 1997a
Australia	Brisbane	27°25′S	A.m.					Mackey 1993
Australia	Subtropical and warm-temperate eastern Australia	27°33′–33°59′S	A.m.		3.07 (SD 1.83)	110–340	Hot summer, uniform rain (climate zone Cfa)	Bunt 1995
Australia	Kooragang Is, NSW	32°51′S	A.m.	3		7.1		Burchett & Pulkownik 1983 in Saenger & Snedaker 1993
Australia	Kooragang Is, NSW	32°51′S	A.m.	7.5		86		Burchett & Pulkownik 1983 in Saenger & Snedaker 1993
Australia	Kooragang Is, NSW	32°51′S	A.m.	10		104		Burchett & Pulkownik 1983 in Saenger & Snedaker 1993
Australia	Kooragang Is, NSW	32°51′S	A.m.	4.3	5.14	21.7		Murray 1985
Australia	Kooragang Is, NSW	32°51′S	A.m.	4.4	5.62	21.8		Murray 1985

Country	Location	Latitude	Species				Notes	Reference
Australia	Merimbula NSW, Bunbury WA, Westernport, Victoria	33°20'–38°25'S	A.m.		4.36 (SD 1.48)		Long, mild summer, cool winter (climate zones Cfb/Csb)	Bunt 1995
Australia	Hawkesbury River, NSW	33°30'S	A.m. A.c.	<3		52.1 54.9	Hypersaline environment	Saintilan 1997a
Australia	Hawkesbury River, NSW	33°30'S	A.m. A.c.	<3		60.1 52.5	Marine environment	Saintilan 1997a
Australia	Hawkesbury River, NSW	33°30'S	A.m.	<3		A.m. 400	Riverine flats	Saintilan 1997a
Australia	Middle Harbour, Sydney	33°46'S	A.m.	5.8		220	Average 79% leaves	Goulter & Allaway 1979
Australia	Lane Cove River, Sydney	33°50'S	A.m.	<0.8		0.2–3.4		Briggs 1977
Australia	Lane Cove River, Sydney	33°50'S	A.m.	6.5–8.2		112.3–144.5	Means of 2 sites (total AGB)	Briggs 1977
Australia	Gulf of St. Vincent, South Australia	34°36'S–34°38'S	A.m.	2–3	7.67 and 11.68		Measured during summer only; 62–85% leaves	Ingraben & Dittmann 2008
Australia	Jervis Bay, NSW	35°07'S	A.m.	5–8	3.67		Fruit and flowers average 9.2% of litter	Clarke 1994
Australia	Westernport Bay, Victoria	38°46'S	A.m.	2				Attiwill & Clough 1974 in Goulter & Allaway 1979
Brazil	Babitonga Bay, Santa Caterina	26°12'S	A.s L.r. R.m.	3.2–5.7 3.1–3.8 2.6–2.8	0.20–0.55 1.02–2.35 0.59–2.36	1.9–11.4 10.2–25.0 4.6–5.9	Ranges for 3 sites across the shore profile	Cunha et al. 2006
New Zealand	Rangaunu Harbour, Northland	34°57'S	A.m.	6.23	6.24		73% leaves	May 1999
New Zealand	Rangaunu Harbour, Northland	34°57'S	A.m.	3.06	3.89			May 1999
New Zealand	Rangaunu Harbour, Northland	34°57'S	A.m.	5.12	4.83			May 1999
New Zealand	Rangaunu Harbour, Northland	34°57'S	A.m.	1.68	1.77			May 1999

(continued on next page)

Table 4 (continued)   Summary of published information on litterfall and above-ground biomass (AGB) of temperate mangroves

Country	Location	Latitude	Species	Height of trees (m)	Litter production (t ha^{-1} yr^{-1} DW)	AGB (t ha^{-1})	Notes	Reference
New Zealand	Whangateau Estuary, Auckland	36°19'S	A.m.	<0.5–1.5 (sapling)	1.68			Oñate-Pacalioga 2005
New Zealand	Whangateau Estuary, Auckland	36°19'S	A.m.	2–4	1.56			Oñate-Pacalioga 2005
New Zealand	Trancar Bay, Auckland	36°19'S	A.m.	2–4	1.3			Oñate-Pacalioga 2005
New Zealand	Tuff Crater, Auckland	36°48'S	A.m.	4	7.12–8.09	130	69% leaves, 12.3% fruits and flowers	Woodroffe 1985a
New Zealand	Tuff Crater, Auckland	36°48'S	A.m.	0.95 (stunted)	2.90–3.65	10	74% leaves, 2.6% fruits and flowers	Woodroffe 1985a
New Zealand	Puhinui Creek, Auckland	37°01'S	A.m.	0.5–1.6 (sapling)	0.11–0.38		Young stands (4–13 yr old)	Burns et al. unpublished
New Zealand	Puhinui Creek, Auckland	37°01'S	A.m.	0.8 (stunted)	0.61		Medium age (13–31 yr old)	Burns et al. unpublished
New Zealand	Puhinui Creek, Auckland	37°01'S	A.m.	2.3–2.6	2.89		Medium age (13–31 yr old)	Burns et al. unpublished
New Zealand	Puhinui Creek, Auckland	37°01'S	A.m.	3.4–4	1.55–4.05		Old stands (31+ yr old)	Burns et al. unpublished
South Africa	Mgeni River, Durban	29°48'S	A.m.	9	6.98	19.82	Leaves 59% of litter (average over 3 yr)	Steinke & Charles 1986
			B.g	7	8.24	74.67		Steinke et al. 1995

South Africa	Wavecrest, Nxaxo-Nqusi Estuary	32°35'S	A.m. + B.g		4.51 (range 0.3–7.3)		Average over 3 yrs; leaves 72% of litter	Steinke & Charles 1990
Various tropical	Various	23°N–23°S	Various	3.9–35	0.01–7.71		Minimum value for *Ceriops tagal* in Andaman Islands; maximum for *Avicennia germinans* in Guyana	Kathiresan & Bingham 2001
Various tropical	Various	23°N–23°S	Various		3.74–18.7	57–436	Average 193 t ha^{-1} biomass, 3.74–14.02 t ha^{-1} yr^{-1} litter; minimum value for *Avicennia* sp. in Sri Lanka; maximum for *Bruguiera* in China; maximum value recorded for *Avicennia* species was 14.0 in Australia and Malaysia	Saenger & Snedaker 1993
Various subtropical	Various	23–30°S and N	Various	1–12.5	1.3–16.31	7.9–164.0	Minimum and maximum both for *Rhizophora* in United States; maximum value for *Avicennia* species was 7.15 in South Africa	Saenger & Snedaker 1993

Note: Data for each country are ordered by increasing latitude. Data for various mangrove species in various tropical and subtropical countries are also shown for comparison. Climate zones refer to the Köppen-Geiger system. A.c. = *Aegiceras corniculatum*, A.m. = *Avicennia marina*; A.s. = *Avicennia schaueriana*; B.g = *Bruguiera gymnorrhiza*; L.r. = *Laguncularia racemosa*; R.m. = *Rhizophora mangle*; NSW = New South Wales; SD = standard deviation; WA = Western Australia.

varied from 0.61 to 8.1 t ha^{-1} yr^{-1}, with the smaller value from stunted (0.8-m tall) plants in Puhinui Creek, Auckland (B.R. Burns, Landcare Research, D.J. Morrisey, NIWA [National Institute of Water and Atmospheric Research], and J. Ellis, NIWA unpublished data) and the larger from 4-m high trees in Tuff Crater, Auckland (Woodroffe 1985a). The higher value approaches the averages for latitudes between 10 and 30°. The average value for full-size, mature trees in temperate stands of *A. marina* (Table 4) was 4.3 (standard deviation [SD] 2.09) t ha^{-1} yr^{-1}. Similar methods of sampling were used in all of these studies.

Burns, Morrisey, & Ellis (unpublished data) measured litter production in each of six stands in Puhinui Creek, near Auckland. Stands were divided into three age groups: those that first developed before 1939 (old stands), those that developed between 1969 and 1987 (medium-age stands) and those that developed between 1987 and 1996 (young stands). Each age class was replicated at two locations along the creek (young stands were generally nearer the creek and old stands highest up the shore). Litterfall in the two young stands was 0.11 and 0.38 t ha^{-1} yr^{-1}, and the stands consisted of saplings with average canopy heights of 0.5 m and 1.6 m, respectively. Of this material, 36% and 71%, respectively, consisted of leaves. One of the medium-age sites contained stunted adult trees (average canopy height 0.8 m) where litterfall was 0.61 t ha^{-1} yr^{-1} (43% wood, 33% leaves). The other stand of this age (average canopy height 2.3 m) produced 2.89 t ha^{-1} yr^{-1} (67% leaves). The two old stands also differed in rate of litter production, with the slightly taller stand (average canopy height 4.0 m) producing less than the shorter stand (average canopy height 3.4 m: litter production 1.55 and 4.05 t ha^{-1} yr^{-1}, respectively). The percentages of leaf material were 72% for the taller and 69% for the shorter stand.

May (1999) measured litterfall at two locations, separated by 250 m, on opposite sides of the mouth of the Awanui River in Rangaunu Harbour, northern New Zealand. At each location, sites were sampled low on the shore near the channel edge, where the trees were tall (3.06 m average canopy height on the northern side of the river, 6.23 m on the southern side), and the upper shore, where the trees were shorter (northern side 1.68 m, southern side 5.12 m). Upper and lower sites were less than 50 m apart. Total annual litterfall for the northern and southern low-shore sites was 3.89 t ha^{-1} yr^{-1} (75% leaf material) and 6.24 t ha^{-1} yr^{-1} (76% leaf material), respectively. Equivalent values for the high-shore sites were 1.77 t ha^{-1} yr^{-1} (86% leaf material) and 4.83 t ha^{-1} yr^{-1} (56% leaf material).

Similarly reduced litterfall at higher-shore sites was reported from Tuff Crater (Woodroffe 1985a), where the stunted (often <0.5 m), sprawling plants produced 2.90–3.65 t ha^{-1} yr^{-1} (average values for two consecutive years: 75% leaf material). Taller (up to 4 m), more erect trees growing along the banks of the major tidal creeks produced 7.12–8.1 t ha^{-1} yr^{-1} (45–69% leaf material).

These data from New Zealand mangroves reveal considerable variation in rates of litter production within and among locations. Highest levels (7–8 t ha^{-1} yr^{-1}) were recorded at the Tuff Crater site near Auckland (36°48′S) rather than at the most northerly site (Rangaunu Harbour, 34°57′S: 2–6 t ha^{-1} yr^{-1}), even though the trees were taller at the latter site. Within locations, such as Puhinui Creek or Rangaunu Harbour, litterfall seems to be broadly proportional to tree height. At Whangateau, however, a newly establishing stand (up to 1.5 m high) produced more litter than established stands (2–4 m high) (Oñate-Pacalioga 2005). Newly establishing areas in Puhinui Creek produced much less litter than those in Whangateau (0.11–0.38 vs. 1.68 t ha^{-1} yr^{-1}).

As May (1999) noted, "The varied topography of creek-dissected mudflats characteristic of northern New Zealand estuaries results in a mosaic of mangrove biomass and litter input across the intertidal". Another factor that may increase spatial and temporal variation in litter production is the variability of populations of herbivores in space and time, and the amount of damage they inflict may influence the amount of litter shed by the trees. The small number of studies of the productivity of temperate mangroves and the large within-location variation recorded by those studies that have been made make it difficult to identify any general trends in productivity. It would be worthwhile to conduct a systematic study of productivity under standardised conditions at a large number of sites throughout the distributional range of temperate mangroves.

Standing crop and tree size

There is a trend of declining mangrove biomass with increasing latitude. Twilley et al. (1992), for example, estimated that the above-ground biomass densities for mangroves in the latitudinal ranges 0–10, 10–20, 20–30 and 30–40° are 283.6, 141.6, 120.6 and 104.2 t ha^{-1}, respectively. Alongi (2009) also compiled data on above-ground biomass and showed a declining trend with increasing latitude (his Figure 2.6). Biomass and tree height correlate reasonably closely (Saenger & Snedaker 1993, Lee 2008), but as with productivity, there is considerable variation in values of biomass at a particular latitude or within a range of latitudes, as Twilley et al. (1992) pointed out. For example, Saenger & Snedaker (1993) gave a range of 57–436 t ha^{-1} for mangrove biomass between the Tropics of Cancer and Capricorn (23°N to 23°S), and 7.9–164 t ha^{-1} between 23 and 30°.

The data representing higher latitudes in the reviews by Twilley et al. (1992) and Alongi (2009), however, omitted a number of the larger published values. Published estimates of biomass for temperate mangrove stands (Table 4) show considerable overlap with values from lower latitudes. For example, biomass was estimated at 130 t ha^{-1} in a stand in Auckland (Woodroffe 1985a) and 220 and 400 t ha^{-1} for stands around Sydney (30°30′ and 30°46′S, Goulter & Allaway 1979 and Saintilan 1997a, respectively). The largest value included by Alongi (2009) for latitudes >30°S, in contrast, was about 125 t ha^{-1}.

Osunkoya & Creese (1997) described a cline of decreasing tree height and propagule size with increasing latitude within New Zealand, and this cline would be expected to impose similar variation on rates of litterfall. There is also, however, variation among locations at the same latitude, as illustrated by the examples described, and this variation obscures or confounds latitudinal gradients.

Reasons for latitudinal trends in productivity and biomass

The physiological basis of this decline in height and productivity with increasing latitude is likely to be complex. In this section, a range of factors is considered that may contribute to the reduced tree height and productivity in temperate mangrove forests compared with more equatorial forests.

Allocation of resources to leaf production may be lower in temperate compared with tropical sites. Patterns of leaf production vary strongly with latitude (Duke 1990). Tropical sites appear to have multiple peaks in leaf fall and leaf initiation compared with the unimodal pattern observed at cooler, southern sites, where leaf flushes are typically in the summer (Duke 1990). Thus, the shorter, compressed growing season of temperate mangroves may be a strong contributing factor in reducing productivity at high latitudes.

Maximum photosynthetic carbon gain and growth (measured as stem extension) was not depressed at higher latitudes compared with tropical sites (Lovelock et al. 2007a). However, leaf nitrogen and phosphorus concentrations increased at higher latitudes compared with lower latitudes, as they do in terrestrial species (Lovelock et al. 2007a). Taken together, these data suggest that per unit leaf photosynthetic rates are maintained at high latitudes through increased investment in the metabolic components of photosynthetic pathways. Declines in forest productivity may therefore be due both to decreases in the length of time when temperatures are suitable for photosynthetic carbon gain and to increases in respiration associated with maintaining thicker, more metabolically active leaves during periods of low potential production.

Anatomical constraints on water transport associated with freezing tolerance (see 'Key drivers' section, p. 51), which may limit canopy development and photosynthesis, are likely to be higher in higher-latitude forests, particularly in the Southern Hemisphere, where freezing temperatures are frequent (Stuart et al. 2007). Xylem vessel diameters were smaller in southern populations of *A. marina* than at more tropical latitudes. A similar trend was not observed in the Northern Hemisphere populations of *A. germinans*, although sample size was small (Stuart et al. 2007). The differences in patterns between biogeographic regions may occur because in Florida freezing is infrequent but severe, which

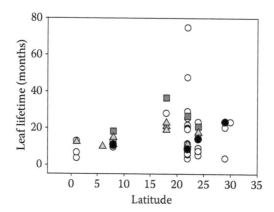

Figure 4 Leaf lifetime of a range of mangrove species over variation in latitude. There is no significant increase in lifetime with increasing latitude. *Avicennia marina* (filled circles), *Rhizophora* spp. (shaded triangles), *Ceriops* spp. (shaded squares) and other species (open circles) are included. (Data are from Saenger 2002.)

may impose different selective pressures on plants compared with those in the Southern Hemisphere (Stuart et al. 2007). In addition, mangrove competition with the saltmarsh grass *Spartina alterniflora*, which occupies similarly low regions of the intertidal zone in Florida and elsewhere, may place a significant selective pressure to maintain high levels of hydraulic conductivity and growth despite the ensuing susceptibility to embolism during freezing temperatures. In *Avicennia germinans* in the Caribbean, leaf characteristics differed over a latitudinal gradient, with leaf size decreasing and leaf mass per area and carbon content increasing with latitude (Mendez-Alonso et al. 2008). These authors suggested that the decrease in leaf area and increase in carbon content was associated with increases in leaf longevity, which helps to balance the higher leaf construction costs (Wright & Westoby 2002, Wright et al. 2002). However, the limited data on longevity of mangrove leaves does not indicate increased leaf lifespans at higher latitudes (Figure 4; Saenger 2002).

Reductions in maximum tree height with increasing latitude could be due to a number of mechanisms. Hydraulic architecture is proposed to limit tree height in tall species, with xylem tension reaching levels that are at the limits of embolism in very tall trees (Koch et al. 2004). However, gravitational forces only account for 0.0098 MPa m^{-1}; thus, in smaller-stature trees (<50 m tall) gravity will only account for about 0.5 MPa contribution to leaf water potential. Another hypothesis is that maximum height is constrained by respiratory demand of non-photosynthetic tissues. Taller trees have more woody tissue (roots and stems) than shorter trees (e.g., Clough et al. 1997). Maximum tree height occurs where photosynthetic carbon gain matches that of respiration losses. Trees may overshoot the balance in some years or seasons but then die back when conditions for photosynthetic carbon gain are less favourable. High soil salinity, which decreases hydraulic conductivity, leaf water potential and photosynthetic rates (Lovelock et al. 2006, Suarez & Medina 2006), will contribute to reducing carbon balances, further decreasing tree heights. There are few direct tests of this hypothesis. In New Zealand, however, loss of twig length was observed over the winter season (Lovelock et al. 2007b). Direct frost damage to leaves, branches and reproductive tissues of trees (Saintilan et al. 2009) and herbivory, particularly by wood-boring insects (I. Feller personal communication), also contributes to limiting height of trees.

Other primary producers in temperate mangrove forests

Although the structural elements of the mangrove forest are dominated by the trees, there is a range of other primary producers that contribute to the overall productivity of the ecosystem. Under the mangrove canopy, well-developed epiphytic algal and microbial communities adhere to roots, stems and sediments (e.g., Beanland & Woelkerling 1983). Macroalgal communities in mangroves are dominated by red algae and characterised as *Bostrychietum* (Phillips et al. 1994, King 1995). Most of the macroalgal species found in temperate mangroves are widespread globally, and species belonging to the genera *Bostrychia*, *Caloglossa* and *Catenella* are most common (Beanland & Woelkerling 1982, Davey & Woelkerling 1985, Eston et al. 1992, King 1995). The taxonomic composition of the algal community in temperate mangroves of Spencer Gulf, South Australia, was distinct from those in nearby Victoria as well as tropical mangroves, but similar to the algal species found on the open coastlines outside the gulf (Beanland & Woelkerling 1982). Several typical tropical macroalgae taxa occurred in the upper Spencer Gulf and are possible relic populations (Beanland & Woelkerling 1982). Subject to inundation, desiccation and shading by the mangrove canopy, the frequency of occurrence can decrease from the seaward to the landward edge of temperate mangroves (Davey & Woelkerling 1985). Different macroalgal species dominate underneath compared with outside the canopy, and algal cover and biomass are higher in the shaded areas (Beanland & Woelkerling 1983).

Ecophysiological adaptations of *Bostrychia* and *Caloglossa* allow them to tolerate salinity changes, desiccation and shading and can thus explain their dominance in mangrove forests. The macroalgae colonise pneumatophores mainly within 10 to 20 cm above the mud surface (Davey & Woelkerling 1985), and although colonisation experiments revealed seasonal changes in the colonisation rate, no successional stages were found (Eston et al. 1992, Gwyther & Fairweather 2002).

This community is highly productive, accounting for up to 15% of gross primary production (Dawes et al. 1999). In temperate locations, biomass can be high (e.g., Dor & Levy 1984, Phillips et al. 1994, Melville et al. 2005), particularly where there is nutrient enrichment (Melville & Pulkownik 2006). The biomass of the epiphytic algal and microbial communities can be heavily grazed (Skilleter & Warren 2000). The high nutrient quality of algal and microbial mat tissue (Lee & Joye 2006) may result in a high contribution to the food web because algae and microbial communities may be preferentially consumed and decomposed compared with the biomass of the trees, which is physically tough and chemically difficult to digest (Bouillon et al. 2008). The photosynthetic activity and nitrogen fixation by epiphytic algae and bacteria on pneumatophores has been discussed as a mutualistic relationship because algal holdfast was detected in the periderm of pneumatophores (Naidoo et al. 2008). Epiphytic algae are grazed on by snails (*Bembicium melanostomum*) (Gwyther & Fairweather 2005) and affect the colonisation of pneumatophores by meiofauna (Gwyther & Fairweather 2002).

Mangrove faunal diversity and abundance

Hutchings & Recher (1982) divided the aquatic faunas of mangroves into an encrusting epifauna (i.e., sessile taxa living on mangrove structures), a mangrove epifauna (i.e., motile species living on the mangroves), a substratum infauna, a substratum epifauna, and a wood-boring infauna. These components of the aquatic fauna are discussed in the following sections, combining substratum infauna and epifauna and encrusting and mangrove epifauna for concision. They are preceded by a review of information on the benthic meiofaunas.

Benthic meiofauna

Introduction

Meiofauna living in the interstitial spaces of sediments has long been known to be a diverse and important component of coastal marine sediments (Fenchel 1978, Giere 2009). They occur in high abundances, have fast response times to disturbances, and are important for decomposer food chains as well as constituting prey for juveniles of commercially important fishes and prawns (Coull 1990, 1999, Coull & Chandler 1992).

Meiofauna in temperate mangroves has been investigated mainly in south-eastern Australia, southern Africa and Brazil. All of these studies used different methods for sampling and extracting meiofauna and focused on different meiofaunal taxa. Although this difference complicates comparisons, some general patterns emerge from the following overview.

Meiofaunal diversity and abundance

Diversity   The meiofauna in sediments of temperate mangroves is dominated by nematodes, comparable with other marine ecosystems (Alongi 1987a, Vanhove et al. 1992, Sasekumar 1994, Olafsson 1995). Nematodes and halacarid mites have been best studied, and the presence of other invertebrate taxa occurring in the interstitial spaces of temperate mangroves has been recorded, but detailed taxonomic work still has to follow, and further undescribed species will be discovered (Nicholas 1996, Marshall & Pugh 2001). The permanent meiofauna reported from temperate mangrove muds and pneumatophores includes Nematoda, Copepoda, Acari, Kinorhynchia, Platyhelmintha, Gastrotricha, Cnidaria and Foraminifera, and temporary meiofauna is represented with Oligochaeta, juvenile polychaetes and insect larvae (Diptera). The occurrence of major meiofaunal taxa in temperate mangroves around the world reflects the current state of knowledge and the level of taxonomic detail, which varies among studies. As mangroves are a transition zone between the land and the sea, meiofauna of not only marine but also terrestrial and freshwater origin can be encountered (Phleger 1970, Hodda & Nicholas 1985, Nicholas et al. 1991, Olafsson 1995, Proches 2004).

Several nematode genera (such as *Microlaimus*, *Spirina*, *Desmodora* and various representatives of the Chromadoridae), and even species, occur in mangroves across temperate latitudes of southern America, Africa and Australia and in tropical mangroves (Hodda & Nicholas 1985, Nicholas et al. 1991, Olafsson 1995, Netto & Gallucci 2003). Yet, other genera that are abundant in temperate mangroves (e.g., *Triplyoides*) are rare in tropical mangroves (Hodda 1990, Olafsson 1995, Gwyther 2003).

Abundance and biomass   Nematoda account for over 70 to 100% of meiofaunal abundances (Hodda & Nicholas 1986a, Gwyther 2000), followed by copepods, ciliates or halacarid mites (Table 5). Meiofaunal abundances in temperate mangrove muds vary within and between regions (Table 5) yet are similar to abundances found in mangrove mud elsewhere or in non-mangrove estuaries or saltmarsh densities (Dye 1983a, Nicholas et al. 1991, Gwyther 2000). Meiofaunal abundances on pneumatophores were about 3 to over 40 times lower than in adjacent sediments (Gwyther 2000, Proches et al. 2001). Mean abundances on 5-cm segments of pneumatophores in South African mangroves ranged from 5 to 90 individuals for copepods and 10 to 160 individuals for acari, with copepods less abundant at the tip of the pneumatophores, where mites (Halacaridae) occurred in highest density (Proches & Marshall 2002). Gwyther (2000) found higher meiofaunal abundances on pneumatophores with algae (52 ± 11 individuals 10 cm^{-2}) than on those with barnacles (39 ± 5 individuals 10 cm^{-2}). These abundances were lower than for epifaunal meiofauna living on seagrass blades or saltmarsh grasses (Gwyther 2000).

Biomass values for nematodes in a New South Wales mangrove estuary ranged from 19 mg dry weight (DW) m^{-2} in the upper intertidal to 888 mg DW m^{-2} in the low intertidal zone, equivalent to 8 and 383 mg carbon m^{-2}, respectively (Nicholas et al. 1991). These values are comparable to estimates made by Dye (1983b) from southern Africa, where the mean biomass was 646 mg DW m^{-2} (range 91–2877 mg DW m^{-2}), but the higher biomass occurred in the upper intertidal zone at most study sites. Biomass also varied over time because Dye (1983a) reported a much higher biomass at most of the sampling sites in the Mngazana Estuary in the following year, with equivalent production of 1.48–9.32 g carbon m^{-2} yr^{-1}.

Table 5 Meiofaunal species numbers (sum of species or phyla recorded in particular studies) and individual density (mean and range of minimum and maximum values) of meiofauna from sediments in temperate mangrove studies

Region/site	Species/ taxa number	Density (ind. cm^{-2}), mean value (minimum–maximum)	Taxa recorded	Most abundant taxa	Reference
New South Wales: Hunter River, Clyde River, Candlagan Creek	21	345 (21–2968)	Nematoda	Nematoda: *Tripyloides* sp., *Desmodora cazca*, *Sabatiera* sp., *Sphaerolaimus* sp. A., *Spirina* sp., *Terschellingia longicaudata*	Hodda 1990
New South Wales: Hunter River	62	225 (63–12,057)	Nematoda, Copepoda, Platyhelminthes, Kinorhyncha, Acari, Polychaeta (Nereididae, Phyllodocidae), Oligochaeta, Insecta (larvae)	Nematoda	Hodda & Nicholas 1985
New South Wales: Hunter River	36	110 (1–9596)	Nematoda, Copepoda, Platyhelminthes, Kinorhyncha, Acari, Polychaeta, Oligochaeta, Insecta (larvae)	Nematoda: *Filipjeva* sp., *Monhystera* sp. B., *Ethmolaimus* sp., *Terschellingia longicaudata*, *Sabatiera* sp., *Desmodora cazca*	Hodda & Nicholas 1986b
New South Wales: Clyde River	40	183 (47–500)	Nematoda, Copepoda, Platyhelminthes, Kinorhyncha, Acari, Ostracoda, Rotifera, Polychaeta (Nereididae, Fabricinae), Oligochaeta, Insecta (larvae)	Nematoda: *Microlaimus capillaris*, *Terschellingia longicaudata*, *Desmodora cazca*	Nicholas et al. 1991
Victoria: Barwon Estuary		2170 (250–8622)	Nematoda, Copepoda, Halacarida, Polychaeta, Oligochaeta, Insecta (larvae), Kinorhynchia, Platyhelminthes, Cnidaria	Nematoda	Gwyther 2000
Transkei: Mngazana River			Nematoda, Ciliata, Oligochaeta, Gastrotricha, Polychaeta, Kinorhyncha, Copepoda, Insecta (larvae), Crustacea (larvae)	Nematoda	Dye 1983b
Transkei: Mngazana River	12	246 (84–530)	Nematoda, Ciliata, Oligochaeta, Platyhelminthes, Kinorhyncha, Copepoda, Polychaeta, Gastrotricha, Crustacea (larvae), Insecta (larvae)	Nematoda	Dye 1983a
South Africa: Beachwood Creek, Bayhead Lagoon	30	2	Copepoda, Insecta (larvae)	Harpacticoid copepods	Proches et al. 2001
South Brazil: Santa Catarina Island	94	(77–1589)	Nematoda, Acari, Oligochaeta, Copepoda, Polychaeta, Kinorhynchia, Platyhelminthes, Ostracoda, Insecta (larvae)	Nematoda	Netto & Gallucci 2003

Note: Studies focused on particular meiofaunal taxa (e.g., Proches et al. 2001 on arthropods) or did not differentiate on species level (e.g., Dye 1983a,b). Insect larvae encountered were mainly Ceratopogonidae and Tipulidae. The recorded taxa are listed in decreasing order of abundance if such information was available. ind. = individuals.

Feeding modes Classification of feeding modes can help elucidate whether the prevalence of particular species in certain habitats may be related to their food supply. For nematodes, classification of feeding modes can be done based on the morphology of the buccal cavity (Wieser 1953, Jensen 1987). Nematodes in mangrove sediments are deposit- or epistratum-feeders, feeding on diatoms, bacteria, plant roots or algae (Hodda & Nicholas 1986a, Nicholas et al. 1991, Netto & Gallucci 2003). Nematodes found on pneumatophores are epigrowth (Gwyther & Fairweather 2002, Gwyther 2003) or epistratum feeders (Nicholas et al. 1991), and non-selective deposit-feeders dominate on mangrove leaf litter (Gwyther 2003). Different nematode feeding modes were found to occur in different tidal zones (Nicholas et al. 1991) and substrata, with epistratum feeders more abundant at sandier sites, while deposit-feeders were dominant at sites with finer, muddier sediments (Hodda & Nicholas 1986a).

Distribution patterns of temperate mangrove meiofauna

Temporal Temperate mangroves experience more pronounced seasons than their tropical counterparts, in particular winter rainfall and dry, hot summers. Dry, hot conditions cause desiccation of the epiphyte cover on pneumatophores, with a subsequent decline in meiofaunal abundances on pneumatophores in summer (Gwyther & Fairweather 2002, 2005). For sediment meiofauna, variations in seasonal patterns of nematodes could be linked to mangrove tree density, which modifies the surface water availability, litterfall and algal blooms on the sediment (Hodda & Nicholas 1986a). Sites with few or no mangroves had highest meiofaunal densities in winter, whereas no seasonal patterns were found in tall and dense mangrove forest regions (Hodda & Nicholas 1986a). Yet, the few nematode species showing consistent seasonal patterns across sampling sites did so in different seasons (e.g., *Filipjeva* sp. highest in winter, *Chromadorina* sp. highest in spring), whereas they all belonged to similar feeding modes (epigrowth or unselective deposit-feeder) (Hodda & Nicholas 1986a). Temporal variation in meiofauna was recorded in most studies, but without clear patterns in relation to seasons or environmental changes over the year (Dye 1983a, Hodda & Nicholas 1986a,b, Gwyther 2000, Proches et al. 2001), and was even seen as stochastic variation (Hodda 1990). Seasonal changes in the species dominating in particular sites complicate the use of meiofauna to detect pollution in temperate mangroves (Hodda & Nicholas 1986b).

Spatial Apart from some taxonomic similarities across global and regional scales (Nicholas et al. 1991, Olafsson 1995, Netto & Gallucci 2003), meiofaunal species compositions in temperate mangroves can differ on a local scale, subject to habitat characteristics along estuaries (Hodda 1990, Proches et al. 2001, Netto & Gallucci 2003) or possibly pollution (Hodda & Nicholas 1985). Differences in meiofaunal assemblages were found along transects within sites, from the high towards the low intertidal zone. Nicholas et al. (1991) reported few differences in nematode diversity along their transects, yet abundances and biomass increased from the high (470×10^3 ind m^{-2}, 19 mg DW m^{-2}) towards the low intertidal zone (5000×10^3 ind m^{-2}, 888 mg DW m^{-2}). Dye (1983a,b), however, found highest meiofaunal abundances in the midintertidal zone where the mangrove forest density was highest, whereas meiofaunal biomass was higher in the upper (231–1835 mg DW m^{-2}) than the lower (91–670 mg DW m^{-2}) intertidal.

Mangroves in both temperate and tropical latitudes create complex habitats with a small-scale heterogeneity of microtopography, biogenic structures (pneumatophores, tree trunks, rootlets, leave litter, algal patches, crab burrows) and respectively varying environmental conditions, which lead to small-scale variability in meiofaunal distributions and complicating the detection of patterns (Hodda 1990, Nicholas et al. 1991, Olafsson 1995). The presence of algal patches was seen as an important determinant of meiofaunal distributions (Dye 1983a,b).

Environmental factors and pollution   Sediment-dwelling meiofauna is subjected to sediment characteristics and porewater qualities. Grain size was important for nematode distributions in mangrove estuaries in New South Wales, where sites with similar grain-size compositions shared similar nematode assemblages, regardless of season or the background pollution of some areas (Hodda & Nicholas 1986b). However, the relevance of particular sediment characteristics can vary between estuaries because Hodda & Nicholas (1985) reported no correlation between grain size and nematode densities. The sediment packing, drainage and content of organic matter and detritus biomass contributed further to differences between sites and were correlated with meiofaunal abundances (Hodda 1990, Netto & Gallucci 2003).

Sediment biogeochemistry shows the greatest changes in the top 15 cm of mangrove sediments; in particular, the oxygen penetration and redox potential profile can affect the vertical distribution of meiofauna within sediments (Dye 1983a,b, Hodda & Nicholas 1985, Hodda 1990) as well as their overall abundance (Hodda & Nicholas 1986b).

Tannins leaching from mangrove leaves have been found to inhibit meiofaunal populations (Alongi 1987b), yet *Avicennia* leaves have a lower tannin content than other mangroves (Robertson 1988), and the relevance of tannins in determining meiofaunal populations and distributions in temperate mangroves may be less than in their tropical counterparts. An indication of lower tannin content can be seen in the higher meiofaunal densities found by Dye (1983b) in the denser parts of the forest.

Pollution by heavy metals in an estuary in New South Wales reduced densities of both nematodes and copepods, so that the nematode/copepod ratio could not be applied to indicate the pollution level of the sediments (Hodda & Nicholas 1985).

Temperature and pH were found to be of less importance in explaining meiofaunal distribution patterns in temperate mangroves, and environmental conditions showed no consistent patterns of seasonal variations between sites and tidal levels (Dye 1983a,b). Yet, the elevation above the low-tide mark can be important for nematode distributions (Hodda & Nicholas 1985, 1986a), and salinity differences along estuaries correlated with copepod distribution (Hodda & Nicholas 1985, 1986a). In mangroves of southern Brazil, the distribution of the most abundant nematodes along the estuary could not be explained by any environmental factors measured (Netto & Gallucci 2003). Clear relations between environmental factors and meiofauna within and between estuaries were also absent in tropical mangroves (Alongi 1987c, Armenteros et al. 2006).

Hodda (1990) argued that the stochastic variation of environmental and habitat factors in temperate mangroves favours adaptability rather than specialisation and can thus account for some of the observed coexistence of related species in mangrove nematodes.

Habitat use by temperate mangrove meiofauna

Sediments   The sediments contain the vast majority of meiofauna occurring in temperate mangroves (Nicholas et al. 1991, Gwyther 2000). Due to the often anoxic conditions, a vertical stratification of meiofauna is found, with highest densities in the top 10 cm of the sediment in South African mangroves (Dye 1983b), but with suspected occurrence of some meiofauna below 60-cm sediment depth. In south-eastern Australian mangroves, nematodes were restricted (85%) to the top 1 cm, above the redox potential discontinuity (RPD) layer (Hodda & Nicholas 1985, Nicholas et al. 1991), with another 10–14% occurring to 5 cm and 2–6% in 5- to 10-cm sediment depth. Some nematode species (including *Terschellingia longicaudata*, *Sabatiera* spp., *Paracyatholaimus* sp.) occurred with >50% of all individuals at depths below the top centimetre, and one species (*Enchodelus coomansi*) was only recorded in greater sediment depths (Nicholas et al. 1991). Nematodes of the same families and genera as found in temperate mangroves have adaptations to suboxic sediments rich in H_2S, including intracellular accumulations of sulphur and symbiotic relationships with bacteria (Nicholas et al. 1987, Polz et al. 1992, Steyaert et al. 2007).

Pneumatophores Pneumatophores constitute a biogenic structure that provides habitat for sessile and mobile epifauna. Although meiofaunal diversity and abundances are lower on pneumatophores than in sediments (Nicholas et al. 1991), they reveal some interesting patterns of microhabitat use and trophic interactions. The species composition and diversity found on pneumatophores (copepods, halacarid mites, tanaids; lowest taxonomic richness) differ from the sediment meiofauna (nematodes dominant; highest taxonomic richness) (Gwyther 2000, Gwyther & Fairweather 2002, Proches & Marshall 2002) and varies between pneumatophores overgrown with barnacles (acari dominant) and algae (harpacticoid copepods, acari, nematodes) (Gwyther 2000, Gwyther & Fairweather 2002). Barnacles are known to harbour a rich halacarid fauna in their crevices, which may explain this pattern (Bartsch 1989). Pneumatophores without fouling were colonised by very few meiofauna (Gwyther 2000, Gwyther & Fairweather 2002, 2005). Insect larvae were found both on pneumatophores and in the sediment (Proches et al. 2001).

Nematode species occurring in the foliaceous brown algae on pneumatophores (*Tylendus* sp.) were different from those inhabiting filamentous green algal mats on the sediment surface (species of *Chromadorina* and *Ethmolaimus*) (Hodda & Nicholas 1985). The length of pneumatophores as such was not correlated with associated meiofaunal abundance (Proches et al. 2001), but vertical distribution on pneumatophores differed, with several species of acari and tanaids increasing in abundance towards the tip (Proches & Marshall 2002). Algal cover and sediment deposition at the base of pneumatophores was positively correlated with some acari, copepods and insect larvae (Proches & Marshall 2002). As algal and barnacle cover and sediment deposition increases towards the low-tide line, this gradient can account for some of the differences in the pneumatophore-associated meiofauna along intertidal transects (Gwyther 2000, Proches & Marshall 2002). Yet, pneumatophores on the seaward fringe of mangroves are subject to more exposure and higher desiccation in summer, and their meiofaunal assemblages had a lower species richness, but higher abundances, and formed a different assemblage compared with the meiofauna on pneumatophores from within the forest (Proches et al. 2001). Desiccation over summer, which seasonally reduces the algal epigrowth on pneumatophores, is followed by a decline in meiofauna (Gwyther & Fairweather 2002).

Colonisation of pneumatophores after experimental removal of all attached algae and sediment was not completed after 25 wk, apart from rapid colonisation by halacarids, harpacticoid copepods and ceratopogonid insect larvae (Proches & Marshall 2002). Proches & Marshall proposed that recovery was related to the recolonisation by epiphytic algae on the pneumatophores and the dispersal ability of the meiofaunal organisms.

The relevance of pneumatophore epigrowth on meiofaunal colonisation was also studied experimentally by Gwyther and Fairweather (2002, 2005), who offered transplanted and mimic pneumatophores. Natural pneumatophores always had the highest meiofaunal numbers, but mimics had a more diverse meiofaunal assemblage. The colonisation was resource dependent, as evident from the nematode feeding modes dominating during colonisation and the differences in colonisation time between experiments in winter and summer (Gwyther & Fairweather 2002, 2005). The importance of algal epiphyte cover on pneumatophores and the varying dispersal abilities of meiofauna (nematodes with sediment dislodgement or floating detritus, copepods through the water column) (Armonies 1988, Hodda 1990, Faust & Gulledge 1996, Gwyther & Fairweather 2005) caused a more ephemeral colonisation rather than a succession (Gwyther & Fairweather 2002, 2005).

Leaf litter Mangrove leaf litter on the forest floor provides another habitat for meiofauna. In a temperate mangrove forest in Victoria, Australia, Gwyther (2003) found 21 nematode species (14 on average per leaf) on leaf litter, where brown leaves and leaf litter in the shade of the canopy were preferred, possibly to avoid desiccation at more exposed sites. Copepods, oligochaetes and acari were also recorded on leaf litter, but in low numbers. Most of the nematodes were non-selective deposit-feeders, followed by epigrowth feeders, which were more common at the exposed sites at the mangrove fringe, where more light could increase microphytobenthic growth.

The diversity of nematodes on leaf litter found by Gwyther (2003) was lower than in the tropics, and whereas one or two feeding groups were dominant at her sites, all nematode feeding groups have been recorded on decaying mangrove leaves in the tropics (Gee & Somerfield 1997).

Trophic interactions of temperate mangrove meiofauna

Although nematode feeding modes in temperate mangroves have been classified using the morphological approach of Wieser (1953) and laboratory observations (Nicholas et al. 1988), detailed analyses of trophic interactions are still lacking. Leduc et al. (2009) demonstrated the usefulness of combining stable isotope, fatty acid and biomarker analyses to elucidate the diet of meiofauna, and this approach could provide valuable insight into mangrove meiofauna.

Meiofauna is a well-known food item for fishes and crustaceans (Gee 1989, Coull et al. 1995, Jenkins et al. 1996, Olafsson & Ndaro 1997), yet the dependence of fishes and prawns on deriving meiofaunal food from temperate mangroves has not yet been explored, apart from observations reported by Nicholas et al. (1991) that juvenile prawns were feeding on meiofauna. Similarly, whether mangrove crabs, which occur in lower diversity and numbers in temperate than tropical mangroves, exert predation pressure on meiofauna in temperate mangroves remains to be studied.

Benthic macrofauna

Introduction

The macrofauna, as in any soft-sediment habitat, consists of a burrowing infauna and an errant or sessile epifauna. In mangroves, in contrast to many other sediment shores, the infauna is generally dominated by crustaceans, and bivalves are usually relatively scarce (Macnae 1968). The motile epifauna is often dominated by gastropods and the sessile fauna by bivalves and barnacles (discussed in the section on encrusting fauna). Most of the species present are marine or estuarine, depending on location along the salinity gradient, but freshwater species may be present in upper estuarine locations (Hutchings & Recher 1982). Information on the composition of mangrove faunas in different temperate regions is reviewed, followed by a discussion of information on their ecology.

Review of regional information

Australia There appears to be little published information on the fauna of mangroves in the Leschenault Inlet Estuary in Western Australia (the only temperate mangroves on the western coast of the continent), although the faunas of other intertidal habitats in the estuary have been described (Semeniuk 2000, Semeniuk & Wurm 2000). The fauna of mangroves in south-eastern Australia, particularly in the Sydney region (33°S), has been the subject of relatively detailed study compared with those of other temperate regions. It is characterised by deposit-feeding oligochaetes, polychaetes (e.g., capitellids and spionids); gastropods; small crustaceans (cumaceans, tanaids, isopods and amphipods); and crabs (ocypodids, grapsids and sesarmids) (Hutchings & Recher 1974, Chapman & Underwood 1995, Chapman 1998, Lindegarth & Hoskin 2001, Chapman & Tolhurst 2004; Table 6). Taxa that were found in at least 33% of core samples from an urbanised mangrove forest in Sydney Harbour included amphipods, insect larvae, oligochaetes, crabs, capitellid, nereid, sabellid and spionid polychaetes and gastropods (Chapman & Tolhurst 2004).

A total of nine species of crabs occurs in mangrove forests in southern Australia, only three of which are common (Chapman & Underwood 1995). Six species of crabs (three ocypodids, two grapsids and a sesarmid), an alpheid prawn and five species of gastropods (including the pulmonate slug *Onchidium damelii*) were among the dominant species of mangrove habitats in the Pittwater,

Table 6 Benthic macrofaunal taxa recorded by studies of temperate mangroves

Region/site	Number of taxa	Total abundance	Taxa recorded	Mangrove species	Reference
Kyushu and Yakushima, Japan (31°22′N)			Gastropods (*Batillaria multiformis, Cerithideopsilla djadjariensis, C. cingulata, Cerithidea rhizophoranum*), crab (*Sesarma dehaani*)	*Kandelia candel*	Wakamatsu & Tomiyama 2000
Matapouri, New Zealand (35°34′S)	29.4 ± 2.7 (SE) per 15 × 15-cm core	43.9 ± 16.1 (SE) per 15 × 15-cm core	Oligochaetes, gastropods (*Diloma subrostrata, Melagraphia aethiops, Turbo smaragdus, Cominella glandiformis, Lepsiella scobina, Zeacumantus* sp.), bivalves (*Austrovenus stutchburyi, Paphies australis*).	*Avicennia marina*	Alfaro 2006
Whitford, New Zealand (36°55′S)	1.0–10.3 (range of means of three 13-cm diameter cores)	2.3–73.3 (range of means of three 13-cm diameter cores)	Polychaetes (*Aquilaspio aucklandica, Nicon* sp., *Scolecolepides* sp., *Scoloplos cylindrifer*), oligochaetes, amphipods (*Corophium* sp., Paracalliopidae), *Helice crassa* (crab)	*Avicennia marina*	Ellis et al. 2004
Various sites, New Zealand			Gastropods (*Amphibola crenata, Diloma subrostrata, Melagraphia aethiops, Turbo smaragdus, Zeacumantus lutulentus*), crabs (*Helice crassa, Scylla serrata*). *Nicon aestuariensis* (polychaete)	*Avicennia marina*	Morton & Miller 1968
Manukau Harbour, New Zealand (37°01′S)	Young stands: 5.5–8.0 Old stands: 2.2–5.8 (ranges of means of six 10-cm diameter cores)	Young stands: 20.7–80.0 Old stands: 17.3–72.6 (ranges of means of six 10-cm diameter cores)	Polychaetes (*Capitella capitata, Nicon aestuariensis, Scolecolepides benhami*), oligochaetes, gastropod (*Potamopyrgus antipodarum*), bivalve spat, crab (*Helice crassa*)	*Avicennia marina*	Morrisey et al. 2003
Sydney, Australia (33°50′S)			Anthozoa (1 sp.), sipunculids, echiurids, gastropods (14 spp. including *Ophicardelus* sp. and *Salinator solida*), bivalves (8 spp., including *Xenostrobus securis*), polychaetes (9 families, including capitellids, nereids, sabellids and spionids), oligochaetes, amphipods (6 spp.), isopods (7 spp.), leptostracans, cumaceans, tanaids, ostracods, maxillipods	*Avicennia marina*	Chapman, 1998, Chapman et al. 2005, Chapman & Tolhurst 2004, Tolhurst & Chapman 2007
Pittwater, Sydney, Australia (33°36′S)			Gastropods (*Littorina scabra, Onchidium damelii, Ophicardelus ornatus, Pyrazus ebeninus, Velacumantus australis*), bivalve (*Crassostrea commercialis*), *Balanus amphitrite* (barnacle), crabs (*Heloegrapsus* (*Helice*) *haswellianus, Heloecius cordiformis, Macrophthalmus carinimanus, M. setosus, Parasesarma erythrodactyla*)	*Aegiceras corniculatum* and *Avicennia marina*	Dakin 1966

Location	Species	Dominant mangrove	Reference
Careel Bay, Sydney, Australia (33°37'S)	Gastropods (trochid, acmaeid, littorinid, *Velacumantus australis*, nassarid, *Ophicardelus* spp, onchidiid), penaeid prawn, mud shrimp (*Laomedia* sp.), crabs (*Scylla serrata*, 2 spp. grapsids, 3 spp. ocypodids), bivalves (arcid, mytiliid, ostreid, *Teredo* sp.)	*Avicennia marina*	Hutchings & Recher 1974
Sydney, Australia (33°50'S)	Polychaetes (*Australonereis ehlersi, Ceratonereis erythraensis, Neanthes vaalii, Marphysa sanguinea, Capitella capitata, Notomastus hemipodus, Nephtys australiensis, Polydora* sp., *Haploscoloplos* sp., *Owenia fusiformis, Phyllodoce novaehollandiae, Chaetopterus* sp.), isopod (*Ligia australiensis*), mud shrimps (*Laomedia healyi*), snapping shrimp (*Alpheus edwardsii*), burrowing prawns (*Callianassa australiensis*), penaeid prawn (*Penaeus plebejus*), crabs (*Helice leachi, Helograpsus haswellianus, Metopograpsus frontalis, Paragrapsus laevis, Parasesarma erythrodactyla, Mictyris longicarpus, Euplax tridentata, Heloecius cordiformis, Macrophthalmus crassipes, Macrophthalmus setosus, Scylla serrata*)	*Aegiceras corniculatum* and *Avicennia marina*	Saenger et al. 1977
South Australia: Port Wakefield (34°10'S), Port Gawler (34°36'S), Port Adelaide (34°48'S)	Gastropods (*Ophicardelus ornatus, Laemodonta ciliata, Maripythia meridionalis, Salinator* spp.), bivalves (*Laternula creccina, Modiolus inconstans*), snapping shrimps (*Alpheus* spp.), burrowing prawns (*Callianassa ceramica, Upogebia simsoni*), crabs (*Cyclograpsus audouini, Helice haswellianus, Paragrapsus gaimardi*)	*Avicennia marina*	Macnae 1966
Various, South Australia	Oligochaetes, nereid polychaetes, gastropods (*Austrocochlea constricta, Bembicium auratum, Cominella lineolata, Niotha (Paracanassa) pauperata, Ophicardelus ornatus, Phasionotrochus* sp., *Salinator solida*), bivalves (*Katelysia peronii, Macomona deltoidalis, Modiolus inconstans*), barnacles (*Balanus amphitrite, Elminius modestus*), crabs (*Eriocheir spinosus, Helograpsus haswellianus, Philyra laevis*)		Butler et al. 1977b
Temperate eastern Australia	Gastropods (*Assiminea tasmanica, Bembicium auratum, Ophicardelus* sp., *Pyrazus ebeninus, Salinator fragile, S. solida, Velacumantus australis*), crabs (*Heloecius cordiformis, Paragrapsus laevis, Parasesarma erythrodactyla*)	*Aegiceras corniculatum* and *Avicennia marina*	Chapman & Underwood 1995

(continued on next page)

Table 6 (continued) Benthic macrofaunal taxa recorded by studies of temperate mangroves

Region/site	Number of taxa	Total abundance	Taxa recorded	Mangrove species	Reference
South Africa: Mtata River (31°57'S), Mngazana River (31°42'S), Isipingo River (30°00'S), Richards Bay (28°50'S)			Polychaete (*Dendronereis arborifera*), gastropods (*Assiminea ?bifasciata, Cassidula labrella, Cerithidea decollata, Littorina scabra, Peronia peronii, Terebralia (Pyrazus) palustris*), bivalve (*Crassostrea cucullata*), barnacle (*Balanus amphitrite*), isopod (*Sphaeroma tenebrans*), snapping shrimp (*Alpheus crassimanus*), burrowing prawns (*Callianassa kraussi, Upogebia africana*), crabs (*Cardisoma carnifex, Clibanarius longitarsus, C. padavensis, Coenobita cavipes, Cyclograpsus punctatus, Eurycarcinus natalensis, Macrophthalmus convexus, M. grandidieri, Metopograpsus gracilipes, Panopeus africanus, Sesarma catenata, S. euillimene, S. guttata, Sesarma meinerti, Uca annulipes, U. chlorophthalmus, U. marionis, U. urvillei*), *Holothuria parva*	*Avicennia marina, Bruguiera gymnorrhiza, Rhizophora mucronata*	Macnae 1963
Mngazana River, South Africa (31°42'S)			Polychaetes (*Ceratonereis keiskama, Marphysa macintoshi, M. depressa, Orbinia angrapaquensis*), gastropods (*Assiminea bifasciata, Cerithidea* sp., *Littorina scabra, Nassarius kraussianus*), burrowing prawn (*Upogebia* sp.), crabs (*Sesarma catenata, S. meinerti, Uca annulipes, U. chlorophthalmus, U. urvillei*)	*Avicennia marina, Bruguiera gymnorrhiza, Rhizophora mucronata*	Branch & Grindley 1979
Santa Caterina Island, Brazil (27°29'S)	17	≤ 7250 m^{-2}	Oligochaetes (especially tubificids and enchytraids) and capitellid polychaetes 82% of the total macrofauna; also other polychaetes (*Laeoneries acuta, Polydora socialis, P. websteri, Typsyllis* sp., *Sigambra grubii*), crabs (*Uca thayeri, Cyrtograpsus altimanus*), tanaid (*Tanais stanfordi*), ostracods, insect larvae; no bivalves or other filter-feeders	*Avicennia schaueriana, Laguncularia racemosa, Rhizophora mangle*	Netto & Gallucci 2003
Bermuda (32°N)			Anthozoa (*Aiptasia pallida*), Scyphozoa (*Cassiopea xamachana*), polychaetes (*Arenicola cristata*, spirorbids, sabellids), gastropods (*Batillaria minima, Cerithium lutosum, Littorina angulifera, Melampus coffeus, Mitrella ocellata*), isopod (*Ligia baudiniana*), crabs (*Pachygrapsus gracilis, Gecarcinus lateralis, Goniopsis cruentata*)	*Avicennia germinans, Conocarpus erectus, Rhizophora mangle*	Thomas & Logan 1992, Thomas 1993

Note: Numbers of taxa and individuals are shown if available, and the mangrove species present are also noted. SE = standard error.

near Sydney (Table 6; Dakin 1966). The commercially harvested portunid crab *Scylla serrata* also occurs in temperate mangroves (Hutchings & Recher 1974), as far south as the Bega River, New South Wales (36°43′; New South Wales Department of Primary Industries 2009). Characteristic, and often very abundant, mangrove gastropods include the potamidids *Pyrazus ebeninus* and *Velacumantus australis* and the amphibolid pulmonates *Salinator* spp. (another amphibolid, *Amphibola crenata*, is a characteristic member of the mangrove fauna in New Zealand). Less-abundant taxa include sipunculids, echiurids, acarids and dipteran larvae.

Bermuda   The mangrove fauna of Bermuda includes about 150 macrofaunal species but is characterised by large variation in species composition among locations around the island (Thomas 1993). Of the 150 species, 73 species were found at only a single location, and only 4 species were found at more than 50% of locations. There were no obvious differences in opportunity for colonisation by macrofauna among stands, and Thomas concluded that variation in macrofaunal composition was likely to be a consequence of effects of environmental factors on settlement, mortality and growth. The most widely distributed benthic species were the gastropods *Cerithium lutosum* and *Mitrella ocelata*, the isopod *Ligia baudiniana* and the grapsid crabs *Goniopsis cruentata* and *Pachygrapsus gracilis* (Table 6). The giant land crab *Cardisoma guanhumi* constructs burrows at the landward edge of some Bermudan mangrove stands, and the land hermit crab *Ceonobita clypeatus* also occurs in this zone (Thomas & Logan 1992).

Japan   There appears to be little published information on the fauna of temperate mangroves in Japan. Wakamatsu & Tomiyama (2000) described seasonal variation in the distribution of the detritivorous gastropods *Batillaria multiformis*, *Cerithideopsilla djadjariensis*, *C. cingulata* and *Cerithidea rhizophoranum* in mangroves at a study site in the southern part of the island of Kyushu (31°S).

New Zealand   A number of studies have been undertaken on the benthic assemblages and species of mangrove forests in New Zealand (Table 6). Taylor (1983) gave a qualitative description of the fauna of mangrove forests in Whangateau Harbour (36°18′S), a relatively pristine embayment north of Auckland. Burrowing animals were rare, apart from the grapsid crab *Helice crassa*. The pulmonate mud snail *Amphibola crenata* was common, along with the gastropods *Diloma subrostrata*, *Zeacumantus lutulentus* and *Z. subcarinatus*. The main predatory species was the whelk *Cominella glandiformis*, feeding on crabs, *Amphibola crenata* and polychaetes.

Ellis et al. (2004) recorded a benthic community at upper-shore mangrove sites in the Whitford embayment (east of Auckland: 36°55′S) dominated by corophiid and paracalliopiid amphipods, oligochaetes and the crabs *Halicarcinus whitei* and *Helice crassa*. Sites closest to intertidal sandflats were differentiated by the limpet *Notoacmea helmsi*; several bivalve species, including *Paphies australis*, *Macomona liliana*, *Austrovenus stutchburyi*, and *Nucula hartvigiana*; and the isopod *Exosphaeroma chilensis*. Subsurface deposit-feeders dominated the benthic community in the mangrove habitats, primarily polychaetes (*Scoloplos cylindrifer*, *Heteromastus filiformis* and other capitellids), oligochaetes and *Helice crassa*. The average number of species per sample was 8.13 in the mangroves and 9.1 in adjacent tidal mudflats.

Morrisey et al. (2003) investigated the effect of mangrove stand age (young 3–12 yr, and old > 60 yr) on associated benthic assemblages at two locations within the Manukau Harbour (on Auckland's western coast: 37°01′S). The overall number of species was generally higher at the younger sites, along with higher numbers of the copepod *Hemicyclops* sp., oligochaetes and *Capitella capitata*. However, the total number of individuals did not differ between the mangrove stands of different ages, largely due to the presence of large numbers of the surface-dwelling gastropod *Potamopyrgus antipodarum* at older sites. The main benthic species found under the mangrove forest were the crab *Helice crassa*, *Hemicyclops* sp., *Potamopyrgus antipodarum*, bivalve spat, oligochaetes and the polychaetes *Nicon aesturiensis*, *Scolecolepides benhami*, and *Capitella capitata*. All taxa varied in their abundance at smaller spatial scales (among sites and plots) apart from bivalve spat and *Helice crassa*, which did not vary at any of the scales examined.

Alfaro (2006) sampled six distinct habitats in Matapouri Estuary (35°34'S), eastern Northland: mangrove stands, the pneumatophore zone, patches of seagrass, channels, banks and sandflats. Each contained distinctive faunal assemblages, with seagrass patches having the highest combined abundance and species diversity per unit area and mangrove forests the lowest. The mangrove fauna contained low numbers of the cockle *Austrovenus stutchburyi*, a variety of deposit-feeding annelid worms, very few crabs, and no shrimps or amphipods. In contrast with the locations sampled by Morrisey et al. (2003) and Ellis et al. (2004), no mud snails (*Amphibola crenata*) or *Potamopyrgus antipodarum* were collected, being found only in saltmarsh areas of the estuary (outside the study area). *Helice crassa* was present in very low densities in mangroves but was common in seagrass. Large volumes of drift brown alga (Neptune's necklace, *Hormosira banksii*) were trapped in the pneumatophore zone, supporting an assemblage of the grazing gastropod snails *Turbo smaragdus*, *Diloma subrostrata* and *Melagraphia aethiops*.

South Africa   The fauna of temperate mangroves in South Africa, like those of Australia and New Zealand, is characterised by polychaetes, gastropod molluscs and decapod crustaceans and is distinctly different from other estuarine habitats (Macnae 1963, Branch & Grindley 1979; Table 6). Although no species occur only in mangroves, several species, such as the crab *Sesarma guttata* and the snails *Cerithidea decollata* and *Pyrazus palustris*, are often only abundant in mangroves (Macnae 1963, Branch & Grindley 1979). These temperate mangrove faunas show stronger affinities with those of lower-latitude, subtropical estuaries than with those further south (Branch & Grindley 1979). In the case of the mangrove fauna of the Mngazana River estuary (31°50'S), 62% of the species extend northwards but not southwards.

Sesarmid crabs are a particularly conspicuous part of the mangrove fauna, and Macnae (1963) recorded four species in mangroves in Richards Bay (28°50'S), together with two species of xanthids, two species of grapsids, six species of ocypodids (including four species of fiddler crabs, genus *Uca*), the portunid *Scylla serrata*, the land crab *Cardisoma carnifex* and three species of hermit crabs. Further south, in the Mngazana Estuary, Branch and Grindley (1979) found two species of sesarmids and three of ocypodids (species of *Uca*, apparently more abundant on the fringe of the mangroves, creek banks and the lower edge of salt marshes; Branch & Grindley 1979). In the Mngazana Estuary, *Sesarma catenata* and *Uca* spp. occur in very large numbers on the mud surface at low tide but retreat to their burrows at high tide.

Other components of the benthic fauna of South African mangroves include alpheid and callianassid prawns; rissoid, potamid and cerithid gastropods; nereid, orbiniid and eunicid polychaetes; and the holothurian *Holothuria parva* (Macnae 1963, Branch & Grindley 1979). Branch & Grindley recorded several species of polychaetes and gastropods from transects through mangrove stands in the Mngazana Estuary, but in an earlier study of the same estuary Macnae reported no polychaetes or molluscs in sediments among the mangroves. This suggests that faunal distributions can be spatially or temporally very variable. The rissoid *Assiminea bifasciata* may occur in very large abundances and is preyed on by fish and mud crabs (*Scylla serrata*) (Branch & Grindley 1979).

United States   No specific information on the fauna of Louisiana mangroves could be found.

Summary   The benthic invertebrate fauna of temperate mangrove forests often appear to be modest in both abundances and species diversity compared with other nearby estuarine habitats. Differences between assemblages from mangroves and adjacent unvegetated sediments are generally identifiable as differences in relative abundance of a largely common suite of species rather than differences in species composition (Morrisey et al. 2003, Alfaro 2006). In Australia, observed lower densities and biodiversity of macrofauna within mangroves relative to other habitats has been attributed (in a tropical context) to the high proportion of tannins from mangrove detritus and mud associated with mangrove habitats (Alongi & Christoffersen 1992, Lee 1999, Alongi et al. 2000). Another contributing factor in older mangrove stands may be the effect of the compacted nature of the sediment on infaunal diversity and abundance.

In general, the fauna is characterised by surface-living detritivorous and predatory gastropods, burrowing crabs, alpheid and callianassid prawns and small crustaceans (such as cumaceans, tanaids, isopods and amphipods). The infauna includes oligochaetes and polychaete worms, but infaunal bivalves are generally absent in mature areas of mangrove stands, presumably because of the compacted nature of the sediment. There is no evidence for a distinct mangrove fauna in any of the other regions considered in this review (Macnae 1963, Butler et al. 1977a, Alfaro 2006). Butler et al. (1977a) noted that even the species occurring most commonly in South Australian mangroves (such as the burrowing crab *Helograpsus haswellianus* and the gastropod *Bembicium auratum*) were rare or absent at some sites, and none was found exclusively in mangroves.

There was spatial variability in taxonomic composition among different locations even within a relatively limited area such as northern New Zealand. For example, Alfaro (2006), working in a small, less-impacted estuarine system (Matapouri Estuary, northern New Zealand), did not record mudsnails (*Amphibola crenata*) or gastropods (*Potamopyrgus antipodarum*) from her mangrove sites and only very low abundances of the crab *Helice crassa*, in contrast to most other New Zealand mangrove forests studied to date. She also recorded an assemblage of the grazing gastropod snails *Turbo smaragdus*, *Diloma subrostrata* and *Melagraphia aethiops*, as did Taylor (1983), working in Whangateau Harbour (another less-impacted system). These species were rare or absent from other mangrove systems that have been sampled in New Zealand (e.g., Morrisey et al. 2003, Ellis et al. 2004). Chapman & Tolhurst (2004) found that variation in faunal diversity and abundances among habitats (open mudflats, shaded and unshaded areas within mangrove stands) was generally small and less than that among sites 10–30 m apart and bays about 2 km apart within each habitat. Again, these observations emphasise the often relatively large variation in diversity among mangrove stands at small or regional scales compared with differences between mangroves and other estuarine habitats.

The ecology of benthic macrofauna in temperate mangroves

Zonation   Mangroves occupy the area of the shore roughly between MSL and mean high water. The fauna of mangrove stands often shows broad patterns of zonation within this tidal range, as they do on other types of shore.

Kaly (1988) examined spatial patterns of distribution of gastropods within mangroves near Sydney. Species found in the upper parts of mangrove forests, such as *Assiminea tasmanica*, *Ophicardelus* sp. and *Salinator solida*, were also found in adjacent salt marshes (Morrisey 1995). Other species (*Bembicium auratum*, *Pyrazus ebeninus* and *Velacumantus australis*) generally occurred in seaward parts of mangrove stands, but *Salinator fragile*, in contrast, occurred throughout the stands. There was overlap in the ranges of tidal height of many of these species and variation in patterns of distribution among different locations. Transplant experiments showed that several species were capable of surviving in other areas of the mangrove stand than those in which they normally occurred (Kaly 1988). Manipulative experiments also indicated that pneumatophores or the algae growing on them were important habitat-components for *Assiminea tasmanica*, *A. solida* and *Ophicardelus*.

In the same region as Kaly's study, the ocypodid *Heloecius cordiformis* did not show clear patterns of distribution with shore height within the mangroves, but was always associated with mounds of well-drained sediment (Warren 1987). Individuals higher on the shore tended to be larger than those in more seaward parts of the forest. *Heloecius cordiformis* hibernated in its burrows during winter (June–September). *Paragrapsus laevis* was generally more abundant in lower parts of the forest and tended to live in burrows in moist or submerged flat areas rather than mounds. A third species, *Parasesarma (Sesarma) erythrodactyla*, was evenly distributed across tidal heights and between mounds and flat areas. This species has also been observed climbing mangrove trees in South Australia (S. Dittmann personal observation).

In the *Kandelia candel* forests in the Atago River estuary, Kyushu, Japan, the cerithid gastropod *Cerithideopsilla rhizophorarum* showed greater tolerance of desiccation than three other, co-occurring members of the same family. Adults of *C. rhizophorarum* occurred at higher tidal levels than those of the other species. Juveniles of all four species, however, were limited to lower tidal levels in the forest. The ability of *C. rhizophorarum* to tolerate desiccation allows it to climb up and hibernate on the trunks of mangrove trees during winter.

Macnae (1963) described zonation of fauna in the Mngazana Estuary in South Africa, from the lower boundary of the mangrove at the level of neap high tides (occupied by a fringe of *Avicennia marina*), through the *Bruguiera gymnorrhiza* thickets further up the shore, a higher band of *Avicennia marina* to the salt marsh in the upper shore. *Sesarma catenata* is the dominant crab from the lower salt marsh downshore to the lowermost *Avicennia marina* pneumatophore zone. *Sesarma meinerti* was also present throughout the mangrove forest but was most abundant at higher levels and in the mangrove fringes. The fiddler crabs *Uca chlorophthalmus* and *U. urvillei* occurred in the salt marsh but only rarely extended in the mangrove thickets.

Animal–habitat relationships   Faunal distribution patterns reflect environmental factors other than, or in combination with, tidal height. Chapman & Tolhurst (2004), working on an urbanised mangrove forest in Sydney Harbour, attempted to relate the invertebrate assemblages that were present to the properties of the sediments. Across three habitats, the benthos showed less variation among mangrove patches with and without leaf litter or mudflats adjacent to mangrove patches than within and among sites in each habitat type. In contrast, biodependent properties of the sediment (water content, water-soluble fraction of carbohydrates, total carbohydrate, chlorophylls *a* and *b*) showed less variation at small scales than among habitats. For the faunal assemblage as a whole, and for all individual taxa examined, most (50–100%) of the variation was at the scale of metres within each habitat. This variation could not be fully explained by tidal inundation, amounts of leaf litter or macroalgal cover. There was no clear correlation between variation in the benthos and variation in the sediment properties at any of the spatial scales. The strongest correlate, albeit still weak (Spearman rank correlation coefficient = −0.10) was chlorophyll *a* concentration.

Chapman & Tolhurst (2007) expanded the approach taken in their earlier work (Chapman & Tolhurst 2004) across three bays in the same system. They found that those sediment properties that contributed most to the differences among habitats, and those that best correlated with the benthos, differed among bays. The single taxon that best correlated with the sediment was spionid polychaetes, but such correlations were generally weak. It was concluded that all spatial scales contributed to variability; that there was little predictability from the patterns shown in one habitat to those in other habitats, or from one component of the sediment to other components; and that such variability suggested structural redundancy in the fauna (i.e., different components of the benthos contributed similar functions in different places). Sediment-related variables showed weaker patterns relating to habitats than the fauna, indicating that the suite of measured, sediment-related variables had failed to capture important environmental differences among sites (Tolhurst & Chapman 2007).

Other environmental variables, however, did appear to influence faunal distribution, and manipulative experiments showed that abundances of the gastropods *Ophicardelus* spp. and *Salinator solida* decreased rapidly in plots where the amount of leaf litter had been reduced (Chapman et al. 2005). Experimental reduction of the heights of pneumatophores did not affect abundances of snails. Responses to manipulations of the amount of macroalgae were inconsistent between two sets of experiments at different times and locations, with no response in one case and reduced densities of both species of snails in the other.

Biogenic structures (including the mangroves themselves; see p. 79) can be important components of the benthic habitat within mangrove forests. In mangroves in south-eastern Australia, abundances of the dominant gastropod *Bembicium auratum* vary with height above low water and are more abundant on oysters (*Crassostrea commercialis*) than other substrata (Underwood & Barrett 1990). Abundances are highest at lower levels of the shore, where oysters are naturally present, and the proportion of small individuals in the population is also highest here. Caging experiments (Branch & Branch 1980) demonstrated that increased population density of *Bembicium auratum* resulted in increased mortality (particularly of juveniles) and reduced growth and food availability (measured as chlorophyll concentrations in the sediment). Growth and mortality (particularly of juveniles) were also reduced in snails caged without access to hard substrata. Branch & Branch (1980) suggested that the observed decline in abundance but increase in body size of *B. auratum* with height on the shore was a result of higher settlement or survival of juveniles on the lower shore, where oysters provide hard substrata. As they grow, individuals become less dependent on hard substrata and may disperse up the shore. The density-dependent relationship between growth or mortality and the availability of food was interpreted as evidence that population density is restricted by intraspecific competition for food, and that predation is unlikely to be important.

This interpretation was disputed by Underwood & Barrett (1990), who found no correlation between the distribution of chlorophyll and that of *B. auratum* and argued that the density-dependent relationship between growth and chlorophyll standing stock could not be interpreted as evidence that population size is limited by availability of food without information on microalgal productivity. Using manipulative experiments, they demonstrated that removal of oysters from areas where they occurred naturally resulted in a rapid decrease in density and mean size of *B. auratum*. More snails, particularly juveniles, emigrated or disappeared from plots without oysters. Conversely, when oysters and snails were transplanted to areas higher up the shore, more snails remained in plots with oysters than in those without; again, the effect was more obvious among juvenile snails. Underwood & Barrett (1990) concluded that the role played by oysters in the distribution, abundance and size of *B. auratum* is a consequence of provision of refuges from predation by crabs or fishes during high tide, although they emphasised that further experimentation is required to support this conclusion.

Oysters also provide a habitat for the limpet *Patelloida mimula* in mangrove forests in south-eastern Australia, and limpets are generally only found on oysters (Minchinton & Ross 1999). Most (98%) oysters with limpets living on them have only one limpet, and >70% of limpets returned to the same scar on their host oyster when tracked for 13 days. Although the distribution of oysters in the mangrove forest sets the limits of distribution of limpets, the abundance of limpets is not directly related to the abundance of oyster habitat present. Oyster abundance decreases with tidal height, both across the shore and with position on the trunks of trees, but this is not reflected in the abundances of limpets.

Warren (1987) experimentally investigated the roles of habitat selection and interspecific interactions in partitioning of mangrove habitat between the crabs *Heloecius cordiformis* (found on well-drained mounds of sediment), *Paragrapsus laevis* (usually found in wetter flats between mounds) and *Parasesarma erythrodactyla* (found equally in flats and mounds). The presence of other species in experimental enclosures had little effect on the habitat a species colonised. Topography and the type of sediment, on the other hand, did influence habitat selection. *Heloecius cordiformis* would colonise artificial mounds if these were composed of sediment obtained from mounds but not from flats. *Paragrapsus laevis* and *P. erythrodactyla* would colonise mounds or flats if artificial burrows were provided but *Heloecius cordiformis* preferred mounds to flats provided with artificial burrows. There was also evidence that *H. cordiformis* was itself at least partly responsible for the presence of hummocky topography, and that by burrowing through the finer surficial sediments into the underlying coarser material, it also made the texture of mound sediments coarser (Warren & Underwood 1986).

Most activities performed by these crabs took place at or near burrow entrances, although collection of mud balls (which were then brought to the burrow to be sifted for food) and courtship took place away from the entrance (at the entrance to a female's burrow in the case of courtship). All three species were active on the surface during daytime low tides, but *Paragrapsus laevis* and *P. erythrodactyla* were also active at night. Burrows played a significant role in reducing mortality from predation by fishes during high tide (Warren 1990).

In South Australian mangroves, field experiments with the crab *Helograpsus haswellianus* showed that crabs ceased digging new burrows in response to experimentally increased population density when the density of burrows reached a critical level (McKillup & Butler 1979). Collapse of the sediment surface occurred when burrow density exceeded this critical value, as determined by experimental manipulations. When the density of crabs exceeded that of holes, crabs entered holes that were already occupied rather than digging new ones, and after a few minutes either the original or new occupant was evicted and entered another hole. McKillup and Butler suggested that, as population density increases, more crabs are exposed to predation as they move from burrow to burrow, and density is reduced back towards the carrying capacity of the sediment. Along tidal channels in South Australian mangroves, burrowing by polychaetes gives rise to mound-covered terraces (Butler et al. 1977b).

Morrisey et al. (2003) investigated the effect of mangrove stand age (young 3–12 yr, and old > 60 yr) on associated benthic assemblages at two locations within the Manukau Harbour (on Auckland's western coast; 37°01′S). The overall number of species was generally higher at the younger sites, along with higher numbers of the copepod *Hemicyclops* sp., oligochaetes and *Capitella capitata*. However, the total number of individuals did not vary between the two ages, largely due to the presence of large numbers of the surface-dwelling gastropod *Potamopyrgus antipodarum* at older sites. All other taxa present varied in their abundance at smaller spatial scales (among sites and plots) apart from bivalve spat and *Helice crassa*, which did not vary at any of the scales examined. It was suggested that, as mangrove stands grow older, the abundance and diversity of the associated fauna shift towards animals living on the mangrove plants themselves (e.g., insects and spiders). This change would correlate with an increase in the size and structural complexity of the plants and perhaps a decrease in the quality of the benthic habitat as the sediment becomes more compacted and the interstitial water more saline and less oxygenated. However, the fauna on the mangrove plants themselves was not sampled.

Ellis et al. (2004) examined the effects of high sedimentation rates on mangrove plant communities and associated benthic community composition, including a comparison with adjacent tidal flats (the Whitford embayment, east of Auckland, New Zealand). Macrofaunal diversity and abundance within the mangrove habitats were lower than expected, and there were clear functional differences along a sedimentation gradient, with lower numbers of suspension-feeders, low macrobenthic diversity and a predominance of deposit-feeding polychaetes and oligochaetes in areas with higher sedimentation rates. All mangroves sites had lower abundance and diversity than nearby sandflats, but heavily sedimented mudflats without mangroves were similar in their benthic composition to mangrove sites. They suggested that this pattern was a response to the increased silt/clay fraction from sedimentation rather than to the presence or absence of the mangroves themselves.

Ellis et al. (2004) concluded that high sediment mud content and rates of deposition were possibly more important than the presence or absence of mangroves in terms of reducing faunal diversity and abundance. Alfaro (2006) also suggested that lower temperatures and lower tidal inundations in New Zealand coastal areas might result in slower organic matter decomposition rates compared with tropical and subtropical mangrove ecosystems, causing reduced productivities. The absence of large crabs from New Zealand mangrove forests, considered to be important sediment bioturbators and consumers of mangrove leaves and detritus in tropical mangroves (Robertson & Daniel 1989, McIvor & Smith 1995, Slim et al. 1997), might also play a role. The dominant crab species in New Zealand mangroves, *H. crassa*, grows to a maximum of around 4 cm carapace length, which is relatively small when compared with tropical species (Alfaro 2006).

Encrusting and motile epifauna of mangrove trees

Introduction

The fauna of mangrove trees includes immobile, encrusting species and motile species that move around on the trees, either as permanent residents or transient visitors. There is also a fauna that bores into the tissues of mangrove trees and woody debris within the forest. Members of each component may derive from adjacent marine or terrestrial environments.

Encrusting marine invertebrates

In general, the encrusting fauna of mangroves is of relatively low diversity compared with that of many other intertidal hard substrata. This low diversity is partly a consequence of the nature of the available substratum, such as the likely inability of limpets to create home scars on mangrove bark (Minchinton & Ross 1999) but also of the general unsuitability of muddy, depositional coastal habitats for sessile organisms because of high suspended sediment loads. In more open coastal situations, the encrusting fauna can be relatively diverse, as in some mangrove stands in Bermuda (see p. 80).

The most common encrusting organisms in temperate mangroves are oysters, mussels, tube-dwelling polychaetes and barnacles, of which barnacles are the most studied. In New Zealand mangroves, rock oysters (*Crassostrea glomerata*) and the introduced Pacific oyster (*Saccostrea glomerata*) are found attached to trunks and pneumatophores; barnacles (*Elminius modestus*) are also characteristic of pneumatophores, trunks and leaves (Taylor 1983). Oysters (*Crassostrea commercialis*) are also common on the trunks and pneumatophores of mangroves in southern Australia (Branch & Branch 1980).

Ross & Underwood (1997) described the distribution and abundance of three barnacle species (*Elminius covertus*, *Hexaminius popeiana* and *H. foliorum*) in *Avicennia marina* forests near Sydney. *Elminius covertus* was more abundant on bark than on leaves or twigs. *Hexaminius popeiana* was only found on bark, whereas *H. foliorum* was most abundant on twigs and did not occur on bark. Densities of all species were higher in seaward parts of the forest than in landward parts, and *H. popeiana* was virtually absent in landward parts. In seaward parts of the forest, barnacles were most abundant at midtidal levels on trunks and more abundant on the lower than upper surfaces of leaves. These patterns of distribution arise primarily from patterns of larval settlement, with subsequent mortality of juveniles and adults modifying density but not distribution (Ross 2001). Cyprid larvae of *Elminius covertus* are most abundant in winter, coinciding with relatively large night-time tides, giving them greater access to landward parts of the forest than the larvae of *Hexaminius popeiana*, which are most abundant in spring and summer. Cyprid abundances are stratified in the water column (Ross 2001), so that variation in density of settling larvae with tidal height is a function of larval abundance at that height and the time available for settlement (which also varies with height).

Two other species of barnacle, *Balanus amphitrite* and *Elminius adelaidae*, occur in *Avicennia marina* forests in South Australia and are present in aggregations on pneumatophores, occurring on some pneumatophores but not others and present in aggregations on those pneumatophores where they do occur (Bayliss 1993). Manipulative experiments with modified pneumatophores demonstrated that cyprids of both species preferred to settle on pneumatophores bearing adults of their own species. Adults of the other species, and juveniles of *E. adelaidae*, did not attract settlers. Aggregations of barnacles may also attract their predators. In Bayliss's study area, large numbers of predatory gastropods, *Bedeva paivae*, occasionally migrate from adjacent mudflats into the seaward part of the mangrove forest and feed on clusters of barnacles on pneumatophores, sometimes eliminating the whole cluster. These aggregations of *B. paivae* are, however, infrequent. The whelk *Lepsiella vinosa* is a more consistent predator on the barnacles but does not form feeding aggregations and rarely consumes all of the barnacles on a pneumatophore (Bayliss 1982).

Poor survival and growth of mangrove seedlings are common in natural and replanted populations of mangroves (e.g., Clarke & Myerscough 1993, Osunkoya & Creese 1997), and this effect has been attributed to the presence of large numbers of fouling organisms, particularly barnacles (e.g., Macnae 1968). Satumanatpan & Keough (1999) found, however, that survival of seedlings of *Avicennia marina* in Westernport, Victoria, Australia, was not influenced by the presence of barnacles on the stems or the upper or lower surfaces of leaves over a 2-yr period following experimental manipulations of barnacle abundances. There was no significant effect on growth. They suggested that other factors were more important, including algal or seagrass cover, smothering by sediments, damage by herbivores and climatic conditions.

In addition to fringing, coastal mangrove stands, Bermuda contains hundreds of landlocked marine ponds, most of which are connected to the sea via underground tunnels and caves (Thomas & Logan 1992). The mangroves in these 'anchialine' ponds have a rich biota compared with other stands on the island. The prop roots of *Rhizophora mangle* and the pneumatophores of *Avicennia germinans* contain some of the richest faunas in the ponds (Thomas & Logan 1992). The fauna also display variation with tidal height, often with a shallow (10 cm below mean low-tide level in the pond) zone of relatively low diversity and a deeper, more diverse assemblage.

Roots in the intertidal zone generally contain spirorbid polychaetes, cyanobacteria and green and red algae (the last of these are largely *Bostrychia montagnei*, which is only found on mangrove roots). Roots in the shallow subtidal are characterised by bryozoans (usually *Bugula neritina*) and spirorbid and sabellid polychaetes. Diversity increases markedly in the deeper zone, but there is also considerable variation in the composition of this fauna among ponds. In some, the roots are dominated by oysters *Isognomon alatus*, which provide a substratum for encrusting sponges and the green alga *Cladophora* sp. Spaces among the oysters may be occupied by sponges (e.g., *Chondrila nucula* and *Terpios aurantiaca*), anemones (*Bartholomea annulata*), bryozoans (including *Schizoporella serialis*, *Bugula neritina* and *B. annulata*) and the solitary ascidian *Styela plicata*.

The ponds with the most diverse assemblages contain species that are rare or absent elsewhere on the island, including several sponges. Large, branching bryozoans (*Amathia vidovici* and *Zoobotryon verticillatum*) form extensive growths, and colonial and solitary ascidians are also abundant, together with the green alga *Caulerpa verticillata* and various anemones.

The high diversity of the fauna living on roots of mangroves in these ponds and the predominant role played by filter-feeders contrast strongly with the epifauna of mangroves in most sheltered coastal environments. It probably reflects the relatively stable temperature and salinity regimes of the ponds and perhaps the availability of suspended organic matter (Thomas & Logan 1992).

Motile marine invertebrates

The pneumatophores and trunks of mangroves in the Whangateau Harbour, New Zealand, are colonised by cyanobacteria and red algae, and the former included active nitrogen fixers (Taylor 1983). Living amongst these are isopods and amphipods. The main grazer at Taylor's study site was a gastropod, the cat's-eye *Turbo smaragdus*, and in the mangrove forest, populations of this species consisted of only large individuals. These animals were up to 45 mm long, with 90% of individuals over 30 mm. Taylor estimated these animals were around 25 yr old, with the nearest younger animals some 500 m away on a rocky shore. The mangrove population of *T. smaragdus* was described as either a relict population or a chance drift event from rafting on a tree trunk, although

T. smaragdus larvae are planktonic for about 12 h, so it would be possible for periodic recruitment to occur (K. Grange, NIWA, personal communication).

In south-eastern Australian mangrove forests, the gastropod *Littoraria luteola* is common on the trunks and branches of the trees. It is able to survive for long periods and occurs on parts of the trees well above the level of high water (Chapman & Underwood 1995). The coffee bean snail *Melampus coffeus* occupies the same habitat in Bermudan mangrove stands (Thomas & Logan 1992).

Several species of crabs in the families Grapsidae and Sesarmidae have evolved a tree-climbing habit in mangrove forests in various parts of the world, and this has apparently occurred on a number of independent occasions (Fratini et al. 2005). The extent of the arboreal habitat varies from species that live predominantly on the ground but also climb roots, through those that live mainly or exclusively on trunks, to those that live in the canopy, sometimes feeding on fresh leaves. The more specialist arboreal species, including the three known to feed on leaves (*Aratus pisonii* on the Atlantic and Pacific coasts of America, *Armases elegans* in West Africa, and *Parasesarma leptosoma* in the western Indo-Pacific), are largely tropical. *Parasesarma leptosoma* has, however, been discovered in the Mngazana Estuary (31°42′S) in South Africa (Emmerson & Ndenze 2007), where they were most abundant on *Rhizophora mucronata* (present on 92% of trees on which traps were deployed), less abundant on *Bruguiera gymnorrhiza* (38%) and absent on *Avicennia marina*. Abundance of crabs was also positively correlated with tree circumference.

Terrestrial invertebrates

The terrestrial invertebrate fauna of tropical mangroves can be abundant and diverse (Kathiresan & Bingham 2001). Common taxa include mites, termites, cockroaches, dragonflies, butterflies and moths, beetles, ants, bees, mosquitoes and spiders. Honey bees living in mangroves produce significant harvests of honey for humans in India, Bangladesh, the Caribbean and Florida. Wood-boring larvae of moths and beetles are common components of the fauna; their tunnels, in turn, provide accommodation for other species. More than 70 species of ants, spiders, mites, moths, cockroaches, termites and scorpions were found in tunnels bored in the wood of mangroves in Belize (see review by Kathiresan & Bingham 2001).

The terrestrial invertebrate fauna of temperate mangroves is poorly known but likely to be less diverse than those of tropical mangroves given the smaller numbers of species of trees and other habitat-related factors. Hutchings and Recher (1974) published a list of 57 taxa of insects and 18 of spiders from a survey of Careel Bay, near Sydney. The insects represented the orders Diptera, Lepidoptera and Hymenoptera, and many belonged to undescribed species. They also cited a study of another site near Sydney in which 35 species of spiders were recorded.

The diversity of spider assemblages in mangroves apparently increases sharply from temperate to tropical regions but, according to Hutchings & Recher (1982), derives from adjacent terrestrial habitats, and none is known to be endemic to mangroves. Orb-weaving species are certainly conspicuous in temperate mangroves (D. Morrisey personal observation), but Hutchings & Recher (1982) suggested that foliage-living species were relatively uncommon. The last observation may simply be a consequence of the lack of appropriate studies. Other species of spiders occur in the lowest vegetation of mangroves in south-eastern Australia or on the substratum among the trees, including wolf spiders (*Geolycosa* spp.) and members of the Pisauridae, such as *Dolomedes* spp. (Hutchings & Recher 1982). Butler et al. (1977b) recorded at least six species of spiders in South Australian mangroves.

In contrast to the marine fauna of mangroves, there appears to be some degree of dependence on mangroves among the terrestrial fauna. Burrows (2003) sampled insects from tropical mangroves at two sites near Townsville, northern Australia, and recorded 61 species, "more than doubling the number of insects recorded from Australian mangroves" and thereby illustrating the scant attention that this component of the mangrove fauna has received to date, even in a relatively thoroughly studied part of the distributional range of mangroves. Among the folivorous species, there was a high level of host specificity, particularly for *Avicennia marina* (largely because of the diversity of gall-forming species on this mangrove). By means of a literature review and by sampling insects on mangroves and adjacent terrestrial trees, Burrows (2003) contested the suggestion that mangroves (at least in this tropical region) do not have depauperate folivore faunas except, perhaps, in comparison to rainforests. Whether the same is true for other regions, including temperate ones, has not been investigated but is not unlikely given the prevalence of *Avicennia* species in temperate mangrove systems.

There are also examples of host specificity among temperate mangroves. Three species of moths, the tortricids *Ctenopseustis obliquana* and *Planotortrix avicenniae* (Cox 1977, Dugdale 1990), the pyralid *Ptyomaxia* sp. (J. Dugdale, Landcare Research, New Zealand [retired], personal communication) and an eriophyid mite, *Aceria avicenniae* (Lamb 1952), have been described from mangroves in New Zealand. *Aceria avicenniae* and the larvae of *Planotortrix avicenniae* are restricted to *Avicennia marina*, whereas *Ctenopseustis obliquana* is distributed throughout New Zealand, and its larvae are polyphagous (Dugdale 1990). All three moths have been collected in Waitemata Harbour, near Auckland, and *Planotortrix avicenniae* and *Ptyomaxia* sp. have been collected from Matakana Island, Tauranga Harbour (J. Dugdale personal communication) on the eastern coast of the central North Island, near the southern limit of mangroves in New Zealand.

The larvae of *Ptyomaxia* sp. cause distinctive distortion of the growing tips of the shoots, and *Aceria avicenniae* cause leaf galls. The larvae of *Ctenopseustis obliquana* cause damage to the leaves, fruit and buds of host plants, including horticultural crops, and presumably do the same to *Avicennia marina*. Young larvae live on the shoot tips or areas of new growth, binding the leaves together with silk and feeding on the inner surface of the leaf, whereas older larvae eat through the leaf (Horticulture and Food Research Institute of New Zealand Ltd. 1998a).

Meades et al. (2002) recorded 252 morphospecies from 13 orders of arthropods in mangrove stands surveyed twice at three locations (separated by kilometres or tens of kilometres) along the coast of southern New South Wales, Australia. Diptera was the most abundant (38% of the total numbers of individuals collected across all three sites) and most diverse (27% of species) order recorded. Contrary to expectations derived from a model of high spatial variability among mangrove patches as a result of natural or anthropogenic disturbance, there was no significant variation among locations in abundance or species composition of the terrestrial arthropod fauna. The results suggested that mangroves in the study area have a common suite of species, with most species occurring in all three locations. Species composition was, however, sometimes variable within a location. The findings were consistent with the hypothesis that the low diversity of mangrove tree species in a geographical region (particularly in temperate areas) contributes to the observed spatial homogeneity of their terrestrial fauna (Farnsworth 1998). The observations did not necessarily support the related hypothesis that low diversity of tree species results in low diversity of terrestrial invertebrates.

In south-eastern Australia, *Avicennia marina* acts as host plant for the larvae of the mangrove fruit fly *Euphranta marina*, which is consequently restricted to mangrove habitats (Hutchings & Recher 1974). The female fly lays its egg within the fruit while it is still on the tree, and the larva bores into and feeds on the developing cotyledons (Minchinton 2006 and references therein). When ready to pupate, the larva burrows to the surface of the fruit, makes an exit hole, but then pupates within the fruit, emerging later through the exit hole as an adult. Individual fruit contained up to six exit holes. In the same geographic region, mangrove fruit are also consumed by the mangrove plume moth *Cenoloba obliteralis*, whose larvae are restricted to the fruit and young shoots of *Avicennia marina* (Hutchings & Recher 1974). The eggs of this species are laid on the outside of the flower cluster or the surface of the fruit, into either of which the larvae burrow on hatching, to feed on the flower buds or cotyledons. The mature larvae leave their galleries within the host tissue to pupate within silk cocoons inside or outside the fruit. At Minchinton's (2006) study site near Sydney, flies and moths emerged from mangrove propagules after being transported tens of kilometres by water currents, potentially representing an important method of dispersal for the insects. Both species are also potential pollinators of mangroves (Minchinton 2006). In addition to fruit flies and plume moths, fruit of mangroves near Sydney contained larvae of other dipterans, lepidopterans and hymenopterans (Minchinton & Dalby-Ball 2001). Larvae of several taxa may occur in the same fruit.

Mosquitoes and biting midges have been studied in relative detail in mangroves generally because of their potential impacts on human health (Hutchings & Recher 1982). *Aedes vigilax* is the major vector of Ross River and Barmah Forest virus in coastal New South Wales. It lays its eggs around drying pools in salt marshes and mangroves, but the adults can disperse tens of kilometres (Department of Medical Entomology, University of Sydney, n.d.). *Aedes alternans* also occurs in estuarine habitats in south-eastern Australia. It is a nuisance to humans but is not known as a vector of disease, although the Ross River virus has been isolated from individuals collected from the southern coast of New South Wales. It breeds in stagnant pools and the larvae feed on larvae of other mosquitoes (Hutchings & Recher 1982).

Small numbers of chironomids and their larvae were collected from sediments among mangroves in Rangaunu and Mahurangi Harbours during a study of fish and their prey in mangrove habitats in several harbours in northern New Zealand (M. Lowe, NIWA, personal communication). Larvae of some tipulid flies feed on intertidal green algae and may exploit this food source that grows on the trunks and pneumatophores of mangroves (J. Dugdale personal communication). These animals, in turn, provide food for fishes and birds.

Ants play a potentially important role in tropical mangrove forests, including deterring herbivorous crabs and sap-feeding scale insects (reviewed by Cannicci et al. 2008), although evidence of long-term effects on the host trees is scarce. Although ants may also have adverse effects on mangroves trees, for example, by reducing longevity of leaves used to form their nests, their positive effects have been estimated to be 3–20 times larger (Cannicci et al. 2008). Ant colonies may establish within the tunnels in mangrove stems created by boring insects in New Zealand (J. Dugdale personal communication), and the ants may perhaps 'farm' the introduced scale insect *Ceroplastes sinensis*, which is also common on mangroves in New Zealand (Brejaart & Brownell 2004). Given their ecological importance in virtually every other terrestrial environment and their abundance and diversity in tropical mangroves (Hutchings & Recher 1982), it seems likely that ants play an ecologically important role in temperate mangroves. There has, however, been little work on them to date.

Animals that bore into mangrove tissues

Worldwide, a number of invertebrate taxa have been reported as burrowing into the living or dead woody tissues of mangroves (Hutchings & Recher 1982). Marine examples include teredinid molluscs (ship worms) and limnorid, sphaeromatid and chelurid isopods, while those from the terrestrial fauna include the larvae of various beetles, such as those of the cerambycid *Oemona hirta* (the lemon-tree borer) in New Zealand (J. Dugdale personal communication).

Fishes

Tropical mangrove systems are well documented as supporting diverse and abundant fish (and prawn) assemblages, including the juveniles of many commercially important species (e.g., Laegdsgaard & Johnson 1995, Vance et al. 1996, Nagelkerken et al. 2000, 2001). Their role as important or critical juvenile nurseries has also been well established, although debate continues regarding exactly how much of total production they contribute relative to alternative nursery habitats (Beck et al. 2001, Dalgren et al. 2006). These tropical mangrove assemblages are usually composed of multiple mangrove species, with very different growth forms and morphologies, including buttress roots. Some mangrove systems are also permanently inundated by water, allowing for continuous access by aquatic organisms (e.g., Curacao, Dutch Antilles, Nagelkerken et al. 2001; Florida, Ley et al.

1994). Until recently, these findings from subtropical and tropical mangroves (high fish abundance and diversity, important nursery role) were uncritically applied to temperate mangroves without any supporting quantitative investigations or data.

However, temperate mangrove systems differ from tropical systems in many ways, including lower mangrove species diversity, less structural complexity and smaller species pools of potentially associated organisms. Only since the year 2000 have scientific studies been directly conducted on fishes in temperate mangrove systems (with the notable exception of Bell et al. 1984). In a review of research on fishes in mangroves over the last 50 years (Faunce & Serafy 2006), only 1 (Bell et al. 1984) of the 111 papers assessed dealt with temperate mangroves. To the best of our knowledge, only one other study of fishes in temperate mangroves existed at that time, that of Clynick & Chapman (2002). Since then, a further 11 studies have been published (10 in temperate Australia, 1 in New Zealand).

As with subtropical and tropical mangrove systems, research on fishes in temperate mangroves has focused strongly on the role of mangroves as fish nurseries, with the types of sampling gear deliberately biased towards the quantification of juvenile or small fishes. More broadly speaking in terms of habitats, only recently have formal definitions been developed of what constitutes a nursery habitat. Beck et al. (2001) suggested that one or more of the following conditions need to be fulfilled for any given habitat to be considered a nursery habitat when contrasted with alternative habitats: (1) greater average densities of juvenile fishes, (2) lower predation rates, (3) higher growth rates or (4) greater than average contributions to adult populations. Of these aspects, most of the temperate mangrove studies have largely focused on condition 1, one study included tethering experiments to address condition 2 (for a fish species widespread across temperate Australasian mangroves), while no work has been done on condition 3. Aspect 4 has been indirectly addressed by Saintilan (2004), who examined commercial catch records (for a range of species) across 55 temperate estuaries along the coast of New South Wales, Australia, with varying proportions of different habitats (e.g., mangroves, seagrass, tidal flats, and deep mud basins). He found that the correlation of catch levels with the proportion of an estuary covered by mangroves was modest at best, and that indeed "as estuaries infill and the area of seagrass and mud basin declines [and mangroves increase], so too does the catch of species dependent upon these habitats".

Additional studies have assessed the relative importance of the proximity of mangroves to other habitats, whether there are artefacts from different fish-sampling methodologies in mangroves and, to a limited extent, connectivity with other adjacent habitats. Table 7 lists the various studies and the aspects that they addressed. Because there were relatively few studies of fish in temperate mangroves, each is briefly discussed individually in the following section.

Individual studies

Bell et al. (1984) sampled the fish assemblage in a temperate, tidal mangrove creek in Botany Bay, near Sydney, New South Wales, using the fish poison rotenone and associated blocking nets. They collected 46 species from 24 families. Six species dominated the assemblage (Table 7), contributing 84% of all individuals. Seasonal variations in abundance were driven largely by restricted recruitment periods of these species, representing young of the year. They suggested that four species (*Gerres ovatus*, the sparid *Acanthopagrus australis*, *Liza argentea* and *Girella tricuspidata*) were almost exclusively restricted to mangrove habitats, and as such these mangrove areas (or more correctly, the tidal creek) were important juvenile nursery areas. Reference was made to earlier studies in other habitats, although none was actually sampled as part of this study.

Clynick & Chapman (2002) sampled small mangrove stands around Sydney Harbour (New South Wales) and found little evidence of mangroves playing an important role as fish nurseries, with the possible exception of one goby species (transparent goby *Gobiopterus semivestitus*), which was more abundant (although highly variable) within the mangrove stands. Overall catches were

Table 7 Summary of information from studies of fish in temperate mangroves

Location	Mangrove species	Method	Design details	Dominant fish species	Reference
Botany Bay, New South Wales, Australia 34°1'S 151°11'E	*Avicennia marina*	Rotenone, block net, and hand-picking recovery	Lower reaches of tidal mangrove creek, sampled every second month between December 1977 and October 1980.	46 species (41 represented by the presence of juveniles), 26 families. Six dominant species: the glassfish *Velambassis jacksoniensis* (31.8%), parore/luderick (*Girella tricuspidata*) (18.6%), blue eye (*Pseudomugil signifer*) (11.5%), the fantail mullet (*Liza argentea*) (9.0%), the common toadfish (*Torquigener hamiltoni*) (6.8%), and the biddy (*Gerres ovatus*) (6.3%); another four relatively common: largemouth goby *Redigobius macrostomus* (2.9%), glass goby *Gobiopterus semivestitus* (2.8%), yellowfin bream *Acanthopagrus australis* (2.4%), exquisite goby *Favonigobius exquisitus* (2.2%)	Bell et al. (1984)
Sydney Harbour, Australia 33°49'S to 33°51'S 151°7'E to 151°119'E	*Avicennia marina*	Fyke nets 3-m wings, 1 m tall, 1-mm mesh	Initial day/night test, then day sampling only, 4 h each side of high tide. Four sites, at each 2 mangrove and 2 mudflat sites (within 50 m of forest), gang of 5 nets at each. Each bay sampled twice. May–June 1999.	17 species, 10 families (mangrove and mudflat samples combined) Dominated by Gobiidae (*Gobiopterus semivestitus*, >90% abundance); juveniles of five commercially important species (*Gerres subfasciatus, Mugil cephalus, Liza argentea, Myxus elongatus,* and one of two species of Sparidae (bream and tarwhine) extremely small, so not identified; numbers of individuals by species not presented by authors	Clynick & Chapman 2002

(continued on next page)

85

Table 7 (continued) Summary of information from studies of fish in temperate mangroves

Location	Mangrove species	Method	Design details	Dominant fish species	Reference
Westernport, and Corner Inlet, Victoria, Australia 38°30'S to 38°50'S 145°30'E to 146°30'E	*Avicennia marina*	Gill nets 1.5 m deep × 30 m long, five panels of different mesh sizes. Fyke nets. Square bag 70 × 70 cm, wings 10 m long by 70 cm deep; 6-mm mesh; beach seine 10 m × 2 m × 1 mm	Three sites within each of two embayments quarterly. Four replicates of gill and fyke in each habitat, 1–2 h before high tide; 3 replicate beach-seine tows adjacent to mangroves at high tide. January 2002–November 2002	41 species (30 in mangroves), 18 of commercial/recreational importance. All methods combined by authors for species numbers: silver fish *Leptatherina presbyteroides* (48.6%), sandy sprat *Hyperlophus vittatus* (14.7%), yellow-eyed mullet *Aldrichetta forsteri* (14.5%), smooth toadfish *Tetractenos glaber* (10.8%), blue spot goby *Pseudogobius olorum* (3.7%), glass goby *Gobiopterus semivestitus* (2.4%)	Hindell & Jenkins 2004
Barwon River, Victoria, Australia 38°16'S 144°30'E	*Avicennia marina*	Gill nets 1.5 m deep × 30 m long, five panels of different mesh sizes. Fyke nets. Square bag 70 × 70 cm, wings 10 m long × 70 cm deep; 6-mm mesh; beach seine 10 m × 2 m × 1 mm. Pop nets 25-m² area, 1.2 m high, 1-mm mesh	Both day and night sampling on 5 occasions in 3 mangrove zones (forest, pneumatophore, channel) with 2 fyke and 2 gill nets in each zone. Day sampling on 5 occasions to compare fyke and pop nets, 1 of each type of net in each of forest and pneumatophore zones. September 2003–April 2004	20 species, 15 families. Across all zones: Australian salmon *Arripis truttacea* and yellow-eyed mullet *Aldrichetta forsteri* represented 63% of catch. Channel, day: *Arripis truttacea* 39.8%, *Favonigobius tamarensis* 15.5%, *Aldrichetta fosteri* 25.6%, total no. fish 246. Channel, night: *Arripis truttacea* 39.4%, *Aldrichetta fosteri* 25.1%, total no. fish 203. Pneumatophore, day: *Galaxias maculatus* 30.0%, *Aldrichetta fosteri* 36.0%, *Tetractenos glaber* 20.0%, total no. fish 50. Pneumatophore, night: *Arripis truttacea* 33.5%, *Aldrichetta fosteri* 33.5%, total no. fish 170. Forest, day: *Tetractenos glaber* 50.0%, total no. fish 4. Forest, night: *Aldrichetta fosteri* 65.0%, total no. fish 20	Smith & Hindell 2005

Location	Species	Method	Sampling	Results	Reference
Port Phillip Bay, Victoria, Australia 38°07'S 144°38'E	*Avicennia marina*	Pop nets 25-m² area, 1.2 m high, 1-mm mesh	Three mangrove zones (forest, edge, mudflat). Sampled on 7 occasions, between October 2003 and January 2004.	15 species (8 in mangroves, 14 on edge), 7 families Forest and edge combined: blue spot goby *Pseudogobius olorum* (59.9%), *Atherinosoma microstoma* (10.9%), half-bridled goby *Areniogobius frenatus* (10.4%), smooth toadfish *Tetractenos glaber* (5.6%), mangrove goby *Mugilogobius paludis* (4.8%), *Favonigobius lateralis* (3.4%)	Hindell & Jenkins 2005
Torrens Island and Barker Inlet, Port River Estuary, Adelaide, South Australia 34°49'S 138°30'E	*Avicennia marina*	Pop nets 9-m² area, 1-mm mesh (Also seine nets [non-vegetated and seagrass habitats only])	Mangroves, seagrass and non-vegetated areas sampled by pop nets, 2 replicate samples from each habitat taken on each of 3 consecutive days. Monthly sampling from May to August and December to February.	19 species sampled by pop-net (10 in mangroves) *Atherinosoma microstoma* (44.5%), unidentified atherinid larvae (37.9%), yellow-eyed mullet *Aldrichetta forsteri* (13.3%), *Sillaginodes punctata* (2.3%), *Favonigobius lateralis* (1.2%)	Bloomfield & Gillanders 2005
Towra Point, Botany Bay, New South Wales 34°1'S 151°10'E	*Avicennia marina*	Pop nets 5.5-m² area, 1.5 m high in mangroves and salt marsh	Four replicate nets in each. Sampled monthly from March 2001 to February 2002.	26 species (25 in mangroves), 12 of 25 species of commercial importance Glassfish *Ambassis jacksoniensis* (27.3%), mangrove goby *Mugilogobius paludis* (19.5%), silver biddy *Gerres subfasciatus* (10.1%), and snakehead goby *Taenioides mordax* (6.5%)	Mazumder et al. 2005
Bicentennial Park, Sydney Towra Point, Allens Creek, Botany Bay, New South Wales 34° 1' S to 34°50'S 151°10'E	*Avicennia marina*	Fyke nets 4 m long funnel-shaped fyke net, one central and two lateral wings, 40 cm wide and 25 cm high entrance; 2-mm mesh size	Four replicate fyke nets in the salt marsh, 2 replicate fyke nets from mangroves. Set before tides flooded habitats; retrieved after tide had fully receded. Fish collected during December 2001, January and August 2002 during spring high tides.	21 species (19 in mangroves), 10 families (8 in mangroves) Mangroves: glassfish *Ambassis jacksonensis* (34.7%), blue spot goby *Pseudogobius olorum* (18.1%), silver biddy *Gerres subfasciatus* (11.3%), glass goby *Gobiopterus semivestitus* (7.3%), blue eye *Pseudomugil signafer* (5.6%), mosquito fish *Gambusia holbrooki* (4.7%), mangrove goby *Mugilogobius paludis* (3.8%), yellowfin bream *Acanthopagrus australis* (2.5%), checkered mangrove goby (*Mugilogobius stigmaticus*) (2.5%)	Mazumder et al. 2006

(continued on next page)

Table 7 (continued)   Summary of information from studies of fish in temperate mangroves

Location	Mangrove species	Method	Design details	Dominant fish species	Reference
Towra Point, Botany Bay, New South Wales 34°1'S 151°10'E	*Avicennia marina*, patches of *Aegiceras corniculatum* to landward	Pop nets 5.5-m² area, 1.5 m high in mangroves and salt marsh, 2 m high in seagrass, 0.6-mm mesh	Three nets per habitat. Monthly sampling of mangroves and salt marsh during spring tides, fortnightly of seagrass during both spring and neap tides.	28 species (24 in mangroves) Dominated by glassfish *Ambassis jacksoniensis* (28%), mangrove goby *Mugilogobius paludis* (20%), silver biddy *Gerres subfasciatus* (10.4%), snakehead goby *Taenioides mordax* (6.7%), yellowfin bream *Rhabdosargus sarba* (5.7%), checkered mangrove goby *Mugilogobius stigmaticus* (5.6%), blue spot goby *Pseudogobius olorum* (4.5%), flat tail mullet *Liza argentea* (2.5%)	Saintilan et al. 2007
Northern New Zealand	*Avicennia marina*	Fyke nets 1.8 m high, 14.5 m extent across wings 9-mm heavy braid mesh	Eight estuaries (2 western coast, 6 eastern coast), 6 sites in each. February–April 2006.	19 species (mangrove sampling only) Yellow-eyed mullet *Aldrichetta forsteri* (65.6%), grey mullet *Mugil cephalus* (17.9%), estuarine triplefin *Grahamina nigripenne* (5.2%), pilchard *Sardinops neopilchardus* (3.8%), smelt *Retropina retropina* (2%), short-finned eel *Anguilla australis* (1.6%)	Morrisey et al. 2007
Port Pirie, Port Broughton, Port Wakefield, South Australia 33°S to 35°51'S 137°50'E to 138°E	*Avicennia marina*	Fyke nets 0.7 × 0.7 m frames, 6-mm mesh, mesh wing of 0.7 × 6 m long each side of frame	Set within 1 h of low tide, samples collected 24 h later, sampling 2 tidal cycles. Six distance classes from mangroves: 10 m within forest, in pneumatophore zone; directly adjacent to pneumatophore zone; 25 m, 200 m, 500 m seaward of mangrove fringe.	26 species (18 in mangroves), 17 families (14 in mangroves) First two distance classes (mangroves) dominated by *Atherinosoma microstoma* (73.7%), yellow-eyed mullet *Aldrichetta forsteri* (19.5%), *Sillaginodes punctata* (1.6%), and *Pelates octolineatus* (1.4%)	Payne & Gillanders 2009

dominated by this one species (>90% of all individuals sampled), and overall species diversity was low (17 species).

Hindell & Jenkins (2004) sampled fish assemblages on the seaward side of mangrove forests and on the adjacent mudflats in Westernport and Corner Inlet, Victoria, Australia. They collected 41 fish species, of which five were found exclusively in the mangrove forest: congolli (*Pseudaphritis urvillii*), atherinid postlarvae, mosaic leatherjacket (*Eubalichthys mosaicus*), parore/luderick (*Girella tricuspidata*) and kahawai/Australian salmon (*Arripis trutta*). An additional six species were found exclusively in the mudflat habitats: hairy pipefish (*Urocampus carinirostris*), garfish (*Hyporhamphus regularis*), the mangrove goby (*Mugilogobius paludis*), sand mullet (*Myxus elongatus*), yank flathead/stargazer (*Platycephalus laevigatus*), and ornate cowfish (*Aracana ornata*). However, many of these 'unique' habitat species were represented by only one to three individuals each. The general fish assemblage was numerically dominated (74% of all individuals) by silver fish (*Leptatherina presbyteroides*), smooth toadfish (*Tetractenos glaber*) and yellow-eyed mullet (*Aldrichetta forsteri*). Overall, fish abundances were always greater in mangroves than mudflats for juveniles, but there were no apparent differences for larger subadult and adult fish. Most of the variability was determined by atherinids, mugulids (mullets), gobiids (gobies), tetraodontids (pufferfishes), pleuronectids (flatfishes) and clupeids. However, there were also strong interactions depending on where and when the three mangrove sites were sampled because the importance of mangroves was both spatially and temporally variable.

Smith & Hindell (2005) sampled the seaward side of mangroves, in the pneumatophore zone and in adjacent subtidal channels during the day and at night in the Barwon River, Victoria, Australia. Overall fish abundance, biomass and species richness were generally lower in the forest than the other two habitats, but varied with date, time of day, and water depth. The general fish assemblage was dominated by yellow-eyed mullet (*Aldrichetta forsteri*). Channel habitats held the highest fish abundances, biomass and species richness (total species pool was 20). Short-finned eels (*Anguilla australis*) and bream (*Acanthopagrus butcheri*) were found across all three habitats, mainly during the night. The authors concluded that the system was relatively low in species richness.

Hindell & Jenkins (2005) collected 15 fish species in and adjacent to mangroves in Port Phillip Bay, Victoria (Table 7). Catches in mangrove forest were dominated by small (<30 mm) gobies (*Pseudogobius olorum*, *Mugilogobius paludis*) and juveniles of the atherinid *Atherinasoma microstoma*. On the forest edges and adjacent mudflats, catches were dominated by King George whiting (*Sillaginodes punctata*), smooth toadfish (*Tetractenos glaber*) and two gobies (the half-bridled goby, *Arenigobius frenatus* and the long-fin goby *Favonigobius lateralis*). Fish densities were highest in the forest (1.98 ± 0.36 m^{-2}, mean ± SE), followed by the forest edge (1.42 ± 0.43 m^{-2}) and the mudflats (0.25 ± 0.19 m^{-2}). Species richness was highest at the mangrove forest edge (0.25 ± 0.19 m^{-2}), followed by the forest (0.17 ± 0.06 m^{-2}) and the mudflats (0.12 ± 0.02 m^{-2}). Fish biomass was highest at the forest edge (4.64 g ± 2.09 m^{-2}), followed by the mudflats (4.06 ± 1.79 m^{-2}) and the forest (1.2 ± 0.38 m^{-2}).

Bloomfield & Gillanders (2005) sampled fishes in seagrass, mangrove (*Avicennia marina*), saltmarsh and non-vegetated habitats in the Barker Inlet–Port River estuary, South Australia. Mangrove forests and non-vegetated habitats had more fish (257 vs. 377) and species (7 vs. 14) than salt marsh (only one fish collected), but less than seagrass (15 species, 590 individuals). Mangrove catches were dominated by unidentified atherinid larvae, the atherinid *Atherinosoma microstoma* and yellow-eyed mullet *Aldrichetta forsteri*, with five other species also caught in low numbers: King George whiting *Sillaginodes punctata*, the long-fin goby *Favonigobius lateralis*, the blue spot goby *Pseudogobius olorum*, a clingfish *Heteroclinus* sp., and an unidentified tetraodontid larva. In contrast, the seagrass samples were dominated by the two goby species and King George whiting (71% of individuals).

Mazumder et al. (2005) sampled small fishes in saltmarsh and mangrove habitats in Botany Bay, New South Wales. Forty-eight samples were collected from each of the two habitats, with 16 species collected from salt marsh at an average total fish density of 0.56 m^{-2} and 25 species at an overall density of 0.76 m^{-2} from mangroves. Twelve species of commercial importance were more common in mangroves, dominated by silver biddy (*Gerres subfasciatus*) and yellow-fin bream (*Acanthopagrus australis*) and including parore/luderick (*Girella tricuspidata*). Overall, in mangrove and salt marsh combined, the dominant species were glass gobies (*Gobiopterus semivestitus*), mangrove gobies (*Mugilogobius paludis*) and glassfish (*Ambassis jacksoniensis*). In a subsequent study in Botany Bay, Mazumder et al. (2006) sampled fishes leaving mangroves with the falling tide and caught 19 species. Species diversity was relatively low and dominated by nine species (Table 7).

Morrison and coworkers (data presented in Morrisey et al. 2007) sampled mangrove forests across eight estuaries in northern New Zealand in the austral summer/autumn, with stations sampled once in each estuary and extending from the upper to the lower reaches of the forest. Seventeen species were caught, but the assemblages were dominated numerically by yellow-eyed mullet (*Aldrichetta forsteri*), grey mullet (*Mugil cephalus*), pilchards and anchovies (the last two species represented by one to two single large catches, each in a different estuarine system). Grey mullet juveniles were generally found only in western coast mangrove forests, while the parore (*Girella tricuspidata*) was common only in eastern coast estuaries. This matched the coast-specific abundance of adult populations. Short-finned eels (*Anguilla australis*) were a common component of the fish–mangrove assemblages on both coasts, and this was the only species to show a positive correlation with the structural complexity of the forests (as measured by the number of mangrove samplings and young trees). No comparisons were made with adjacent habitats.

Finally, Payne & Gillanders (2009) sampled three estuaries (Port Wakefield, Port Broughton, and Port Pirie) in South Australia for small fishes. Neither total abundance nor species richness was found to vary between the mangrove and mudflat locations. Twenty-six species were sampled overall, with 18 of these occurring within mangrove habitats (inside forest or pneumatophore zones). However, only three of these were classified as mangrove residents (defined as total abundance more than five individuals sampled, >70% of these in mangroves): yellow-eyed mullet (*Aldrichetta forsteri*), *Sillago schomburgkii* and *Arripis georgiana*, although the abundances of the last two species were modest. In mangrove habitats, total fish abundance, mangrove residents and *Aldrichetta forsteri* were positively associated with pneumatophore density, indicating that structural complexity probably influences the distribution of some fish species.

Connectivity of temperate fish assemblages with surrounding habitat mosaics

Work by Nagelkerken et al. (2001) on tropical systems has shown that the presence of mangroves significantly increases species richness and abundance of fish assemblages in adjacent seagrass beds relative to seagrass beds without adjacent mangroves. Jelbart et al. (2007), working in the Pittwater Estuary just north of Sydney, sampled three seagrass (*Zostera capricorni*) beds close to mangroves (*Avicennia marina*) (<200 m) and three seagrass beds further away (>500 m). They found seagrass beds closer to mangroves had greater fish densities and diversities than more distant beds, especially for juveniles. Six species followed this pattern: the half-bridled goby *Arenigobius frenatus*, bridled leatherjacket *Acanthaluteres spilomelanurus*, parore/luderick *Girella tricuspidata*, *Pelates sexlineatus*, tarwhine (a sparid) *Rhabdosargus sarba* and hairy pipefish *Urocampus carinirostris*. Conversely, the density of those fish species in the seagrass at low tide that were also found in mangroves at high tide was negatively correlated with the distance of the seagrass bed from the mangroves. This finding showed the important daily connectivity that exists through tidal movements between mangrove and seagrass habitats.

Saintilan et al. (2007) sampled mangroves (*Avicennia marina*, although *Aegiceras corniculatum* was also present in the wider area) at Towra Point, Botany Bay, New South Wales. They sampled three habitats: seagrass, mangroves and salt marsh. Samples were taken monthly during spring tides in the mangrove and saltmarsh habitats and fortnightly in the seagrass, covering both spring and neap tides (salt marsh dries out during neap tides, and only the seaward sections of the mangrove forest are inundated). Twenty-eight species were sampled overall (24 from mangroves), with each habitat containing consistently different fish assemblages. The mangrove assemblage was dominated by gobies, including the mangrove goby *Mugilogobius paludis*, the checkered mangrove goby *Mugilogobius stigmaticus*, the snakehead goby *Taenioides mordax* and the blue spot goby *Psuedogobius olorum*, along with the glassfish *Ambassis jacksoniensis*, the silver biddy *Gerres sub-fasciatus* and the yellow-fin bream *Acanthopagrus australis*. Contrasts of the spring and neap tide assemblages (across habitats) found that seagrass assemblages had greater fish abundances during neap tides, especially of those species that visited the adjacent habitats (saltmarsh, mangroves) when they were available during spring tides. This was interpreted as evidence that fish were moving from seagrass into these adjacent habitats during spring tides, to exploit high abundances of zooplankton, and retreating to seagrass habitats as a refuge during low tides.

Why are juvenile fish in mangroves?

A number of hypotheses have been advanced regarding why mangrove habitat might be dispropor-tionately important as nurseries for juvenile fish in areas where this has been shown to be the case. The main hypotheses are that they provide protection from predation (e.g., larger fish and birds) and elevated foraging opportunities through high prey abundances. Using experimental manipulations, Laegdsgaard & Johnson (2001) looked at the interactions between fish predation and root densi-ties of the mangrove species *Avicennia marina*. They concluded that (1) habitat complexity regu-lates predation, (2) not all prey species use structurally complex habitats in the absence of predators and (3) the use of structurally complex habitats decreases with fish size. Further support for these findings comes from surveys documenting higher densities and biomass of demersal fishes in shal-low, inland mangroves with dense pneumatophores relative to prop-root habitats, where almost all predatory fish collected were found to inhabit the more open, seaward sites (Vance et al. 1996, Rönnbäck et al. 1999). However, these studies were undertaken in tropical mangroves. Smith & Hindell (2005) undertook tethering experiments with small yellow-eyed mullet (*Aldrichetta for-steri*) across a range of temperate mangrove microhabitats (mangrove forest, pneumatophore zone and adjacent channels) in the Barwon River, Victoria, Australia. They found low rates of daytime predation across all of these habitats. They suggested that predation refuges provided by mangroves might be less important in temperate systems, and that the lower number of fish in temperate man-grove forests was likely to be due to a lack of food, in agreement with models of lower productivity in temperate mangrove forests (Alongi et al. 2002). They argued that more attention should be given to assessing changes in the distribution of invertebrate prey across such microhabitats. Such studies are rare in general for mangrove systems (Faunce & Serafy 2006).

Summary of studies of fish in temperate mangroves

Collectively, the studies discussed, all undertaken in temperate Australasia, encompass many shared species given their bias towards the southern and south-eastern Australian seaboard, and northern New Zealand. They all focused on one common mangrove species (the grey mangrove *Avicennia marina*) and a range of temperate estuarine fish species. Diversity of fish species was consistently low relative to subtropical and tropical mangroves, and a few key species consistently dominated the fish assemblages, notably members of the families Gobiidae, Atherinidae and Mullidae (especially the yellow-eyed mullet *Aldrichetta forsteri*). Although a range of commercial species was found in mangroves across these studies, the species were often present in low numbers and were species known to occur in schools (e.g., species from the families Arripidae, Sparidae and Sillaginidae), making them more prone to large random variations in abundance during sampling. This was also true for a broader range of non-commercial species. The general conclusion from these temperate mangrove studies is that, although mangrove habitats do provide habitat for fishes, many of the species involved are small bodied, of little or no commercial value, and often equally abundant in alternative habitats. No temperate species appeared to be dependent solely on mangrove habitat.

Reptiles and amphibians

According to Hutchings & Recher (1982), reptiles are uncommon in temperate mangroves but common in tropical ones, although for most species mangroves are marginal habitats. In some situations, where other types of forest are scarce, they may serve as corridors for movement of individuals. Nagelkerken et al. (2008) suggested that mangroves may be important to marine turtles, including providing habitat for algae on which turtles feed, although there is a lack of information. Given that most of the examples of use of mangroves by turtles and other reptiles in their review were from tropical regions, this lack seems to be particularly severe for temperate areas.

Thomas & Logan (1992) listed the lizard *Anolis grahami*, the green turtle (*Chelonia mydas*) and the diamondback terrapin *Malaclemys terrapin* as present in mangrove forests in Bermuda. In North America, *M. terrapin* occurs in coastal and estuarine marshes, flats and lagoons (Tortoise & Freshwater Turtle Specialist Group 1996). Green and loggerhead (*Caretta caretta*) turtles also enter mangrove waterways in Australia (Milward 1982).

Two amphibians were also recorded in Bermudan mangroves, the frog *Eleutherodactylus johnstoni* and the cane toad *Bufo marinus*. *Eleutherodactylus johnstoni* is an invasive species found in disturbed habitats. It is native to several Caribbean islands but introduced to parts of north-eastern South America (Hedges et al. 2008) and possibly also to Bermuda. *Bufo marinus* is a notoriously invasive species native to Central and South America but introduced to islands in the Caribbean (among many other places) and, presumably, Bermuda (Solís et al. 2008).

Crisp et al. (1990) noted that various geckos have been found among mangroves in northern harbours of New Zealand (particularly Rangaunu and Hokianga), most commonly Pacific and forest geckos (*Hoplodactylus pacificus* and *H. granulatus*), but did not reference their sources of information. They also noted that sea snakes (*Laticauda colubrina*, *L. laticordata* and *Pelamis platurus*) sometimes occur in New Zealand mangroves as far south as Tauranga Harbour (37°40′S), but these are likely to be rare and chance events.

Birds

Worldwide, mangroves harbour a moderate number of species of birds (Nagelkerken et al. 2008). Among these, however, are a surprisingly small number of mangrove specialists. No species have been recorded exclusively in mangroves in Africa and only one in north-eastern South America (Surinam) and the Caribbean (Trinidad). Even the relatively diverse mangrove avifauna of Australia (Schodde et al. 1982) includes only 13 mangrove endemics among more than 200 species that occur in this habitat (Saenger et al. 1977).

Schodde et al. (1982) provided one of the most detailed discussions to date of the composition, structure and origin of assemblages of birds in mangroves, focusing on Australia. They concluded that, across Australia, the avifauna of mangroves has developed relatively recently and mainly from rainforest sources in Australo-Papua. The number and diversity of mangrove endemics or mangrove-dependent species decreases with the number of tree species but has been limited, at least partly, by historical factors.

Saenger et al. (1977) listed 242 species of birds recorded from mangroves in Australia, of which 13 species (5%) were found exclusively in mangroves and 60 species (25%) used mangroves as an integral part of their habitat. The total number of species in this list that have been recorded in south-eastern Australia is 131, including 4 that occur only in mangroves and 20 for which mangroves are an integral part of their habitat. The remainder visit mangroves opportunistically, for example, to take advantage of food resources provided by flowering mangroves and the insects that they attract (Schodde et al. 1982).

The south-eastern Australian species found exclusively in mangroves, as reported by Saenger et al. (1977), were the mangrove heron (*Butorides striatus*), the mangrove warbler (*Gerygone laevigaster*) and the mangrove honeyeater (*Meliphaga fasciogularis*). Species closely associated with mangroves include a heron, two species of ibis, a sea eagle, a kite, an osprey, a rail, an oyster catcher, a plover, a godwit, a whimbrel, a stone-curlew, a dove, a cuckoo, a triller (family Grallinidae), a flycatcher (family Muscicapidae), the red-browed finch (family Meliphagidae), an oriole (family Oriolidae) and a wood swallow (family Artamidae).

In Careel Bay, near Sydney, relatively few birds were recorded as feeding in mangroves or salt marshes compared with adjacent intertidal seagrass beds (Hutchings & Recher 1974). Species that did feed in the mangroves included black cormorants (*Phalacrocorax carbo*), white-faced herons (*Ardea novaehollandiae*), white egrets (*Egretta alba*), mangrove herons white and straw-necked ibis (*Threskiornis molucca* and *T. spinocollis*), black-billed spoonbills (*Platalea regia*), yellow-billed spoonbills (*P. flavipes*), eastern curlews (*Numenius madagascariensis*) and bar-tailed godwits (*Limosa lapponica*). The mangroves and salt marshes also provided an important high-tide refuge, and a number of species, including herons, nested in the mangroves.

Saenger et al. (1977) listed Australian pelicans (*Pelicanus conspicillatus*) as visitors to mangroves, although they do not nest in this habitat (G. Johnston, Flinders University of South Australia, personal communication). In the United States in Louisiana, however, brown pelicans (*Pelicanus occidentalis*) nest in mangroves (Lester et al. 2005).

Raines et al. (2000) included one site where mangroves were present in a series of waterbird surveys of the Leschenault Inlet Estuary in south-western Australia. They recorded a "small variety of species in small numbers for feeding", but noted that, because mangroves were underrepresented among the habitats in their surveys, the actual importance of mangroves to waterbirds in the inlet may be greater than the surveys suggested. The site was used as a dry-season refuge by many species, and the authors suggested that, in general, mangroves may be of moderate conservation value to waterbirds.

There is relatively little published information on the use of mangroves by birds in New Zealand, and some of what is available consists of chance observations (e.g., Miller & Miller 1991). In the most detailed such study to date, Cox (1977) investigated use by birds of a mangrove stand in the Kaipara Harbour (north-west of Auckland) over 2 years and made one-off surveys of other locations. The Kaipara site consisted of tall (5- to 6-m) trees along the seaward fringe, backed by a broad, flat area of stunted (1.5-m) trees, bounded by a dyke at the top of the shore. Cox (1977) recorded 22 species at the Kaipara site, of which 12 occurred regularly within the mangroves, and 6–7 bred. The 11 species regularly recorded were white-faced heron (*Ardea novaehollandiae*), harrier (*Circus approximans*), chaffinch (*Fringilla coelebs*), grey warbler (*Gerygone igata*), Australian magpie (*Gymnorhina tibicen*), kingfisher (*Halcyon sancta*), welcome swallow (*Hirundo tahitica neoxena*), house sparrow (*Passer domesticus*), pukeko (*Porphyrio porphyrio*), blackbird (*Turdus merula*) and silvereye (*Zosterops lateralis*). The species breeding in the mangroves were grey warbler, silvereye, fantail (*Rhipidura fuliginosa*), house sparrow and shining cuckoo (*Chrysococcyx lucidus*) (in the nest of a grey warbler). A further five species, including roosting colonies of little black shags (*Phalacrocorax sulcirostris*) and pied shags (*P. varius*), were recorded in mangroves at other locations (Parengarenga, Hatea, Kaipara, Manukau, Waitemata and Ohiwa Harbours). Cox (1977) concluded that mangroves are generally a marginal habitat for birds and in no case were they a major habitat, even though surveys of the invertebrate fauna of the Kaipara site indicated that prey was abundant in the mangroves.

Although all the species recorded at Cox's Kaipara site were either common natives or introduced species (such as house sparrows), other studies have documented use of mangroves by less abundant species. Miller & Miller (1991) reported bitterns (*Botaurus poiciloptilus*) using mangroves near

Whangarei in the north-east of the North Island. Royal spoonbills (*Platalea regia*) used mangroves on a small island as their principle roost site in the sediment settlement ponds at Port Whangarei (Beauchamp & Parrish 1999). White-faced herons and various species of shags also roosted in these mangroves, and there were resident populations of grey warbler, blackbirds, song thrushes (*Turdus philomelos*) and dunnock (*Prunella modularis*) and transient silvereyes, shining cuckoos and fantails. Thousands of starlings (*Sturnus vulgaris*) and hundreds of house sparrows and chaffinches roosted in mangroves in other parts of the settlement ponds. Cox (1977) cited information from other studies indicating that banded rail (*Gallirallus philippensis assimilis*) were the "only New Zealand bird typically described as associated with mangrove swamp", but that this was probably only the case where the mangroves were adjacent to suitable high-tide habitat. Beauchamp (n.d.) noted that mangroves are "the only northern habitat of the banded rail ... and are a substantial breeding habitat for New Zealand kingfisher". Of the species that occurred in both Australia and New Zealand (including royal or black-billed spoonbills, banded rails and bitterns), all those found in mangroves in New Zealand were also found in this habitat in Australia.

Crisp et al. (1990) provided a list of 48 species of native or introduced birds using mangroves in New Zealand (no references were given, but parts of their commentary appeared to derive from Cox's thesis). Like Cox (1977), they concluded that mangroves are a marginal habitat for birds. Available evidence therefore suggests that there are no New Zealand birds that are exclusively found in mangroves, but that many species make extensive use of them for roosting, feeding or breeding. Given the difference in total numbers of birds found in mangroves in Australia and New Zealand and the small percentage of Australian species found exclusively in this habitat, it is perhaps not surprising that New Zealand does not appear to have any mangrove-dependent species.

Other than as pollinators, the ecological importance of birds in mangroves has received little attention. One exception, however, is a comparison of growth rates of stands of *Rhizophora mangle* in Florida with and without breeding colonies of pelicans and egrets (Onuf et al. 1977). Growth began earlier in the year and was faster at a site with colonies of birds than at a nearby site without birds. The former site also exhibited greater production of leaves, propagules, branches and rates of growth of existing branches, and growth showed two maxima per year, in contrast to one at the site without birds. The authors attributed these differences to higher inputs of nutrients, in the form of guano, to the site with birds. Unfortunately, there was no replication of sites with or without bird colonies, so that the influence of birds may be confounded with other differences between the two sites.

Mammals

The use of mangroves by mammals appears to be at least as opportunistic, and possibly even less studied, than that by birds. Other than their role as pollinators, there appears to be little or no information on the ecological roles of mammals in mangroves.

No Australian mammals are restricted to mangroves (Hutchings & Recher 1982), but various species of native and introduced mammals have been recorded in them, including bandicoots, wallabies, possums, various rodents and feral pigs (Lovelock 1993). Given the ubiquity of the introduced brush-tailed possum (*Trichosurus vulpecula*) in New Zealand, it seems likely that they also make use of mangroves in that country. Weasels (*Mustela nivalis*) have occasionally been sighted in mangrove forests in New Zealand (Blom 1992).

The presence of rats (*Rattus* spp.) in mangroves is sometimes cited as a reason for clearing mangroves, and there is plenty of anecdotal evidence of their occurrence. We are not aware, however, of any quantitative evidence for their use of mangroves. It is very likely that rats would use mangroves as habitats even at high tide because they can swim and climb well. Mature stands of large trees are more likely to provide refuges for rats in the form of holes in their trunks and a firmer ground among the trees. Mangroves are likely to provide a good source of food for rats, as they do for birds, in the form of invertebrates and plant material, such as propagules. Cox (1977) noted rat footprints and droppings at his study site in the Kaipara Harbour (New Zealand) and concluded that they were feeding on vegetable matter. As far as is known, there is no information concerning feeding on mangrove propagules by rats or on the palatability of mangrove material.

There does not seem to be any reason to presume that rats would occur in mangroves in larger population densities than other similarly vegetated habitats. It is possible that if mangroves colonise more open habitats, such as mud- or sandflats, they will locally increase the area of suitable habit for rats by raising the height of the ground and reducing frequency of tidal flooding and provide a source of food and shelter. Abundances of rats in mangrove areas are likely to vary with the nature of adjoining habitats (open pasture, freshwater wetland, urban or industrial areas) and probably reflect the relative abundance of rats in these habitats.

In temperate parts of South Africa, vervet monkeys [*Chlorocebus* (*Cercopithecus*) *aethiops*] are common in the canopy and on the ground in mangrove forests, and blue duiker (*Cephalophus monticola*), reedbuck (*Redunca arundinum*) and bushbuck (*Tragecephalus scriptus*) visit mangrove forests from adjacent habitats (Hughes & Hughes 1992).

Bats also use mangroves as feeding and roosting habitats. In south-eastern Australia, the grey-headed flying fox *Pteropus poliocephalus* feeds on the pollen and nectar of mangrove flowers and also roosts in mangrove forests (Hutchings & Recher 1982). This species shows strong fidelity to roosting sites and may occupy the same site for years. In southern Brazil, the fishing bat *Noctilio leporinus* forages in and around mangroves. Analysis of faeces indicated that its main food items were fish (90% of samples examined), moths and beetles (Bordignon 2006).

Primary consumption

Mangrove material may be consumed fresh by direct herbivory or indirectly by consumption of dead leaves and other tissues by detritivores. The latter trophic pathway has been studied in considerable detail in tropical mangrove systems (reviewed by Alongi 2009) but much less so in temperate systems. Direct consumption has, in general, been less studied, but again the emphasis has been on tropical systems.

Direct consumption of mangrove material by animals

The role of herbivores in mangrove forests has, until relatively recently, been considered minimal compared with their role in terrestrial forests (e.g., Macnae 1968, Tomlinson 1986). However, the importance of crabs in mangrove trophic pathways has become evident (reviewed by Cannicci et al. 2008, Alongi 2009), particularly as processors of leaf litter but also, in the case of arboreal species in the New World and the Indo-Pacific, as herbivores. Hartnoll et al. (2002) suggested that biomass of crabs in mangroves is highest in warm temperate regions, although abundances may decrease with latitude. They also argued that a shift from ocypodids (which are deposit-feeders) to grapsids (which are herbivores) with increasing latitude may result in latitudinal changes in patterns of transfer of primary production through trophic pathways.

Herbivory: consumption of mangrove foliage

Several species of arboreal crabs live on the trunks and in the canopies of mangroves and feed on living plant tissues, including leaves and propagules. The large majority are tropical, but the exclusively arboreal *Parasesarma leptosoma* also occurs in mangroves in warm temperate parts of South Africa (Emmerson & Ndenze 2007). Intensity of browsing varied with height on the shore and with species of mangrove; from 100% of *Rhizophora mucronata* and 52% of *Bruguiera gymnorrhiza* near the low-tide creek showing evidence of browsing, 25.7% of *Rhizophora mucronata* but 0% of *Bruguiera gymnorrhiza* browsed away from the creek and no browsing occurred on

Avicennia marina (near or away from the creek). These differences reflected relative palatability as both *Rhizophora mucronata* and *Bruguiera gymnorrhiza* are salt excluders, and *Avicennia marina* secretes salt from its leaves. Leaves of *Rhizophora mucronata* also had a higher nutrient content, particularly nearer the creek. The average area of leaf damaged was 1.7–2.6% for *R. mucronata* and 0–1.76% for *Bruguiera gymnorrhiza*. There was a significant correlation between the number of crabs caught in traps on the trees and the amount of leaf damage for both species of trees.

Insect herbivores have been even more overlooked than crabs, with their perceived lack of abundance and diversity in mangroves relative to terrestrial forests leading, in turn, to an assumption that their ecological role is similarly minimal (Burrows 2003, Cannicci et al. 2008). Again, however, more detailed study has revealed their previously underestimated importance. In the most thorough study to date, Burrows (2003) showed that the diversity of folivorous insects on *Rhizophora stylosa* and *Avicennia marina* in northern Queensland, Australia, was similar to that of other tropical species of trees.

Insects and other herbivores may cause a number of types of damage to mangrove trees, including loss of leaf area, premature abscission of leaves, leaf mining, leaf deformation or stunting and gall formation (Burrows 2003). Indirect damage may result from destruction of growing tips and branches, causing a loss of leaf biomass, or from and sap feeding, leading to necrosis. In northern Queensland, *Rhizophora stylosa* and *Avicennia marina* contained specialised but different assemblages of herbivores, with gall-forming species comprising nearly a third of the fauna on *A. marina* but none occurring on *Rhizophora stylosa* (Burrows 2003). Temperate examples are scarce (probably because of a lack of research) but include the distinctive distortion of the growing tips of the shoots of *Avicennia marina* caused by larvae of the moth *Ptyomaxia* sp. and leaf galls caused by the mite *Aceria avicenniae* in New Zealand (Lamb 1952, J. Dugdale personal communication).

The large majority of studies of herbivory in mangroves have been based on measures of damage to leaves sampled at a single point in time (Burrows 2003). When Burrows (2003) compared such 'discrete' data with those from long-term methods, he found that the former failed to account for loss of leaves that were entirely consumed by herbivores or that were prematurely abscised as a result of damage. Loss of leaf biomass as a result of premature abscission caused by insect damage was equal to or greater than direct damage or consumption. Consequently, discrete methods underestimated consumption by herbivores by three to six times, and in Burrows's study in northern Queensland, discrete measurements gave estimates of loss of leaf area for *Avicennia marina* of 6–7%, compared with 28–36% from long-term studies. Susceptibility to herbivore damage varied between the two mangrove species studied, with 5–8% of the leaves of *Rhizophora stylosa* either completely consumed or prematurely abscised, compared with 19–29% of those of *Avicennia marina*.

Although most of the studies of insect herbivory reviewed by Burrows were of tropical mangrove systems, these conclusions may well apply to temperate systems, where the insect fauna has been equally neglected. The only study of herbivore damage to leaves of temperate mangroves of which we are aware is that by Johnstone (1981), who recorded 0–2.6% of leaf area loss in *Avicennia marina* in Auckland, New Zealand, using discrete measurements.

In their study of stands of *Rhizophora mangle* with and without breeding colonies of birds in Florida, Onuf et al. (1977) found that, in addition to higher rates of growth, the stand with birds experienced a much higher level of insect herbivory. Five lepidopteran species and a scolytid beetle were either more abundant or only occurred at the site with birds. The authors hypothesised that the higher rate of herbivory was due to higher nutritive value of leaves at this site. The difference in relative levels of herbivory disappeared when the birds migrated away from the site after breeding (although, of course, this could simply reflect a response of both groups of organisms to seasonality of some other environmental factor).

Animals that bore into plant tissues may lead to death of mangrove limbs and loss of leaves. In some areas of mangroves in New Zealand, such as Puhinui Creek, Manukau Harbour west of Auckland (D. Morrisey personal observation) and Puhoi Estuary north of Auckland (Kronen 2001), damage to woody mangrove tissue by boring insects is common. The insect responsible is a cerambycid beetle, the lemon-tree borer *Oemona hirta* (J. Dugdale personal communication), which occurs on a wide range of species of trees throughout New Zeland. The larvae excavate long tunnels throughout the woody tissue, with side tunnels leading to holes to the outside, through which waste materials are ejected (Horticulture and Food Research Institute of New Zealand Ltd. 1998b).

Frugivory: predation of mangrove propagules

In a worldwide survey of herbivore damage to mangrove propagules (germinated seeds) prior to dispersal from the parent tree, Farnsworth & Ellison (1997) concluded that predispersal herbivory is a ubiquitous feature of mangrove forests worldwide and must be accounted for in estimates of reproductive output. Clarke & Allaway (1993) commented that herbivory of postdispersal mangrove propagules by crabs was less important in temperate than tropical parts of Australia. Consistent with this suggestion, individuals of the temperate Australian mangrove crab *Helograpsus haswellianus* did not consume propagules of *Avicennia marina* when offered them in laboratory experiments (Imgraben & Dittmann 2008).

In contrast to herbivory of leaves, there have been several studies of herbivory of propagules in temperate mangroves. Farnsworth & Ellison (1997) also noted that crabs consumed the largest number of propagules and at the widest range of locations around the world. Other important consumers of propagules were scolytid beetles and, to a lesser extent, lepidopteran larvae. The last of these groups was relatively more important in the Southern Hemisphere.

Farnsworth & Ellison's study (1997) surveyed propagules for predispersal herbivory across 10 species of mangrove and 42 sites around the world, including 2 near Durban, South Africa (29°48′ and 29°53′S). Across these two sites, their samples showed no damage to propagules of *Bruguiera gymnorrhiza* ($n = 85$ propagules) or *Rhizophora mucronata* ($n = 7$), but damage to *Avicennia marina* propagules ranged from 10 to 90% ($n = 80$) and was caused by lepidopteran larvae.

In the Sydney region and at Westernport, Victoria, larvae of phycitine moths attack up to 60% of fruit, and several cohorts may feed on an individual fruit as it develops (Clarke 1992). The larvae also feed on flower buds, as do cantherid beetles, but exclusion experiments showed that they not reduce the proportion of buds that survive to become fruits. Exclusions did, however, result in a doubling of fruit survival, but Clarke concluded that the impact of predation was minor in comparison with that of maternal regulation of fruit survival, which was responsible for 75% of mortality. Of the fruit that showed evidence of damage by herbivores, only 1–16% were consequently found not to be viable.

Fruit of *Avicennia marina* are attacked by insect herbivores throughout their development, and at a site near Sydney, Minchinton & Dalby-Ball (2001) recorded larval exit holes of the mangrove fruit fly *Euphranta marina* and the mangrove plume moth *Cenoloba obliteralis* in the cotyledons of 53% of predispersal fruit, 69% of abscised propagules and 80% of established seedlings, with the number of holes increasing with time since abscission. Because the larvae attacked only the cotyledons and not the embryonic axis, they did not affect establishment of seedlings, but by reducing the food supply of the developing seedlings, they did reduce their rates of growth. Mortality of early seedlings was, nevertheless, minimal and not related to the level of frugivory. Whether decreased size may lead to reduced competitive ability as seedlings wait for release from the 'seedling bank' (Burns & Ogden 1985) is not yet known.

In the United States, Louisiana mangroves (*Avicennia germinans*) are constrained in their distribution to higher levels of the shore than the saltmarsh grass *Spartina alterniflora*. The controlling factors include dispersal of propagules by tidal movement and relative rates of desiccation, decay and herbivory (Patterson et al. 1997). Herbivory was more intense in the *Spartina* zone, where 40% (±4%) of experimentally placed propagules were attacked compared with 5% (±4%) in the mangrove zone. Few propagules were consumed completely, but damage to the cotyledons resulted in increased rates of decay.

Summary

The available evidence suggests that the recently recognised importance of herbivores in tropical mangroves may also apply to their temperate equivalents, at least in the case of feeding on propagules. In contrast to the detritivorous trophic pathway described next, much of this direct herbivory may be exported to adjacent terrestrial systems through migration of adult insects or their predation by birds.

Detritivory

Whether or not insects and other herbivores play a significant role as consumers of mangrove primary production, it is likely that the largest component of production enters the food web as detritus (Bouillon et al. 2008, Alongi 2009).

Decomposition of mangrove and other plant material

The nutritional value of detrital material to consumers increases over time as it is colonised and broken down by microbial organisms, with concurrent increase in nitrogen content (mainly in the form of mucopolysaccharides produced by bacteria but also in the bodies of bacteria and fungi; Steinke et al. 1990) and decreases in carbon content and net weight (Robertson 1988). Concentrations of tannins also decrease rapidly during the early stages of decomposition (Robertson 1988). Leaves of *Avicennia* and *Kandelia* are inherently rapidly decomposed relative to other mangrove species because of their relatively high nitrogen content, low carbon-to-nitrogen ratio, low content of structural lignocellulose and low tannin content (Robertson 1988, Alongi 2009). This relative rapidity may suggest that decomposition processes are faster in temperate compared with tropical forests given the predominant role of these genera in temperate forests, although lower temperatures in temperate regions may offset enhanced rates of tissue decomposition. Concentrations of nitrogen in the leaves of young (0.6–1.6 m tall, 2.72% DW) and mature (1.88% DW) *Avicennia marina* in Auckland (Morrisey et al. 2003) are slightly higher than the concentration reported by Robertson (1988) in Queensland (~0.9% DW from his Figure 2). The percentage of nitrogen in senescent leaves of *A. marina* was 0.7% DW in both Westernport Bay, Victoria, Australia (van der Valk & Attiwill 1984), and the Mgeni Estuary, South Africa (Steinke et al. 1983).

Rates of decay vary with climate and latitude (Figure 5 and see Mackey & Smail 1996). However, effects of latitude can apparently be obscured by local differences in position within the

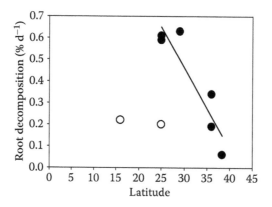

Figure 5 Variation in root decomposition of *Avicennia* sp. with latitude in mineral soils (filled circles) and in peat soils (open circles). The line of best fit for mineral soils is described by $y = -0.038x + 1.595$, $R^2 = 0.85$, $p = 0.0084$. (Data are from Albright 1976, van der Valk & Attiwill 1984, Middleton & McKee 2001, Perry & Mendelssohn 2009, D. Salomone & C.E. Lovelock, unpublished data.)

intertidal area and associated period of immersion and by seasonal effects. Middleton & McKee (2001) reviewed published rates of decomposition (mainly from tropical and subtropical locations), and among species of *Avicennia*, the average daily per cent loss of DW on the low/middle shore was 0.42 (±0.05 SE). The two temperate examples included in Middleton & McKee's (2001) review (Woodroffe 1982, van der Valk & Attiwill 1984) were below this average value but were not the lowest values for *Avicennia*.

This variation in rates of decomposition with latitude is illustrated more clearly by comparison of studies of decomposition of leaves of *A. marina* at different latitudes in eastern Australia (Table 8). Robertson (1988) recorded a time of 11 days for 50% reduction in ash-free dry weight (AFDW) of leaves of *A. marina* submerged in small mangrove creeks in northern Queensland (latitude 19°17′S) and 90 days for leaves in the midintertidal area (leaves of *A. marina* decomposed considerably faster than those of other mangrove species at the same locations). Further south, near Brisbane (27°24′S), Mackey & Smail (1996) recorded times for 50% loss of AFDW of 44 days low on the shore (inundated by 100% of high tides) and 59 days higher on the shore (inundated by 20% of high tides) in summer. Both of these rates are faster than the intertidal value reported by Robertson (1988). Equivalent times for winter were 78 days (low shore) and 98 days (high shore). In Sydney Harbour (33°46′S), newly senesced leaves of *A. marina* lost 50% of their oven-dried weight in about 56 days in winter (May–July) (Goulter & Allaway 1979), although changes in AFDW and DW are not strictly comparable. The rate of decomposition at a location near Adelaide, South Australia (34°36′), in summer (February–April) was even faster than near Brisbane, with 50% loss of air-dried weight in the first 14 days (Imgraben & Dittmann 2008). Rates of decomposition varied among heights on the shore, being highest at the seaward fringe, and the rates at this location were significantly higher than at another 7 km away (50% loss of DW in about 6 wk). In Westernport Bay, Victoria (latitude 38°20′S), times for 50% loss of oven-dried weight in summer (beginning in October and November) were about 70 days for leaves placed on the sediment surface in mesh bags and about 40 days for unbagged leaves (van der Valk & Attiwill 1984). Unbagged leaves lost weight faster because of loss of small fragments that were retained by the bags.

In New Zealand, 50% loss of leaf weight occurred after about 53 days in early summer (October–December) for leaves placed on the mud surface in Whangateau Harbour (latitude 36°19′S; Albright 1976). The rate of decomposition was much slower for leaves buried 20 cm below the sediment surface, where decomposition processes would have been anaerobic. Woodroffe (1982) measured rates of decomposition of mangrove leaves in the Tuff Crater (Auckland) of 6–8 wk for 50% reduction in oven-dried weight, with little variation between summer and winter or between a site in tall mangroves along a creek bank and a site in low mangroves on the intertidal flats (there was no measure of variation in rates of decay in summer because only one sample was measured each time). A subsequent study at the same location resulted in 50% reduction in AFDW over 10 wk (time of year not stated; Woodroffe 1985a). The lack of difference between summer and winter in the first of these studies is surprising given that decomposition rates are generally considered to be temperature dependent (Mackey & Smail 1996). The lack of seasonal difference in Woodroffe's study contrasts with that of Oñate-Pacalioga (2005) near Auckland. She found that 50% loss of AFDW occurred after 8 wk in mature stands in autumn but took more than 12 wk in winter and spring. Equivalent times for newly establishing stands were 12 wk in autumn and more than 12 wk in winter and spring.

In the Mgeni Estuary, South Africa (29°48′S), senescent leaves of *A. marina* lost 50% of their oven-dried weight over 3 wk (starting in spring to late October), whereas those of *Bruguiera gymnorrhiza* took about 6 wk to lose the same amount (Steinke et al. 1983). Stems of these two species were more recalcitrant, with only 20% loss after 6 mo. Similar rates of decomposition (~2 wk and 9 wk for 50% loss of oven-dried weight by *Avicennia marina* and *Bruguiera gymnorrhiza*, respectively) were recorded in a later study at the same site but starting in February (Steinke et al. 1990).

Table 8   Rates of decomposition (as number of days for 50% loss of weight) of mangrove leaf litter at temperate locations

Location	Latitude	Species	Season	Time for 50% loss (d)	Notes	Reference
North Queensland, Australia	19°17'S	A.m.		11	Creek	Robertson 1988
				90	Midintertidal	
					AFDW, senescent	
Brisbane, Australia	27°24'S	A.m.	Summer	44	Low shore	Mackey &
				59	Midshore	Smail 1996
					AFDW, senescent	
Brisbane, Australia	27°24'S	A.m.	Winter	78	Low shore	Mackey &
				98	Midshore	Smail 1996
					AFDW, senescent	
Newcastle, Australia	32°52'S	A.m.	Summer	<30	Downstream of floodgate	Dick &
				>180	Upstream of floodgate	Osunkoya
					DW, fresh	2000
Sydney, Australia	33°46'S	A.m.	Winter	56	Submerged by all high tides; same times for 1- and 7-mm bags	Goulter & Allaway 1979
					DW, senescent	
Middle Beach, South Australia	34°36'S	A.m.	Summer	11	Low and high shore	Imgraben &
				14	Midshore	Dittmann 2008
					DW, senescent	
Port Gawler, South Australia	34°38'S	A.m.	Summer	42	Low shore	Imgraben &
				—	Mid- and high shore sites did not reach 50% reduction within 8-wk duration of study	Dittmann 2008
					DW, senescent	
Westernport Bay, Australia	38°20'S	A.m.	Summer	70	Mesh bags	Van der Valk &
				40	Unbagged	Attiwill 1984
					Midshore, DW, senescent	
Whangateau Harbour, New Zealand	36°19'S	A.m.	Summer	53	DW, fresh?	Albright 1976
Auckland, New Zealand	36°48'S	A.m.	Summer	42	Creek bank	Woodroffe 1982
				56	Tidal flat	
					AFDW, fresh	
Auckland, New Zealand	36°48'S	A.m.	Winter	35–70	Creek bank	Woodroffe 1982
				39–42	Tidal flat	
					AFDW, fresh	
Auckland, New Zealand	36°48'S	A.m.	Not stated	70	Creek bank and tidal flat AFDW, fresh?	Woodroffe 1985a
Auckland, New Zealand	36°19'S	A.m.	Autumn	56	AFDW	Oñate-Pacalioga
			Spring	>84		2005
Mgeni Estuary, South Africa	29°48'S	A.m.	Spring	21	DW, senescent	Steinke et al.
		B.g.		42		1983
Mgeni Estuary, South Africa	29°48'S	A.m.	Summer	14	DW, senescent	Steinke et al.
		B.g.		63		1990

Note: Examples of rates from lower latitudes in Australia are also shown for comparison. Whether fresh or senescent leaves were used also noted if known. A.m. = *Avicennia marina*; B.g. = *Bruguiera gymnorrhiza*; AFDW = ash-free dry weight; DW = air- or oven-dried weight.

These comparisons of the decomposition rates reported by Robertson (1988) and Mackey & Smail (1996) with those for temperate Australia, New Zealand and South Africa (Table 8) suggest that rates are not necessarily slower at higher latitudes.

Variation among locations at a given latitude, and among times at a given location, seems to be as large as variation among latitudes. There also seems to be variation with salinity (Steinke & Charles 1986) and among species of mangrove, as indicated by the comparison of leaves of *Avicennia marina* and *Bruguiera gymnorrhiza* in South Africa (Steinke et al. 1990). As noted, leaves of *Avicennia* and *Kandelia* decompose relatively rapidly because of their high nitrogen content, low carbon-to-nitrogen ratio and low tannin content compared with other species (Robertson 1988, Alongi 2009). Because temperate mangrove stands tend to contain only one or two species, differences in the dominant species present in different temperate regions (e.g., *Avicennia marina* in southern Australia and New Zealand, *Kandelia candel* in Japan, *Bruguiera gymnorrhiza* in South Africa) may give rise to large variations in the rates at which mangrove litter enters trophic pathways.

Experimental differences among the studies cited may also contribute to apparent variation in rates of decomposition. Most studies use litter bags to contain the decomposing material, but the size of the mesh varies among studies and may influence decay rates. The evidence for the effects of mesh size is equivocal because Goulter & Allaway (1979) found no difference in rates of decay measured in litter bags of 1-mm and 7-mm mesh. The majority of studies also used senescent (yellow) leaves picked from the tree before abscission, but since up to 61% of the nutrients in leaves of *Avicennia marina* may be absorbed prior to abscission (Ochieng & Erftemeijer 2002), there may be a difference in nutritive value between leaves still on the tree and those that have abscised naturally, as Imgraben & Dittmann (2008) pointed out.

Decomposition rates of fine roots of *Avicennia* sp. published in the literature appear to decline with increasing latitude in mineral soils (Figure 5), although few studies have been published from tropical latitudes. Decomposition of fine roots of *Avicennia* spp. in peat soils is slower than in mineral soils, but data are limited.

Decomposition is carried out by assemblages of bacteria and fungi (Steinke et al. 1990, Singh et al. 1991, Kristensen et al. 2008) and often proceeds relatively rapidly at first, followed by a slower rate of decrease or an increase in DW (Woodroffe 1982, 1985a, Steinke et al. 1990, Imgraben & Dittmann 2008). Although concentrations of nutrients such as nitrogen, phosphorus, potassium, magnesium and calcium decline steadily during decomposition (Steinke et al. 1983, van der Valk & Attiwill 1984, Steinke & Charles 1986), concentrations of nitrogen subsequently increase, reflecting colonisation of the detritus by bacteria, fungi, cyanobacteria and diatoms (Steinke et al. 1983).

Microbial colonisation of litter during decomposition increases its nutritive value for larger organisms, and in turn, the rate of decomposition of litter is increased by the activities of animals that break it down into smaller fragments, providing a larger surface area for microbial activity. In some parts of the world, crabs are particularly important in this role (northern Australia, Robertson & Daniel 1989; South Africa, Emmerson & McGwynne 1992). Kristensen et al. (2008) estimated consumption of mangrove litter as a percentage of litterfall from data collected in various locations, including the Mngazana Estuary in South Africa (Emmerson & McGwynne 1992). Although the biomass of sesarmids at this location was low relative to other locations, even this value (and other conservative assumptions) suggested that the crabs were capable of removing up to 30% of litter from the surface of the sediment in mangrove stands.

Lee (2008) suggested that gastropods may provide a similar service in parts of the world where crabs are a less-dominant part of the macrofauna. This may be the case in New Zealand, where the mangrove crab fauna is of relatively low diversity, but as yet, there is little information available. In Florida, the gastropod *Melampus coffeus* is capable of assimilating mangrove leaf material, but its grazing also increases rates of leaf decomposition (Proffitt & Devlin 2005). In the presence of snails, DW of decomposing leaves decreased by 90% in 4 wk in the case of *Avicennia germinans* and 7 wk in the case of *Rhizophora mucronata*, compared with 12–26 wk and more than 26 wk, respectively, when snails were excluded.

In New Zealand, Oñate-Pacalioga (2005) showed that decomposition rates of mangrove leaves in the laboratory were 8–12% slower when macrofauna were removed from the sediment on which the leaves were placed (leaves and sediment were collected from Whangateau Harbour, near Auckland). It is not known whether *Helice crassa* or other crabs found in New Zealand mangrove habitats have the same effect on rates of decomposition as crabs in tropical mangroves. A study in subtropical Japanese mangroves (Mchenga et al. 2007) suggested that a related species, *H. formosensis*, is an important bioturbator, increasing oxygenation of sediments at the landward edge of the mangroves and significantly influencing the distribution and rate of decomposition of organic matter. Amphipods and deposit-feeding snails may also be important, but their abundance appears to vary among locations. The amphipod *Orchestia* sp. was abundant in litter decomposition bags deployed in the Tuff Crater (Woodroffe 1985a), but amphipods were generally absent in mangrove and pneumatophore habitats in Matapouri Estuary (Alfaro 2006). Because most studies of rates of decomposition of mangrove litter use mesh bags to prevent the litter being washed away by the tide, the bags may exclude larger macrofauna and, by preventing them from breaking up the litter, underestimate rates of decomposition, as most studies have acknowledged.

Uptake of mangrove-derived material within the mangrove forest and export to adjacent habitats

Nutrients derived from mangrove material, either through direct grazing or via detritus, may be recycled within the mangrove system or exported. Material may be exported in the form of leaves, twigs, fragmented detritus, dissolved organic matter or inorganic matter or as living organisms. The proportion of detrital material that is retained in the sediment within the mangroves relative to the proportion exported by water movement is not known. Although net primary production is lower in temperate compared with tropical mangroves, knowledge of how temperate mangroves differ from tropical mangroves in other components of the carbon budget and how carbon and nutrients derived from mangroves is incorporated into marine food webs through inwelling and outwelling is less well documented. Generalised carbon budgets for mangroves, based mainly on tropical forests, have been developed most recently by Alongi (2009). Carbon fixation by the trees is the largest input of carbon to the system, and respiration by trees is the largest efflux of carbon. This is unlikely to differ between temperate and tropical forests, but other components of carbon budgets may well do so.

The proportional contribution of benthic primary producers may be higher in temperate than in tropical mangroves. Many temperate forests are short, scrub forests where light penetration to the sediment is relatively high, leading to substantial rates of primary production of algae and microbial communities (Joye & Lee 2004). Increased production by these components may increase the availability of labile carbon for consumption and outwelling in temperate compared with tropical mangroves.

Carbon allocation between shoots and roots may differ between temperate and tropical mangroves. Shorter scrub forests typical of temperate sites tend to allocate a greater proportion of carbon below ground than taller forests (Lovelock 2008). Allocation to roots may contribute to enhancing microbial carbon mineralization and thus carbon availability for export from temperate compared with tropical mangrove estuaries. Few data are available to allow this hypothesis to be assessed.

Woodroffe (1985a) noted that the organic content of sediments in mangrove areas in the Tuff Crater (Auckland) was high, and although not all of it necessarily derived from mangroves, at least a portion of detrital production was clearly retained and recycled *in situ*. Morrisey et al. (2003) also measured high percentage cover of the sediment surface by detrital material (4.5–72.5%) and high proportions of organic matter (7.9–17.2% DW) in the sediment in mature stands of mangroves in Puhinui Creek, Auckland, but not in newly establishing stands (0–1.3% and 4.4–5.8%, respectively).

However, concentrations of particulate organic matter (POM) were also high in the water in the tidal creeks of Tuff Crater, indicating that some of this material is exported (Woodroffe 1985a).

The interplay of factors influencing the relative rates of litter decomposition and export from temperate mangroves is illustrated by a study of the effects of feeding by *Sesarma meinerti* in the Mgeni Estuary, South Africa (Steinke et al. 1993). In the field, leaf litter represented 75% of the diet of the crabs, and they showed differential preferences for different mangrove species and in different states of decomposition, in the order yellow *Bruguiera gymnorrhiza* > yellow *Avicennia marina* > green *Bruguiera gymnorrhiza* > green *Avicennia marina*. Laboratory studies showed a correlation between the relative amount consumed and the degree of decomposition. The role of the crabs varied with tidal height and therefore between the two mangroves species, being relatively more important in *Bruguiera gymnorrhiza* stands. In *Avicennia marina* stands, there was more detritus present on the sediment surface, and microbial decomposition may play a larger role. At the same location, *Rhizophora mucronata* lines the banks of creeks, and this may result in relatively large export of detritus from these stands. In stands of *Bruguiera gymnorrhiza*, the feeding behaviour of *Sesarma meinerti* results in removal of mangrove leaves from the sediment surface and their retention within the forest. It also leads to increased rates of degradation of detrital particle size, so that the material is more rapidly decomposed by detritivores and microbes.

Although it has been assumed for a long time that mangroves export detritus and faunal biomass to adjacent habitats and offshore (see discussion in Lee 1995), this hypothesis has only been tested relatively recently. Mass balance studies tend to support the hypothesis, but estimates of net imports and exports from such studies are strongly dependent on the method of calculation (Murray & Spencer 1997). Studies using stable isotopes to track the fate of organic matter originating within mangrove habitats have suggested that this 'outwelling' may be less extensive and ecologically significant than previously assumed (Lee 1995, Loneragan et al. 1997, Kathiresan & Bingham 2001, Bouillon et al. 2008).

Guest & Connolly (2004, 2006) found that the ^{13}C signatures of crabs (*Parasesarma erythrodactyla* and *Australoplax tridentata*) and slugs (the pulmonate *Onchidina australis*) in salt marshes in south-east Australia reflected that of the dominant saltmarsh plant (*Sporobolus virginica*). The signatures of the same animals living in mangroves reflected those of *Avicennia marina*. The sharply defined zone of transitional ^{13}C values between the two habitats suggested that movement and assimilation of carbon from one habitat to the other is limited to 5–7 m. Animals living in large (>0.4-ha) patches of salt marsh had signatures matching those of *Sporobolus virginica*, while the signatures of those in smaller patches indicated that they had assimilated carbon from both *S. virginica* and mangroves.

In Westernport, Victoria, the stable isotope signatures of a deposit-feeding callianassid shrimps (*Biffarius arenosus* and *Callianassa* (*Trypea*) *australiensis*) indicated that mangroves (*Avicennia marina*) and salt marshes (dominated by *Sarcocornia quiqeflora*) were not major sources of food, even though these habitats occurred close to the intertidal flats on which the shrimps lived (Boon et al. 1997).

Alfaro et al. (2006) used lipid biomarkers and stable isotopes to identify the trophic pathways of an estuarine food web in Matapouri Estuary, Northland. Mangroves are the dominant habitat in the estuary, but seagrass beds, sandflats and salt marshes are also present. Mangroves and brown algae were identified as important contributors of suspended organic matter in the creeks draining the harbour, but seagrass detritus is also likely to be important. Suspension of bacteria associated with detritus in surficial sediments probably represents another route by which mangroves and seagrasses contribute to suspended organic material. Biomarkers characteristic of bacteria were dominant in sediments. Biomarkers for fresh mangrove material were present in sediments collected adjacent to mangrove stands but not in those from a sandflat further down the estuary, suggesting that distribution of fresh mangrove organic matter may be quite localised.

Mangrove detritus, however, is probably more widely distributed because detrital biomarkers found at the sandflat site were likely to have derived from both seagrass and mangroves. This detrital material then becomes available to the infauna of the sandflats, particularly filter-feeders. This was confirmed by the presence of mangrove biomarkers in three filter-feeding bivalve species (cockles *Austrovenus stutchburyi*, pipis *Paphies australis*, and particularly oysters *Crassostrea gigas*, which were collected from trunks and roots of mangroves). Two grazing gastropods (*Turbo smaragdus* and *Nerita atramentosa*) both contained relatively large amounts of diatom, seagrass and fresh mangrove biomarkers. *Turbo smaragdus* occurs in association with mangrove pneumatophores, in addition to brown algae and seagrass, while *Nerita atramentosa* occurs on mangrove trunks; hence, the presence of fresh mangrove material in their diets is not surprising. Stable isotope signatures suggested that *N. atramentosa* consumes equal amounts of mangrove and seagrass material, whereas *Turbo smaragdus* consumes relatively more brown algae and seagrass.

The predatory whelk *Lepsiella scobina* contained relatively large amounts of mangrove biomarkers, possibly because it preys on oysters, which in turn consume mangrove material in suspension. Biomarkers found in glass shrimps (*Palaemon affinis*) indicated that they consume a wide range of organic matter, but with diatoms and mangrove material predominant. The mud crab *Helice crassa* is abundant in mangroves in some parts of New Zealand (e.g., Morrisey et al. 2003), although not in Matapouri, where it has been found only in small numbers (Alfaro 2006). Its diet includes diatoms, macroalgae, sediment-related bacteria and meiofauna (Morton 2004), and this dietary diversity was reflected by the diversity of biomarkers found in these crabs in Matapouri. However, it appears to consume little fresh mangrove or seagrass material, in contrast to the grapsid crabs of tropical mangroves.

Alfaro et al. (2006) concluded that the food web in the Matapouri Estuary thus incorporates several sources of organic matter and a range of trophic pathways. The various consumers (primary, secondary and higher order) appear to exploit different sources to different degrees, with none of the food sources being obligatory for the dominant organisms studied. Little fresh mangrove material appears to be incorporated directly into adjacent habitats, but mangrove detritus, in contrast, appears to be important to a range of organisms via the detrital food web. May (1999) also deduced that mangrove detritus was potentially important to deposit-feeding organisms in northern New Zealand estuaries, as did Knox (1983) in a study of the Waitemata Harbour, Auckland, although neither of these studies could provide the resolution achieved by use of stable isotopes or other biomarkers. The brown alga *Hormosira banksii* was found to contribute a relatively large amount of organic material to the estuarine system (Alfaro et al. 2006). *Hormosira banksii* is present all year and is very abundant among mangrove pneumatophores, presumably because they trap the alga and provide a substratum for it. Mangroves may therefore provide a further, indirect contribution to trophic pathways in the estuary. Like mangroves, the contribution of seagrass as a direct food source was smaller than expected but did provide material to the detrital food chain. May (1999) noted that in Rangaunu Harbour (35°00′S), a large estuarine system with high water clarity, extensive seagrass beds and mangrove forests, prevalent south-easterly winds often imported drift seagrass into mangrove forests on the windward side of the harbour, implying that primary production by seagrasses can subsidise mangrove forests in some situations. Observations in late April 2007 of the mangrove forest in the most southern arm of this harbour found dried seagrass drift festooned across the mangrove trees, up to the high-tide mark (M. Morrison personal observation). The seagrass meadows present extended from the low-tide channel up and into the pneumatophore zone.

Work on the trophic role mangroves play in supporting fish production has also shown them to be less critical than traditionally thought. Melville & Connolly (2003) examined the carbon and nitrogen stable isotope signatures of three commercially important fish species (yellow-fin bream *Acanthopagrus australis*, sand whiting *Sillago ciliata* and winter whiting *Sillago maculata*) over bare mudflats in a subtropical estuary (Morton Bay, southern Queensland). They undertook this both at the whole estuary scale and through finer-scale 'spatial tracking' by sampling nine separate locations and looking for spatial trends in the isotope signatures. The primary producer signatures identified were mangroves, seagrass, seagrass epiphytes, saltmarsh grass and saltmarsh succulent plants, POM and microphytobenthos. For yellow-fin bream, seagrass, salt marsh and POM were important trophic sources at the whole estuary scale, but the use of spatial tracking also identified mangroves as an additional important source dependent on spatial location (up to 33% of the carbon used—upper 95% confidence limit of the mean contribution derived from an isotope mixing model). Similarly, for sand whiting, only POM appeared to be important at the whole-estuary scale. At more localised scales, mangroves and microphytobenthos were also important, with up to 25% (upper 95% confidence interval) of carbon contributed by mangroves. Relative contributions for different producers could not be assigned for winter whiting, possibly because of either site-specific diet selection or movement of individuals among sites.

Subsequent to their 2003 study, Melville & Connolly (2005) examined ^{13}C isotope markers in 22 species of estuarine fishes collected from bare mudflats in the same estuary. They pooled the similar isotopic signatures of three mangrove species (*Avicennia marina*, *Aegiceras corniculatum* and *Rhizophora stylosa*) and those of three seagrass species (*Zostera capricorni*, *Halophila ovalis*, and *H. spinulosa*). Other primary producers identifiable by their isotopic signatures included epiphytes on seagrass, saltmarsh plants and microphytobenthos. The majority of the carbon in the fishes caught over mudflats was clearly derived from adjacent habitats, with seagrass contributions dominating strongly. The authors found it difficult to separate the contributions of mangroves from that of saltmarsh succulents and microphytobenthos, but for most fishes the maximum mangroves could have contributed was 30%. For five species, however, the contribution from mangroves may be up to 50% and possibly even higher for yellow perchlet (*Ambassis jacksonienis*).

In summary, evidence worldwide for the suggested importance of mangrove productivity to estuarine and coastal food webs is equivocal (Lee 1995). Kristensen et al. (2008) concluded that globally mangroves are an important net source of detritus to adjacent coastal waters, with an average export rate equivalent to 50% of total litter production. Utilisation of this material is, however, inconsistent and dependent on local conditions, including net primary production, the abundance of litter-collecting fauna and the local tidal range.

Information on nutrient and energy flows between mangroves and other habitats in temperate estuaries is relatively limited, but recent studies, such as that in Matapouri Estuary, northern New Zealand (Alfaro et al. 2006), suggested that estuarine consumers exploit a range of sources of primary production (seagrasses, mangroves, benthic microalgae and macroalgae) rather than being dependent on one particular source. Fresh mangrove and seagrass material appears to play a relatively minor, local role in the overall estuarine food web, but detritus derived from these plants and exported via tidal movement may play a more significant role via the detrital pathway. Macroalgae, such as *Hormosira banksii* in southern Australia and New Zealand, can be an important source of organic material to estuarine food webs and is abundant year round among mangrove pneumatophores and seagrass beds, indicating an additional, indirect role for these habitats.

Mangroves and sediments

Geomorphic distribution of temperate mangroves

Globally, the largest mangrove forests occur on deltas and muddy coasts of the tropical Indo–West Pacific (IWP) (Thom et al. 1975, Robertson et al. 1991, Walsh & Nittrouer 2004) and Atlantic–East Pacific (AEP) (Wells & Coleman 1981, Woodroffe & Davies 2009). Examples include the Ganges–Brahmaputra–Meghna, Irrawaddy, Mekong, Red and Pearl Rivers (IWP) and the mangrove coast of the Amazon River mouth (AEP) (Araújo da Silva et al. 2009, Woodroffe & Davies 2009). These rivers supply billions of tonnes of terrigenous sediment to tropical continental margins each year (Syvitski et al. 2005), some of which is deposited in shallow nearshore environments, including mangrove forests (Allison et al. 1995, Walsh & Nittrouer 2004). Extensive tropical mangrove forests develop on meso- and macrotidal coasts, where deposition of fine sediments builds extensive deltas and low-gradient intertidal mudflats (Schaeffer-Novelli et al. 2002, Ellison 2009, Woodroffe & Davies 2009). Large mudflat systems also develop in mesotidal (tidal range 2–4 m) settings with large terrigenous sediment supply (Healy 2002) and on coasts subjected to high wave energy (Wells & Coleman 1981, Mathew & Baba 1995).

In contrast to tropical systems, the temperate mangrove forests of south-eastern Australia, northern New Zealand and South Africa's eastern coast predominantly occur in estuaries, rather than on river deltas or muddy coasts. These Southern Hemisphere forests account for about 95% of temperate mangrove habitat worldwide. The estuary classification of Roy et al. (2001) for south-eastern Australian estuaries, which is generally applicable to New Zealand, provides a framework to describe the geomorphic distribution of temperate mangroves in these systems. In this scheme, estuaries are classified into types based on their geological setting, rate of sediment infilling and the relative dominance of tides and waves in controlling water circulation and sediment transport. This scheme has similarities with Thom's (1982) geomorphic classification of mangroves, which describes coastal environments dominated by terrigenous sediment supply (i.e., deltas, barrier islands, spits, lagoons and estuaries), although the vast majority of temperate mangrove forests are found in estuaries.

Temperate mangroves occur in bays with small freshwater inflow, drowned river valleys, tidal basins, barrier estuaries and lagoons. Bays, river valleys and basins have tidal ranges similar to the ocean, and circulation is dominated by tidal exchange except during floods, which episodically discharge freshwater. Tidal ranges in barrier estuaries and lagoons are reduced by sand deposits at their inlets, so that tidal currents are weak and wind waves and wind-driven circulation control sediment transport (Roy et al. 2001). In northern New Zealand, many drowned river valley estuaries have reached an advanced stage of infilling, and sediment transport on intertidal flats is controlled by wind waves rather than tidal currents (Green et al. 1997, Hume 2003, Swales et al. 2004, Green & Coco 2007). Along South Africa's eastern coast, temperate mangroves occupy tidal flats in narrow, bedrock-confined drowned river valleys and river-dominated barrier estuaries and lagoons (Cooper et al. 1999, Steinke 1999). Many of these estuaries have infilled with terrigenous sediments and have limited tidal prisms so that sediment transport is largely driven by river discharge rather than tides. River-dominated estuaries display cyclic sedimentation and erosion due to episodic large floods, and sediment deposits maybe completely eroded down to bedrock. River-dominated lagoons are also subject to episodic inlet closure during periods of low freshwater discharge. Inlets are re-established by high river flows during the early summer wet season (Cooper et al. 1999, Schumann et al. 1999). Thus, intertidal flat habitats suitable for mangroves are limited by the inherent morphology and dynamics of South Africa's river-dominated estuaries.

During the last 6000–7000 years since sea level has been stable, estuaries have filled with sediment at different rates and have reach different stages of evolution (Cooper et al. 1999, Roy et al. 2001, Hume 2003). Stages of development range from youthful systems that retain much of their original tidal volume to mature estuaries dominated by river discharge. In semimature estuaries, sediment infilling is usually associated with the seaward expansion of accreting intertidal flats that progressively displace subtidal basins. The rate of estuary infilling primarily reflects the original volume of the tidal basin and rate of sediment supply.

In south-eastern Australia, the largest mangrove forests occur in the mature drowned river valley and barrier estuaries of New South Wales, which reflects the larger intertidal flat areas found in these systems (Roy et al. 2001). Drowned river valley estuaries in particular display a trend, independent of estuary size, of expanding mangrove and saltmarsh habitat with increasing estuary maturity. Port Stephens is an exception to this pattern, and it remains at a relatively youthful stage of development due to its relatively small catchment; it contains the largest area of mangrove forest (2330 ha) in New South Wales (Roy et al. 2001, their Appendix 1). Mangrove forests also occur in ocean embayments with small river catchments, such as Westernport Bay (Victoria), where *Avicennia marina* forests fringe 40% of the shoreline (Bird 1986). Drowned river valley estuaries are also numerous along the coast of northern New Zealand, within the climatic range of *A. marina* (Hume 2003). However, the largest temperate mangrove forests occur in estuarine embayments with large terrigenous sediment supply. Examples include the Firth of Thames (1100 ha) and Rangaunu Harbour (2415.5 ha) as well as large barrier-enclosed estuaries such as the Kaipara (6167 ha) and Tauranga (623 ha) Harbours (Hume et al. 2007 and see Table 9). The area of temperate mangrove forest in these four estuaries accounts for 47% of the New Zealand total (data in Table 9 and from S. Park, Environment Bay of Plenty, personal communication). In South Africa, the largest temperate mangrove forests occur in the drowned river valley estuaries of northern Kwazulu-Natal that have formed in unconsolidated Tertiary–Pleistocene sediments (Cooper et al. 1999). Shoreline erosion has produced relatively large, shallow, sandy estuaries with extensive intertidal flats. The Umhlatuze (Richards Bay), St. Lucia, Kosi and Mfolozi estuaries of Kwazulu-Natal account for about 75% (~734 ha) of South Africa's temperate mangrove forests (Steinke 1999, Bedin 2001).

Sediment sources

Sediment supply to mangrove forests can be classified as either external (allochthonous) associated with transport and deposition of inorganic fine sediments or *in situ* (autochthonous) production of organic peat. Mangrove peat sediments occur in systems isolated from terrigenous sediment sources and are most commonly associated with isolated oceanic low islands in the tropical Pacific and Caribbean (Woodroffe & Davies 2009). With the exception of the peat-forming *Avicennia germinans* and *Rhizophora mangle* stands of Bermuda (Ellison 1993), most temperate mangroves occur on sediment-rich continental margins and high islands. Thus, sedimentation in temperate mangrove systems is characterised by accumulation of terrigenous mineral sediments that are ultimately derived from the erosion of catchment soils and coastal margins (Bird 1986, Ellison 2009).

Terrigenous sediment yields have increased by an order of magnitude or more as a consequence of land use changes associated with human activities in catchments (Walling 1999). The impact of humans on the terrigenous suspended sediment flux to the coastal ocean has been evaluated by Syvitski et al. (2005) based on long-term hydrological records and modelling of the world's major river systems. In the warm temperate zone (10–25°C), the annual flux to the coastal ocean of about $8 \times 10^6 \pm 1.25$ (observational uncertainty) $\times 10^6$ t yr^{-1} is about 10% lower than the prehuman flux. A similar trend is observed for the global sediment flux and was attributed to sediment storage in floodplains and reservoirs constructed in large river basins during the twentieth century. However, the sediment loads delivered by small river systems that drain islands on tectonically active plate margins, such as occur in New Zealand, may be similar to loads delivered by large rivers on passive margins (Milliman & Syvitski 1992). The average size of catchments draining to estuaries with temperate mangroves in northern New Zealand is 239 km^2 (range 6–4194 km^2; Hume et al. 2007). These small, steep basins are subject to episodic intense rainstorms and have relatively small capacity to store eroded sediments, and sediment delivery to estuaries and coasts will generally be higher than for large basins (Griffiths & Glasby 1985, Milliman & Syvitski 1992, Elliot et al. 2009), such as those in south-eastern Australia (Wasson & Galloway 1986, Wasson 1994). For example, catchments draining to New South Wales estuaries with temperate mangrove habitat are an order of magnitude larger (average 2545 km^2, range 9–22,400 km^2; Roy et al. 2001) than occur in northern New Zealand.

The natural process of estuary infilling in New Zealand and south-eastern Australia has accelerated over the last approximately 200 years following catchment deforestation by European settlers. Environmental changes (i.e., catchment deforestation, drainage of wetlands, clearing of riparian vegetation and land conversion for pastoral agriculture and cropping) occurred rapidly and on a large scale. Widespread urbanisation of coastal margins has also occurred since the 1950s. At the peak of

Table 9 Summary of historical changes in temperate mangrove habitat in southern Australia, New Zealand and South Africa compiled from published sources

Location	Species	Time period	Habitat area (ha)	Habitat area change %	% yr^{-1}	Reference
Southern Australia						
Light River and Swan Alley, South Australia	A.m.	1949–1979	624–1221	96	3.2	Burton 1982
Jackson-Paramata River, NSW	A.m., A.c.	1930–1951	182–217.5	20	0.95	Thorogood 1985
Minnamurra River, NSW	A.m., A.c.	1938–1997	56.5–95.4	69	1.2	Chafer 1998
North Arm Creek, South Australia	A.m.	1979–1993	7.4–8.8	19	1.4	Coleman 1998
Tweed River, NSW	A.m.	1930–1994	144.1–281.9	96	1.5	Saintilan 1998
Kurnell Peninsula, Botany Bay, NSW	A.m., A.c.	1956–1996	80.8–120.3	49	1.3	Evans & Williams 2001
Currambene Creek, Jervis Bay, NSW	A.m.	1949–1993	34.1–44.9	32	0.7	Saintilan & Wilton 2001
Pittwater, NSW	A.m., A.c.	1940–1996	2–12.2	510	9.1	Wilton 2001
Kooweerup, Westernport Bay, Victoria	A.m.	1939–1999	11.3–18.1	60	1	Rogers et al. 2005b
Rhyll, Westernport Bay, Victoria	A.m.	1939–1999	40–62.3	56	0.9	Rogers et al. 2005b
New Zealand						
Houhora, Northland	A.m.	1944–1979	52.3–56.4	8	0.2	NCC 1984
Rangaunu, Northland	A.m.	1944–1981	1821.5–2415.5	33	0.9	NCC 1984
Ngunguru, Northland	A.m.	1942–1979	139.9–175.5	25	0.7	NCC 1984
Whangapae, Northland	A.m.	1939–1993	280–338.8	21	0.4	Creese et al. 1998
Tauranga, Bay of Plenty (8 subestuaries)	A.m.	1940–1999	13–168.2	1193	20.2	Park 2004
Whangapoua, Great Barrier Island, Auckland	A.m.	1960–1999	88–162.8	85	2.2	Morrisey et al. 1999
Puhinui, Auckland	A.m.	1939–1996	14.4–37.7	162	2.8	Morrisey et al. 2003
Puhoi, Auckland	A.m.	1960–2007	46.6–59.7	28	0.6	Swales et al. 2009
Waiwera, Auckland	A.m.	1960–2007	7.5–37.8	404	8.6	Swales et al. 2009
Orewa, Auckland	A.m.	1960–2007	17.9–41.9	134	2.9	Swales et al. 2009

Location	Species	Years	Range			Reference
Okura, Auckland	A.m.	1959–2007	12.1–25	106	2.2	Swales et al. 2009
Lucas Creek, Auckland	A.m.	1950–1996	47.5–50.5	6	0.14	Morrisey et al. 1999
Central Waitemata, Auckland	A.m.	1959–2007	722.5–665	–8	–0.17	Swales et al. 2009
Shoal Bay, Auckland	A.m.	1959–2007	149–139	–7	–0.15	Swales et al. 2009
Orakei, Auckland	A.m.	1959–2007	18.7–26	39	0.8	Swales et al. 2009
Whitford, Auckland	A.m.	1955–2007	72.4–112.1	55	1.1	Swales et al. 2009
Wairoa, Auckland	A.m.	1955–2007	124.7–161.7	30	0.6	Swales et al. 2009
Te Matuku, Auckland	A.m.	1940–2007	33–47.4	44	0.7	Swales et al. 2009
Wharekawa, Coromandel	A.m.	1983–2008	20.1–40	99	4	NCC 1984, EW 2009
Tairua, Coromandel	A.m.	1983–2008	12.1–35.6	194	7.8	NCC 1984, EW 2009
Manaia, Coromandel	A.m.	1971–2008	58.6–168.7	187.7	5.1	NCC 1984, EW 2009
Whangamata, Coromandel	A.m.	1978–2007	73.8–95.8	30	1	NCC 1984, EW 2009
Otahu, Coromandel	A.m.	1978–2007	1.9–10.2	448	15.4	NCC 1984, EW 2009
Coromandel Harbour	A.m.	1971–1999	27.5–61.6	124	4.4	NCC 1984, EW 2009
Te Kouma, Coromandel	A.m.	1971–1999	9.1–15.1	67	2.4	NCC 1984, EW 2009
Whitianga, Coromandel	A.m.	1970–1999	444.9–488.6	10	0.3	NCC 1984, EW 2009
Raglan, Waikato	A.m.	1979–2005	6.3–27.4	336	12.9	NCC 1984, EW 2009
Firth of Thames (south shore)	A.m.	1952–2006	75–741	888	16.4	Swales et al. 2007b
Firth of Thames (total)	A.m.	1983–2004	662–1100	66	3.1	NCC 1984, Brownell, 2004
South Africa						
Umhlatuze (Richards Bay), Kwazulu-Natal	A.m., B.g., R.m.	1974–1999	197–489	148	5.9	Bedin 2001, Ward & Steinke 1982
Mtamvuna, Eastern Cape	B.g.	1982–1999	1–0.25	–133	–7.8	Adams et al. 2004
Mzamba, Eastern Cape	B.g.	1982–1999	1–0.15	–118	–6.9	Adams et al. 2004
Mnyameni, Eastern Cape	A.m., B.g.	1982–1999	3–0	–100	–5.9	Adams et al. 2004
Mtentu, Eastern Cape	A.m., B.g.	1982–1999	1–2	100	5.9	Adams et al. 2004

(continued on next page)

Table 9 (continued) Summary of historical changes in temperate mangrove habitat in southern Australia, New Zealand and South Africa compiled from published sources

Location	Species	Time period	Habitat area (ha)	Habitat area change		Reference
				%	% yr^{-1}	
Mzintlava, Eastern Cape	B.g.	1982–1999	1.5–1.75	17	1	Adams et al. 2004
Mntafufu, Eastern Cape	Am., B.g., R.m.	1982–1999	10–12.4	24	1.4	Adams et al. 2004
Mzimvubu, Eastern Cape	A.m., B.g.	1982–1999	1–0	-100	-5.9	Adams et al. 2004
Mngazana, Eastern Cape	Am., B.g., R.m.	1982–1999	150–145	-3	-0.2	Adams et al. 2004
Mtakatye, Eastern Cape	Am., B.g., R.m.	1982–1999	7.5–9	20	1.2	Adams et al. 2004
Mdumbi, Eastern Cape	A.m.	1982–1999	1–0.5	-50	-2.9	Adams et al. 2004
Mtata, Eastern Cape	Am., B.g., R.m.	1982–1999	34–42	24	1.4	Adams et al. 2004
Bulungula, Eastern Cape	Am., B.g., R.m.	1982–1999	3.5–0	-100	-5.9	Adams et al. 2004
Xora, Eastern Cape	Am., B.g.	1982–1999	16–16.5	3	0.2	Adams et al. 2004
Mbashe, Eastern Cape	Am., B.g.	1982–1999	12.5–14	12	0.7	Adams et al. 2004
Nqabara, Eastern Cape	A.m.	1982–1999	9–8.5	6	0.35	Adams et al. 2004
Nxaxo/Ngqusi, Eastern Cape	Am., B.g., R.m.	1982–1999	14–15	7	0.4	Adams et al. 2004
Kobonqaba, Eastern Cape	A.m.	1982–1999	6–3.5	-42	-2.5	Adams et al. 2004
Total, Eastern Cape (17 estuaries)		1982–1999	272–270.6	-0.4	-0.02	Adams et al. 2004

Note: A.c. = *Aegiceras corniculatum*; A.m. = *Avicennia marina*; B.g. = *Bruguiera gymnorhiza*; R.m. = *Rhizophora mucronata*; NCC = Nature Conservation Council; EW = Environment Waikato.

these catchment land use changes in the mid- to late 1800s, catchment sediment yields were as much as several orders of magnitude higher than pre-European values of several tonnes $km^{-2} yr^{-1}$ (Wasson & Galloway 1986, Prosser et al. 2001, Olley & Wasson 2003). Sediment yields have declined over the last century due to reforestation, destocking or reduced sediment supply but remain several times higher than pre-European values (Wasson 1994, Wilmshurst 1997, Erskine et al. 2002, Healy 2002, Swales et al. 2002, Olley & Wasson 2003). In South Africa, human activities in catchments had similar environmental effects: increased soil erosion due to overgrazing and poor farming methods and destruction of wetlands and riparian vegetation (Kakembo & Rowntree 2003, Adams et al. 2004, Foster et al. 2009) that have accelerated estuary infilling (Morant & Quinn 1999). Modern rates of terrigenous sediment supply to South Africa's eastern coast, associated with human activities in catchments, remain 10–20 times higher than geological rates of 15–25 t $km^{-2} yr^{-1}$ (Martin 1987).

In temperate estuaries, these increased terrigenous sediment loads have resulted in increased water turbidity and accelerated sedimentation, with a shift to increasingly muddy intertidal systems and changing plant and animal communities (Chenhall et al. 1995, Morant & Quinn 1999, Nichol et al. 2000, Roy et al. 2001, Swales et al. 2002, Thrush et al. 2004). Although such studies have quantified the effects of human activities in catchments on temperate estuaries, there is a general paucity of data on rates of sediment delivery to receiving estuaries.

Sediment delivery to temperate mangrove forests

Mechanisms controlling delivery of terrigenous sediments to temperate mangrove forests in estuaries are spatially and temporally variable and reflect the relative importance of sediment transport by rivers, tides and waves. Fine suspended sediments are delivered directly or indirectly to mangrove forests by buoyant silt plumes discharged from rivers (Geyer et al. 1991), remobilisation and transport of muds stored in intertidal and subtidal deposits in estuaries by waves or currents (Green et al. 1997, Christie et al. 1999, Perry 2007), erosion of mangrove-fringed muddy coasts (Semeniuk 1980, J.T. Wells 1983, Mathew & Baba 1995, Walsh & Nittrouer 2004) and erosion and retreat of low-lying shorelines (Bird 1986).

Terrigenous sediments are largely delivered to estuaries during storms, and infrequent large-magnitude storms can account for years or decades of the average annual sediment load (Nichols 1977, Schubel & Pritchard 1986, Stern et al. 1986). Most of the terrigenous sediment load delivered to New Zealand and south-eastern Australian estuaries is composed of silt and clay particles transported in suspension (Griffiths & Glasby 1985, Roy et al. 2001, Healy 2002). By contrast, rivers deliver sand-rich sediments to many South African estuaries (Cooper et al. 1999). The cohesive behaviour of clay and fine silt particles differs markedly from sand (Dyer 1986). Mangrove forests preferentially accumulate these fine sediments in temperate (Woodroffe 1992, Kathiresan & Bingham 2001, Swales et al. 2002, Ellis et al. 2004, Swales et al. 2007b) and tropical environments (Bird 1972, Walsh & Nittrouer 2004).

Along tidal rivers and creeks in the upper reaches of estuaries, fine-sediment delivery to mangrove forests is primarily controlled by riverine and tidal forcing. Key mechanisms include tidal pumping, density-driven circulation and flocculation (Dyer 1986, Friedrichs & Aubrey 1988, Chant & Stoner 2001, Wolanski et al. 2001, Mehta 2002, Uncles 2002). In meso- and macrotidal estuaries, turbulent flows, density-driven circulation and tidal pumping maintains high suspended sediment concentrations (SSCs) in their upper reaches. Within this turbidity maximum, SSCs are higher than in the river system and receiving estuary (Dyer 1997), which supplies fine suspended sediments to riverine mangrove forests. Stormwater discharge alters the hydrodynamics of estuaries, such as the vertical stratification of the water column that develops due to differences in the density of freshwater run-off and saline estuarine waters (Fugate & Chant 2005, Traynum & Styles 2007). Under conditions of high freshwater discharge, baroclinic (salt-wedge) circulation can develop (Dyer 1997). Silt-laden stormwater in the seaward-flowing surface layer is delivered to riverine mangroves fringing the creeks during stratified-flow conditions when the water spills over the creek banks. Deposition of cohesive fine suspended sediments in mangrove forests is enhanced by flocculation (Furukawa et al. 1997), which produces aggregates composed of clay and fine silt-size particles. Settling velocities of these constituent particles, typically less than 0.005 cm s^{-1}, are orders of magnitude lower than for flocs (Wolanski et al. 2002). The process of floc formation and decay is modulated by tidal cycle variations in flow turbulence (Dyer 1986, Wolanski 1995, Augustinus 2002).

Seawards of the tidal creeks, temperate mangroves occupy intertidal flats fringing the central subtidal basin of semimature estuaries. The extent of these forests increases with tidal range and system maturity. In large estuaries, mangrove forests fringe extensive intertidal flats where sediment transport is controlled by waves (Green et al. 1997, Roy et al. 2001, Swales et al. 2007b). Tidal currents are relatively less important in mature infilled estuaries largely due to the small tidal volume exchanged (i.e., tidal prism) (Friedrichs & Aubrey 1988, Roy et al. 2001, Hume 2003). The short-period waves that typify fetch-limited estuaries are rapidly attenuated within the upper few metres of the water column. Consequently, sediment resuspension by waves is strongly modulated by tidal variations in water depth (Dyer 1986, Uncles 2002). Seasonal variations in biofilm production by benthic diatoms and fauna also influence tidal flat accretion and erosion (Grant et al. 1986, Meadows et al. 1990, Andersen 2001). In large muddy estuaries, wave attenuation may also occur in hyperconcentrated suspensions (i.e., SSC ~ 10 g l^{-1}) due to viscous-energy dissipation in fluid-mud wave-boundary layers (Mehta 2002, Gratiot et al. 2007). Once suspended, fine sediments can be transported long distances due to their low settling velocities (Lee & Mehta 1997), and transport velocities are generally much lower than required for sediment erosion (Postma 1961). Coupled with wind-driven surface flows and mass transport, these high-concentration suspensions are preferentially transported onshore during flood tides to mangrove forests fringing the intertidal flats (Bird 1986, Gratiot et al. 2007).

The role of mangroves in trapping sediments

The influence of temperate and tropical mangroves on sedimentation processes and geomorphic evolution of estuaries and coasts has been the focus of research since the early 1900s (Vaughan 1909, Davis 1940, Chapman & Ronaldson 1958, Stephens 1962, Scholl 1968, Scoffin 1970, Bird 1971, Carlton 1974). The question is "whether mangroves promote sedimentation, and thereby the evolution of depositional landforms, or whether they simply occupy sites that have become ecologically suitable, moving in to colonise (and possibly thereafter to stabilise and protect) an intertidal morphology that would have formed independently in their absence" (Bird 1986).

Reconstructions of the evolution of tidal flats/mangrove forests based on analysis of sediment cores and historical records have demonstrated the strong dependence of mangrove forest ecology on coastal geomorphology (Walsh & Nittrouer 2004, Swales et al. 2007b, J.C. Ellison 2008). In temperate and tropical mangrove forests, changes in habitat extent are closely linked to substratum elevation relative to sea level. Mangrove habitat expansion occurs on accreting tidal flats when the bed elevation has increased sufficiently for mangrove seedlings to colonise the substratum (Chapman & Ronaldson 1958, Thom 1967, Bird 1986, Panapitukkul et al. 1998, Bedin 2001, Swales et al. 2002, 2007b, Walsh & Nittrouer 2004, Carvalho do Amaral et al. 2006). Mangrove habitat loss has followed relative declines in surface elevation due to factors such as decreased sediment supply, tidal flat erosion, subsidence and relative SLR (Semeniuk 1980, Bird 1986, Ellison 1993, Lebigre 1999). These observations reflect the present consensus that mangroves are not in themselves 'land builders' (Mitsch et al. 2009).

Few physical process studies have been undertaken in temperate mangrove forests, and the following descriptions include studies done in tropical systems. Mangrove trees enhance the settling and accumulation of fine suspended sediments delivered by river plumes, tidal currents and waves in a number of ways.

In tidal rivers and creeks, suspended sediments are transported into mangrove forests by tidal currents and river plumes. Where the volume of flow through the mangrove forest is similar to or exceeds the tidal channel volume, mangrove forests can substantially influence creek hydrodynamics (Wolanski 1995). At low current velocities (i.e., less than about 5 cm s^{-1}) laminar flow prevails in mangrove forests. Turbulent flow is generated as current velocities increase, with the trunks, stems, leaves and root structures of mangrove trees and seedlings enhancing friction through pressure (form) drag and vegetation-induced viscous drag (i.e., eddy viscosity) (Wolanski et al. 1980, Mazda et al. 1995, Kobashi & Mazda 2005, Mazda et al. 2005, Mazda & Wolanski 2009). Pressure drag is generated by flow separation around the irregular-shaped structural elements of mangrove trees. The total drag force F_D induced by mangrove trees can be described as a function of the projected area of obstacles within the flow and a representative drag coefficient C_D for the particular mangrove species (Struve et al. 2003, Mazda et al. 2005). The F_D will also vary with tide level due to the changes in the projected area of trees with height above the bed. The eddy-viscosity coefficient f encompasses the effects of small-scale turbulence (eddies) on energy dissipation in the flow. Both F_D and f are enhanced by the presence of mangrove trees, the net effect of which is to reduce current velocities within mangrove forests. Current velocities within mangrove forests are typically on the order of 1–10 cm s^{-1} due to this drag-induced current dampening (Furukawa et al. 1997, Kitheka et al. 2002), in comparison with current velocities of 10–100 cm s^{-1} in tidal channels.

Tidal pumping is a key mechanism controlling the flux of suspended sediment to mangrove forests in riverine and tidal creek environments. Tidal pumping describes the asymmetry between the peak flood- and ebb-tide current velocities that develops in estuarine channels due to frictional deformation of the tide as it propagates in shallow water. Flood-tide durations are appreciably shorter, so current velocities are higher than for ebb tides. The degree of tidal asymmetry also varies over the spring–neap cycle (Dyer 1986). Bryce et al.'s (2003) multiyear study of a mangrove creek system demonstrated that changes in flood- and ebb-tide dominance over the spring–neap cycle controlled sediment delivery to mangrove forests. During neap tides, flows were contained with the channel with net flood-tide transport of suspended sediments to the head of the creek, where deposition occurred. These fine-sediment deposits were subsequently resuspended from the channel bed and transported into the mangrove during overbank spring tides.

In systems where mangrove forests occupy a substantial proportion of the tidal prism, current asymmetry in channels is reversed during overbank high spring tides. Under these conditions, water surface gradients induced by vegetation-enhanced friction result in shorter ebb tides with peak current velocities substantially higher than during flood tides (Wolanski et al. 1980, Woodroffe 1985b, Lessa & Masselink 1995). This ebb-tide dominance provides a mechanism to maintain a deep tidal channel with net export of creek sediments (Wolanski et al. 1980). In degraded mangrove systems, with reduced vegetation density, flood-tide dominance in channels is maintained, with net landward transport of suspended sediment (Kitheka et al. 2002). Baroclinic circulation developed during periods of relatively high freshwater discharge also enhances the landward transport of flocs in the near-bottom saline water (Wolanski 1995).

Sedimentation in riverine and tidal creek mangrove forests during overbank tides is dominated by the settling of clay-rich flocs. Settling is also enhanced by the stagnation zones of tree trunks and other flow obstructions (Furukawa & Wolanski 1996, Furukawa et al. 1997, Kitheka et al. 2002). Sedimentation rapidly declines with distance from distributory creeks due to the combined effects of current dampening and flocculation. Furukawa et al. (1997) found that about 40% of the suspended sediment advected into the forest was deposited within tens of metres of the creek edge, and 80% of the suspended sediment imported from coastal waters was trapped in the mangrove forest. In tidal creek systems at catchment outlets, stormwater may displace estuarine water from channels. Under these conditions, suspended silts are deposited as levees in mangroves fringing the channel banks (Swales et al. 2002).

On intertidal flats with sufficient fetches for wave generation, sediment trapping in fringing mangrove forests is enhanced by wave attenuation (Othman 1994, Brinkman et al. 1997, Massel 1999, Massel et al. 1999, Phuoc & Massel 2006, Quartel et al. 2007). Wave attenuation by plants has similarities with the physical mechanisms of vegetation drag-induced current dampening previously described for mangroves; the degree of wave attention varies with the submerged projected area of trees, and the drag profile is species dependent because of differences in tree morphology.

Although the spectral characteristics of waves propagating into mangrove forests are documented, little is known about sedimentation under attenuating waves. Suspended sediment influxes to mangrove forests typically increase with incident wave height (Phuoc & Massel 2006), which is consistent with wave-driven sediment resuspension on tidal flats (Green et al. 1997, Christie et al. 1999). Flocculation is also likely to be important (Furukawa et al. 1997), and the bulk of sedimentation will occur in the wave attenuation zone at the seaward margins of mangrove forests. This is consistent with sediment trap, marker bed and optical sensor data (Young & Harvey 1996, Ellis et al. 2004, Van Santen et al. 2007) and long-term sedimentation patterns derived from cores for mangrove forests in wave-dominated environments (Bird 1972, 1986, Lynch et al. 1989, Walsh & Nittrouer 2004, Swales et al. 2007b).

Moreover, hydroperiod (i.e., frequency and duration of inundation) also rapidly declines with surface elevation in the mangrove forest, so that sediment delivery is also modulated by the spring–neap tidal cycle. In mature mangrove forests where surface elevations are close to the upper tidal limit, delivery of this wave-driven suspended sediment influx only occurs episodically during high spring or storm tides (Swales et al. 2007b).

Sedimentation along the margins of mangrove forests may also be enhanced by the aerial roots of mangrove trees. Studies have included measurements of sedimentation rates in pneumatophore fields for several mangrove species, including *Avicennia marina*, under natural conditions as well as experiments using artificial pneumatophore arrays of various densities (10^2–10^5 m^{-2}). Environments have included wave-exposed mudflat/forest fringes and sheltered sites within forests (Bird 1971, Spenceley 1977, 1987, Bird 1986, Young & Harvey 1996). The results of these studies indicated differences in the pattern and rate of sedimentation that depended on the environmental conditions. Spenceley (1977) undertook experiments using artificial pneumatophore arrays with densities similar to the mangrove trees (400–1600 m^{-2}). His data suggested that local scour by waves occurred on mudflats and forest fringe during episodic storms, with local deposition occurring between these events. Bird (1986) mimicked the radial network of *A. marina* pneumatophore on a wave-exposed sand flat using pegs (density > 400 m^{-2}). Mud deposition was observed, and erosion followed the removal of the pegs. Similar experiments on a wave-cut platform and open mudflat did not induce sedimentation, which Bird (1986) attributed to local variations in sediment supply.

Manipulative experiments done within the sheltered interiors of mangrove forests also suggested that the influence of root structures on sedimentation depends on factors other than root density. Young & Harvey (1996) measured sedimentation rates over 3 mo around pneumatophore arrays of varying densities (100–350 m^{-2}) within an *A. marina* forest. Sedimentation was strongly and positively correlated with pneumatophore density. Differences between Young & Harvey's (1996) and Spenceley's (1977) results likely reflect differences in wave exposure (forest interior vs. fringe) and density of the artificial pneumatophore arrays. Krauss et al. (2003) extended this approach by considering the effect of root structure (i.e., pneumatophores, prop roots and root knees) on sedimentation and surface elevation changes in a Micronesian mangrove forest. Their 2.5-year field experiment included fringing, riverine and basin forests composed of *Rhizophora* sp., *Bruguiera gymnorrhiza* and *Sonneratia alba*. Treatments were prepared by removal of natural roots rather than the addition of artificial arrays, with the specific root area (cm^2 m^{-2}) used to account for differences in root density. Measurements enabled the contribution of subsurface processes, such as root biomass changes and sediment compaction, on net surface elevation changes to be constrained. The results indicated that annual sedimentation rates were significantly higher within the prop root treatments than for pneumatophore treatments or bare substrata controls in the riverine forest. By contrast, there were no significant differences in elevation changes between root types in all three forest types. Unlike Young & Harvey's (1996) study, Krauss et al. (2003) did not find a strong relationship between root density and sedimentation. This difference may reflect the greater 3-dimensional complexity of the natural root structure in the Micronesian forest and resulting hydrodynamic conditions.

The overall impression gained from these studies is that physical processes exert a strong control on sediment trapping in mangrove forests along with structural differences in the forest that relate to species composition and sedimentary environment. Present understanding of sediment dynamics and their role in the long-term development of temperate mangrove forests is limited.

Mangrove forest sediment budgets

Sediment budget studies of riverine mangrove forests have sought to quantify the role these systems play in sequestering terrigenous sediments in coastal and estuarine environments. Small-scale or short-term (i.e., tidal cycle or months) measurements of suspended sediment fluxes (Woodroffe 1985b, Furukawa et al. 1997, Kitheka et al. 2002) may not be representative of longer-term conditions. Over timescales of weeks to months, the direction of net suspended sediment transport in mangrove creek systems can be highly variable (Bryce et al. 2003). Historical data, reconstructions using dated sediment cores that integrate longer timescales (i.e., years to millennia) and modelling enable the geomorphic evolution of mangrove systems to be considered (Walsh & Nittrouer 2004, Swales et al. 2007b, J.C. Ellison 2008).

Sedimentation rates in temperate mangrove forests

Despite the importance of mangrove forests as sinks for terrigenous sediments, few detailed studies of sedimentation processes have been undertaken in modern mangrove systems (Woodroffe 1992, Walsh & Nittrouer 2004). By comparison, studies of mainly tropical autochthonous mangrove systems on sediment-starved oceanic low islands are more numerous (Lynch et al. 1989, Ellison & Stoddart 1991, Woodroffe 1992, Ellison 1993, Parkinson et al. 1994, Snedaker 1995, Cahoon & Lynch 1997, Valiela et al. 2001, Cahoon et al. 2006, McKee et al. 2007). This research interest in part reflects the vulnerability of these remote island mangrove systems to SLR associated with climate warming.

Sediment accumulation rates (SARs) have been estimated for temperate mangrove forests using a variety of techniques: marker beds (Chapman & Ronaldson 1958, Bird 1971, 1986), sedimentation pegs, artificial pneumatophores and plates (Spenceley 1977, Young & Harvey 1996, Coleman 1998, Perry 2007) and sediment-core geochronology. The radioisotopes lead-210 (^{210}Pb) and caesium-137 (^{137}Cs) have proved particularly useful to quantify SAR over annual-to-decadal timescales (Ellis et al. 2004, Rogers et al. 2005b, Perry 2007, Swales et al. 2007b). Radiocarbon (^{14}C) and pollen dating have also been used to estimate sedimentation rates in temperate mangrove forests over hundreds to thousands of years (Nichol et al. 2000, Carvalho do Amaral et al. 2006).

Short-term measurements (i.e., weeks to months) of sedimentation are unlikely to be representative of longer-term conditions in temperate mangrove forests. This is because large-scale ecosystem processes, such as mangrove forest development, occur over annual-to-decadal timescales. Morphological reconstructions of mangrove forest development based on long-term records provided by dated sediment cores, aerial photography and historical archives can overcome limitations of short-term studies (Dahdouh-Guebas & Koedam 2008, J.C. Ellison 2008). These data integrate the effects of biological and physical processes operating over annual-to-decadal timescales, which include the effects of episodic infrequent high-magnitude events, such as storms, that perturb the system. The temporal resolution of sediment cores is a potential limitation of these reconstructions, which depends mainly on sediment-mixing characteristics and SAR. Sediment-mixing effects become less pronounced as SAR increase (Valette-Silver 1993). High-resolution

Table 10 Summary of sedimentation rates in modern temperate mangrove forests from short- and long-term measurements (approximately last 100 yr)

Location	Environment	SAR (mm yr^{-1})	Method	Duration	Reference (species)
Waitemata Harbour, Auckland, New Zealand	Basin	~1.7	mb	14 mo	Chapman & Ronaldson (1958) (*A. marina* subsp. *australasica*)
Westernport Bay, Victoria, Australia	Fringe	1.3–15	mb	3 yr	Bird (1986) (*A. marina* subsp. *australasica*)
	Basin	3–8			
	High basin	0.7–2.3			
Firth of Thames, New Zealand	Fringe/basin	16–100	ap	3 mo	Young & Harvey (1996) (*A. marina* subsp. *australasica*)
		5.5	pegs	5 mo	
		6.4	tfs	~50 yr	
Pakuranga Creek, New Zealand	Tidal creek	3.5–26	^{137}Cs	~35 yr	Swales et al. (2002) (*A. marina* subsp. *australasica*)
		3.3–33	pollen		
Whitford Bay, New Zealand	Tidal creek	1.1–23	^{137}Cs	~50 yr	Ellis et al. (2004) (*A. marina* subsp. *australasica*)
Westernport Bay, Victoria, Australia	Fringe/basin	5–9	mb	3 yr	Rogers et al. (2005b) (*A. marina* subsp. *australasica*)
	Mangrove in salt marsh	1.4–2.5	^{210}Pb	~100 yr	
Bayou Larfourche, Louisiana, USA	Basin with salt marsh	0–17.7	mb	1 yr	Perry (2007) (*A. germinans*)
		5.3	^{137}Cs	~40 yr	
Firth of Thames, New Zealand	Fringe	30–100	^{210}Pb	~50 yr	Swales et al. (2007) (*A. marina* subsp. *australasica*)
	High basin	7–12	^{210}Pb		
	Average	21–36	^{137}Cs		

Note: Types of environment generally follow the classification of Lugo et al. (1976). High basin denotes mangroves near the upper-tidal limit. Sediment accumulation rates (SARs) from radioisotopes are time-averaged values and average annual values from upscaling of short-term (<1 yr) rates. Methods: marker beds (mb); artificial pneumatophores (ap); measuring pegs (pegs); pollen dating (pollen); tidal flat slope and historical aerial photography (tfs); caesium-137 (^{137}Cs); lead-210 (^{210}Pb). Radioisotope SARs are time-averaged values.

sediment cores are preserved in modern mangrove forests that are rapidly accumulating terrigenous sediments (i.e., 10–100 mm yr^{-1}) near large rivers and in systems with enhanced sediment supply following catchment disturbance (Allison et al. 1995, Augustinus 1995, Swales et al. 2002, 2007b, Walsh & Nittrouer 2004).

Table 10 summarises sedimentation rates in modern temperate mangrove forests and includes data from short-term experiments as well as longer-term SAR estimates derived from cores. With one exception (Louisiana; Perry 2007), these studies solely relate to the estuarine temperate mangrove forests (*Avicennia marina* subsp. *australasica*) of northern New Zealand and south-eastern Australia. Estimates of short-term sedimentation rates (0–18 mm yr^{-1}) over 1- to 3-yr timescales integrate some seasonal variability. Longer-term SARs derived from dated cores indicate differences based on geomorphic setting, system maturity or sediment supply. In relatively sediment-poor systems, sedimentation rates in mangrove forests are of the order of several millimetres per year (Rogers et al. 2005b, Perry 2007). High SAR measured in New Zealand mangrove forests (10–100 mm yr^{-1}) reflect large terrigenous sediment supply or close proximity to catchment outlets (e.g., tidal creeks). Sedimentation rates decrease over time as mangrove forests mature due to a progressive reduction in hydroperiod and sediment delivery (Swales & Bentley 2008). Patterns and long-term rates of sedimentation in tropical mangrove forests are similar to those reported for temperate systems (Lynch et al. 1989, Cahoon & Lynch 1997, Smoak & Patchineelam 1999, Walsh & Nittrouer 2004, Van Santen et al. 2007, Kamaruzzaman & Ong 2008).

The ecological significance of sedimentation in temperate mangrove systems primarily relates to the process of tidal flat accretion as a precursor to mangrove habitat expansion (Bird 1986, Bedin

2001, Roy et al. 2001, Swales et al. 2007b) and maintenance of surface elevation relative to sea level (Cahoon et al. 2006, McKee et al. 2007). The delivery of nutrients associated with sediment inputs to mangrove forests is a secondary factor (Saintilan & Williams 1999, Lovelock et al. 2007a). Mangroves and saltmarsh systems are particularly vulnerable to the effects of relative SLR because they occupy a relatively narrow elevation zone in the mid- to upper intertidal zone. The vulnerability of mangrove forests to inundation by rising sea levels has been evaluated based on long-term SAR data, with forest stability assumed to depend on sedimentation keeping pace with relative SLR (Parkinson et al. 1994, Alongi 2008, J.C. Ellison 2008). However, this approach does not account for the effects of *in situ* processes such as sediment compaction and subsurface root production, which also influence local surface elevation changes in mangrove forests (Cahoon et al. 2006, McKee et al. 2007).

The surface elevation table (SET) and marker bed technique (Boumans & Day 1993, Cahoon et al. 1995) enable substratum elevation changes to be apportioned to sedimentation and subsurface processes. This method has been applied to temperate (Rogers et al. 2005a,b, 2006, Rogers & Saintilan 2008, Stokes et al. 2009) and tropical mangrove forests (Cahoon & Lynch 1997, McKee et al. 2007). Cahoon et al. (2006) analysed SET data from a wide range of temperate and tropical mangrove forests in a range of geomorphic settings characterised by terrigenous and organic sediments. Sedimentation rates exceeded relative SLR at all sites, but net elevation changes were mostly negative due to shallow subsidence. Similar trends were observed in *Avicennia* and *Rhizophora* mangrove forests. Subsidence was also highest at sites with the highest sedimentation rates. These types of studies have highlighted the complex biotic and physical feedbacks that drive surface elevation dynamics in mangrove forests. However, records of surface elevation changes are relatively short and as yet do not provide a long-term perspective. Swales and Bentley (2008) reconstructed the morphodynamics of an *Avicennia marina* forest over the prior 50 yr based on high-resolution ^{210}Pb profiles preserved in low-permeability, rapidly depositing muds. The effects of sediment compaction were negligible, and SAR could be used to estimate surface elevation changes. Ultimately, the time resolution of stratigraphic records declines due to sediment compaction and lithification (J.C. Ellison 2008). However, under high-SAR regimes it would appear that detailed reconstructions of modern mangrove forest development can be obtained from sediment cores (Walsh & Nittrouer 2004, Swales & Bentley 2008).

Capacity of mangroves to mitigate coastal hazards

The presence of mangrove forests along low-lying shorelines has been credited with the mitigation of erosion and inundation hazards caused by local wind and swell waves (Othman 1994, Brinkman et al. 1997, Massel 1999, Massel et al. 1999, Phuoc & Massel 2006, Quartel et al. 2007, Alongi 2008), storms (Granek & Ruttenberg 2007, Alongi 2008) and tsunami (Harada & Kawata 2004, Dahdouh-Guebas et al. 2005, Danielsen et al 2005, Kathiresan & Rajendran 2005, Alongi 2008).

Mangrove forests are effective at attenuating the heights of wind waves and swell (wave periods < 30 s) due to vegetation-induced drag. Brinkman et al. (1997) measured a 25–50% reduction in wave energy within about 200 m of the seaward edge of a *Rhizophora* forest. Wave attenuation was also pronounced in a mixed *Avicennia* and *Rhizophora* forest, with a 50–70% reduction in wave energy occurring within 20 m of the forest edge (Phuoc & Massel 2006). The capacity of mangrove forests to attenuate waves depends on tree density, tree morphology, bed slope, spectral characteristics of the waves and water depth (Massel et al. 1999, Alongi 2008). The relative effect of these factors varies with mangrove species. For example, the drag induced by mangroves varies with water depth due to vertical differences in the submerged projected area of trees. The drag profile of *Avicennia* sp., with their peg-like pneumatophores, differs markedly from *Rhizophora* sp., with their large prop roots, so that wave attenuation by *Rhizophora* is more effective when the prop roots are partially submerged (Massel et al. 1999). Unlike *Rhizophora* sp., *Kandelia* sp. lack prop roots or pneumatophores, so that the largest flow obstruction is the leaf canopy, and wave attenuation is less effective when the canopy is exposed (Quartel et al. 2007).

Although mangrove forests effectively attenuate wind and swell waves and thereby protect soft-sediment shorelines from erosion, this may not be the case during severe storms (Massel et al. 1999, Alongi 2008). The erosive capacity of storm waves is much greater due to the quadratic increase in total energy with wave height. Penetration of storm waves into shallow-water coastal and estuarine environments is enhanced by elevated sea levels (storm tides) related to meteorological drivers (i.e., inverse barometric effect and wind setup) and spring tides. Important attributes of mangrove forests that mitigate storm erosion and inundation hazards include forest width, degree of sediment compaction, tree density and tree morphology (height, root structure, ratio of above- to below-ground biomass) (Alongi 2008). The capacity of mangrove trees to withstand bed erosion varies between species largely due to differences in root structure. For example, *Avicennia* spp. are anchored by a radial network of horizontal cable roots and vertical anchor roots within about the top 0.5 m of the substratum (Janssen-Stelder et al. 2002). Mudflat erosion during storms can detach these shallow-rooted mangrove trees, resulting in tree loss on the exposed seaward edge of mangrove forest (Othman 1994, Swales et al. 2007a). Bed erosion will also depend on the degree of substratum compaction, so that substrata in older forest stands will be more resistant to wave erosion (Othman 1994). Mangrove forests that occupy a wide band across the intertidal zone will also be more resilient to damage by episodic storms and may also reduce the risk of inundation of low-lying coastal land by protecting coastal defence structures from direct wave attack and dynamic run-up (Swales et al. 2007a).

The capacity of mangrove forests to mitigate tsunami hazard will depend on the tsunami characteristics, environmental setting and mangrove forest characteristics. Tsunami characteristics include source type (tectonic, volcanic, subsidence, underwater landslide); ocean volume displaced; size and speed; distance from source (remote, regional, local) and approach angle to the coast (de Lange 2003, Alongi 2008). The transformation of tsunamis as they propagate into the near shore will also differ markedly between open coasts and estuarine environments due to differences in shoreline bathymetry (i.e., seabed slope) and estuary shape, tidal volume, intertidal flat area and elevation. Tsunamis have unique hydraulic characteristics due to their much longer periods (0.1–1 h; de Lange 2003). Tsunamis propagate into estuaries in a similar way to tidal bores, with a large water mass behind the wave crest and momentum increasing as shoaling occurs (de Lange 2003, Alongi 2008). Transformations of a tsunami in estuaries are largely controlled by the tidal inlet characteristics, basin shape and volume, proportion of intertidal area, bed elevation and slope. For example, amplification of a tsunami in funnel-shaped estuaries occurs due to landward reduction in basin volume, whereas substantial attenuation in estuaries with extensive intertidal flats is due to friction.

Attributes of mangrove forests that determine their value in tsunami hazard mitigation are similar to those described for storm waves. The degree of forest modification by humans may also be a factor (Dahdouh-Guebas et al. 2005, Danielsen et al. 2005, Kathiresan & Rajendran 2005). In the aftermath of the 2004 Indian Ocean tsunami, several thousand hectares of tropical mangrove forest were uprooted due to bed erosion or died due to prolonged inundation (Alongi 2008). Mangrove species with wide prop roots or knee roots, such as *Rhizophora* and *Bruguiera* sp., withstood the tsunami surge, and forests with *Rhizophora* along their seaward fringes experienced less damage. Mangrove species with shallow, subsurface cable roots, such as *Avicennia* sp., were less effective at withstanding the tsunami (Kathiresan & Rajendran 2005). Modelling of tsunami flow attenuation by mangroves also indicated that species-related differences in drag coefficient are important (Tanaka et al. 2007, Alongi 2008).

Temperate mangrove forests are composed of relatively few species (dominated by *A. marina*), so that their capacity to mitigate tsunami hazard will be less than reported for tropical mangroves. Monospecific stands of *A. marina* occur in northern New Zealand, Victoria, South Australia and

South Africa's Eastern Cape Province, which are susceptible to bed erosion and will provide minimal protection in many cases. In northern and central NSW and South Africa, mangrove forests include more robust *Rhizophora* and *Bruguiera* species. These mixed-species forests are likely to be more effective at mitigating tsunami hazard (Alongi 2008). However, the protective function and resilience of *Avicennia marina* will be enhanced where forests occupy wide intertidal zones that have the capacity to provide an erosion buffer as well as attenuate the tsunami flow. The recovery time of mangrove forests from large-scale disturbances will vary from years to decades depending on event magnitude (Alongi 2008). Thus, large forests will generally have the highest capacity to mitigate storm and tsunami hazards.

The role of mangroves in the ageing of temperate estuaries

The role of biology in long-term landscape evolution is an emerging theme in geomorphology. The concept of 'biomorphodynamics' encapsulates the two-way couplings between biological processes and the physical processes of sediment transport and morphological evolution (Murray et al. 2009). Dietrich & Perron (2006) posed the following question in relation to landscape evolution: "If life had not arisen, would the tectonic and climatic processes that drive uplift and erosion of landscapes be significantly different?" In the context of the present review, it also is apparent that the physical processes that control sediment transport are fundamental drivers of temperate mangrove ecology. Here, we ask the more specific question: What is the role of temperate mangroves in the natural ageing of estuaries?

As has already been seen, estuaries are long-term sinks for terrigenous and marine sediments. How rapidly estuaries have infilled or 'aged' has mainly depended on the original shape and volume of their tidal basins, the rate of sediment supply and their ability to flush sediment to the sea. These factors largely explain observed regional differences in estuarine maturity (Roy et al. 2001).

The geomorphic evolution of estuaries exerts a direct control on temperate mangrove ecology through the process of sedimentation and tidal flat development. Mangroves colonise intertidal flats only after they have become "ecologically suitable" (Bird 1986). More specifically, the physical process of estuarine sedimentation builds intertidal flats that provide potential habitat for mangroves. On intertidal flats in temperate regions, a lower elevation threshold for growth of mangrove forests occurs at about MSL. This threshold primarily relates to the physiological requirement of mangroves, particularly seedlings, for regular emersion (Clarke & Hannon 1970). Although tidal flat accretion provides the opportunity for mangrove forest development or expansion, the actual pattern of seedling recruitment also depends on physical factors, such as wave energy at the site and dispersal of propagules by currents, in addition to biotic factors, such as predation of propagules (Clarke & Kerrigan 2002, and see discussion on p. 124).

Thus, rapid expansion of temperate and tropical mangrove forests has occurred mainly in estuaries and deltas with large terrigenous sediment supply and rapidly accreting intertidal flats (Thom 1967, Panapitukkul et al. 1998, Bedin 2001, Walsh & Nittrouer 2004, Carvalho do Amaral et al. 2006, Swales et al. 2007b). These patterns of mangrove forest expansion are also consistent with long-term reconstructions of estuarine development over centuries or millennia. These reconstructions showed that, as estuaries infill with fluvial sediments, tidal flats build seaward and mangroves initially colonise tidal flats in the upper reaches of the estuary. As the accommodation space is filled, tidal flats expand into the central basin, where the most extensive mangrove forests occur. Eventually, the estuary becomes largely intertidal, and mangrove forests occupy the entire tidal flat area save for areas drained by narrow, sinuous, fluvial channels (Swales et al. 1997, 2002, Roy et al. 2001, Hume 2003, Carvalho do Amaral et al. 2006).

Humans have accelerated the natural process of estuary infilling by increasing terrigenous sediment delivery to estuaries. Average SARs are typically an order of magnitude higher today than prior to catchment deforestation (Wasson & Galloway 1986, Martin 1987, Wasson 1994, Wilmshurst

1997, Morant & Quinn 1999, Prosser et al. 2001, Erskine et al. 2002, Healy 2002, Swales et al. 2002, Kakembo & Rowntree 2003, Olley & Wasson 2003, Adams et al. 2004, Foster et al. 2009). Estuarine sedimentation has also been exacerbated by engineering structures that reduce the sediment accommodation space (e.g., reclamations) and restrict tidal flows and wave fetch (e.g., causeways). It can be argued that the accelerated ageing of estuaries due to human activities largely explains the rapid expansion of temperate mangrove forests that has occurred in many New Zealand, south-eastern Australian and South African estuaries.

Are temperate mangrove forests therefore merely a symptom of estuarine maturity rather than agents of geomorphic change, or do they exert other, subtle effects on the natural process of estuarine ageing? Mangrove forests are highly effective at dampening currents (e.g., Wolanski et al. 1980, Furukawa et al. 1997, Mazda & Wolanski 2009) and attenuating waves (e.g., Brinkman et al. 1997, Massel et al. 1999, Phuoc & Massel 2006), so that they preferentially accumulate suspended fine particles and flocs composed of silt and clay particles (Woodroffe 1992, Kathiresan & Bingham 2001, Swales et al. 2002, 2007b, Ellis et al. 2004, Walsh & Nittrouer 2004) delivered by river plumes and tidal currents or resuspended from tidal flats. In doing so, temperate mangrove forests may mitigate some of the adverse effects of fine terrigenous sediments on estuarine and coastal ecosystems (Thrush et al. 2004). For instance, this sequestration function may add to the fine-sediment budget of large, infilled, wave-dominated estuaries (e.g., Swales et al. 2007b) that would otherwise export a large fraction of their terrigenous sediment load to the coast. In doing so, the mangrove forests enhance the sediment-trapping function of estuaries, even in relatively mature systems. Ultimately, this sediment trapping will be limited by the available accommodation space for sediment in an estuary, with sedimentation rates proportional to relative SLR (Woodroffe & Davies, 2009).

Patterns and causes of changes in the distribution of temperate mangroves

The rapid deforestation of terrestrial ecosystems over the last century has been mirrored in coastal marine ecosystems with large-scale loss of mangrove habitat. An estimated 75% of the world's tropical coasts were originally fringed by mangrove forests compared with 25% today. About one-third of this deforestation has occurred since the 1980s (Chapman 1976, Rönnbäck et al. 1999, Valiela et al. 2001). The current global rate of mangrove habitat destruction is estimated at 2.1%, which exceeds the rate for tropical rainforest. Major causes of tropical mangrove habitat loss associated with human activities include reclamations for industrial, residential and tourism developments, aquaculture and salt production (United Nations Environment Program, World Conservation Monitoring Centre [UNEP-WCMC] 2006). Despite this global trend of net mangrove habitat loss, expansion of tropical mangrove forests has occurred in some systems, particularly in forests growing on river deltas (Neil 1998, Panapitukkul et al. 1998, Lebigre 1999, Lacerda et al. 2007).

Historical loss of temperate mangroves in New Zealand and south-eastern Australian estuaries has followed the construction of causeways and structures that restrict tidal flows or elevate water levels, marina development, stock grazing, stormwater pollution, dredging, reclamations for ports, waste landfills, airports, agriculture, industrial and urban development, harvesting for timber and charcoal manufacture and excessive sedimentation (Chapman & Ronaldson 1958, Nature Conservation Council 1984, Thorogood 1985, Bird 1986, Crisp et al. 1990, Thomas 1993, Williams & Watford 1997, Wright et al. 1997, Steinke 1999, Adams et al. 2004). Large-scale loss of mangrove habitat occurred in New Zealand and south-eastern Australia prior to the 1960s (Glanville 1947, Morrisey et al. 2007). The introduction of new legislation, such as New Zealand's Harbours Amendment Act 1977, made it illegal to reclaim seabed for agricultural purposes, so that mangrove habitat loss has substantially reduced since that time. More recently, New Zealand coastal communities have lobbied regional and central government agencies for the controlled removal of mangroves to restore tidal flat habitats and recreational amenities. In other estuaries, mangrove removal has occurred illegally (Morrisey et al. 2007). In South Africa, extensive mangrove habitat loss during the 1960s–1970s was related to harbour development, bridge construction and land development impacts. Natural events such as lagoon-mouth closures and tidal flat erosion by storm flows in river-dominated estuaries have also resulted in loss of mangrove habitat (Cooper et al. 1999, Steinke 1999). Long-term habitat loss has occurred in Bermuda due to relative increases in sea level that have outpaced peat sediment accretion in mangrove stands (Ellison 1993). The total extent of temperate mangrove habitat loss in New Zealand, Australia and South Africa cannot be accurately quantified because much of this loss occurred before aerial photographic surveys began in the 1930s.

Despite these historical losses, increases in temperate mangrove distribution have occurred in New Zealand, south-eastern Australian (*Avicennia marina*) and South African (*A. marina*, *Bruguiera gymnorrhiza*, *Rhizophora mucronata*) estuaries during the last 50–70 years. Mangrove colonisation of tidal flats as well as encroachment into saltmarsh habitats has been documented (Burton 1982, Burns & Ogden 1985, Mitchell & Adam 1989, Coleman 1998, Creese et al. 1998, Saintilan 1998, Saintilan & Hashimoto 1999, Saintilan & Williams 1999, Bedin 2001, Evans & Williams 2001, Saintilan & Wilton 2001, Wilton 2001, Adams et al. 2004, Ellis et al., 2004, Rogers et al. 2005b, Morrisey et al. 2007, Swales et al. 2007a,b, 2009). The black mangrove (*Avicennia germinans*) has also re-established in the *Spartina* salt marshes of Louisiana following dieback due to severe winter frosts in the late 1980s (Perry 2007, Saintilan et al. 2009).

Historical records also indicate a longer-term pattern of mangrove habitat expansion in south-eastern Australian estuaries since the time of European settlement (McLoughlin 1987, Saintilan & Williams 1999, McLoughlin 2000). In New Zealand, estuaries with relatively large catchments infilled with terrigenous sediment and had been colonised by mangroves by the early twentieth century. This process of estuary infilling was accelerated by large-scale catchment deforestation in the mid- to late 1800s (e.g., Swales et al. 1997, 2002). Extensive areas of mature mangrove forest, infrequently inundated by high spring tides, had developed in the Kaipara Harbour (Northland) before the 1920s (Ferrar 1934). However, in other estuaries with large sediment supply, such as the Firth of Thames, mangrove habitat expansion has only occurred in recent decades. Captain James Cook, who visited the Firth in 1769, recorded that mangroves were present on the delta of the Waihou River (Beaglehole 1968). Aerial photographs from the 1940s showed that these deltaic mangrove forests had not substantially increased their distribution since Cook's time. Large-scale mangrove habitat expansion did not occur until the 1950s (Swales et al. 2007b).

Mangrove habitat expansion

Causes of expansion

Recent expansion of temperate mangrove habitat observed in New Zealand and south-eastern Australian estuaries has been attributed to estuary infilling and vertical accretion of tidal flats (Chapman & Ronaldson 1958, Bird 1986, Young & Harvey 1996, Creese et al. 1998, Saintilan & Williams 1999, Ellis et al. 2004, Swales et al., 2007b), increased nutrient inputs (Saintilan & Williams 1999), climate warming (Burns & Ogden 1985), changes in relative sea level due to sedimentation or subsidence (Burton 1982, Rogers et al. 2005a,b, 2006) or a combination of all or some of the these factors. Seaward expansion of mangrove habitat onto accreting intertidal flats is a feature of New Zealand estuaries (Burns & Ogden 1985, Deng et al. 2004, Morrisey et al. 2007, Swales et al. 2007a), whereas landward encroachment into salt marshes is a notable trend in south-eastern Australian estuaries (Saintilan & Williams 1999, 2000, Wilton 2002, Williams & Meehan 2004). Mangrove encroachment of salt marshes in NSW and Victoria is correlated with declines in tidal flat surface elevations and resulting increases in tidal inundation (Rogers et al. 2005b, 2006). Tidal creeks and drainage channels provide conduits for mangrove propagules to penetrate into saltmarsh habitats (Saintilan & Williams 1999, Rogers et al. 2005b). Mangrove encroachment into salt marsh is not well documented in New Zealand estuaries (Morrisey et al. 2007). Temperate mangroves have also increased their distribution in some South Africa estuaries in recent decades. This change has been attributed to increased sediment delivery to estuaries and tidal flat accretion, although increases in mangrove habitat have been modest in most cases. Future increases in mangrove habitat are also likely to be constrained by the limited extent of suitable habitat, cyclic erosion of river-dominated estuaries and limited number of permanently open barrier estuaries (Wright et al. 1997, Cooper et al. 1999, Steinke 1999, Bedin 2001, Adams et al. 2004). In southern Brazil (e.g., Baia da Iiha Grande), mudflat accretion and seaward progradation of mangrove forests has been limited by low sediment supply (Vann 1980).

Patterns and extent of expansion

Table 9 summarises information on changes in the extent of mangrove habitat in New Zealand, south-eastern Australian and South African estuaries compiled from journal articles and published reports. The studies included data from small mangrove stands as well as large forests (10^0–10^3 ha) and major estuary types (i.e., drowned river valleys, barriers, embayments and coastal lagoons). Mangrove habitat mapping is primarily based on analysis of time series of aerial photographs taken since the 1940s–1950s. The majority of studies came from New Zealand and south-eastern Australia and related to monospecific stands of the *Avicennia marina* subsp. *australasica*. South African studies included mixed-species forests, although these are also predominantly composed of *A. marina* stands. The reliability of the habitat change data varied from study to study due to the methods employed, image scale and quality, habitat classification and accuracy of habitat boundary digitization. The historical periods covered by these studies also varied in length and timing. Few, if any, of the studies satisfied recent protocols for mangrove habitat mapping (Wilton & Saintilan 2000). Although the dataset is not exhaustive, it is sufficient to enable the general patterns and scale of changes in temperate mangrove habitat over the last several decades to be identified.

Rate of mangrove habitat expansion in south-eastern Australian estuaries (average 2.1% yr^{-1}, range 0.7–9.1% yr^{-1}) are substantially lower and less variable than occur in New Zealand estuaries (average 4.1% yr^{-1}, range −0.2 to 20.2% yr^{-1}; Table 9). In South Africa's Eastern Cape Province, the total area of mangrove habitat in estuaries was relatively stable during the period 1982–1999 (−0.02% yr^{-1}). This statistic masks a pattern of habitat expansion and declines for individual estuaries (Adams et al. 2004). Many of these mangrove forests are also small, with 60% being less than 10 ha in size (Table 9). In Kwazulu-Natal, rapid mangrove habitat expansion (5.9% yr^{-1}) in Umhlatuze Estuary (Richards Bay) during 1974–1999 followed the construction of a berm to reduce sedimentation in the port. The berm isolated the Umhlatuze River from the northern half of the estuary and initiated a sequence of large-scale deltaic sedimentation, rapid vertical tidal flat accretion and mangrove habitat expansion in the estuary. The Umhlatuze forest represents about 50% of the temperate mangrove habitat in South Africa.

In northern New Zealand, rapid mangrove habitat expansion in many estuaries reflects high rates of sediment delivery from relatively large catchments. For example, the Wairoa Estuary (high-tide area 3 km^2) represents a mature end member of drowned river valley estuaries on Auckland's eastern coast. The sediment load delivered by its 311-km^2 steep-land catchment had formed extensive intertidal flats that were colonised by mangroves before the 1940s (Swales et al. 2009). A regional study of Auckland estuaries confirmed that the largest increases in mangrove habitat over the last 50–60 years have occurred in the smallest (i.e., <5-km^2) systems (Swales et al. 2009). During the same time, there were virtually no increases in mangrove habitat in the largest estuaries, such as the 65-km^2 Waitemata Harbour, which accounts for 30% of the 2700 ha of mangrove habitat in these estuaries today. In fact, substantial habitat loss (−8%; Table 9) occurred in the Waitemata

Harbour due to reclamations associated with motorway construction, industrial development and refuse landfills in the 1950s–1970s.

Controls on expansion

Temperate mangrove habitats in New Zealand, south-eastern Australia and South Africa are dominated by the genus *Avicennia*, which displays the widest tolerance to environmental factors, such as water salinity, low air and water temperatures, frost frequency and severity and day length (Chapman 1976, Duke 1990, Augustinus 1995, Stuart et al. 2007, Krauss et al. 2008, Woodroffe & Davies 2009). Within its latitudinal range, factors controlling mangrove habitat expansion primarily relate to tidal flat elevation and seedling dispersal, establishment and survival. In this section, the physical processes that influence the spatial distribution of temperate mangroves within estuaries are reviewed.

Tidal amplitude and intertidal flat slope exert first-order controls on the spatial distribution and extent of intertidal flat habitat above MSL elevation. Tides in northern New Zealand are mesotidal (range 2–4 m), whereas those in south-eastern Australia and along South Africa's eastern coast are microtidal (range <2 m) (Bird 1984, Schumann et al. 1999, de Lange et al. 2003). The most extensive intertidal flats suitable for mangroves occur in mature estuaries with large tidal amplitudes and terrigenous sediment supply, such as occur in northern New Zealand. The relationship between relative catchment size and degree of estuary infilling has been demonstrated for the large drowned valley systems of NSW (Roy et al. 2001) and the drowned river valley and barrier estuaries of Auckland's eastern coast (Swales et al. 2009). In the New Zealand study, estuaries with relatively large catchments, which are most at risk of infilling, have also experienced the largest increases in mangrove habitat over the last 50 years (Swales et al. 2009).

Temperate mangrove species occupy the mid- to upper intertidal zone between high tide and MSL elevation (Galloway 1982, Bird 1986, Clarke & Myerscough 1993, Ellison 1993, Steinke 1999, Schaeffer-Novelli et al. 2002), although the exact elevation limits of temperate mangroves species have not been adequately determined (Clarke & Myerscough 1993, Ellison 2009). An important distinction to make in evaluating information on the lower elevation of temperate mangrove forests is between dynamic sea level and MSL attributed to a fixed vertical datum. The latter is typically defined on the basis of historical sea-level records from tide gauges. For example, the fixed MSL datum at Auckland (Auckland Vertical Datum 1946, AVD-46) is based on sea-level measurements during the 1920s–1930s. Likewise, the Australian Height Datum (AHD) is based on MSL measured during the period 1966–1968. By contrast, the dynamic or actual MSL is a varying level that accounts for the effects of long-period fluctuations in sea level. These include the annual heating and cooling cycle, multiyear El Niño-Southern Oscillation (ENSO) cycle, longer-term 20- to 30-yr Interdecadal Pacific Oscillation (IPO) and progressive SLR associated with climate warming. These annual–decadal cycles result in decimetre-scale variations in MSL from year to year (Goring & Bell 1999, Hannah 2004) and in doing so may influence seedling recruitment in a given year by elevating or depressing sea levels. In the long term, SLR associated with climate warming also has ecological significance because of the potential for mangrove habitat loss. For example, relative SLR at Auckland has averaged 1.4 mm yr^{-1} over the last century (Hannah 2004), so that actual MSL today is now about 0.13 m higher than in the 1920s–1930s. This increase in relative SLR translates into a potential 3–250 m landward retreat of the MSL elevation contour for the range of tidal flat slopes in Auckland estuaries. This predicted retreat does not take into account increases in tidal flat elevations due to estuary sedimentation, which has averaged 3.8 mm yr^{-1} over the last 50 years (Swales et al. 2009). The relative importance of short-term sea-level variations and progressive long-term relative SLR on the LEL of present-day temperate mangrove forests is not known.

The LEL for temperate mangroves at about MSL elevation primarily reflects the fact that mangrove seedlings are intolerant to continuous submersion (Clarke & Hannon 1970). Regular daily exposure enables seedlings to maintain an adequate oxygen supply to their tissues, particularly as mangroves typically grow in anaerobic muddy substrata. Laboratory experiments to determine the influence of submergence on the growth and development of *Avicennia marina* seedling indicated a statistically significant effect of inundation time on root mass (Curran et al. 1986, Hovenden et al. 1995). Although aerobic respiration in roots can be maintained throughout a 6-h period of tidal inundation (Curran et al. 1986), a large reduction in root growth rates occurred for inundation periods in excess of about 4 h (Hovenden et al. 1995). Semidiurnal tides with periods of 12.4 h prevail in northern New Zealand, south-eastern Australia and South Africa's eastern coast (Schumann et al. 1999, de Lange et al. 2003), so that tidal flats above MSL are submerged for no more than about 6 h per tidal cycle.

The distribution of *A. marina* was surveyed by Clarke and Myerscough (1993) along four transects in Botany Bay and Jervis Bay (NSW), with mangroves observed down to MSL (AHD). In Tauranga Harbour (Bay of Plenty, New Zealand), grey mangroves extend down to MSL in sheltered creeks and bays and 0.2 m above MSL on open tidal flats in the harbour (Park 2004). Swales et al. (2009) surveyed the LEL for *Avicennia* forests in terms of actual MSL (2007) along 81 transects in 17 Auckland estuaries. The average LEL for adult mangrove trees varied between estuaries (−0.05 to 0.76 m MSL), with an overall average of 0.35 m MSL. For seedlings recruited during the summer of 2007/2008, the LEL varied between −0.41 and 0.21 m MSL, with an overall average of −0.15 m MSL. Mapping of the MSL elevation contour from aerial photography using the waterline method (George & Gulliard 2003) also indicated that the seaward edge of mangrove forests occurred down to −0.33 m MSL in several estuaries (Swales et al. 2009).

In the fetch-limited, infilled estuaries that characterise temperate mangrove habitats, sediment transport on intertidal flats is largely driven by small short-period waves (Green et al. 1997, Roy et al. 2001, Hume 2003, Green & Coco 2007). Thus, these estuarine waves influence substratum stability, mangrove seedling recruitment and the lower elevation of mangroves on intertidal flats (Chapman & Ronaldson 1958, Clarke & Allaway 1993, Clarke & Myerscough 1993, Augustinus 1995, Osunkoya & Creese 1997, Park 2004, Swales et al. 2007b, 2009), although few studies have considered the ecological significance of these physical processes in detail.

Clarke & Myerscough (1993) studied *A. marina* propagule and seedling establishment in Botany Bay and Jervis Bay (NSW). Mangrove seedlings occurred above and below the elevation limits of adult trees. However, seedlings growing on the mudflats displayed substantially higher mortality after 2 yr than seedlings growing in the forest. Less than 5% of seedlings remained on the mudflats after 3.5 yr in comparison with 50% of seedlings in the mangrove forest. The low rate of seedling recruitment on the mudflats as well as the low density of seedlings on mudflats in other south-eastern Australian estuaries was attributed to "the mechanical effects of tides and currents on establishment". Osunkoya & Creese (1997) evaluated the survival and growth of self-established and transplanted *A. marina* seedlings in three New Zealand estuaries (Bay of Islands, Whangateau and Tauranga). Seedlings on lower intertidal flats had disproportionately (but not significantly) higher mortality rates after 18 mo, although higher growth rates, than seedlings on mid- to high intertidal flats. Seedling density declined with decreasing substratum elevation, which Osunkoya and Creese attributed to increased substratum erosion by tides and waves. The survival of recently settled *A. marina* propagules and seedlings on an energetic wave-exposed mudflat was also monitored by Swales et al. (2007b). Propagule and seedling numbers declined rapidly and exponentially, with fewer than 10% remaining after 2 mo. Seedling mortality was not significantly influenced by site or distance from the forest fringe, and surviving seedlings were actively growing. Loss of propagules and seedlings was due to frequent and deep erosion (~7 cm) of the mudflat by waves. Swales et al. (2007b) hypothesized that major recruitment events were likely to coincide with infrequent periods of calm weather lasting 3–5 wk, which is the time required for *A. marina* propagules to strike roots and establish (Clarke 1993). Apart from the realisation that waves influence mangrove seedling recruitment in estuaries, present understanding of this process is largely qualitative.

Regional-scale and long-term effects of wave exposure on mangrove seedling recruitment in estuaries were considered by Swales et al. (2009). Overall, mangrove forests occupied only 58% (2710 ha) of their potential habitat above actual MSL. Within individual estuaries, the area above MSL occupied by mangroves varied from 22% to 74% and was generally higher in small estuaries (high-tide area, $A < 5$ km^2) with large catchments ($A \geq 10$ km^2). Substantial increases in mangrove habitat have occurred in these estuaries since the 1950s (28–400%). However, the historical record suggests substantial time lags (i.e., decades) between tidal flat accretion and subsequent colonisation by mangroves. Modelling of long-term wave climate and sediment entrainment indicated that seedling establishment was more likely to occur in smallest estuaries ($A < 5$ km^2), which was consistent with the historical patterns of mangrove habitat expansion.

In contrast to the dominant role of physical processes influencing the lower elevation of temperate mangrove forests, factors influencing the upper elevation limit (UEL) are more varied. Maximum tide height represents a first-order control on the UEL. Tidal flows control propagule delivery to the upper intertidal zone (Saintilan & Williams 1999, Rogers et al. 2005b) and maintain substratum porewater salinity, preventing establishment of freshwater plants (Chapman 1976, Gillanders & Kingsford 2002, Mitsch et al. 2009). Porewater salinities in mangrove forest sediments are also enhanced by evapotranspiration, with salinities of about 50 or more being substantially higher than seawater (Galloway 1982, Swales et al. 2007b, Mitsch et al. 2009). In NSW estuaries, the upper limit of *A. marina* stands occurs at about mean high-water neap (MHWN) tide elevation. Mixed stands of *A. marina* and the river mangrove *A. corniculatum* occur at higher elevations in salt marsh (Clarke & Myerscough 1993). In Westernport Bay (Victoria), the upper limit of *A. marina* occurs between mean high-water spring (MHWS) and MHWN tide levels (Bird 1986). In the estuaries of northern New Zealand and South Africa's eastern coast, mangrove forests generally occur up to MHWS tide elevation (Chapman & Ronaldson 1958, Steinke 1999, Swales et al. 2007b). Mature forests of *A. marina* also occur in New Zealand estuaries above MHWS elevation, and episodic storm tides can substantially increase the hydroperiod in these forests (Swales et al. 2007b).

In south-eastern Australia, the pattern of mangrove encroachment into saltmarsh habitat is not ubiquitous, and in many estuaries the boundaries between saltmarsh and mangrove habitats have remained stable over the last 50 years or more (Saintilan & Williams 1999). Where landward expansion of mangroves has occurred, propagules have used tidal creeks and artificial drainage channels as conduits to enter saltmarsh habitats (Saintilan & Williams 1999, Rogers et al. 2005b). At elevated saltmarsh sites, *A. marina* propagules may fail to establish due to desiccation after settlement (Clarke & Myerscough 1993). In Westernport Bay (Victoria), mangrove encroachment was strongly correlated with decreases in saltmarsh surface elevation because mangroves generally establish at lower tidal elevations (Rogers et al. 2005b). Declines in marsh surface elevation coincided with El Niño drought conditions, as was also observed in NSW estuaries, so that groundwater recharge may play a significant role in controlling marsh surface elevation (Rogers et al. 2005a, 2006). The effects of progressive SLR on mangrove encroachment into salt marshes may also be exacerbated during droughts. However, subsidence rates in salt marshes and the exact causal mechanisms vary from site to site (Cahoon et al. 2006, Rogers et al. 2006).

In south-eastern Louisiana, the observed south-to-north expansion of *A. germinans* into *Spartina alterniflora* (cordgrass) salt marsh has coincided with two decades of mild winter air temperatures (Perry 2007, Saintilan et al. 2009). Mangrove and saltmarsh zonation differs from south-eastern Australia in that *Avicennia germinans* generally occupies the upper intertidal zone, whereas *Spartina* salt marsh occurs at lower elevations. Patterson et al. (1993) did greenhouse and field-transplant experiments to study factors controlling mangrove seedling establishment in *Spartina* salt marsh. The survival and growth of mangrove seedlings in salt marsh were reduced in comparison with the mangrove forest. Seedling mortality in the lower salt marsh followed submergence and sediment deposition on leaves. Sulphide concentrations were also about eight times higher in saltmarsh sediments. Patterson et al. (1993) attributed the lower success of mangrove seedling recruitment in salt marsh to less-than-optimal physicochemical conditions as well as competition. Where mangrove encroachment has occurred, substratum physicochemical conditions, long-term (^{137}Cs) sedimentation rates and above- and below-ground productivity are not significantly different from pure saltmarsh habitats. Environmental changes may emerge as larger mangrove stands and forests develop in these saltmarsh systems (Perry 2007).

In many estuaries, engineering structures restrict the landward expansion of mangrove forests by physical exclusion, restricting tidal flows and controlling water levels (Williams & Watford 1997). Structures that exclude mangroves include embankments, reclamations and sea walls (Burton 1982, Thorogood 1985, Coleman 1998, Saintilan 1998, Saintilan & Williams 1999, Steinke 1999, Bedin 2001, Wilton 2001, Rogers et al. 2005a, Swales et al. 2007b, 2009). Structures that restrict or enhance tidal flows or control water levels include causeways, navigation, drainage and access canals, floodgates, culverts and weirs (Thorogood 1985, Evans & Williams 2001, Rogers et al. 2005a, Perry 2007).

Future changes in the distribution of temperate mangroves

Environmental stressors

The historical record of temperate mangrove habitat loss is closely related to human activities in estuaries and their catchments. These activities include reclamation, modification of tidal flows and water levels associated with dredging and flow control structures, wood harvesting and sedimentation. Environmental legislation and management practices now curtail many of these earlier practices. However, the potential effects of climate warming and increasing human population in the coastal zone are likely to increase pressure on temperate mangrove habitats. Key stressors include catchment urbanisation, armouring of shorelines to mitigate erosion and inundation hazards associated with SLR, increasing stormwater and sewage contamination and sediment loads (Crisp et al. 1990, Morant & Quinn 1999, Steinke 1999, Morrisey et al. 2003).

The quality of temperate mangrove habitat is also likely to deteriorate as a consequence of future land-use intensification, particularly in estuaries downstream of urban and urbanising catchments. Environmental changes associated with urbanisation may occur gradually, and the ecological consequences are often poorly perceived (Pearce 1995). Over time, mangrove forests accumulate stormwater contaminants, such as trace metals, due to preferential trapping of fine sediments (Ellis et al. 2004). However, field and laboratory studies showed no clear link between trace metal concentrations and mangrove growth or mortality rates (Tam & Wong 1997, Wong et al. 1997a,b, Duke 2008). The effects of earlier management practices, such as reclamations, landfills and flow control structures, will also continue to degrade mangrove habitat. For example, landfill leachates may locally affect the diversity of benthic communities in mangrove forest sediments (Blom 1992). Herbicide accumulation in estuarine sediments may also adversely affect mangrove health, as documented for *Avicennia marina* seedlings and trees (Duke 2008). Despite the protected status of mangroves in New Zealand, illegal reclamations and structures, removal of trees to enhance visual and recreational amenity, dumping of waste and stock grazing still occur (Kronen 2001, Morrisey et al. 2007). Wood harvesting and stock grazing also continue to threaten mangroves in South Africa's Eastern Cape Province (Steinke 1999).

Mangroves are vulnerable to excessive and rapid sedimentation due to smothering of seedlings and the aerial roots of trees. Effects of sedimentation on mangroves include reduced vigour and growth rates and tree loss (Ellison 2009). Sedimentation induced die-off events have been associated with catchment floods, hurricanes, changes in deltaic sedimentation patterns, dredge spoil, reclamations and mining (Terrados et al. 1997, Ellison 1999, 2009). The sensitivity of mangrove species to burial depends on root structure because adult trees die when pneumatophores

(*Avicennia*) and knee roots (*Bruguiera*) are rapidly buried. Adverse effects are also more likely to occur in anaerobic muddy sediments than in sandy substrata. Death of *Avicennia* sp. trees occurs when sedimentation depths exceed 10 cm (Ellison 1999), although the burial rate also appears to be a determining factor. For example, large-scale seaward expansion of *A. marina* forests occurs on rapidly accreting (50–100 mm yr^{-1}) tidal flats (Swales et al. 2007b). In high-sedimentation environments, mangroves may respond by upward extension of pneumatophores or development of higher root arches or knee roots (Ellison 1999).

Climate change and sea-level rise

Changes in the distribution of tropical and temperate mangroves over geological timescales have been reconstructed from sedimentary records of mangrove peat deposits and pollen (Ellison 2009). Mangrove sediments can also provide reliable sea-level markers because they occupy a relatively narrow elevation band in the intertidal zone. At high latitudes, changes in their distribution inferred from sedimentary records also indicate range extension during periods of climate warming. Such retrospective studies also provide a basis to inform predictions of future changes in mangrove habitat distribution associated with the rapid climate warming and SLR that is expected to occur this century and beyond.

The long-term fate of mangrove forests depends on surface (i.e., substratum) elevations increasing at a rate equal to or exceeding SLR so that they maintain their position in the tidal frame (Cahoon et al. 2006, McKee et al. 2007, Alongi 2008, Woodroffe & Davies 2009). Mangrove forest response to sea-level fluctuations can be complex depending on local relative SLR (including vertical landmass movement), regional departures from global mean eustatic SLR, climate change effects on storm surges and waves, and tidal flat accretion and subsidence (Alongi 2008). Paleoenvironmental studies have been used to reconstruct how mangrove forests have responded to SLR over geological timescales (Woodroffe & Davies 2009). Early Holocene mangrove forests were inundated by rapid SLR (5–15 mm yr^{-1}), and the former existence of these forests is preserved in shelf sediments (Pocknall et al. 1989, Woodroffe 1990, McKee et al. 2007, J.C. Ellison 2008, Woodroffe & Davies 2009). Where rates of SLR were slightly lower, mangroves re-established at higher landward positions on the shelf. As the rate of SLR slowed, mangrove forests were able to keep pace with sea level. In the Southern Hemisphere, sea levels stabilised about 6500 years ago, with decimetre variations since that time (Gibb 1986, Roy et al. 2001, Ellison 2009). Temperate mangrove forests subsequently developed in estuaries as they progressively infilled with sediment.

The response of modern mangrove forests to historical sea-level changes has been commonly inferred from dated sediment cores, with SARs used as a surrogate measure of surface elevation change (Cahoon et al. 2006). For example, Alongi (2008) compared relative SLR data with mangrove forest SAR. The analysis indicated a linear relationship between SLR and SAR, with sedimentation exceeding SLR in most cases. Furthermore, sedimentation rates in mangrove forests with large terrigenous sediment supply have exceeded SLR by an order of magnitude or more (Walsh & Nittrouer 2004, Swales et al. 2007b). In south-eastern Australian estuaries, sedimentation rates in mangrove forests, measured over several years, generally exceeded rates of actual surface elevation change due to factors such as sediment compaction, root production and groundwater hydrology (Saintilan & Williams 1999, Rogers et al. 2006). Despite these local and regional differences in drivers, the general pattern of mangrove habitat expansion suggests that over the long term surface elevations in modern temperate forests have kept pace with rates of relative SLR of 0.9–1.4 mm yr^{-1} over the last century (Schumann et al. 1999, Hannah 2004, Rogers et al. 2005b, Saintilan & Williams, 1999). This is not the case in Bermuda, where long-term loss of temperate mangrove habitat has occurred. Rates of peat sediment accretion have not compensated for relative SLR of 2.8 mm yr^{-1} since the 1930s (Ellison 1993).

The effects of climate warming are now evident from increases in global average atmospheric and ocean temperatures over the last century, widespread glacier ice and snow melt and rising eustatic MSL (Pfeffer et al. 2008). These observed effects are very likely due to increases in anthropogenic greenhouse gas concentrations in the atmosphere (IPCC 2007). Global surface temperatures have risen by 0.74 ± 0.18°C during the last century, and the likely range of projected increase this century is 1.1–6.4°C for different emission scenarios. Global mean (eustatic) sea levels have increased by 1.7 ± 0.5 mm yr^{-1} during the twentieth century. Sea levels will continue to rise primarily due to thermal expansion effects and loss of ice sheets and glaciers on land. Global eustatic rates of SLR are projected to be between 2 and 6 mm yr^{-1} by the 2090s (2090–2099) relative to the 1980–1999 average for the basic set of six future emission scenarios (IPCC 2007). A further sea-level increase of 0.1–0.2 m would arise from increased ice flow from Greenland and Antarctica if its contribution grew linearly with global surface temperature change. This effect would increase the projected rate of SLR up to 8 mm yr^{-1} by the 2090s. Actual rates of SLR could be even higher if discharges from the Greenland and West Antarctic ice sheets occur more rapidly than projected, with a possible increase in sea level of 1 m (i.e., 10 mm yr^{-1}) or more by 2100 (Hansen 2007, Rahmstorf 2007, Pfeffer et al. 2008, Rignot et al. 2008). Furthermore, sea level will continue to rise for several centuries due to the long time lag in the response of the deep ocean to climate warming, as well as future greenhouse gas emissions (IPCC 2007).

The future effects of climate warming on temperate mangrove forests will vary due to local differences in climate, relative SLR, increases in water temperatures, tidal regimes, sediment supply and species-dependent tolerance to environmental stressors (e.g., Alongi 2008). Climate warming effects on temperate mangroves include changes in seedling recruitment and tree growth, maintenance of surface elevation relative to sea level and potential for habitat expansion to higher-latitude environments. Climate warming has the potential to alter the distribution of temperate mangroves because these systems are sensitive to climate-related drivers such as increasing air, ocean and estuary temperatures, reduced frost frequency, increased storm frequency or intensity, changing rainfall patterns, river discharges and terrigenous sediment and nutrient loads and accelerated SLR (Field 1995, McKee & Rooth 2008, Ellison 2009).

How will temperate mangrove forests respond to accelerated SLR during the twenty-first century? To avoid submergence, surface elevation increases in mangrove forests will need to equal future rates of relative SLR. Eustatic SLR rates of 2–8 mm yr^{-1} and potentially as high as about 10 mm yr^{-1} projected to the end of this century are within the low to midrange of SLR experienced during the early Holocene. The regional effects of tectonics, postglacial isostatic adjustments and sediment loading of continental margins will locally offset or exacerbate SLR. For example, average relative SLR of 1.4 mm yr^{-1} in northern New Zealand over the last century was about 0.45 mm yr^{-1} lower than the eustatic rate due to isostatic adjustment (Hannah 2004).

Mangrove forests that occupy oceanic low-relief islands with human habitation are most vulnerable to inundation by rising sea levels. Causal factors include subsidence, low rates of sediment supply and engineering structures that prevent landward migration of mangrove forests (Alongi 2008). Based on these criteria, the temperate mangrove stands of Bermuda are unlikely to survive in the long term, with accelerated SLR exacerbating the historical trend of forest loss (Ellison 1993). The temperate mangrove forests of Australasia and South Africa that exist in sediment-rich infilled estuaries are likely to be among the most resilient to the future effects of climate change (Schaeffer-Novelli et al. 2002, Alongi 2008). Terrigenous and marine sediments that have accumulated in estuaries during the Holocene provide a buffer against the potential inundation by rising sea levels. Mangrove forests in rapidly infilling meso–macrotidal drowned river valley estuaries are particularly resilient (Schaeffer-Novelli et al. 2002) and will provide refuges for temperate mangroves.

The relative vulnerability of temperate mangrove forests can be evaluated by considering the likelihood of intertidal flat inundation by rising sea levels using SAR as a proxy for surface elevation change. This approach recognises that estuary inundation is a precursor to mangrove forest retreat (Ellison 1993). Sedimentation data also provide a long-term perspective, and where indicated by uniform bulk-density profiles, sediment compaction effects are negligible (Hancock & Pietsch

2006, Swales et al. 2007b). Recent sedimentation rates in New Zealand and south-eastern Australian estuaries have typically averaged 1–5 mm yr^{-1} over the last 50–100 years (Barnett 1994, Chenhall et al. 1995, Hancock & Hunter 1999, Hancock 2000, Hancock et al. 2001, Jones & Chenhall 2001, Swales et al. 2002, 2009, Ellis et al. 2004, Rogers et al. 2005b, Hancock & Pietsch 2006). These sediment records suggest that increases in tidal flat elevation have outpaced historical rates of relative SLR of less than 1.5 mm yr^{-1} along the temperate mangrove coasts of northern New Zealand, south-eastern Australia and South Africa. This net increase in elevation is consistent with the seaward expansion of mangrove forests observed in New Zealand estuaries over the last 50–70 years. Estuarine sedimentation rates are similar to the range of eustatic SLR (2–8 mm yr^{-1}) projected to occur this century, so that temperate mangrove forests are likely to persist in most estuaries, assuming similar rates of sediment delivery to estuaries, which may not be valid if the trend towards more stringent controls on sediment run-off continues. Under present worst-case SLR scenarios, a 2- to 5-fold increase in estuary sedimentation rates would be required to keep pace with about a 10 mm yr^{-1} average rate of eustatic SLR or more if subsidence is occurring. Such an increase in sedimentation has not occurred since the early Holocene. The implied increase in sediment delivery to estuaries is unlikely to occur unless there are major shifts in rainfall patterns or land use that drastically increase catchment soil erosion. A large sediment deficit means that mangrove forests are likely to retreat landwards towards the end of this century and beyond as SLR continues (Swales et al. 2009). Temperate mangrove forests with low rates of sediment delivery (e.g., estuarine bays with small catchments) or high rates of sediment compaction are unlikely to be sustainable (Rogers et al. 2005b). High-sedimentation environments, such as tidal creeks (e.g., Swales et al. 2002), could provide refuge for mangroves.

In many estuaries, artificial structures such as embankments, reclamations, rock revetments and sea walls will limit the landward retreat of mangrove forests as sea levels rise. Similar restrictions will also occur on estuarine shorelines backed by steep bedrock cliffs. This process, referred to as *coastal squeeze* (French 1997), will affect mangrove forests in many New Zealand, south-eastern Australia and South African estuaries, with their long history of coastal development (Crisp et al. 1990, Williams & Watford 1997, Morant & Quinn 1999). Coastal squeeze will particularly affect highly modified urban estuaries and estuaries with low-lying shorelines that are already vulnerable to inundation by storm tides. The effects of coastal squeeze on temperate mangrove forests are already evident in some estuaries. For example, in the Firth of Thames (New Zealand), rapid vertical accretion of mangrove sediments seaward of a stop bank (levee) has raised tidal flat surface elevations well above land levels (Swales et al. 2007a).

Temperate mangrove forests occur at the latitudinal limits of mangroves globally, the most southern of which are composed of monospecific stands of *Avicennia marina*. The distribution of *Avicennia* appears to be constrained by its physiological limitations to low temperatures and freezing (Steinke & Naidoo 1991, Walbert 2002, Beard 2006, Stuart et al. 2007). Geological evidence indicates that temperate mangroves have extended beyond their latitudinal range in the past. For example, *Avicennia* pollen is preserved in early Holocene sediments at Poverty Bay (New Zealand) about 150 km south of their present-day extent and coincides with a period of warmer climate (Mildenhall & Brown 1987, Mildenhall 1994). Thus, future climate warming may also enable temperate mangroves to extend their present distribution.

Extension of temperate mangrove forests to higher latitudes also depends on propagule dispersal between estuaries and the suitability of intertidal habitats. The buoyancy of *Avicennia* propagules is limited to a few days after shedding of the pericarp on exposure to seawater (Steinke 1975, 1986, Burns 1982, de Lange & de Lange 1994). The hydrodynamic characteristics of an estuary, such as degree of tidal asymmetry, will also determine the distance and direction of net transport (de Lange & de Lange 1994) and thus the likelihood that propagules are exported to the open coast. The propagules of trees fringing tidal channels are also less likely to be stranded in the immediate vicinity of their parent tree (Clarke & Myerscough 1991). In south-eastern Australia, *Avicennia* propagules may be transported up 50 km along shore but most propagules stranded on beaches occur within 1 km of estuaries with *Avicennia* forests (Clarke 1993). In northern New Zealand, longshore dispersal of propagules is limited by the coastal current velocities and alongshore wave drift, so that net transport distances for propagules are no more than ~175 km over several days. Most viable propagules establish in the immediate vicinity of their parent tree (de Lange & de Lange 1994). In the Southern Hemisphere, the potential for temperate mangrove forests to extend their latitudinal range is also limited by the small number of estuaries south of their present distributions. In Louisiana, the potential for increases in *A. germinans* distribution to higher latitudes appears to be less dependent on propagule viability as propagules remain buoyant for long periods (Rabinowitz 1978). Viable *A. germinans* propagules stranded on beaches along the Gulf coast of Texas have been found many kilometres from potential source populations (Gunn & Dennis 1973). However, as observed for *A. marina*, most *A. germinans* propagules establish in the immediate vicinity of their parent trees, with a small fraction dispersing large distances (Sousa et al. 2007). This limited dispersal pattern of propagules is consistent with the considerable genetic variability observed between local *Avicennia* populations (Duke et al. 1998, Dodd et al. 2002).

Management of temperate mangroves

Background

Attitudes towards mangroves in the temperate regions considered in this review have undergone a number of changes, principally since the arrival of Europeans. Pre-European inhabitants generally held the mangrove, and its associated habitat, in high regard as a source of food, fuels and medicines (Crisp et al. 1990, Tomlinson 1986). Following European settlement, however, mangroves were generally considered of little use and with little in the way of aesthetic or economic value. More recently, the value of mangroves and the services that they provide has again been recognised (Duke et al. 2007). In developed countries, these services often relate to their role in promoting biodiversity, supporting fisheries and providing coastal protection rather than more traditional, artisanal uses (A.M. Ellison 2008). Nevertheless, some clearance continues and, in Australia, tends to affect small areas in subtropical or temperate regions or involves concentrated clearance for infrastructure projects (Bridgewater & Cresswell 1999).

Even in the absence of clearance, chronic, low-intensity human disturbance (e.g., trampling, dumping of rubbish and diffuse inputs of contaminants) continues to affect temperate mangroves, particularly around urban areas. In mangrove stands in Sydney, trampling resulted in alteration of the structure of the benthic habitat (decreased numbers and size of pneumatophores and reduced algal biomass) and loss of macrofaunal species associated with these structures (Ross 2006). More generally, however, Lindegarth and Hoskin (2001) failed to detect differences in benthic macrofaunal assemblages between urban and non-urban mangrove areas around Sydney. Human activities may affect mangrove assemblages even when their intention is protective management, as exemplified by changes in benthic assemblages around boardwalks (Kelaher et al. 1998a,b).

The following discussion uses New Zealand as a case study of recent management issues. In New Zealand, adverse attitudes to mangroves post-European settlement were particularly prevalent where mangrove occupied potential areas for new farmland or where they had encroached and transformed the environment into vegetated swamplands. Consequently, little or no consideration was given to the conservation of mangroves, and there were few restrictions preventing the widespread and large-scale destruction of these plants that followed in many regions. Clearance and reclamation of intertidal areas continued for almost a century, and in some harbours (e.g., the Hokianga), approximately 34% of mangrove were destroyed (Chapman 1978).

By the late 1970s, perceptions of mangrove in New Zealand changed again, a move championed largely by Professor V.J. Chapman. Chapman's work, along with that of others, emphasised the unique ecological and economical values of mangrove based largely on work done on mangroves overseas because there was little supporting evidence from New Zealand at that time. A subsequent shift occurred to preserve mangrove, and several reserves (e.g., Waitangi National Reserve in the Bay of Islands) were established on the basis of their recommendations. At the present time, a total of 16 mangrove reserve areas have been established in New Zealand, covering approximately 2000 ha (or about 10% of total mangrove cover) (Mom 2005).

New Zealand's principal environmental legislation, the Resource Management Act (1991), allows governing bodies to uphold protection of mangroves against indiscriminate destruction or reclamation. However, concerns over recent expansion of mangrove areas, coupled with a push to preserve the ecology of adjacent habitats (e.g., salt marsh, seagrass beds and open mudflats), have resulted in increased pressure on regional councils and environmental agencies to provide information about the causes of, and possible resolutions to, this perceived problem.

Meanwhile, the public view of mangroves remains polarised, with some groups advocating protection at all costs, while others see mangroves as a nuisance and a loss to the economic and aesthetic values of the harbours and estuaries in which they grow. In some cases, management initiatives have been put in place with governing agencies, research scientists, community groups and traditional Māori owners (iwi) working closely to find a balance between mangrove and other estuarine habitats. One such programme in the Waikaraka Estuary in Tauranga Harbour has been very successful (Wildland Consultants 2003). However, despite these initiatives, protective legislation and due process, several groups and private individuals in other parts of New Zealand have removed mangroves from estuaries in protest at controls and perceived inaction.

Management initiatives

The concept of 'mangrove management' in New Zealand is increasingly associated with some form of control measure involving mangrove removal. However, management actually encompasses a broader range of possible actions and corresponding outcomes.

At one end of this range, a low-impact 'non-intervention' approach to mangrove management may be taken, allowing mangroves to remain intact and natural processes to take their course. This approach does not necessarily result in expansion of areas occupied by mangroves, but it does infer that people need to adapt to, and accept, the changes that take place in the mangrove habitat over time. This style of management may be more suited to relatively stable mangrove areas where little change has occurred in the populations over several decades (Mom 2005).

A similar approach may also be applied in preserving mangrove areas. In New Zealand, preservation has largely been achieved through the formation of a number of marine reserves that encompass areas of the ocean and foreshore, including mangroves, and are managed for scientific and preservation reasons. Examples of such marine reserves in New Zealand where mangroves form a significant component of the protected foreshore vegetation are Motu Manawa (Pollen Island) marine reserve in the Waitemata Harbour, Auckland, and Te Matuku marine reserve, Waiheke Island, in the nearby Hauraki Gulf; both are managed by the New Zealand Department of Conservation. Reserves have added advantages in that they provide opportunities to enhance appreciation of the mangrove ecosystem and ecology through education (by way of access and interpretative signage) and recreation. For instance, at Waitangi and Paihia in the Bay of Islands and Waikareo Estuary in the Bay of Plenty, mangroves are being managed in a way that allows people access right into the tidal forest habitat by way of boardwalks and tracks.

A middle-road approach to mangrove management, and one that also allows adult plants to remain intact, is the prevention of their further expansion into areas where they have been identified as potentially decreasing or removing existing values (aesthetic, ecological or economic). This approach involves the annual removal of first-year seedlings and requires ongoing and active management, often coupled with large-scale participation by local community groups. Recent consents have been granted by Bay of Plenty and Waikato regional councils to allow such activity in Whangamata and Tauranga Harbours, by which seedling mangrove plants may be removed from newly colonised mudflats (a *seedling* being defined as a mangrove plant with 2–12 leaves and one stem and between 5 and about 55 cm tall) (Maxwell 2006). Removal must be undertaken by hand to avoid unnecessary disturbance of the estuarine sediments.

In contrast, a relatively high-impact control measure, and one that is increasingly being considered as a method of mangrove management in New Zealand, is the large-scale removal of all adult plants, saplings and seedlings back to a predetermined baseline. The main aims of this approach are to preserve the ecology of habitats threatened by mangrove encroachment (e.g., salt marsh, eelgrass beds, open mudflat), to restore aesthetic values in an estuary (i.e., to open up views and to allow built-up sediment to shift following removal of the binding and accumulation properties provided by mangrove roots and stems) and to maintain access ways to, and throughout, a harbour or estuary.

A number of different approaches to large-scale removal have been trialled to date, including removal of all above- and below-ground mangrove material (including crowns, stems, roots and pneumatophores), removal of above-ground material only (also including pneumatophores) and cutting all to the level of the substratum surface, and removing above-ground crowns and stems but leaving pneumatophores and roots intact (Coffey 2001, 2002, 2004, Wildland Consultants 2003). Mangrove debris is either stockpiled, dried and eventually burned within the intertidal area or removed and disposed of outside the coastal marine boundaries.

Management focus has now moved towards catchments because there is a general acceptance that mangrove expansion is a response to increased sediment input into harbours and estuaries, and this perspective on mangrove management is also being adopted in other temperate regions (e.g., Harty & Cheng 2003). Many catchment areas have been greatly modified over the last 200 years, and the native vegetation that would have once slowed the flow of water from hillsides and helped to prevent erosion has been cleared for agriculture, forestry and urban development. These activities have resulted in significant changes in sediment quantities within the coastal marine environment. River and catchment programmes of the regional councils are focused on providing physical works, services and advice to landowners to reduce the risk of soil erosion and flooding, reduce the amount of sediment getting into waterways, and improve water quality, river stability and river environments. Reducing sediment and nutrient inputs will ultimately limit growth and expansion of mangroves in New Zealand harbours and estuaries (Nichol et al. 2000, Mom 2005).

Effects of mangrove removal

Removal may be considered an effective management option for mangroves in some harbours or estuaries in New Zealand, although relatively little is known of the short- and long-term effects of these activities on the immediate and wider environment. In other regions, research indicated that anthropogenic disturbance to the structure of mangrove forests alters physical processes and has ongoing effects on the associated assemblages of plants and animals (Gladstone & Schreider 2003, Prosser 2004, Ross 2006). For example, a study of damaged mangrove habitats in northern Queensland, Australia, revealed that changes linked to human disturbance were largely due to the loss of biological function and to other physical effects. A decline in abundance and diversity of associated mangrove fauna (such as sediment-dwelling crabs) was evident in areas where mangroves had been removed (Kaly et al. 1997). Losses of this nature may have negative effects, such as reduced soil aeration and bioturbation, which in turn can affect productivity and reproductive outputs of mangroves (if they remain) and other organisms. Use of vehicles, machinery and human traffic during the process of mangrove removal inevitably results in mechanical perturbation or compaction of soft sediments. These processes affect the ability of organisms to re-establish in the substratum following disturbance (Kaly et al. 1997).

Mangrove clearance has very significant impacts on vegetation communities and habitats for some fauna. For example, removal of mangrove cover radically alters the habitat for birds. For some species, such as the banded rail, a species commonly associated with mangroves in New Zealand, this removal results in loss of a major part of their foraging, feeding and breeding habitat. However, other species may benefit from mangrove clearance, particularly those that feed over open mudflats (e.g., white-faced heron, reef heron, pied stilt and oystercatcher). Areas cleared of mangroves in Panama developed higher algal biomass and diversity than uncleared areas, and the algal communities in the cleared areas included species that were rare within mangrove stands (Granek & Ruttenberg 2008).

Sediment grain size may also be altered following mangrove removal as a result of changes to run-off and current and tidal flows brought about by the absence of the trees and pneumatophores. Clearance of mangroves in Panama did not affect rates of sedimentation, but the sediments that accumulated in the mangrove habitat contained more organic matter than that accumulating in nearby cleared areas (Granek & Ruttenberg 2008). Sediments in highly altered mangrove areas in northern Australia showed smaller fractions of clay and a higher index of compaction compared with mangrove forests where no human disturbance had occurred (Kaly et al. 1997). Changes in forest nutrient status also occurred via altered processes of run-off and leaching and resulted in decreases of phosphorus and clay particles in disturbed areas.

Removal or slow physical breakdown of root material in the substratum following mangrove clearance may increase the possibility of erosion and transport of sediments to other areas, which in turn could have potentially significant impacts on water circulation, drainage patterns and flooding within an estuary. Removal may also result in the remobilisation not only of previously bound sediments but also of sediment-associated contaminants, thus increasing the potential for bioaccumulation and other effects of chemical contaminants in organisms.

Ongoing monitoring of mangrove areas before and after removal in two New Zealand harbours (Tauranga and Whangamata) has revealed trends similar to overseas studies, with measurable effects of tree removal on the composition and movement of sediment, benthic infauna, mobile epibenthos (crustaceans and gastropods) and birds. In addition, activities associated with mangrove removal, such as physical access, use of vehicles and machinery, trampling and disposal of mangrove debris, also contribute to disturbance of existing plant and animal communities and to some physical changes within and adjacent to mangrove habitat (Coffey 2001, 2002, 2004, Stokes & Healy 2005, Wildland Consultants 2005).

Mangrove restoration and enhancement

Mangrove habitats around the world have long been exploited for fuel, fishing and construction purposes and have also been subject to various forms of pollution from industrial waste, mining, oil exploration and eutrophication. From a worldwide standpoint, they are now counted as one of the most threatened natural community types, with approximately 50% of their global area destroyed or degraded since 1900 (Gilman et al. 2006). Widespread recognition of this global decline, and a growing appreciation of mangrove values in coastal protection, water quality, wildlife or fisheries habitat and tourism, has led to increasing efforts in many countries to restore, conserve and sustainably manage mangrove areas (Field 1999, Saenger 2002, Lewis 2005, Walton et al. 2006, Bosire et al. 2008). Of the approximately 90 countries that have mangrove vegetation, around 20 have undertaken rehabilitation initiatives (Field 1999), establishing nurseries and attempting aforestation of previously uncolonised mudflats and replanting in degraded areas (Erftemeijer & Lewis 1999).

Rehabilitation, restoration and planting of mangrove areas is not, and has not been, common practice in New Zealand. As recently as 1970, the preferred option for many mangrove areas was actually reclamation for various types of land development, including marinas, roading, oxidation ponds, agriculture and tip sites. This practice still continues, albeit on a much smaller scale and under the control of the Resource Management Act (Crisp et al. 1990). Even though mangroves continue to support ecological, community and traditional Māori values in New Zealand and despite historical losses, the recent and ongoing expansion of mangrove in many harbours and the lack of any major industry based on this vegetation has encouraged management initiatives that focus largely on removal rather than restoration.

Mangroves have been introduced to a few areas in New Zealand with a view to controlling erosion (e.g., Mohakatino, Mokau and Urenui River mouths at 38°59′ to 38°44′S, slightly beyond their natural latitudinal limit), but these attempts were largely unsuccessful due to plant mortality (Crisp et al. 1990). Successful establishment of mangroves, or enhancement of degraded areas, can only be achieved if the stresses (or actions) that initially caused their decline or absence are removed or discontinued. In some cases, mangrove wetlands will then repair themselves if the necessary natural processes, such as seedling recruitment and hydrology, are still intact. Otherwise, given appropriate environmental conditions (e.g., wave energy, salinity, pH, nutrient concentrations, substratum composition, inundation), successful rehabilitation may be a long-term process dependent on human assistance and ongoing active management (including replanting and weed control) (Gilman et al. 2006).

Effectiveness of mangrove management initiatives

Thus far, few conclusions have been reached regarding the most effective or ecologically sound method of mangrove removal in terms of sediment remobilization and impacts on other organisms, including other vegetation types, benthic fauna, shellfish, fishes and birds. However, in a number of New Zealand coastal areas, ongoing monitoring and research of both intact mangrove systems, and those where mangroves have been removed, are helping to answer some of these questions. Conclusions have also yet to be drawn in regard to economics because large-scale clearance of this nature can be costly in terms of equipment and labour and require ongoing and active management to prevent seedlings re-establishing.

Research has established that, regardless of which approach is decided on, sustainable management can only be achieved if evaluation of mangrove areas is undertaken on a site-by-site basis. Processes and effects vary according to the type of mangrove community, whether it is stable or dynamic and site-specific physical and ecological characteristics defined by a range of factors, including geomorphology, climate, sediment input, nutrient status and hydrodynamics.

Thorough research, provision of information and communication are crucial components of any management initiative. The recent debate about values of New Zealand mangrove, particularly their ecological role in coastal ecosystems, has highlighted the need for more comprehensive information than has been available up to very recent times. Much of the information on which New Zealand mangrove values were based was gleaned from a small number of isolated studies, anecdotal evidence and comparisons with overseas mangrove systems. This information proved inadequate not only for communities seeking guidance or action on mangrove management but also for the governing agencies responsible for providing those services.

Conclusions and directions for future research

Public interest in mangroves in temperate regions has waxed and waned over time and with it the pressure to manage them. Current concern is relatively high, but there are markedly conflicting viewpoints. Much of the basic information required to address concerns and manage mangroves is lacking. This review has identified variation in ecological values of temperate mangroves, such as levels of litter production and decomposition, faunal abundance or importance to fishes, at a range of scales among studies and among locations. This in turn highlights the need to assess the appropriate management actions for a given area of mangroves or a given estuary on a case-by-case basis.

The information that would allow us to make the assessments is, however, often lacking. The usefulness of applying general conclusions from the wider body of mangrove research, dominated as it is by studies in tropical regions, needs to be treated with caution. Important differences between the ecology of tropical and temperate mangroves, such as the roles played by crabs in processing mangrove material and the relative importance of mangroves as fish habitat, have been identified in the discussion. A critical eye is needed even when comparing information from studies of different temperate regions, as illustrated by the differing patterns of mangrove spread between eastern Australia and New Zealand. There is therefore a strong need for local studies to provide information that will allow understanding and management.

Current work addressing some of the issues has been discussed, but there are many other aspects that still need to be investigated. For example, our present knowledge of relative productivity of mangroves across the range of latitude, estuarine characteristics, tidal elevation, tree size and age indicates that there is considerable variation but is not sufficient to allow us to predict productivity at a particular site based on these factors. Systematic studies of productivity and incorporation of mangrove material into local food webs along these gradients are needed. Similarly, although we have a reasonably good knowledge of the benthic fauna of mangroves and how it varies with stand age and height on the shore, our knowledge of other components of faunal and floral diversity (particularly terrestrial invertebrates) is extremely limited.

Subsequent work has addressed some of the sampling deficiencies identified by Faunce and Serafy (2006) in earlier studies of fishes in mangroves, including not sampling alternative habitats or measuring environmental conditions beyond the basics of water temperature and salinity (e.g., structural complexity and landscape measures). However, many fundamental issues remain unresolved in both the sampling of fishes in mangrove habitats and in assessing the role of mangroves for fishes relative to alternative habitats. Some of these gaps in our knowledge are related to funding cycles, practical constraints and issues of scale rather than lack of awareness of the issues. Limitations include the use of gears that do not provide estimates of density per unit area (e.g., fyke and gill nets), the short timescale of most studies (generally only 1 or 2 yr at most), sampling of only mangroves and immediate adjacent habitats rather than the spectrum of habitats present in estuarine systems and not accounting for where mangroves are located in the wider habitat landscape mosaic. Of the four measures of nursery function proposed by Beck et al. (2001), no work has been done to date on whether mangrove habitats enhance fish growth relative to other habitats or whether they ultimately contribute more to final adult populations on a habitat area basis than alternative nursery habitats. As with work on other habitat types, such as seagrass meadows and coral reefs, future work will need to be directed towards larger-scale studies that include mangroves as one of a number of habitat types contributing to the support and production of fish populations in estuarine and coastal ecosystems.

Description of patterns and mechanisms of change in the distribution of mangroves themselves are hampered by lack of physical process studies of temperate mangrove forests, and this review has had to draw on studies done in tropical mangrove systems. There is a general paucity of information on sedimentation in temperate regions, including present rates of sediment accumulation in mangrove forests (see Walsh & Nittrouer 2004, p. 228), and sedimentation processes within fringing mangroves in wave-dominated environments. Furthermore, there have been few long-term physical studies and none in temperate mangrove systems, so the extent to which the information currently available is representative is not known. Precise elevation limits have not been adequately described for most species, including *Avicennia marina* (Clarke & Myerscough 1993, p. 307). The influence of wave exposure on elevation limits and the relative effects of short-term sea-level variations versus progressive, long-term relative SLR on the LEL of present-day temperate mangrove forests is not known. Apart from recognition that waves influence mangrove seedling recruitment in estuaries, our present understanding of this process is largely qualitative.

Our ability to predict the long-term development of mangrove forests is limited by the lack of deterministic morphodynamic models to simulate these systems. However, models have been developed to improve understanding of the dynamics of mangrove forests themselves (Berger et al. 2008). Furthermore, morphodynamic models have been developed to predict the effects of sediment supply, subsidence and SLR on saltmarsh ecosystems (e.g., Allen 1990, French 1993, van Wijnen & Bakker 2001, Morris et al. 2002, Temmerman et al. 2004, French 2006, Kirwan et al. 2008, Craft et al. 2009). The absence of similar morphodynamic models for mangrove systems is surprising given the similarity of the physical drivers and dynamics of mangrove and saltmarsh communities. The need for morphodynamic models of estuaries that incorporate mangrove ecosystems will become pressing as the combined environmental effects of burgeoning human populations in the coastal zone and climate change, such as SLR and changes in terrigenous sediment supply, become apparent (probably within the present century). Such models should explicitly incorporate the feedbacks among hydrodynamics, sediment processes, geomorphology and mangrove ecology and be underpinned by process measurements at a wide range of temporal and spatial scales.

As our understanding of different estuarine habitats and their assemblages (plants, invertebrates, fishes, birds) increases, the next obvious step is to start assessing how changes in the spatial habitat landscape (including the pelagic environment) might influence the overall biological/ecological functioning of the estuary. This information is especially relevant to the potential influence of human activities, which speed up the 'ageing' and infilling of estuaries. For instance, Saintilan (2004) showed that as NSW (Australia) estuaries infill and 'age', the relative proportion of different habitats change (e.g., seagrasses decline, mangroves expand), and the production of many fish species valuable to humans declines. However, in addition to the total habitat extents, 'habitat landscape' factors are also important. These factors include spatial configuration (e.g., the ratio of area to edge, the proximity of habitat patches to each other and distance from the harbour mouth) and habitat quality (e.g., age, health). Mangroves are part of the landscape dynamics of this estuarine habitat and need to be assessed in this context as new information becomes available.

Acknowledgements

We are very grateful to the NIWA librarians, in particular Hannah Russell, for library services and for compiling the list of references, Max Oulton for the map of mangrove distribution, the Auckland Regional Council (particularly Dominic McCarthy and Megan Stewart) for the original impetus for the study, Terry Hume and Giovanni Coco (NIWA) for reviewing parts of the text, and to Atsuko Fukunaga for translating information in Japanese. We thank Gary Hancock of the Commonwealth Scientific and Industrial Research Organisation (CSIRO; Land and Water Division) for information on sedimentation rates in south-eastern Australian estuaries, Jacquie Reed (Northland Regional Council, New Zealand), Grant Barnes (Auckland Regional Council) and Stephen Park (Environment Bay of Plenty) for providing access to reports and data. We particularly thank Rob Bell (NIWA) for his careful review of the discussion of climate change effects.

Information for the original New Zealand study was kindly provided by numerous people, and some of it has been included in the present report: Andrea Alfaro (Auckland University of Technology), Oliver Ball (Northland Polytechnic), Phil Battley (Massey University), Tony Beauchamp (Department of Conservation), Bruce Burns (Landcare Research), John Dugdale (Landcare Research, retired), Bruce Howse (Northland Regional Council), Joe Lee (Griffith University), David Melville, Stephen Park (Environment Bay of Plenty), Ray Pierce (Eco Oceania Ltd.), Greg Skilleter (University of Queensland), Brian Sorrell (NIWA), Debra Stokes (University of Waikato), Gillian Vaughan (Miranda Naturalists Trust), Robert Williams (NSW Department of Primary Industries), Kylee Wilton (NSW Department of Natural Resources) and Rosalind Wilton (Environment Waikato).

Any errors, omissions or misinterpretations contained in this review, despite the best efforts of those named, remain our responsibility. Finally, given the breadth of the published and unpublished information on temperate mangroves, we apologise to any authors whose contributions may have been omitted.

D.M., M.M. and A.S. were partly funded by NIWA Capability Fund Contract CRBE094 during the preparation of this review.

References

Adams, J.B., Colloty, B.M. & Bate, G.C. 2004. The distribution and state of mangroves along the coast of Transkei, Eastern Cape Province, South Africa. *Wetlands Ecology and Management* **12**, 531–541.

Albright, L.J. 1976. *In situ* degradation of mangrove tissues. *New Zealand Journal of Marine and Freshwater Research* **10**, 385–389.

Alfaro, A.C. 2006. Benthic macro-invertebrate community composition within a mangrove/seagrass estuary in northern New Zealand. *Estuarine Coastal and Shelf Science* **66**, 97–110.

Alfaro, A.C., Thomas, F., Sergent, L. & Duxbury, M. 2006. Identification of trophic interactions within an estuarine food web (northern New Zealand) using fatty acid biomarkers and stable isotopes. *Estuarine Coastal and Shelf Science* **70**, 271–286.

Allen, J.R.L. 1990. Salt-marsh growth and stratification: a numerical model with special reference to the Severn Estuary, southwest Britain. *Marine Geology* **95**, 77–96.

Allison, M.A., Nittrouer, C.A. & Faria, L.E.C., Jr. 1995. Rates and mechanisms of shoreface progradation and retreat downdrift of the Amazon river mouth. *Marine Geology* **125**, 373–392.

Alongi, D.M. 1987a. Intertidal zonation and seasonality of meiobenthos in tropical mangrove estuaries. *Marine Biology* **95**, 447–458.

Alongi, D.M. 1987b. The influence of mangrove-derived tannins on intertidal meiobenthos in tropical estuaries. *Oecologia* **71**, 537–540.

Alongi, D.M. 1987c. Inter-estuary variation and intertidal zonation of free-living nematode communities in tropical mangrove systems. *Marine Ecology Progress Series* **40**, 103–114.

Alongi, D.M. 2008. Mangrove forests: resilience, protection from tsunamis, and responses to global climate change. *Estuarine Coastal and Shelf Science* **76**, 1–13.

Alongi, D.M. 2009. *The Energetics of Mangrove Forests*. Berlin: Springer.

Alongi, D.M. & Christoffersen, P. 1992. Benthic infauna and organism-sediment relations in a shallow, tropical coastal area: influence of outwelled mangrove detritus and physical disturbance. *Marine Ecology Progress Series* **81**, 229–245.

Alongi, D.M., Tirendi, E. & Clough, B.F. 2000. Below-ground decomposition of organic matter in forests of the mangroves *Rhizophora stylosa* and *Avicennia marina* along the arid coast of Western Australia. *Aquatic Botany* **68**, 97–122.

Alongi, D.M., Trott, L.A., Wattayakorn, G. & Clough, B.F. 2002. Below-ground nitrogen cycling in relation to net canopy production in mangrove forests of southern Thailand. *Marine Biology* **140**, 855–864.

Andersen, T.J. 2001. Seasonal variation in erodibility of two temperate, microtidal mudflats. *Estuarine Coastal and Shelf Science* **53**, 1–12.

Araújo da Silva, C., Souza-Filho, P.W.M. & Rodrigues, S.W.P. 2009. Morphology and modern sedimentary deposits of the macrotidal Marapanim Estuary (Amazon, Brazil). *Continental Shelf Research* **29**, 619–631.

Armenteros, M., Martin, I., Williams, J.P., Creagh, B., Gonzalez-Sanson, G. & Capetillo, N. 2006. Spatial and temporal variations of meiofaunal communities from the western sector of the Gulf of Batabano, Cuba. I. Mangrove systems. *Estuaries and Coasts* **29**, 124–132.

Armonies, W. 1988. Active emergence of meiofauna from intertidal sediment. *Marine Ecology Progress Series* **43**, 151–159.

Arnaud-Haond, S., Teixeira, S., Massa, S.I., Billot, C., Saenger, P., Coupland, G., Duarte, C.M. & Serrao, E.A. 2006. Genetic structure at range edge: diversity and high inbreeding in southeast Asian mangrove (*Avicennia marina*) populations. *Molecular Ecology* **15**, 3515–3525.

Augustinus, P. 2002. Biochemical factors influencing deposition and erosion of fine grained sediment. In *Muddy Coasts of the World: Processes, Deposits and Function,* T. Healy et al. (eds). Amsterdam: Elsevier Science, 203–228.

Augustinus, P.G.E.F. 1995. Geomorphology and sedimentology of mangroves. In *Geomorphology and Sedimentology of Estuaries,* G.M.E. Perillo (ed.). Amsterdam: Elsevier Science, 333–357.

Ball, M.C. 1988. Salinity tolerance in the mangroves *Aegiceras corniculatum* and *Avicennia marina.* I. Water use in relation to growth, carbon partitioning, and salt balance. *Australian Journal of Plant Physiology* **15**, 447–464.

Ball, M.C. 1998. Mangrove species richness in relation to salinity and waterlogging: a case study along the Adelaide River floodplain, northern Australia. *Global Ecology and Biogeography Letters* **7**, 73–82.

Barnett, E.J. 1994. A Holocene paleoenvironmental history of Lake Alexandria. *Journal of Paleolimnology* **12**, 259–268.

Bartsch, I. 1989. Marine mites (*Halacaroidea: Acari*): a geographical and ecological survey. *Hydrobiologia* **178**, 21–42.

Bayliss, D.E. 1982. Switching by *Lepsiella vinosa* (Gastropoda) in South Australian mangroves. *Oecologia* **54**, 212–226.

Bayliss, D.E. 1993. Spatial distribution of *Balanus amphitrite* and *Elminius adelaidae* on mangrove pneumatophores. *Marine Biology* **116**, 251–256.

Beaglehole, J.C. 1968. *The Journals of Captain James Cook on His Voyages of Discovery. The Voyage of the Endeavour 1968–1771.* Cambridge, United Kingdom: Cambridge University Press.

Beanland, W.R. & Woelkerling, W.J. 1982. Studies on Australian mangrove algae: II. Composition and geographic distribution of communities in Spencer Gulf, South Australia. *Proceedings of the Royal Society of Victoria* **94**, 89–106.

Beanland, W.R. & Woelkerling, W.J. 1983. *Avicennia* canopy effects on mangrove algal communities in Spencer Gulf, South Australia. *Aquatic Botany* **17**, 309–313.

Beard, C.M. 2006. *Physiological constraints on the latitudinal distribution of the mangrove Avicennia marina (Forsk.) Vierh. subsp. australasica (Walp.) J. Everett (Avicenniaceae) in New Zealand.* PhD thesis, University of Waikato, Hamilton, New Zealand.

Beauchamp, A.J. & Parrish, G.R. 1999. Bird use of the sediment settling ponds and roost areas at Port Whangarei. *Notornis* **46**, 470–483.

Beauchamp, T. (n.d.). Statement of evidence in the matter of the Resource Management Act 1991 and in the matter of submissions and further submissions by the Director General of Conservation on proposed Plan Change No. 3 (Mangrove Management) to the Regional Coastal Plan for Northland. Whangarei, New Zealand: Department of Conservation. 13 p.

Beck, M.W., Heck, K.L., Jr., Able, K.W., Childers, D.L., Eggleston, D.B., Gillanders, B.M., Halpern, B., Hays, G.G., Hoshino, K., Minello, T.J., Orth, R.J., Sheridan, P.F. & Weinstein, M.P. 2001. The identification, conservation, and management of estuarine and marine nurseries for fish and invertebrates. *BioScience* **51**, 633–641.

Bedin, T. 2001. The progression of a mangrove forest over a newly formed delta in the Umhlatuze Estuary, South Africa. *South African Journal of Botany* **67**, 433–438.

Bell, J.D., Pollard, D.A., Burchmore, J.J., Pease, B.C. & Middleton, M.J. 1984. Structure of a fish community in a temperate tidal mangrove creek in Botany Bay, New South Wales. *Australian Journal of Marine and Freshwater Research* **35**, 33–46.

Berger, U., Rivera-Monroy, V.H., Doyle, T.W., Dahdouh-Guebas, F., Duke, N.C., Fontalvo-Herazo, M.L., Hildenbrandt, H., Koedam, N., Mehlig, U., Piou, C. & Twilley, R.R. 2008. Advances and limitations of individual-based models to analyze and predict dynamics of mangrove forests: a review. *Aquatic Botany* **89**, 260–274.

Bhat, N.R., Suleiman, M.K. & Shahid, S.A. 2004. Mangrove, *Avicennia marina*, establishment and growth under the arid climate of Kuwait. *Arid Land Research and Management* **18**, 127–139.

Bird, E.C.F. 1971. Mangroves as land builders. *Victorian Naturalist* **88**, 189–197.

Bird, E.C.F. 1972. Mangroves and coastal morphology in Cairns Bay, North Queensland. *Journal of Tropical Geography* **35**, 11–16.

Bird, E.C.F. 1984. *Coasts. An Introduction to Coastal Geomorphology.* Oxford, United Kingdom: Blackwell, 3rd edition.

Bird, E.C.F. 1986. Mangroves and intertidal morphology in Westernport Bay, Victoria, Australia. *Marine Geology* **69**, 251–271.

Blom, C.M. 1992. *Anthropogenic impacts on the benthic fauna and forest structure of the New Zealand mangrove Avicennia marina var.* resinifera. MSc thesis, University of Auckland, Auckland, New Zealand.

Bloomfield, A.L. & Gillanders, B.M. 2005. Fish and invertebrate assemblages in seagrass, mangrove, saltmarsh, and nonvegetated habitats. *Estuaries* **28**, 63–77.

Boon, P.I., Bird, F.L. & Bunn, S.E. 1997. Diet of the intertidal callianassid shrimps *Biffarius arenosus* and *Trypea australiensis* (Decapoda : Thalassinidea) in Western Port (southern Australia), determined with multiple stable-isotope analyses. *Marine and Freshwater Research* **48**, 503–511.

Bordignon, M.O. 2006. Diet of the fishing bat *Noctilio leporinus* (Linnaeus) (Mammalia, Chiroptera) in a mangrove area of southern Brazil. *Revista Brasileira de Zoologia* **23**, 256–260.

Bosire, J.O., Dahdouh-Guebas, F., Walton, M., Crona, B.I., Lewis, R.R., III, Field, C., Kairo, J.G. & Koedam, N. 2008. Functionality of restored mangroves: a review. *Aquatic Botany* **89**, 251–259.

Bouillon, S., Borges, A.V., Castaneda-Moya, E., Diele, K., Dittmar, T., Duke, N.C., Kristensen, E., Lee, S.Y., Marchand, C., Middelburg, J.J., Rivera-Monroy, V.H., Smith, T.J., III & Twilley, R.R. 2008. Mangrove production and carbon sinks: a revision of global budget estimates. *Global Biogeochemical Cycles* **22**, GB2013.

Boumans, R. & Day, J.W. 1993. High precision measurements of sediment elevation in shallow coastal areas using a sedimentation-erosion table. *Estuaries* **16**, 375–380.

Branch, G.M. & Branch, M.L. 1980. Competition in *Bembicium auratum* (Gastropoda) and its effect on microalgal standing stock in mangrove muds. *Oecologia* **46**, 106–114.

Branch, G.M. & Grindley, J.R. 1979. Ecology of southern African estuaries. Part XI. Mgazana: a mangrove estuary in Transkei. *South African Journal of Zoology* **14**, 149–170.

Brejaart, R. & Brownell, B. 2004. Mangroves: the cornerstone of a dynamic coastal environment. In *Muddy Feet: Firth of Thames Ramsar Site Update 2004*, B. Brownell (ed.). Pokeno, New Zealand: Ecoquest Education Foundation, 34–52.

Bridgewater, P. & Cresswell, I.D. 1999. Biogeography of mangrove and saltmarsh vegetation: implications for conservation and management in Australia. *Mangroves and Salt Marshes* **3**, 117–125.

Briggs, S.V. 1977. Estimates of biomass in a temperate mangrove community. *Australian Journal of Ecology* **2**, 369–373.

Brinkman, R.M., Massel, S.R., Ridd, P.V. & Furukawa, K. 1997. *Surface wave attenuation in mangrove forests*. Christchurch, New Zealand: Centre for Advanced Engineering, University of Canterbury.

Brownell, B. 2004. *Muddy Feet: Firth of Thames Ramsar Site Update 2004*. Kaiaua, New Zealand: Ecoquest Education Foundation.

Bryce, S., Larcombe, P. & Ridd, P.V. 2003. Hydrodynamic and geomorphological controls on suspended sediment transport in mangrove creek systems, a case study: Cocoa Creek, Townsville, Australia. *Estuarine Coastal and Shelf Science* **56**, 415–431.

Bunt, J.S. 1995. Continental scale patterns in mangrove litter fall. *Hydrobiologia* **295**, 135–140.

Burchett, M.D., Field, C.D. & Pulkownik, A. 1984. Salinity, growth and root respiration in the grey mangrove, *Avicennia marina. Physiologia Plantarum* **60**, 113–118.

Burns, B.R. 1982. *Population biology of Avicennia marina var.* resinifera. MSc thesis, University of Auckland, Auckland, New Zealand.

Burns, B.R. & Ogden, J. 1985. The demography of the temperate mangrove (*Avicennia marina* (Forsk.) Vierh.) at its southern limit in New Zealand. *Australian Journal of Ecology* **10**, 125–133.

Burrows, D.W. 2003. *The role of insect leaf herbivory on the mangroves Avicennia marina and Rhizophora stylosa*. PhD thesis, James Cook University, Townsville, Australia.

Burton, T. 1982. Mangrove development north of Adelaide. *Transactions of the Royal Society of South Australia* **106**, 183–189.

Butler, A.J., Depers, A.M., McKillup, S.C. & Thomas, D.P. 1977a. Distribution and sediments of mangrove forests in South Australia. *Transactions of the Royal Society of South Australia* **101**, 35–44.

Butler, A.J., Depers, A.M., McKillup, S.C. & Thomas, D.P. 1977b. A survey of mangrove forests in South Australia. *South Australian Naturalist* **51**, 34–49.

Cahoon, D.R., Hensel, P.F., Spencer, T., Reed, D.J., McKee, K.L. & Saintilan, N. 2006. Coastal wetland vulnerability to relative sea-level rise: wetland elevation trends and process controls. In *Wetlands and Natural Resource Management*, J.T.A. Verhoeven et al. (eds). Berlin: Springer-Verlag, 271–292.

Cahoon, D.R. & Lynch, J.C. 1997. Vertical accretion and shallow subsidence in a mangrove forest of southwestern Florida, USA. *Mangroves and Salt Marshes* **1**, 173–186.

Cahoon, D.R., Reed, D.R. & Day, J.W. 1995. Estimating shallow subsidence in microtidal salt marshes of the southeastern United States: Kaye and Barghoorn revisited. *Marine Geology* **128**, 1–9.

Cannicci, S., Burrows, D., Fratini, S., Smith, T.J., III, Offenberg, J. & Dahdouh-Guebas, F. 2008. Faunal impact on vegetation structure and ecosystem function in mangrove forests: a review. *Aquatic Botany* **89**, 186–200.

Carlton, J.M. 1974. Land-building and stabilization by mangroves. *Environmental Conservation* **1**, 285–294.

Carvalho do Amaral, P.G., Ledru, M.P., Branco, F.R. & Giannini, P.C.F. 2006. Late Holocene development of a mangrove ecosystem in southeastern Brazil (Itanhaem, state of Sao Paulo). *Palaeogeography Palaeoclimatology Palaeoecology* **241**, 608–620.

Chafer, C.J. 1998. A spatio-temporal analysis of estuarine vegetation change in the Minnamurra River 1938–1997. Report prepared for the Minnamurra Estuary Management Committee, Kiama, New South Wales, Australia.

Chant, R.J. & Stoner, A.W. 2001. Particle trapping in a stratified flood-dominated estuary. *Journal of Marine Research* **59**, 29–51.

Chapman, M.G. 1998. Relationships between spatial patterns of benthic assemblages in a mangrove forest using different levels of taxonomic resolution. *Marine Ecology Progress Series* **162**, 71–78.

Chapman, M.G., Michie, K. & Lasiak, T. 2005. Responses of gastropods to changes in amounts of leaf litter and algae in mangrove forests. *Journal of the Marine Biological Association of the United Kingdom* **85**, 1481–1488.

Chapman, M.G. & Tolhurst, T.J. 2004. The relationship between invertebrate assemblages and bio-dependent properties of sediment in urbanized temperate mangrove forests. *Journal of Experimental Marine Biology and Ecology* **304**, 51–73.

Chapman, M.G. & Tolhurst, T.J. 2007. Relationships between benthic macrofauna and biogeochemical proper-ties of sediments at different spatial scales and among different habitats in mangrove forests. *Journal of Experimental Marine Biology and Ecology* **343**, 96–109.

Chapman, M.G. & Underwood, A.J. 1995. Mangrove forests. In *Coastal Marine Ecology of Temperate Australia*, A.J. Underwood & M.G. Chapman (eds). Sydney: University of New South Wales Press, 187–204.

Chapman, V.J. (ed.). 1976. *Mangrove Vegetation*. Lehre, Germany: J. Cramer Verlag.

Chapman, V.J. (ed.). 1977. *Wet Coastal Ecosystems*. Amsterdam: Elsevier.

Chapman, V.J. 1978. *Mangroves and Salt Marshes of the Herekino, Whangape and Hokianga Harbours: A Study with Proposals for Preservation of Areas Supporting the Harbour Ecosystem*. Auckland, New Zealand: Department of Lands and Survey.

Chapman, V.J. & Ronaldson, J.W. 1958. The mangrove and saltmarsh flats of the Auckland isthmus. New Zealand Department of Scientific and Industrial Research Bulletin No. 125. Wellington, New Zealand.

Chenhall, B.E., Yassini, I., Depers, A.M., Caitcheon, G., Jones, B.G., Batley, G.E. & Ohmsen, G.S. 1995. Anthropogenic marker evidence for accelerated sedimentation in Lake Illawarra, New South Wales, Australia. *Environmental Geology* **26**, 124–135.

Christie, M.C., Dyer, K.R. & Turner, P. 1999. Sediment flux and bed level measurements from a macrotidal mudflat. *Estuarine Coastal and Shelf Science* **49**, 667–688.

Clarke, L.D. & Hannon, N.J. 1970. The mangrove swamp and salt marsh communities of the Sydney district. III. Plant growth in relation to salinity and waterlogging. *Journal of Ecology* **58**, 351–369.

Clarke, P.J. 1992. Predispersal mortality and fecundity in the grey mangrove (*Avicennia marina*) in southeast-ern Australia. *Australian Journal of Ecology* **17**, 161–168.

Clarke, P.J. 1993. Dispersal of grey mangrove (*Avicennia marina*) propagules in southeastern Australia. *Aquatic Botany* **45**, 195–204.

Clarke, P.J. 1994. Baseline studies of temperate mangrove growth and reproduction: demographic and litterfall measures of leafing and flowering. *Australian Journal of Botany* **42**, 37–48.

Clarke, P.J. & Allaway, W.G. 1993. The regeneration niche of the grey mangrove (*Avicennia marina*): effects of salinity, light and sediment factors on establishment, growth and survival in the field. *Oecologia* **93**, 548–556.

Clarke, P.J. & Kerrigan, R.A. 2002. The effects of seed predators on the recruitment of mangroves. *Journal of Ecology* **90**, 728–736.

Clarke, P.J., Kerrigan, R.A. & Westphal, C.J. 2001. Dispersal potential and early growth in 14 tropical man-groves: do early life history traits correlate with patterns of adult distribution? *Journal of Ecology* **89**, 648–659.

Clarke, P.J. & Myerscough, P.J. 1991. Buoyancy of *Avicennia marina* propagules in south-eastern Australia. *Australian Journal of Botany* **39**, 77–83.

Clarke, P.J. & Myerscough, P.J. 1993. The intertidal distribution of the grey mangrove (*Avicennia marina*) in southeastern Australia: the effects of physical conditions, interspecific competition, and predation on propagule establishment and survival. *Australian Journal of Ecology* **18**, 307–315.

Clough, B.F. 1984. Growth and salt balance of the mangroves *Avicennia marina* (Forsk.) Vierh. and *Rhizophora stylosa* Griff. in relation to salinity. *Australian Journal of Plant Physiology* **11**, 419–430.

Clough, B.F., Dixon, P. & Dalhaus, O. 1997. Allometric relationships for estimating biomass in multi-stemmed mangrove trees. *Australian Journal of Botany* **45**, 1023–1031.

Clynick, B. & Chapman, M.G. 2002. Assemblages of small fish in patchy mangrove forests in Sydney Harbour. *Marine and Freshwater Research* **53**, 669–677.

Coffey, B.T. 2001. Resource Consent 103475: trial clearance of mangroves Patiki Place Reserve, Whangamata Harbour. Final monitoring report—January 2001. Hamilton, New Zealand: Environment Waikato (Waikato Regional Council).

Coffey, B.T. 2002. *Resource Consent 102475: Trial Clearance of Mangroves Patiki Place Reserve, Whangamata Harbour*. Hamilton, New Zealand: Environment Waikato (Waikato Regional Council).

Coffey, B.T. 2004. *Mangrove Clearance Whangamata Harbour, October 2004: Sediment Monitoring Programme to Meet Condition 10 of Resource Consent 107665*. Hamilton, New Zealand: Environment Waikato (Waikato Regional Council).

Coleman, P.S.J. 1998. Changes in a mangrove/samphire community, North Arm Creek, South Australia. *Transactions of the Royal Society of South Australia* **122**, 173–178.

Cooper, A., Wright, I. & Mason, T. 1999. Geomorphology and sedimentology. In *Estuaries of South Africa*, B.R. Allanson & D. Baird (eds). Cambridge, United Kingdom: Cambridge University Press, 5–26.

Coull, B.C. 1990. Are members of the meiofauna food for higher trophic levels? *Transactions of the American Microscopical Society* **109**, 233–246.

Coull, B.C. 1999. Role of meiofauna in estuarine soft-bottom habitats. *Australian Journal of Ecology* **24**, 327–343.

Coull, B.C. & Chandler, G.T. 1992. Pollution and meiofauna: field, laboratory and mesocosm studies. *Oceanography and Marine Biology An Annual Review* **30**, 191–271.

Coull, B.C., Greenwood, J.G., Fielder, D.R. & Coull, B.A. 1995. Subtropical Australian juvenile fish eat meiofauna: experiments with winter whiting *Sillago maculata* and observations on other species. *Marine Ecology Progress Series* **125**, 13–19.

Cox, G.J. 1977. *Utilization of New Zealand mangrove swamps by birds*. MSc thesis, University of Auckland, Auckland, New Zealand.

Craft, C., Clough, J., Ehman, J., Jove, S., Park, R., Pennings, S., Guo, H. & Machmuller, M. 2009. Forecasting the effects of accelerated sea-level rise on tidal marsh ecosystem services. *Frontiers in Ecology and the Environment* **7**, 73–78.

Creese, B., Nichol, S., Gregory, M., Augustinus, P., Horrocks, M. & Mom, B. 1998. *Siltation of Whangape Harbour, Northland. Implications for local iwi*. Auckland, New Zealand: James Henare Maori Research Centre, University of Auckland/Foundation for Research, Science and Technology.

Crisp, P., Daniel, L. & Tortell, P. 1990. *Mangroves in New Zealand—Trees in the Tide*. Wellington, New Zealand: GP Books.

Cunha, S.R., Tognella-De Rosa, M.M.P. & Costa, C.S.B. 2006. Structure and litter production of mangrove forests under different tidal influences in Babitonga Bay, Santa Catarina, southern Brazil. *Journal of Coastal Research* **Special Issue 39**, 1169–1174.

Curran, M., Cole, M. & Allaway, W.G. 1986. Root aeration and respiration in young mangrove plants (*Avicennia marina* (Forsk.) Vierh.). *Journal of Experimental Botany* **37**, 1225–1233.

Dahdouh-Guebas, F., Jayatissa, L.P., Di Nitto, D., Bosire, J.O., Lo Seen, D. & Koedam, N. 2005. How effective were mangroves as a defence against the recent tsunami? *Current Biology* **15**, R443–R447.

Dahdouh-Guebas, F. & Koedam, N. 2008. Long-term retrospection on mangrove development using transdisciplinary approaches: a review. *Aquatic Botany* **89**, 80–92.

Dakin, W.J. 1966. *Australian Seashores*. Sydney, Australia: Angus & Robertson Ltd., revised edition.

Dalgren, C.P., Kellison, G.T., Adams, A.J., Gillanders, B.M., Kendall, M.S., Layman, C.A., Ley, J.A., Nagelkerken, I. & Serafy, J.E. 2006. Marine nurseries and effective juvenile habitats: concepts and applications. *Marine Ecology Progress Series* **312**, 291–295.

Danielsen, F., Sorensen, M.K., Olwig, M.F., Selvam, V., Parish, F., Burgess, N.D., Hiraishi, T., Karunagaran, V.M., Rasmussen, M.S., Hansen, L.B., Quarto, A. & Suryadiputra, N. 2005. The Asian tsunami: a protective role for coastal vegetation. *Science* **310**, 643.

Davey, A. & Woelkerling, W.J. 1985. Studies on Australian mangrove algae. III. Victorian communities: structure and recolonization in Western Port Bay. *Journal of Experimental Marine Biology and Ecology* **85**, 177–190.

Davis, J.H. 1940. Ecology and geologic role of mangroves in Florida. Carnegie Institution of Washington Publication No. 517. Carnegie Institute of Washington, Washington, D.C., USA.

Dawes, C.J., Siar, K. & Marlett, D. 1999. Mangrove structure, litter and macroalgal productivity in a northernmost forest of Florida. *Mangroves and Salt Marshes* **3**, 259–267.

de Lange, W.P. 2003. Tsunamis and storm-surge hazard in New Zealand. In *The New Zealand Coast: Te Tai O Aotearoa*, J.R. Goff et al. (eds). Palmerston North, New Zealand: Dunmore Press, 79–95.

de Lange, W.P., Bell, R.G., Gorman, R. & Reid, S. 2003. Physical oceanography of New Zealand waters. In *The New Zealand Coast: Te Tai O Aotearoa*, J.R. Goff et al. (eds). Palmerston North, New Zealand: Dunmore Press, 59–78.

de Lange, W.P. & de Lange, P.J. 1994. An appraisal of factors controlling the latitudinal distribution of mangrove (*Avicennia marina* var. *resinifera*) in New Zealand. *Journal of Coastal Research* **10**, 539–548.

Delgado, P., Hensel, P.F., Jimenez, J.A. & Day, J.W. 2001. The importance of propagule establishment and physical factors in mangrove distributional patterns in a Costa Rican estuary. *Aquatic Botany* **71**, 157–178.

Deng, Y., Ogden, J., Horrocks, M., Anderson, S.H. & Nichol, S.L. 2004. The vegetation sequence at Whangapoua Estuary, Great Barrier Island, New Zealand. *New Zealand Journal of Botany* **42**, 565–588.

Department of Medical Entomology, University of Sydney. (n.d.). *Saltwater wetlands*. Online. Available HTTP: http://www.medent.usyd.edu.au/fact/saltwet.htm (accessed 2 September 2009).

Dick, T.M. & Osunkoya, O.O. 2000. Influence of tidal restriction floodgates on decomposition of mangrove litter. *Aquatic Botany* **68**, 273–280.

Dietrich, W.E. & Perron, J.T. 2006. The search for a topographic signature of life. *Nature* **439**, 411–418.

Dodd, R.S. & Afzal Rafii, Z. 2002. Evolutionary genetics of mangroves: continental drift to recent climate change. *Trees Structure and Function* **16**, 80–86.

Dodd, R.S., Afzal-Rafii, Z., Kashini, N. & Budrick, J. 2002. Land barriers and open oceans: effects on gene diversity and population structure in *Avicennia germinans* L. (*Avicenniaceae*). *Molecular Ecology* **11**, 1327–1338.

Dor, I. & Levy, I. 1984. Primary productivity of the benthic algae in the hard-bottom mangal of Sinai. In *Hydrobiology of the Mangal: The Ecosystem of the Mangrove Forests*, F.D. Por & I. Dor (eds). The Hague: Dr. W. Junk, 179–191.

Downton, W.J.S. 1982. Growth and osmotic relations of the mangrove *Avicennia marina*, as influenced by salinity. *Australian Journal of Plant Physiology* **9**, 519–528.

Dugdale, J.S. 1990. Reassessment of *Ctenopseustis* Meyrick and *Planotortrix* Dugdale with descriptions of two new genera (Lepidoptera: Tortricidae). *New Zealand Journal of Zoology* **17**, 437–465.

Duke, N.C. 1990. Phenological trends with latitude in the mangrove tree *Avicennia marina*. *Journal of Ecology* **78**, 113–133.

Duke, N.C. 1991. A systematic revision of the mangrove genus *Avicennia* (Avicenniaceae) in Australia. *Australian Systematic Botany* **4**, 299–324.

Duke, N.C. 2006. *Australia's Mangroves. The Authoritative Guide to Australia's Mangrove Plants*. Brisbane: MER.

Duke, N.C. 2008. Corrections and updates to the article by Duke et al. (2005) reporting on the unusual occurrence and cause of dieback of the common mangrove species, *Avicennia marina*, in NE Australia. *Marine Pollution Bulletin* **56**, 1668–1670

Duke, N.C., Benzie, J.A.H., Goodall, J.A. & Ballment, E.R. 1998. Genetic structure and evolution of species in the mangrove genus *Avicennia* (Avicenniaceae) in the Indo-West Pacific. *Evolution* **52**, 1612–1626.

Duke, N.C., Meynecke, J.O., Dittmann, S., Ellison, A.M., Anger, K., Berger, U., Cannicci, S., Diele, K., Ewel, K.C., Field, C.D., Koedam, N., Lee, S.Y., Marchand, C., Nordhaus, I. & Dahdouh-Guebas, F. 2007. A world without mangroves? *Science* **317**, 41–42.

Dye, A.H. 1983a. Composition and seasonal fluctuations of meiofauna in a southern African mangrove estuary. *Marine Biology* **73**, 165–170.

Dye, A.H. 1983b. Vertical and horizontal distribution of meiofauna in mangrove sediments in Transkei, southern Africa. *Estuarine Coastal and Shelf Science* **16**, 591–598.

Dyer, K. 1986. *Coastal and Estuarine Sediment Dynamics*. Chichester, United Kingdom: Wiley.

Dyer, K.R. 1997. *Estuaries—Physical Introduction*. Chichester, United Kingdom: Wiley, 2nd ed.

Elliot, A.H., Shaanker, U., Hicks, D.M., Woods, R.A. & Dymonds, J.R. 2009. SPARROW regional regression for sediment yields in New Zealand rivers. In *Sediment Dynamics in Changing Environments (Proceedings of a Symposium Held in Christchurch, New Zealand, December 2008),* J. Schmidt et al. (eds). Wallingford, U.K.: IAHS Press, 242–249.

Ellis, J., Nicholls, P., Craggs, R., Hofstra, D. & Hewitt, J. 2004. Effects of terrigenous sedimentation on mangrove physiology and associated macrobenthic communities. *Marine Ecology Progress Series* **270**, 71–82.

Ellison, A.M. 2002. Macroecology of mangroves: large-scale patterns and processes in tropical coastal forests. *Trees Structure and Function* **16**, 181–194.

Ellison, A.M. 2008. Managing mangroves with benthic biodiversity in mind: moving beyond roving banditry. *Journal of Sea Research* **59**, 2–15.

Ellison, J.C. 1993. Mangrove retreat with rising sea-level, Bermuda. *Estuarine Coastal and Shelf Science* **37**, 75–87.

Ellison, J.C. 1999. Impacts of sediment burial on mangroves. *Marine Pollution Bulletin* **37**, 420–426.

Ellison, J.C. 2008. Long-term retrospection on mangrove development using sediment cores and pollen analysis: a review. *Aquatic Botany* **89**, 93–104.

Ellison, J.C. 2009. Geomorphology and sedimentology of mangroves. In *Coastal Wetlands—An Integrated Ecosystem Approach,* G.M.E. Perillo et al. (eds). Amsterdam: Elsevier, 565–591.

Ellison, J.C. & Stoddart, D.R. 1991. Mangrove ecosystem collapse during predicted sea-level rise: Holocene analogues and implications. *Journal of Coastal Research* **7**, 151–165.

Emmerson, W.D. & McGwynne, L.E. 1992. Feeding and assimilation of mangroves leaves by the crab *Sesarma meinerti* de Man in relation to leaf-litter production in Mgazana, a warm-temperate southern African mangrove swamp. *Journal of Experimental Marine Biology and Ecology* **157**, 41–53.

Emmerson, W.D. & Ndenze, T.T. 2007. Mangrove tree specificity and conservation implications of the arboreal crab *Parasesarma leptosoma* at Mngazana, a mangrove estuary in the Eastern Cape, South Africa. *Wetlands Ecology and Management* **15**, 13–25.

Environment Waikato. 2009. *Estuarine Vegetation Survey, Extent of Coastal Habitats.* Hamilton, New Zealand: Environment Waikato (Waikato Regional Councl). Online. Available HTTP: http://www.ew.govt.nz/Environmental-information/REDI/628521/ (accessed 6 July 2009).

Erftemeijer, P.L.A. & Lewis, R.R., III. 1999. Planting mangroves on intertidal mudflats: habitat restoration or habitat conversion? In *Enhancing Coastal Ecosystem Restoration for the 21st Century. Proceedings of Regional Seminar for East and Southeast Asian Countries: ECOTONE VIII, Ranong and Phuket, Thailand, 23–28 May 1999,* V. Sumantakul (ed.). Bangkok, Thailand: Royal Forest Department of Thailand, 156–165.

Erskine, W.D., Mahmoudzadeh, A. & Myers, C. 2002. Land use effects on sediment yields and soil loss rates in small basins of Triassic sandstone near Sydney, NSW, Australia. *Catena* **49**, 271–287.

Eston, V.R., Braga, M.R.A., Cordeiro-Marinao, M., Fujii, M.T. & Yokoya, N.S. 1992. Macroalgal colonization patterns on artificial substrates inside southeastern Brazilian mangroves. *Aquatic Botany* **42**, 315–325.

Evans, M.J. & Williams, R.J. 2001. Historical distribution of estuarine wetlands at Kurnell Peninsula, Botany Bay. *Wetlands (Australia)* **19**, 61–71.

Farnsworth, E.J. 1998. Issues of spatial, taxonomic and temporal scale in delineating links between mangrove diversity and ecosystem function. *Global Ecology and Biogeography Letters* **7**, 15–25.

Farnsworth, E.J. & Ellison, A.M. 1997. Global patterns of pre-dispersal propagule predation in mangrove forests. *Biotropica* **29**, 318–330.

Farnsworth, E.J. & Farrant, J.M. 1998. Reductions in abscisic acid are linked with viviparous reproduction in mangroves. *American Journal of Botany* **85**, 760–769.

Faunce, C.H. & Serafy, J.E. 2006. Mangroves as fish habitat: 50 years of field studies. *Marine Ecology Progress Series* **318**, 1–18.

Faust, M.A. & Gulledge, R.A. 1996. Associations of microalgae and meiofauna in floating detritus at a mangrove island, Twin Cays, Belize. *Journal of Experimental Marine Biology and Ecology* **197**, 159–175.

Feller, I.C. 1995. Effects of nutrient enrichment on growth and herbivory of dwarf red mangrove (*Rhizophora mangle*). *Ecological Monographs* **65**, 477.

Fenchel, T.M. 1978. The ecology of micro- and meiobenthos. *Annual Review of Ecology and Systematics* **9**, 99–121.

Ferrar, H.T. 1934. The geology of the Dargaville—Rodney Subdivision, Hokianga and Kaipara Divisions. Wellington, New Zealand: Geology Survey Branch, Department of Scientific and Industrial Research, Bulletin No. 34.

Field, C.D. 1995. Impact of expected climate change on mangroves. *Hydrobiologia* **295**, 75–81.

Field, C.D. 1999. Rehabilitation of mangrove ecosystems: an overview. *Marine Pollution Bulletin* **37**, 383–392.

Foster, I.D.L., Boardman, J. & Gates, J.B. 2009. Reconstructing historical sediment yields from the infilling of farm reservoirs, Eastern Cape, South Africa. In *Sediment Dynamics in Changing Environments (Proceedings of a Symposium Held in Christchurch, New Zealand, December 2008)*, J. Schmidt et al. (eds). Wallingford, U.K.: IAHS Press, 440–447.

Fratini, S., Vannini, M., Cannicci, S. & Schubart, C.D. 2005. Tree-climbing mangrove crabs: a case of convergent evolution. *Evolutionary Ecology Research* **7**, 219–233.

French, J. 2006. Tidal marsh sedimentation and resilience to environmental change: exploratory modelling of tidal, sea-level and sediment supply forcing in predominantly allochthonous systems. *Marine Geology* **235**, 119–136.

French, J.R. 1993. Numerical simulation of vertical marsh growth and adjustment to accelerated sea-level rise, north Norfolk, UK. *Earth Surface Processes and Landforms* **18**, 63–81.

French, P.W. 1997. *Coastal and Estuarine Management*. London: Routledge.

Friedrichs, C.T. & Aubrey, D.G. 1988. Non-linear tidal distortion in shallow well-mixed estuaries: a synthesis. *Estuarine Coastal and Shelf Science* **27**, 521–545.

Fugate, D.C. & Chant, R.J. 2005. Near-bottom shear stresses in shallow well-mixed estuaries: a synthesis. *Journal of Geophysical Research* **110**, C03022, doi:03010.01029/02004JC002563.

Furukawa, K. & Wolanski, E. 1996. Sedimentation in mangrove forests. *Mangroves and Salt Marshes* **1**, 3–10.

Furukawa, K., Wolanski, E. & Mueller, H. 1997. Currents and sediment transport in mangrove forests. *Estuarine Coastal and Shelf Science* **44**, 301–310.

Galloway, R.W. 1982. Distribution and physiographic patterns of Australian mangroves. In *Mangrove Ecosystems in Australia: Structure, Function and Management*, B.F. Clough (ed.). Townsville, Australia: Australian Institute of Marine Science, 31–54.

Gee, J.M. 1989. An ecological and economic review of meiofauna as food for fish. *Zoological Journal of the Linnean Society* **96**, 243–261.

Gee, J.M. & Somerfield, P.J. 1997. Do mangrove diversity and leaf litter decay promote meiofaunal diversity? *Journal of Experimental Marine Biology and Ecology* **218**, 13–33.

George, K. & Gulliard, F. 2003. Application of the waterline method to the bathymetry of a drying macrotidal bay. *Hydrographic Journal* **108**, 3–9.

Geyer, W.R., Beardsley, R.C., Candela, J., Castro, B.M., Legeckis, R.V., Lentz, S.J., Limeburner, R., Miranda, L.B. & Trowbridge, J.H. 1991. The physical oceanography of the Amazon outflow. *Oceanography* **4**, 8–14.

Gibb, J.G. 1986. A New Zealand regional Holocene eustatic sea-level curve and its application to determination of vertical tectonic movements. A contribution to IGCP-Project 200. *Royal Society of New Zealand Bulletin* **24**, 377–395.

Giere, O. 2009. *Meiobenthology. The Microscopic Motile Fauna of Aquatic Sediments*. Berlin: Springer.

Gillanders, B.M. & Kingsford, M.J. 2002. Impact of changes in flow of freshwater on estuarine and open coastal habitats and the associated organisms. *Oceanography and Marine Biology An Annual Review* **40**, 233–309.

Gilman, E., Van Lavieren, H., Ellison, J., Jungblut, V., Wilson, L., Areki, F., Brighouse, G., Bungitak, J., Dus, E., Henry, M., Kilman, M., Matthews, E., Sauni, I., Teariki-Ruatu, N., Tukia, S. & Yuknavage, K. 2006. Pacific island mangroves in a changing climate and rising sea. Nairobi: United Nations Environment Programme, Regional Seas Programme, Regional Seas Report and Studies No. 179.

Gladstone, W. & Schreider, M.J. 2003. Effects of pruning a temperate mangrove forest on the associated assemblages of macroinvertebrates. *Marine and Freshwater Research* **54**, 683–690.

Glanville, E.B. 1947. Reclamation of tidal flats. *New Zealand Journal of Agriculture* **74**, 49–58.

Goring, D.G. & Bell, R.G. 1999. El Niño and decadal effects on sea-level variability in northern New Zealand: a wavelet analysis. *New Zealand Journal of Marine and Freshwater Research* **33**, 587–598.

Goulter, P.F.E. & Allaway, W.B. 1979. Litterfall and decomposition in a mangrove stand, *Avicennia marina* (Forsk.) Vierh., in Middle Harbour, Sydney. *Australian Journal of Marine and Freshwater Research* **30**, 541–546.

Granek, E. & Ruttenberg, B.I. 2008. Changes in biotic and abiotic processes following mangrove clearing. *Estuarine Coastal and Shelf Science* **80**, 555–562.

Granek, E.F. & Ruttenberg, B.I. 2007. Protective capacity of mangroves during tropical storms: a case study from 'Wilma' and 'Gamma' in Belize. *Marine Ecology Progress Series* **343**, 101–105.

Grant, J., Bathmann, U.V. & Mills, E.L. 1986. The interaction between benthic diatom films and sediment transport. *Estuarine Coastal and Shelf Science* **23**, 225–238.

Gratiot, N., Gardel, A. & Anthony, E.J. 2007. Trade-wind waves and mud dynamics on the French Guiana coast, South America: input from ERA-40 wave data and field investigations. *Marine Geology* **236**, 15–26.

Green, M.O., Black, K.P. & Amos, C.L. 1997. Control of estuarine sediment dynamics by interactions between currents and waves at several scales. *Marine Geology* **144**, 97–116.

Green, M.O. & Coco, G. 2007. Sediment transport on an estuarine intertidal flat: measurements and conceptual model of waves, rainfall and exchanges with a tidal creek. *Estuarine Coastal and Shelf Science* **72**, 553–569.

Green, P.S. 1994. *Oceanic Islands*. Canberra: Australian Government Printing Service.

Griffiths, G.A. & Glasby, G.P. 1985. Input of river-derived sediment to the New Zealand continental shelf: I. Mass. *Estuarine Coastal and Shelf Science* **21**, 773–787.

Guest, M.A. & Connolly, R.M. 2004. Fine-scale movement and assimilation of carbon in saltmarsh and mangrove habitat by resident animals. *Aquatic Ecology* **38**, 599–609.

Guest, M.A. & Connolly, R.M. 2006. Movement of carbon among estuarine habitats: the influence of saltmarsh patch size. *Marine Ecology Progress Series* **310**, 15–24.

Gunn, C.R. & Dennis, J.V. 1973. Tropical and temperate stranded seeds and fruits from the Gulf of Mexico. *Contributions in Marine Science* **17**, 111–121.

Gwyther, J. 2000. Meiofauna in phytal-based and sedimentary habitats of a temperate mangrove ecosystem—a preliminary survey. *Proceedings of the Royal Society of Victoria* **112**, 137–151.

Gwyther, J. 2003. Nematode assemblages from *Avicennia marina* leaf litter in a temperate mangrove forest in south-eastern Australia. *Marine Biology* **142**, 289–297.

Gwyther, J. & Fairweather, P.G. 2002. Colonisation by epibionts and meiofauna of real and mimic pneumatophores in a cool temperate mangrove habitat. *Marine Ecology Progress Series* **229**, 137–149.

Gwyther, J. & Fairweather, P.G. 2005. Meiofaunal recruitment to mimic pneumatophores in a cool-temperate mangrove forest: spatial context and biofilm effects. *Journal of Experimental Marine Biology and Ecology* **317**, 69–85.

Hancock, G.J. 2000. Identifying resuspended sediment in an estuary using the 228Th/232Th activity ratio: the fate of lagoon sediment in the Bega River estuary, Australia. *Marine and Freshwater Research* **51**, 659–667.

Hancock, G.J. & Hunter, J.R. 1999. Use of excess 210Pb and 228Th to estimate rates of sediment accumulation and bioturbation in Port Phillip Bay, Australia. *Marine and Freshwater Research* **50**, 533–545.

Hancock, G.J., Olley, J.M. & Wallbrink, P.J. 2001. Sediment transport and accumulation in the Western Port. Report on Phase I of a study to determine the sources of sediment to Western Port. CSIRO Land and Water Technical Report No. 47/01. CSIRO Land and Water, Canberra, Australia.

Hancock, G.J. & Pietsch, T. 2006. Sedimentation in the Gippsland Lakes as determinded from sediment cores. CSIRO Land and Water Science Report No. 40/06. CSIRO Land and Water, Canberra, Australia.

Hannah, J. 2004. An updated analysis of long-term sea level change in New Zealand. *Geophysical Research Letters* **31**, L03307, doi:03310.01029/02003GL019166.

Hansen, J.E. 2007. Scientific reticence and sea level rise. *Environmental Research Letters* **2**, 024002.

Harada, K. & Kawata, Y. 2004. Study on the effect of coastal forest to tsunami reduction. *Annals of the Disaster Prevention Research Institute, Kyoto University* **47C**, 161–166.

Hartnoll, R.G., Cannici, S., Emmerson, W.D., Fratini, S., Macia, A., Mgaya, Y., Porri, F., Ruwa, R.K., Shunula, J.P., Skov, M.W. & Vannini, M. 2002. Geographic trends in mangrove crab abundance in East Africa. *Wetlands Ecology and Management* **10**, 203–213.

Harty, C. & Cheng, D. 2003. Ecological assessment and strategies for the management of mangroves in Brisbane Water—Gosford, New South Wales, Australia. *Landscape and Urban Planning* **62**, 219–240.

He, B., Lai, T., Fan, H., Wang, W. & Zheng, H. 2007. Comparison of flooding-tolerance in four mangrove species in a diurnal tidal zone in the Beibu Gulf. *Estuarine Coastal and Shelf Science* **74**, 254–262.

Healy, T. 2002. Muddy coasts of mid-latitude oceanic islands on an active plate margin—New Zealand. In *Muddy Coasts of the World: Processes, Deposits and Function,* T. Healy et al. (eds). Amsterdam: Elsevier Science, 347–374.

Hedges, B., Ibene, B., Koenig, S., La Marca, R.I. & Hardy, J. 2008. *Eleutherodactylus johnstonei.* In *IUCN Red List of Threatened Species.* International Union for Conservation of Nature and Natural Resources, Red List Unit, Cambridge, U.K. Online. Available HTTP: http://www.iucnredlist.org (accessed 2 May 2009).

Hindell, J.S. & Jenkins, G.P. 2004. Spatial and temporal variability in the assemblage structure of fishes associated with mangroves (*Avicennia marina*) and intertidal mudflats in temperate Australian embayments. *Marine Biology* **144**, 385–395.

Hindell, J.S. & Jenkins, G.P. 2005. Assessing patterns of fish zonation in temperate mangroves, with emphasis on evaluating sampling artefacts. *Marine Ecology Progress Series* **290**, 193–205.

Hodda, M. 1990. Variation in estuarine littoral nematode populations over three spatial scales. *Estuarine Coastal and Shelf Science* **30**, 325–340.

Hodda, M. & Nicholas, W.L. 1985. Meiofauna associated with mangroves in the Hunter River estuary and Fullerton Cove, south-eastern Australia. *Australian Journal of Marine and Freshwater Research* **36**, 41–50.

Hodda, M. & Nicholas, W.L. 1986a. Temporal changes in littoral meiofauna from the Hunter River estuary. *Australian Journal of Marine and Freshwater Research* **37**, 729–741.

Hodda, M. & Nicholas, W.L. 1986b. Nematode diversity and industrial pollution in the Hunter River estuary, NSW, Australia. *Marine Pollution Bulletin* **17**, 251–255.

Hogarth, P.J. 2007. *The Biology of Mangroves.* New York, USA: Oxford University Press.

Horticulture and Food Research Institute of New Zealand Ltd. 1998a. *HortFACT factsheet. Brownheaded leafroller, Ctenopseustis obliquana (Walker).* Plant and Food Research, Auckland, New Zealand. Online. Available HTTP: http://www.hortnet.co.nz/publications/hortfacts/hf401027.htm (accessed 20 July 2009).

Horticulture and Food Research Institute of New Zealand Ltd. 1998b. *HortFACT factsheet. Lemon tree borer, Oemona hirta (Fabricius).* Plant and Food Research, Auckland, New Zealand. Online. Available HTTP: http://www.hortnet.co.nz/publications/hortfacts/hf401033.htm (accessed 20 July 2009).

Hovenden, M.J., Curran, M., Cole, M.A., Goulter, P.F.E., Skelton, N.J. & Allaway, W.G. 1995. Ventilation and respiration in roots of one-year-old seedlings of grey mangrove *Avicennia marina* (Forsk.) Vierh. *Hydrobiologia* **295**, 23–29.

Hughes, R.H. & Hughes, J.S. 1992. *A Directory of African Wetlands.* Gland, Switzerland/Cambridge, United Kingdom/Nairobi, Kenya: IUCN/UNEP/WCMC.

Hume, T.M. 2003. Estuaries and tidal inlets. In *The New Zealand Coast: Te Tai O Aotearoa,* J.R. Goff et al. (eds). Palmerston North, New Zealand: Dunsmore Press, 191–213.

Hume, T.M., Snelder, T., Weatherhead, M. & Liefting, R. 2007. A controlling factor approach to estuary classification. *Ocean and Coastal Management* **50**, 905–929.

Hutchings, P. & Saenger, P. 1986. *Ecology of Mangroves.* Brisbane, Australia: Queensland University Press.

Hutchings, P.A. & Recher, H.F. 1974. The fauna of Careel Bay with comments on the ecology of mangrove and sea-grass communities. *Australian Zoologist* **18**, 99–128.

Hutchings, P.A. & Recher, H.F. 1982. Fauna of Australian mangroves. *Proceedings of the Linnean Society of New South Wales* **106**, 703–712.

Imgraben, S. & Dittmann, S. 2008. Leaf litter dynamics and litter consumption in two temperate South Australian mangrove forests. *Journal of Sea Research* **59**, 83–93.

IPCC 2007. *Climate change 2007: the physical basis. Contributions of Working Group I to the Fourth Assessment Report of the Intergovernmental Panel for Climate Change.* S. Solomon et al. (eds). Cambridge: Cambridge University Press.

Janssen-Stelder, B.M., Augustinus, P.G.E.F. & van Santen, W.A.C. 2002. Sedimentation in a coastal mangrove system, Red River Delta, Vietnam. In *Fine Sediment Dynamics in the Marine Environment,* J.C. Winterwerp & C. Kranenberg (eds). Amsterdam: Elsevier Science, 455–467.

Jelbart, J.E., Ross, P.M. & Connolly, R.M. 2007. Fish assemblages in seagrass beds are influenced by the proximity of mangrove forests. *Marine Biology* **150**, 993–1002.

Jenkins, G.P., Wheatley, M.J. & Poore, A.G.B. 1996. Spatial variation in recruitment, growth, and feeding of postsettlement King George whiting, *Sillaginoides punctata,* associated with seagrass beds of Port Phillip Bay, Australia. *Canadian Journal of Fisheries and Aquatic Sciences* **53**, 350–359.

Jensen, P. 1987. Feeding ecology of free-living aquatic nematodes. *Marine Ecology Progress Series* **35**, 187–196.

Johnstone, I.M. 1981. Consumption of leaves by herbivores in mixed mangrove stands. *Biotropica* **13**, 252–259.

Jones, B.G. & Chenhall, B.E. 2001. Lagoonal and estuarine sedimentation during the past 200 years. In *Australian Institute of Nuclear Science and Engineering: Environment Workshop on Archives of Human Impact of the Last 200 Years*, H. Heijnis & K. Harle (eds). Lucas Heights, New South Wales, Australia: Australian Institute of Nuclear Science and Engineering, 42–46.

Joye, S.B. & Lee, R.Y. 2004. Benthic microbial mats: important sources of fixed nitrogen and carbon to the Twin Cays, Belize ecosystem. *Atoll Research Bulletin*, 1–24.

Kakembo, V. & Rowntree, K.M. 2003. Relationship between land use and soil erosion in the communal lands near Peddie Town, Eastern Cape, South Africa. *Land Degradation and Development* **14**, 39–49.

Kaly, U.L. 1988. *Distribution, abundance and size of mangrove and saltmarsh gastropods*. PhD thesis, University of Sydney, Sydney, Australia.

Kaly, U.L., Eugelink, G. & Robertson, A.I. 1997. Soil conditions in damaged North Queensland mangroves. *Estuaries* **20**, 291–300.

Kamaruzzaman, B.Y. & Ong, M.C. 2008. Recent sedimentation rate and sediment ages determination of Kemaman-Chukai mangrove forest, Terengganu, Malaysia. *American Journal of Agricultural and Biological Science* **3**, 522–525.

Kangas, P.C. & Lugo, A.E. 1990. The distribution of mangroves and saltmarsh in Florida. *Journal of Tropical Ecology* **31**, 32–39.

Kathiresan, K. & Bingham, B.L. 2001. Biology of mangroves and mangrove ecosystems. *Advances in Marine Biology* **40**, 84–254.

Kathiresan, K. & Rajendran, N. 2005. Coastal mangrove forests mitigated tsunami. *Estuarine Coastal and Shelf Science* **65**, 601–606.

Kelaher, B.P., Chapman, M.G. & Underwood, A.J. 1998a. Changes in benthic assemblages near boardwalks in temperate urban mangrove forests. *Journal of Experimental Marine Biology and Ecology* **228**, 291–307.

Kelaher, B.P., Underwood, A.J. & Chapman, M.G. 1998b. Effect of boardwalks on the semaphore crab *Heloecius cordiformis* in temperate urban mangrove forests. *Journal of Experimental Marine Biology and Ecology* **227**, 281–300.

King, R.J. 1995. Mangrove macroalgae: a review of Australian studies. *Proceedings of the Linnean Society of New South Wales* **115**, 151–161.

Kirwan, M.L., Murray, A.B. & Boyd, W.S. 2008. Temporary vegetation disturbance as an explanation for permanent loss of tidal wetlands. *Geophysical Research Letters* **35**, L05403.

Kitheka, J.U., Ongwenyi, G.S. & Mavuti, K.M. 2002. Dynamics of suspended sediment exchange and transport in a degraded mangrove creek in Kenya. *Ambio* **31**, 580–587.

Knox, G.A. 1983. *Estuarine Ecology*. Auckland, New Zealand: Auckland Regional Authority.

Kobashi, D. & Mazda, Y. 2005. Tidal flow in riverine-type mangroves. *Wetlands Ecology and Management* **13**, 615–619.

Koch, G.W., Sillet, S.C., Jennings, G.M. & Davis, S. 2004. Limits to tree growth. *Nature* **428**, 851–854.

Krauss, K.W., Allen, J.A. & Cahoon, D.R. 2003. Differential rates of vertical accretion and elevation change among aerial root types in Micronesian mangrove forests. *Estuarine Coastal and Shelf Science* **56**, 251–259.

Krauss, K.W., Lovelock, C.E., McKee, K.L., Lopez-Hoffman, L., Ewe, S.M.L. & Sousa, W.P. 2008. Environmental drivers in mangrove establishment and early development: a review. *Aquatic Botany* **89**, 105–127.

Kristensen, E., Bouillon, S., Dittmar, T. & Marchand, C. 2008. Organic carbon dynamics in mangrove ecosystems: a review. *Aquatic Botany* **89**, 201–219.

Kronen, M. 2001. *Dynamics of mangroves (Avicennia marina) in the Puhoi Estuary, New Zealand*. MSc thesis, University of Auckland, Auckland, New Zealand.

Lacerda, L.D., Menezes, M.O.T. & Molisani, M.M. 2007. Changes in mangrove extension at the Pacoti River estuary, CE, NE Brazil due to regional environmental changes between 1958 and 2004. *Biota Neotropica* **7**, 67–72.

Laegdsgaard, P. & Johnson, C.R. 1995. Mangrove habitats as nurseries: unique assemblages of juvenile fish in subtropical mangroves in eastern Australia. *Marine Ecology Progress Series* **126**, 67–81.

Laegdsgaard, P. & Johnson, C. 2001. Why do juvenile fish utilise mangrove habitats? *Journal of Experimental Marine Biology and Ecology* **257**, 229–253.

Lamb, K.P. 1952. New plant galls: 1—Mite and insect galls. *Transactions of the Royal Society of New Zealand* **79**, 349–362.

Lebigre, J.M. 1999. Natural spatial dynamics of mangals through their margins: diagnostic elements. *Hydrobiologia* **413**, 103–113.

Leduc, D., Probert, P.K. & Duncan, A. 2009. A multi-method approach for identifying meiofaunal trophic interactions. *Marine Ecology Progress Series* **383**, 95–111.

Lee, R.Y. & Joye, S.B. 2006. Seasonal patterns of nitrogen fixation and dentrification in oceanic mangrove habitats. *Marine Ecology Progress Series* **307**, 127–141.

Lee, S.C. & Mehta, A.J. 1997. Problems in characterizing dynamics of mud shore profiles. *Journal of Hydraulic Engineering* **123**, 351–360.

Lee, S.Y. 1995. Mangrove outwelling: a review. *Hydrobiologia* **295**, 203–212.

Lee, S.Y. 1999. The effect of mangrove leaf litter enrichment on macrobenthic colonization of defaunated sandy substrates. *Estuarine Coastal and Shelf Science* **49**, 703–712.

Lee, S.Y. 2008. Mangrove macrobenthos: assemblages, services, and linkages. *Journal of Sea Research* **59**, 16–29.

Lessa, G. & Masselink, G. 1995. Morphodynamic evolution of a macrotidal barrier estuary. *Marine Geology* **129**, 25–46.

Lester, G.D., Sorensen, S.G., Faulkner, P.L., Reid, C.S. & Maxit, I.E. 2005. *Louisiana Comprehensive Wildlife Conservation Strategy*. Baton Rouge, Louisiana, USA: Louisiana Department of Wildlife and Fisheries.

Lewis, R.R., III. 2005. Ecological engineering for successful management and restoration of mangrove forests. *Ecological Engineering* **24**, 403–418.

Ley, J.A., Montague, C.L. & Mcivor, C.C. 1994. Food habits of mangrove fishes: a comparison along estuarine gradients in northeastern Florida Bay. *Bulletin of Marine Science* **54**, 881–899.

Li, M.S. & Lee, S.Y. 1997. Mangroves of China: a brief review. *Forest Ecology and Management* **96**, 241–259.

Lindegarth, M. & Hoskin, M. 2001. Patterns of distribution of macro-fauna in different types of estuarine, soft sediment habitats adjacent to urban and non-urban areas. *Estuarine Coastal and Shelf Science* **52**, 237–247.

Loneragan, N.R., Bunn, S.E. & Kellaway, D.M. 1997. Are mangroves and seagrasses sources of organic carbon for penaeid prawns in a tropical Australian estuary? A multiple stable-isotope study. *Marine Biology* **130**, 289–300.

Lovelock, C. 1993. *Field Guide to the Mangroves of Queensland*. Townsville, Australia: Australian Institute of Marine Science.

Lovelock, C.E. 2008. Soil respiration and belowground carbon allocation in mangrove forests. *Ecosystems* **11**, 342–354.

Lovelock, C.E., Ball, M.C., Feller, I.C., Engelbrecht, B.M.J. & Ewe, M.L. 2006. Variation in hydraulic conductivity of mangroves: influence of species, salinity, and nitrogen and phosphorus availability. *Physiologia Plantarum* **127**, 457–464.

Lovelock, C.E., Ball, M.C., Martin, K.C. & Feller, I.C. 2009. Nutrient enrichment increases mortality of mangroves. *PLoS One* **4**, e5600.

Lovelock, C.E., Feller, I.C., Ball, M.C., Ellis, J. & Sorrell, B.K. 2007a. Testing the growth rate vs. geochemical hypothesis for latitudinal variation in plant nutrients. *Ecology Letters* **10**, 1154–1163.

Lovelock, C.E., Feller, I.C., Ellis, J., Schwarz, A.M., Hancock, N., Nichols, P. & Sorrell, B. 2007b. Mangrove growth in New Zealand estuaries: the role of nutrient enrichment at sites with contrasting rates of sedimentation. *Oecologia* **153**, 633–641.

Lugo, A.E., Sell, M. & Snedaker, S.C. 1976. Mangrove ecosystem analysis. In *Systems Analysis and Simulation in Ecology*, B.C. Pattern (ed.). New York, USA: Academic Press, 113–145.

Lynch, J.C., Meriwether, J.R., McKee, B.A., Vera-Herrera, F. & Twilley, R.R. 1989. Recent accretion in mangrove ecosystems based on super(137)Cs and super(210)Pb. *Estuaries* **12**, 284–299.

Mackey, A.P. 1993. Biomass of the mangrove *Avicennia marina* (Forssk.) Vierh. near Brisbane, south-eastern Queensland. *Australian Journal of Marine and Freshwater Research* **44**, 721–725.

Mackey, A.P. & Smail, G. 1996. The decomposition of mangrove litter in a subtropical mangrove forest. *Hydrobiologia* **332**, 93–98.

Macnae, W. 1963. Mangroves swamps in South Africa. *Journal of Ecology* **51**, 1–25.

Macnae, W. 1966. Mangroves in eastern and southern Australia. *Australian Journal of Botany* **14**, 67–104.

Macnae, W. 1968. A general account of the fauna and flora of mangrove swamps and forests in the Indo-West-Pacific region. *Advances in Marine Biology* **6**, 73–270.

Maguire, T.L., Edwards, K.J., Saenger, P. & Henry, R. 2000. Characterisation and analysis of microsatellite loci in a mangrove species, *Avicennia marina* (Forsk.) Vierh. (*Avicenniaceae*). *Theoretical and Applied Genetics* **101**, 279–285.

Maguire, T.L., Peakall, R. & Saenger, P. 2002. Comparative analysis of genetic diversity in the mangrove species *Avicennia marina* (Forsk.) Vierh. (*Avicenniaceae*) detected by AFLPs and SSRs. *Theoretical and Applied Genetics* **104**, 388–398.

Marshall, D.J. & Pugh, P.J.A. 2001. Two new species of *Schusteria* (Acari: Oribatida: Ameronothroidea) from marine shores in southern Africa. *African Zoology* **35**, 201–205.

Martin, A.K. 1987. Comparison of sedimentation rates in the Natal Valley, south-west Indian Ocean, with modern sediment yields in east coast rivers of Southern Africa. *South African Journal of Science* **83**, 716–724.

Martin, K.C. 2007. *Interactive effects of salinity and nutrients on mangrove physiology: implications for mangrove forest structure and function.* PhD thesis, Australian National University, Canberra.

Massel, S.R. 1999. *Fluid Mechanics for Marine Ecologists.* Berlin: Springer-Verlag.

Massel, S.R., Furukawa, K. & Brinkman, R.M. 1999. Surface wave propagation in mangrove forests. *Fluid Dynamics Research* **24**, 219–249.

Mathew, J. & Baba, M. 1995. Mudbanks of the southwest coast of India. II: Wave-mud interactions. *Journal of Coastal Research* **11**, 179–187.

Maxwell, G.S. 2006. *The Removal of Mangrove Seedlings from Whangamata Harbour—An Assessment of Environmental Effects.* Hamilton, New Zealand: Environment Waikato (Waikato Regional Council).

May, J.D. 1999. Spatial variation in litter production by the mangrove *Avicennia marina* var. *australasica* in Rangaunu Harbour, Northland, New Zealand. *New Zealand Journal of Marine and Freshwater Research* **33**, 163–172.

Mazda, Y., Kanazawa, N. & Wolanski, E. 1995. Tidal asymmetry in mangrove creeks. *Hydrobiologia* **295**, 51–58.

Mazda, Y., Kobashi, D. & Okada, S. 2005. Tidal-scale hydrodynamics within mangrove swamps. *Wetlands Ecology and Management* **13**, 647–655.

Mazda, Y. & Wolanski, E. 2009. Hydrodynamics and modelling of water flow in mangrove areas. In *Coastal Wetlands—An Integrated Ecosystem Approach*, G.M.E. Perillo et al. (eds). Amsterdam: Elsevier, 231–261.

Mazumder, D., Saintilan, N. & Williams, R.J. 2005. Temporal variations in fish catch using pop nets in mangrove and saltmarsh flats at Towra Point, NSW, Australia. *Wetlands Ecology and Management* **13**, 457–467.

Mazumder, D., Saintilan, N. & Williams, R.J. 2006. Fish assemblages in three tidal saltmarsh and mangrove flats in temperate NSW, Australia: a comparison based on species diversity and abundance. *Wetlands Ecology and Management* **14**, 201–209.

Mchenga, I.S.S., Mfilinge, P.L. & Tsuchiya, M. 2007. Bioturbation activity by the grapsid crab *Helice formosensis* and its effects on mangrove sedimentary organic matter. *Estuarine Coastal and Shelf Science* **73**, 316–324.

McIvor, C.C. & Smith, T.J., III. 1995. Differences in the crab fauna of mangrove areas at a southwest Florida and a northeast Australia location: implications for leaf litter processing. *Estuaries* **18**, 591.

McKee, K.L. 1995a. Interspecific variation in growth, biomass partitioning, and defensive characteristics of neotropical mangrove seedlings: response to light and nutrient availability. *American Journal of Botany* **82**, 299–307.

McKee, K.L. 1995b. Mangrove species distribution and propagule predation in Belize: an exception to the dominance-predation hypothesis. *Biotropica* **27**, 334–345.

McKee, K.L., Cahoon, D.R. & Feller, I.C. 2007. Caribbean mangroves adjust to rising sea level through biotic controls on change in soil elevation. *Global Ecology and Biogeography* **16**, 545–556.

McKee, K.L. & Rooth, J.E. 2008. Where temperate meets tropical: multi-factorial effects of elevated CO_2, nitrogen enrichment, and competition on a mangrove-salt marsh community. *Global Change Biology* **14**, 971–984.

McKillup, S.C. & Butler, A.J. 1979. Cessation of hole-digging by the crab *Helograpsus haswellianus*: a resource-conserving adaptation. *Marine Biology* **50**, 157–161.

McLoughlin, L. 1987. Mangroves and grass swamps: changes in the shoreline vegetation of the Middle Lane Cover River, Sydney, 1780s–1880s. *Wetlands (Australia)* **7**, 13–24.

McLoughlin, L. 2000. Estuarine wetlands distribution along the Paramatta River, Sydney, 1788–1940: implications for planning and conservation. *Cunninghamia* **6**, 579–610.

McTainsh, G., Iles, B. & Saffinga, P. 1986. Spatial and temporal patterns of mangroves at Oyster Point Bay, South East Queensland. *Proceedings of the Royal Society of Queensland* **99**, 83–91.

Meades, L., Rodgerson, L., York, A. & French, K. 2002. Assessment of the diversity and abundance of terrestrial mangrove arthropods in southern New South Wales, Australia. *Austral Ecology* **27**, 451–458.

Meadows, P.S., Tait, J. & Hussain, S.A. 1990. Effects of estuarine infauna on sediment stability and particle sedimentation. *Hydrobiologia* **190**, 263–266.

Mehta, A.H. 2002. Mud-shore dynamics and controls. In *Muddy Coasts of the World: Processes, Deposits and Function,* T. Healy et al. (eds). Amsterdam, Elsevier Science, 19–60.

Melville, A.J. & Connolly, R.M. 2003. Spatial analysis of stable isotope data to determine primary sources of nutrition for fish. *Oecologia* **136**, 499–507.

Melville, A.J. & Connolly, R.M. 2005. Food webs supporting fish over subtropical mudflats are based on transported organic matter not in situ microalgae. *Marine Biology* **148**, 363–371.

Melville, F. & Pulkownik, A. 2006. Investigation of mangrove macroalgae as bioindicators of estuarine contamination. *Marine Pollution Bulletin* **52**, 1260–1269.

Melville, F., Pulkownik, A. & Burchett, M. 2005. Zonal and seasonal variation in the distribution and abundance of mangrove macroalgae in the Parramatta River, Australia. *Estuarine Coastal and Shelf Science* **64**, 267–276.

Mendez-Alonzo, R., Lopez-Portillo, J. & Rivera-Monroy, V.H. 2008. Latitudinal variation in leaf and tree traits of the mangrove *Avicennia germinans* (*Avicenniaceae*) in the central region of the Gulf of Mexico. *Biotropica* **40**, 449–456.

Middleton, B.A. & McKee, K.L. 2001. Degradation of mangrove tissues and implications for peat formation in Belizean island forests. *Journal of Ecology* **89**, 818–828.

Mildenhall, D.C. 1994. Early to mid Holocene pollen samples containing mangrove pollen from Sponge Bay, East Coast, North Island, New Zealand. *Journal of the Royal Society of New Zealand* **24**, 219–230.

Mildenhall, D.C. & Brown, L. 1987. An early Holocene occurrence of the mangrove *Avicennia marina* in Poverty Bay, North Island, New Zealand. *New Zealand Journal of Botany* **25**, 281–294.

Miller, P. & Miller, K. 1991. Bitterns using mangroves. *Notornis* **38**, 79.

Milliman, J.D. & Syvitski, J.P.M. 1992. Geomorphic/tectonic control of sediment discharge to the ocean: the importance of small mountainous rivers. *Journal of Geology* **100**, 525–544.

Milward, N.E. 1982. Mangrove-dependent biota. In *Mangrove Ecosystems in Australia: Structure, Function and Management*, B.F. Clough (ed.). Townsville, Australia: Australian Institute of Marine Science, 121–139.

Minchinton, T.E. 2006. Consequences of pre-dispersal damage by insects for the dispersal and recruitment of mangroves. *Oecologia* **148**, 70–80.

Minchinton, T.E. & Dalby-Ball, M. 2001. Frugivory by insects on mangrove propagules: effects on the early life history of *Avicennia marina*. *Oecologia* **129**, 243–252.

Minchinton, T.E. & Ross, P.M. 1999. Oysters as habitats for limpets in a temperate mangrove forest. *Australian Journal of Ecology* **24**, 157–170.

Mitchell, M.L. & Adam, P. 1989. The relationship between mangrove and saltmarsh communities in the Sydney region. *Wetlands (Australia)* **8**, 37–46.

Mitsch, W.J., Gosselink, J.G., Anderson, C.J. & Zhang, L. 2009. *Wetland Ecosystems*. Chichester, United Kingdom: Wiley.

Mom, G.A.T. 2005. *Effects of management on the ecology of mangrove communities in New Zealand*. PhD thesis, University of Auckland, Auckland, New Zealand.

Morant, P. & Quinn, N. 1999. Influence of man and management of South African estuaries. In *Estuaries of South Africa*, B.R. Allanson & D. Baird (eds). Cambridge, United Kingdom: Cambridge University Press, 289–320.

Morris, J.T., Sundareshwar, P.V., Nietch, C.T., Kjerfve, B. & Cahoon, D.R. 2002. Responses of coastal wetlands to rising sea level. *Ecology* **83**, 2869–2877.

Morrisey, D. 1995. Saltmarshes. In *Coastal Ecology of Temperate Australia*, A.J. Underwood & M.G. Chapman (eds). Sydney: University of New South Wales Press, 205–220.

Morrisey, D., Beard, C., Morrison, M., Craggs, R. & Lowe, M. 2007. The New Zealand mangrove: review of the current state of knowledge. Auckland Regional Council Technical Publication No. 325. Auckland, New Zealand: Auckland Regional Council.

Morrisey, D.J., Hill, A.F., Kemp, C.L.S. & Smith, R.K. 1999. Changes in abundance and distribution of coastal and estuarine vegetation in the Auckland region. NIWA Client Report No. ARC90232/1. Hamilton, New Zealand: National Institute of Water and Atmospheric Research Ltd (NIWA).

Morrisey, D.J., Hume, T.M., Swales, A., Hewitt, J.E., Hill, A.F., Ovenden, R., Smith, R.K., Taylor, M.D. & Wilkinson, M.R. 1995. *Ecological monitoring for potential effects of forestry activities on the intertidal habitats of Whangapoua Harbour. Annual report 1991–1995*. Hamilton, New Zealand: NIWA.

Morrisey, D.J., Skilleter, G.A., Ellis, J.I., Burns, B.R., Kemp, C.E. & Burt, K. 2003. Differences in benthic fauna and sediment among mangrove (*Avicennia marina* var. *australasica*) stands of different ages in New Zealand. *Estuarine Coastal and Shelf Science* **56**, 581–592.

Morton, B. 2004. Predator-prey interactions between *Lepsiella vinosa* (Gastropoda : Muricidae) and *Xenostrobus inconstans* (Bivalvia : Mytilidae) in a southwest Australian marsh. *Journal of Molluscan Studies* **70**, 237–245.

Morton, J. & Miller, M. 1968. *The New Zealand Sea Shore*. Auckland, New Zealand: Collins.

Murray, A.B., Lazarus, E., Ashton, A., Baas, A., Coco, G., Coulthard, T., Fonstad, M., Haff, P., McNamara, D., Paola, C., Pelletier, J. & Reinhardt, L. 2009. Geomorphology, complexity and the emerging science of the Earth's surface. *Geomorphology* **103**, 496–505.

Murray, A.L. & Spencer, T. 1997. On the wisdom of calculating annual material budgets in tidal wetlands. *Marine Ecology Progress Series* **150**, 207–216.

Murray, F. 1985. Cycling of fluoride in a mangrove community near a fluoride emission source. *Journal of Applied Ecology* **22**, 277–285.

Nagelkerken, I., Blaber, S.J.M., Bouillon, S., Green, P., Haywood, M., Kirton, L.G., Meynecke, J.O., Pawlik, J., Penrose, H.M., Sasekumar, A. & Somerfield, P.J. 2008. The habitat function of mangroves for terrestrial and marine fauna: a review. *Aquatic Botany* **89**, 155–185.

Nagelkerken, I., Kleijnen, S., Klop, T., van den Brand, R.A., de la Moriniere, E.C. & van der Velde, G. 2001. Dependence of Caribbean reef fishes on mangroves and seagrass beds as nursery habitats: a comparison of fish faunas between bays with and without mangroves/seagrass beds. *Marine Ecology Progress Series* **214**, 225–235.

Nagelkerken, I., van der Velde, G., Gorissen, M.W., Meijer, G.J., vant Hof, T. & den Hartog, C. 2000. Importance of mangroves, seagrass beds and the shallow coral reef as a nursery for important coral reef fishes, using a visual census technique. *Estuarine Coastal and Shelf Science* **51**, 31–44.

Naidoo, Y., Steinke, T.D., Mann, F.D., Bhatt, A. & Gairola, S. 2008. Epiphytic organisms on the pneumatophores of the mangrove *Avicennia marina*: occurrence and possible function. *African Journal of Plant Science* **2**, 12–15.

Nakasuga, T., Oyama, H. & Haruki, M. 1974. Studies on the mangrove community. 1. The distribution of the mangrove community in Japan. *Japanese Journal of Ecology* **24**, 237–246.

Nature Conservation Council. 1984. *Strategies for the management of mangrove forests in New Zealand, Nature Conservation Council discussion document*. Auckland, New Zealand: Nature Conservation Council.

Neil, D.T. 1998. Moreton Bay and its catchment: seascape and landscape, development and degradation. In *Moreton Bay and Catchment*, I.R. Tibbetts et al. (eds). Brisbane, Australia: University of Queensland, 3–54.

Netto, S.A. & Gallucci, F. 2003. Meiofauna and macrofauna communities in a mangrove from the island of Santa Catarina, South Brazil. *Hydrobiologia* **505**, 159–170.

New South Wales Department of Primary Industries. 2009. *Mud crab Scylla serrata*. Cronulla, Australia: New South Wales Department of Primary Industries. Online. Available HTTP: http://www.dpi.nsw.gov.au/fisheries/recreational/saltwater/sw-species/mud-crab (accessed 13 June 2009).

Nichol, S.L., Augustinus, P.C., Gregory, M.R., Creese, R. & Horrocks, M. 2000. Geomorphic and sedimentary evidence of human impact on the New Zealand coastal landscape. *Physical Geography* **21**, 109–132.

Nicholas, W.L. 1996. *Robustnema fosteri* sp. nov., gen. nov. (Xyalidae, Monhysterida, Nematoda), a common nematode of mangrove mudflats in Australia. *Transactions of the Royal Society of South Australia* **120**, 161–165.

Nicholas, W.L., Elek, J.A., Stewart, A. & Marples, T.G. 1991. The nematode fauna of a temperate Australian mangrove mudflat; its population density, diversity and distribution. *Hydrobiologia* **209**, 13–27.

Nicholas, W.L., Goodchild, D.J. & Stewart, A. 1987. The mineral composition of intracellular inclusions in nematodes from thiobiotic mangrove mud-flats. *Nematologica* **33**, 167–179.

Nicholas, W.L., Stewart, A.C. & Marples, T.G. 1988. Field and laboratory studies of *Desmodora cazca* Gerlach, 1956 (Desmodoridae: Nematoda) from mangrove mud-flats. *Nematologica* **34**, 331–349.

Nichols, M.M. 1977. Response and recovery of an estuary following a river flood. *Journal of Sedimentary Petrology* **47**, 1171–1186.

Ochieng, C.A. & Erftemeijer, P.L.A. 2002. Phenology, litterfall and nutrient resorption in *Avicennia marina* (Forssk.) Vierh in Gazi Bay, Kenya. *Trees Structure and Function* **16**, 167–171.

Olafsson, E. 1995. Meiobenthos in mangrove areas in eastern Africa with emphasis on assemblage structure of free-living marine nematodes. *Hydrobiologia* **312**, 47–57.

Olafsson, E. & Ndaro, S.G.M. 1997. Impact of the mangrove crabs *Uca annulipes* and *Dotilla fenestrata* on meiobenthos. *Marine Ecology Progress Series* **158**, 225–231.

Olley, J.M. & Wasson, R.J. 2003. Changes in the flux of sediment in the Upper Murrumbidgee catchment, Southeastern Australia, since European settlement. *Hydrological Processes* **17**, 3307–3320.

Oñate-Pacalioga, J.A. 2005. *Leaf litter production, retention, and decomposition of Avicennia marina var. australasica at Whangateau Estuary, Northland, New Zealand.* MSc thesis, University of Auckland, Auckland, New Zealand.

Onuf, C.P., Teal, J.M. & Valiela, I. 1977. Interactions of nutrients, plant growth and herbivory in a mangrove ecosystem. *Ecology* **58**, 514–526.

Osunkoya, O.O. & Creese, R.G. 1997. Population structure, spatial pattern and seedling establishment of the grey mangrove, *Avicennia marina* var. *australasica*, in New Zealand. *Australian Journal of Botany* **45**, 707–725.

Othman, M.A. 1994. Value of mangroves in coastal protection. *Hydrobiologia* **285**, 277–282.

Panapitukkul, N., Duarte, C.M., Thampanya, U., Kheowvongsri, P., Srichai, N., Geertz-Hansen, O., Terrados, J. & Boromthanarath, S. 1998. Mangrove colonization: mangrove progression over the growing Pak Phanang (SE Thailand) mud flat. *Estuarine Coastal and Shelf Science* **47**, 51–61.

Park, S. 2004. Aspects of mangrove distribution and abundance in Tauranga Harbour. Tauranga, New Zealand: Environmental Bay of Plenty (Bay of Plenty Regional Council) Environmental Publication No. 2004/16.

Parkinson, R.W., DeLaune, R.D. & White, J.R. 1994. Holocene sea-level rise and the fate of mangrove forests within the wider Caribbean region. *Journal of Coastal Research* **10**, 1077–1086.

Patterson, C.S., McKee, K.L. & Mendelssohn, I.A. 1997. Effects of tidal inundation and predation on *Avicennia germinans* seedling establishment and survival in a sub-tropical mangal/salt marsh community. *Mangroves and Salt Marshes* **1**, 103–111.

Patterson, C.S., Mendelssohn, I.A. & Swenson, E.M. 1993. Growth and survival of *Avicennia germinans* seedlings in a mangal/salt marsh community in Louisiana, USA. *Journal of Coastal Research* **9**, 801–810.

Payne, J.L. & Gillanders, B.M. 2009. Assemblages of fish along a mangrove-mudflat gradient in temperate Australia. *Marine and Freshwater Research* **60**, 1–13.

Pearce, J.B. 1995. Urban development and marine and riparian habitat quality. *Marine Pollution Bulletin* **30**, 496–499.

Peel, M.C., Finlayson, B.L. & McMahon, T.A. 2007. Updated world map of the Koppen-Geiger climate classification. *Hydrology and Earth System Sciences* **11**, 1633–1644.

Perry, C.L. 2007. *Ecosystem effects of expanding populations of Avicennia germinans in a southeastern Louisiana Spartina alterniflora saltmarsh.* MSc thesis, Louisiana State University, Baton Rogue, Louisiana.

Perry, C.L. & Mendelssohn, I.A. 2009. Ecosystem effects of expanding populations of *Avicennia germinans* in a Louisiana salt marsh. *Wetlands* **29**, 396–406.

Pfeffer, W.T., Harper, J.T. & O'Neel, S. 2008. Kinematic constraints on glacier contributions to 21st-century sea-level rise. *Science* **321**, 1340–1343.

Phillips, A., Lambert, G., Granger, J.E. & Steinke, T.D. 1994. Horizontal zonation of epiphytic algae associated with *Avicennia marina* (Forssk) Vierh pneumatophores at Beachwood Mangroves Nature Reserve, Durban, South Africa. *Botanica Marina* **37**, 567–576.

Phleger, F.B. 1970. Foraminiferal populations and marine marsh processes. *Limnology and Oceanography* **15**, 522–536.

Phuoc, V.L.H. & Massel, S.R. 2006. Experiments on wave motion and suspended sediment concentration at Nang Hai, CanGio mangrove forest, Southern Vietnam. *Oceanologia* **48**, 23–40.

Pocknall, D.T., Gregory, M.R. & Greig, D.A. 1989. Palynology of core 80/20 and its implications for understanding Holocene sea level changes in the Firth of Thames, New Zealand. *Journal of the Royal Society of New Zealand* **19**, 171–179.

Polz, M.F., Felbeck, H., Novak, R., Nebelsick, M. & Ott, J.A. 1992. Chemoautotrophic, sulfur-oxidizing symbiotic bacteria on marine nematodes: morphological and biochemical characterization. *Microbial Ecology* **24**, 313–329.

Postma, H. 1961. Transport and accumulation of suspended matter in the Dutch Wadden Sea. *Netherlands Journal of Sea Research* **1**, 148–190.

Proches, S. 2004. Ecological associations between organisms of different evolutionary history: mangrove pneumatophore arthropods as a case study. *Journal of the Marine Biological Association of the United Kingdom* **84**, 341–344.

Proches, S. & Marshall, D.J. 2002. Epiphytic algal cover and sediment deposition as determinants of arthropod distribution and abundance on mangrove pneumatophores. *Journal of the Marine Biological Association of the United Kingdom* **82**, 937–942.

Proches, S., Marshall, D.J., Ugrasen, K. & Ramcharan, A. 2001. Mangrove pneumatophore arthropod assemblages and temporal patterns. *Journal of the Marine Biological Association of the United Kingdom* **81**, 545–552.

Proffitt, C.E. & Devlin, D.J. 2005. Grazing by the intertidal gastropod *Melampus coffeus* greatly increases mangrove leaf litter degradation rates. *Marine Ecology Progress Series* **296**, 209–218.

Prosser, A.J. 2004. *Faunal community change following mangrove dieback in Moreton Bay, Australia.* BSc Hons thesis, University of Queensland, Brisbane, Australia.

Prosser, I.P., Rutherfurd, I.D., Olley, J.M., Young, W.J., Wallbrink, P.J. & Moran, C.J. 2001. Large-scale patterns of erosion and sediment transport in river networks, with examples from Australia. *Marine and Freshwater Research* **52**, 81–99.

Quartel, S., Kroon, A., Augustinus, P.G.E.F., Van Santen, P. & Tri, N.H. 2007. Wave attenuation in coastal mangroves in the Red River Delta, Vietnam. *Journal of Asian Earth Sciences* **29**, 576–584.

Rabinowitz, D. 1978. Dispersal properties of mangrove propagules. *Biotropica* **10**, 47–57.

Rahmstorf, S. 2007. A semi-empirical approach to projecting future sea-level rise. *Science* **315**, 368–370.

Raines, J., Youngson, K. & Unno, J. 2000. Use of the Leschenault Inlet estuary by waterbirds. *Journal of the Royal Society of Western Australia* **83**, 503–512.

Rignot, E., Bamber, J.L., Van Den Broeke, M.R., Davis, C., Li, Y., Van De Berg, W.J. & Van Meijgaard, E. 2008. Recent Antarctic ice mass loss from radar interferometry and regional climate modelling. *Nature Geoscience* **1**, 106–110.

Robertson, A.I. 1988. Decomposition of mangrove leaf litter in tropical Australia. *Journal of Experimental Marine Biology and Ecology* **116**, 235–248.

Robertson, A.I. & Daniel, P.A. 1989. The influence of crabs on litter processing in high intertidal mangrove forests in tropical Australia. *Oecologia* **78**, 191–198.

Robertson, A.I., Daniel, P.A. & Dixon, P. 1991. Mangrove forest structure and productivity in the Fly River estuary, Papua New Guinea. *Marine Biology* **111**, 147–155.

Rogers, K. & Saintilan, N. 2008. Relationships between surface elevation and groundwater in mangrove forests of southeast Australia. *Journal of Coastal Research* **24**, 63–69.

Rogers, K., Saintilan, N. & Cahoon, D. 2005a. Surface elevation dynamics in a regenerating mangrove forest at Homebush Bay, Australia. *Wetlands Ecology and Management* **13**, 587–598.

Rogers, K., Saintilan, N. & Heijnis, H. 2005b. Mangrove encroachment of salt marsh in Western Port Bay, Victoria: the role of sedimentation, subsidence, and sea level rise. *Estuaries* **28**, 551–559.

Rogers, K., Wilton, K.M. & Saintilan, N. 2006. Vegetation change and surface elevation dynamics in estuarine wetlands of southeast Australia. *Estuarine Coastal and Shelf Science* **66**, 3–4.

Rönnbäck, P., Troell, M., Kautsky, N. & Primavera, J.H. 1999. Distribution pattern of shrimps and fish among *Avicennia* and *Rhizophora* microhabitats in the Pagbilao mangroves, Philippines. *Estuarine Coastal and Shelf Science* **48**, 223–234.

Ross, P.M. 2001. Larval supply, settlement and survival of barnacles in a temperate mangrove forest. *Marine Ecology Progress Series* **215**, 237–249.

Ross, P.M. 2006. Macrofaunal loss and microhabitat destruction: the impact of trampling in a temperate mangrove forest, NSW Australia. *Wetlands Ecology and Management* **14**, 167–184.

Ross, P.M. & Underwood, A.J. 1997. The distribution and abundance of barnacles in a mangrove forest. *Australian Journal of Ecology* **22**, 37–47.

Roy, P.S., Williams, R.J., Jones, A.R., Yassini, I., Gibbs, P.J., Coates, B., West, R.J., Scanes, P.R., Hudson, J.P. & Nichol, S. 2001. Structure and function of south-east Australian estuaries. *Estuarine Coastal and Shelf Science* **53**, 351–384.

Saenger, P. 1998. Mangrove vegetation: an evolutionary perspective. *Marine and Freshwater Research* **49**, 277–286.

Saenger, P. 2002. *Mangrove Ecology, Silviculture and Conservation*. Dordrecht, The Netherlands: Kluwer.

Saenger, P. & Snedaker, S.C. 1993. Pan-tropical trends in mangrove above-ground biomass and litter fall. *Oecologia* **96**, 293–299.

Saenger, P., Specht, M.M., Specht, R.L. & Chapman, V.J. 1977. Mangal and coastal salt-marsh communities in Australasia. In *Wet Coastal Ecosystems*, V.J. Chapman (ed.). Amsterdam: Elsevier, 293–339.

Saintilan, N. 1997b. Above- and below-ground biomass of mangroves in a sub-tropical estuary. *Marine and Freshwater Research* **48**, 601–604.

Saintilan, N. 1997a. Above- and below-ground biomasses of two species of mangrove on the Hawkesbury River Estuary, New South Wales. *Marine and Freshwater Research* **48**, 147–152.

Saintilan, N. 1998. Photogrammetric survey of the Tweed River wetlands. *Wetlands (Australia)* **17**, 74–82.

Saintilan, N. 2004. Relationships between estuarine geomorphology, wetland extent and fish landings in New South Wales estuaries. *Estuarine Coastal and Shelf Science* **61**, 591–601.

Saintilan, N. & Hashimoto, T.R. 1999. Mangrove-saltmarsh dynamics on a bay-head delta in the Hawkesbury River estuary, New South Wales, Australia. *Hydrobiologia* **413**, 95–102.

Saintilan, N., Hossain, K. & Mazumder, D. 2007. Linkages between seagrass, mangrove and saltmarsh as fish habitat in the Botany Bay estuary, New South Wales. *Wetlands Ecology and Management* **15**, 277–286.

Saintilan, N., Rogers, K. & McKee, K.L. 2009. Salt marsh-mangrove interactions in Australasia and the Americas. In *Coastal Wetlands—An Integrated Ecosystem Approach*, G.M.E. Perillo et al. (eds). Amsterdam: Elsevier, 855–883.

Saintilan, N. & Williams, R.J. 1999. Mangrove transgression into saltmarsh environments in south-east Australia. *Global Ecology and Biogeography* **8**, 117–124.

Saintilan, N. & Williams, R.J. 2000. The decline of saltmarsh in southeast Australia: results of recent surveys. *Wetlands (Australia)* **18**, 49–54.

Saintilan, N. & Wilton, K. 2001. Changes in the distribution of mangroves and saltmarshes in Jervis Bay, Australia. *Wetlands Ecology and Management* **9**, 409–420.

Sakai, A., Paton, D. & Wardle, P. 1981. Freezing resistance of trees of the south temperate zone, especially subalpine species of Australasia. *Ecology* **62**, 563–570.

Sakai, A. & Wardle, P. 1978. Freezing resistance of New Zealand trees and shrubs. *New Zealand Journal of Ecology* **1**, 51–61.

Sasekumar, A. 1994. Meiofauna of a mangrove shore on the west coast of peninsular Malaysia. *Raffles Bulletin of Zoology* **42**, 901–915.

Satumanatpan, S. & Keough, M.J. 1999. Effect of barnacles on the survival and growth of temperate mangrove seedlings. *Marine Ecology Progress Series* **181**, 189–199.

Schaeffer-Novelli, Y., Cintron-Molero, G., Adaime, R.R. & de Camargo, T.M. 1990. Variability of mangrove ecosystems along the Brazilian coast. *Estuaries* **13**, 204–218.

Schaeffer-Novelli, Y., Cintron-Molero, G. & Soares, M.L.G. 2002. Mangroves as indicators of sea level change in the muddy coasts of the world. In *Muddy Coasts of the World: Processes, Deposits and Function*, T. Healy et al. (eds). Amsterdam: Elsevier Science, 245–262.

Schodde, R., Mason, I.J. & Gill, H.B. 1982. The avifauna of the Australian mangroves—a brief review of composition, structure and origin. In *Mangrove Ecosystems in Australia: Structure, Function and Management*, B.F. Clough (ed.). Townsville, Australia: Australian Institute of Marine Science, 141–150.

Scholl, D.W. 1968. Mangrove swamps: geology and sedimentology. In *Encyclopedia of Geomorphology*, F.R. Fairbridge (ed.). New York, USA: Reinhold, 683–688.

Schubel, J.R. & Pritchard, D.W. 1986. Responses of upper Chesapeake Bay to variations in discharge of the Susquehanna River. *Estuaries* **9**, 236–249.

Schumann, E.H., Largier, J.L. & Slinger, J.H. 1999. Estuarine hydrodynamics. In *Estuaries of South Africa*, B.R. Allanson & D. Baird (eds). Cambridge, United Kingdom: Cambridge University Press, 27–52.

Schwarzbach, A.E. & McDade, L.A. 2002. Phylogenetic relationships of the mangrove family Avicenniaceae based on chloroplast and nuclear ribosomal DNA sequences. *Systematic Botany* **27**, 84–98.

Scoffin, T.P. 1970. Trapping and binding of subtidal carbonate sediments by marine vegetation in Bimimi Lagoon, Bahamas. *Journal of Sedimentary Petrology* **40**, 249–273.

Semeniuk, T.A. 2000. Small benthic Crustacea of the Leschenault Inlet estuary. *Journal of the Royal Society of Western Australia* **83**, 429–441.

Semeniuk, V. 1980. Mangrove zonation along an eroding coastline in King Sound, north-western Australia. *Journal of Ecology* **68**, 789–812.

Semeniuk, V. & Wurm, P.A.S. 2000. Molluscs of the Leschenault Inlet estuary: their diversity, distribution, and population dynamics. *Journal of the Royal Society of Western Australia* **83**, 377–418.

Singh, N., Steinke, T.D. & Lawton, J.R. 1991. Morphological changes and the associated fungal colonization during decomposition of leaves of a mangrove, *Bruguiera gymnorrhiza* (Rhizophoraceae). *South African Journal of Botany* **57**, 151–155.

Skilleter, G.A. & Warren, S. 2000. Effects of habitat modification in mangroves on the structure of mollusc and crab assemblages. *Journal of Experimental Marine Biology and Ecology* **244**, 107–129.

Slim, F.J., Hemminga, M.A., Ochieng, C., Jannink, N.T., Cocheret de la Morinière, E. & van der Velde, G. 1997. Leaf litter removal by the snail *Terebralia palustris* (Linnaeus) and sesarmid crabs in an East African mangrove forest (Gazi Bay, Kenya). *Journal of Experimental Marine Biology and Ecology* **215**, 35–48.

Smith, T.J., III. 1987. Effects of light and intertidal position on seedling survival and growth in tropical tidal forests. *Journal of Experimental Marine Biology and Ecology* **110**, 133–146.

Smith, T.M. & Hindell, J.S. 2005. Assessing effects of diel period, gear selectivity and predation on patterns of microhabitat use by fish in a mangrove dominated system in S.E. Australia. *Marine Ecology Progress Series* **294**, 257–270.

Smoak, J.M. & Patchineelam, S.R. 1999. Sediment mixing and accumulation in a mangrove ecosystem: evidence from 210Pb, 234Th and 7Be. *Mangroves and Salt Marshes* **3**, 17–27.

Snedaker, S.C. 1995. Mangroves and climate change in the Florida and Caribbean region: scenarios and hypotheses. *Hydrobiologia* **295**, 43–49.

Solís, F., Ibanez, R., Hammerson, G., Hedges, B., Diesmos, A., Matsui, J.-M.H., Richards, S., Coloma, L.A., Ron, S., La Marca, E., Hardy, J., Powell, R., Bolanos, F. & Chaves, G. 2008. *Rhinella marina*. In: *IUCN Red List of Threatened Species*. Online. Available HTTP: http://www.iucnredlist.org (accessed 2 May 2009).

Sousa, W.P., Kennedy, P.G., Mitchell, B.J. & Ordonez, B.M. 2007. Supply-side ecology in mangroves: do propagule dispersal and seedling establishment explain forest structure? *Ecological Monographs* **77**, 53–76.

Spalding, M., Blasco, F. & Field, C. (eds). 1997. *World Mangrove Atlas*. Okinawa, Japan: International Society for Mangrove Ecosytems.

Spenceley, A.P. 1977. The role of pneumatophores in sedimentary processes. *Marine Geology* **24**: M31–M37.

Spenceley, A.P. 1987. Mangroves and intertidal morphology in Westernport Bay, Victoria, Australia—comment. *Marine Geology* **77**, 327–330.

Steinke, G.W., Naidoo, G. & Charles, L.M. 1983. Degradation of mangrove leaf and stem tissues in situ in Mgeni Estuary, South Africa. In *Biology and Ecology of Mangroves*, H.J. Teas (ed.). The Hague, The Netherlands: Dr. W. Junk, 141–149.

Steinke, T. 1999. Mangroves in South African estuaries. In *Estuaries of South Africa*, B.R. Allanson & D. Baird (eds). Cambridge, United Kingdom: Cambridge University Press, 119–140.

Steinke, T.D. 1975. Some factors affecting the dispersal and establishment of propagules of *Avicennia marina* (forsk.) Vierh. In *Proceedings of the International Symposium on the Biology and Management of Mangroves, University of Gainsville, Florida*, G.E. Walsh et al. (eds), Gainsville, Florida, USA: University of Florida, 404–414.

Steinke, T.D. 1986. A preliminary study of buoyancy behaviour in *Avicennia marina* propagules. *South African Journal of Botany* **52**, 559–565.

Steinke, T.D. 1995. A general review of the mangroves of South Africa. In *Wetlands of South Africa*, G.I. Cowan (ed.). Pretoria, South Africa: South African Wetlands Conservation Programme, Department of Environmental Affairs and Tourism, 53–73.

Steinke, T.D., Barnabas, A.D. & Somaru, R. 1990. Structural changes and associated microbial activity accompanying decomposition of mangrove leaves in Mgeni Estuary. *South African Journal of Botany* **56**, 39–48.

Steinke, T.D. & Charles, L.M. 1986. In vitro rates of decomposition of leaves of the mangrove *Bruguiera gymnorrhiza* as affected by temperature and salinity. *South African Journal of Botany* **52**, 39–42.

Steinke, T.D. & Charles, L.M. 1990. Litter production by mangroves. III. Wavecrest (Transkei) with predictions for other Transkei estuaries. *South African Journal of Botany* **56**, 514–519.

Steinke, T.D. & Naidoo, Y. 1991. Respiration and net photosynthesis of cotyledons during establishment and early growth of propagules of the mangrove, *Avicennia marina*, at three temperatures. *South African Journal of Botany* **57**, 171–174.

Steinke, T.D., Rajh, A. & Holland, A.J. 1993. The feeding behaviour of the red mangrove crab *Sesarma meinerti* de Man, 1887 (Crustacea: Decapoda: Grapsidae), and its effect on the degradation of mangrove leaf litter. *South African Journal of Marine Science* **13**, 151–160.

Steinke, T.D., Ward, C.J. & Rajh, A. 1995. Forest structure and biomass of mangroves in the Mgeni Estuary, South Africa. *Hydrobiologia* **295**, 159–166.

Stephens, W.M. 1962. Trees that make land. *Sea Frontiers* **8**, 219–230.

Stern, M.K., Day, J.W. & Teague, K.G. 1986. Seasonality of materials transport through a coastal freshwater marsh: riverine versus tidal forcing. *Estuaries* **9**, 301–308.

Steyaert, M., Moodley, L., Nadong, T., Moens, T., Soetaert, K. & Vincx, M. 2007. Responses of intertidal nematodes to short-term anoxic events. *Journal of Experimental Marine Biology and Ecology* **345**, 175–184.

Stokes, D. & Healy, T. 2005. Mangrove expansion and their human removal in Tauranga Harbour, New Zealand. In *Proceedings of the 17th Coastal and Ocean Engineering Conference and the 10th Australasian Port and Harbour Conference, Adelaide, Australia*. M. Townsend & D. Walker (eds). Barton, ACT, Australia: Australian Institution of Engineers, 577–582.

Stokes, D.J., Healy, T. & Cooke, P.J. 2009. Surface elevation changes and sediment characteristics of intertidal surfaces undergoing mangrove expansion and mangrove removal, Waikaraka Estuary, Tauranga Harbour, New Zealand. *International Journal of Ecology and Development* **12**, 88–106.

Struve, J., Falconer, R.A. & Wu, Y. 2003. Influence of model mangrove trees on the hydrodynamics in a flume. *Estuarine Coastal and Shelf Science* **58**, 163–171.

Stuart, S., Choat, B., Martin, K., Holbrook, N. & Ball, M. 2007. The role of freezing in setting the latitudinal limits of mangrove forests. *New Phytologist* **173**, 576–583.

Suarez, N. & Medina, E. 2006. Influence of salinity on Na+ and K+ accumulation, and gas exchange in *Avicennia germinans*. *Photosynthetica* **44**, 268–274.

Swales, A., Bell, R.G., Ovenden, R., Hart, C., Horrocks, M., Hermansphan, N. & Smith, R.K. 2007a. Mangrove-habitat expansion in the southern Firth of Thames: sedimentation processes and coastal-hazards mitigation. Environment Waikato Technical Report No. 2008/13. Hamilton, New Zealand: Environment Waikato (Waikato Regional Council).

Swales, A. & Bentley, S.J. 2008. Recent tidal-flat evolution and mangrove-habitat expansion: application of radioisotope dating to environmental reconstruction. In *Sediment Dynamics in Changing Environments (Proceedings of a Symposium Held in Christchurch, New Zealand, December 2008)*, J. Schmidt et al. (eds). Wallingford, U.K.: IAHS Press, 76–84.

Swales, A., Bentley, S.J., Lovelock, C. & Bell, R.G. 2007b. Sediment processes and mangrove-habitat expansion on a rapidly-prograding muddy coast, New Zealand. In *Coastal Sediments '07. Proceedings of the Sixth International Conference on Coastal Engineering and Science of Coastal Sediment Processes, New Orleans, May 2007*. Reston, Virginia, USA: American Society of Civil Engineers, 1441–1454.

Swales, A., Gorman, R., Oldman, J.W., Altenberger, A., Hart, C., Bell, R.G., Claydon, L., Wadhwa, S. & Ovenden, R. 2009. Potential future changes in mangrove habitat in Auckland's east-coast estuaries. Auckland, New Zealand: Auckland Regional Council, Technical Report No. 2009/079.

Swales, A., Hume, T.M., Oldman, J.W. & Green, M.O. 1997. Holocene sedimentation and recent human impacts on infilling in a drowned valley estuary. In *Proceedings of the 13th Australasian Coastal and Ocean Engineering Conference and the 6th Australasian Port and Harbour Conference, Christchurch, New Zealand*, J. Lumsden (ed.). Christchurch, New Zealand: Centre for Advanced Engineering, University of Canterbury, 895–900.

Swales, A., MacDonald, I.T. & Green, M.O. 2004. Influence of wave and sediment dynamics on cordgrass (*Spartina anglica*) growth and sediment accumulation on an exposed intertidal flat. *Estuaries* **27**, 225–243.

Swales, A., Williamson, R.B., Van Dam, L.F., Stroud, M.J. & McGlone, M.S. 2002. Reconstruction of urban stormwater contamination of an estuary using catchment history and sediment profile dating. *Estuaries* **25**, 43–56.

Syvitski, J.P.M., Vorosmarty, C.J., Kettner, A.J. & Green, P. 2005. Impact of humans on the flux of terrestrial sediment to the global coastal ocean. *Science* **308**, 376–380.

Tam, N.F.Y. & Wong, Y.-S. 1997. Accumulation and distribution of heavy metals in a simulated mangrove system treated with sewage. *Hydrobiologia* **352**, 67–75.

Tanaka, N., Sasaki, Y., Mowjood, M.I.M., Jindasa, K.B.S.N. & Homchuen, S. 2007. Coastal vegetation structures and their functions in tsunami protection: experience of the recent Indian Ocean tsunami. *Landscape Ecology and Engineering* **3**, 33–45.

Taylor, F.J. 1983. The New Zealand mangrove association. In *Biology and Ecology of Mangroves*, H.J. Teas (ed.). The Hague, The Netherlands: Dr. W. Junk, 77–79.

Teas, H.J. (ed.). 1983. *Biology and Ecology of Mangroves*. The Hague, Netherlands: Dr. W. Junk.

Temmerman, S., Govers, G., Wartel, S. & Meire, P. 2004. Modelling estuarine variations in tidal marsh sedimentation: response to changing sea level and suspended sediment concentrations. *Marine Geology* **212**, 1–19.

Terrados, J., Thampanya, U. & Duarte, C.M. 1997. The effect of increased sediment accretion on the survival and growth of *Rhizophora apiculata* seedlings. *Estuarine Coastal and Shelf Science* **45**, 697–701.

Thom, B.G. 1967. Mangrove ecology and deltaic geomorphology, Tabasco, Mexico. *Journal of Ecology* **55**, 301–343.

Thom, B.G. 1982. Mangrove ecology—a geomorphological perspective. In *Mangrove Ecosystems in Australia: Structure, Function and Management*, B.F. Clough (ed.). Townsville, Australia: Australian Institute of Marine Science, 3–17.

Thom, B.G., Wright, L.D. & Coleman, J.M. 1975. Mangrove ecology and deltaic-estuarine geomorphology: Cambridge Gulf-Ord River, Western Australia. *Journal of Ecology* **63**, 203–232.

Thomas, M.L. & Logan, A. 1992. A guide to the ecology of the shoreline and shallow-water marine communities of Bermuda. Bermuda Biological Station for Research Special Publication No. 30. Bermuda: Bermuda Biological Station for Research.

Thomas, M.L.H. 1993. Mangrove swamps in Bermuda. *Atoll Research Bulletin* **386**, 1–17.

Thorogood, C.A. 1985. Changes in the distribution of mangroves in the Port Jackson-Paramata River Estuary from 1930 to 1985. *Wetlands (Australia)* **5**, 91–93.

Thrush, S.F., Hewitt, J.E., Cummings, V.J., Ellis, J.I., Hatton, C., Lohrer, A. & Norkko, A. 2004. Muddy waters: elevating sediment input to coastal and estaurine habitats. *Frontiers in Ecology and the Environment* **2**, 299–306.

Tolhurst, T.J. & Chapman, M.G. 2007. Patterns in biogeochemical properties of sediments and benthic animals among different habitats in mangrove forests. *Austral Ecology* **32**, 775–788.

Tomlinson, P.B. 1986. *The Botany of Mangroves*. Cambridge, United Kingdom: Cambridge University Press.

Tortoise and Freshwater Turtle Specialist Group 1996. *Malaclemys terrapin*. In *IUCN Red List of Threatened Species*. International Union for Conservation of Nature and Natural Resources, Red List Unit, Cambridge, U.K. Online. Available HTTP: http://www.iucnredlist.org (accessed 2 May 2009).

Traynum, S. & Styles, R. 2007. Flow, stress and sediment resuspension in a shallow tidal channel. *Estuaries and Coasts* **30**, 94–101.

Twilley, R.R., Chen, R.H. & Hargis, T. 1992. Carbon sinks in mangrove forests and their implications to the carbon budget of tropical coastal ecosystems. *Water Air and Soil Pollution* **64**, 265–288.

Uncles, R.J. 2002. Estuarine physical processes research: some recent studies and progress. *Estuarine Coastal and Shelf Science* **55**, 829–856.

Underwood, A.J. & Barrett, G. 1990. Experiments on the influence of oysters on the distribution, abundance and sizes of the gastropod *Bembicium auratum* in a mangrove swamp in New South Wales, Australia. *Journal of Experimental Marine Biology and Ecology* **137**, 25–45.

United Nations Environment Program, World Conservation Monitoring Centre (UNEP-WCMC). 2006. *In the Front Line: Shoreline Protection and Other Ecosystem Services from Mangroves*. Cambridge, United Kingdom: United Nations Environment Program, World Conservation Monitoring Centre.

Valette-Silver, N.J. 1993. The use of sediment cores to reconstruct historical trends in contamination of estuarine and coastal sediments. *Estuaries* **16**, 577–588.

Valiela, I., Bowen, J.L. & York, J.K. 2001. Mangrove forests: one of the world's threatened major tropical environments. *BioScience* **51**, 807–815.

Vance, D.J., Haywood, M.D.E., Heales, D.S., Kenyon, R.A., Loneragan, N.R. & Pendrey, R.C. 1996. How far do prawns and fish move into mangroves? Distribution of juvenile banana prawns *Penaeus merguiensis* and fish in a tropical mangrove forest in northern Australia. *Marine Ecology Progress Series* **131**, 115–124.

Van der Valk, A.G. & Attiwill, P.M. 1984. Decomposition of leaf and root litter of *Avicennia marina* at Westernport Bay, Victoria, Australia. *Aquatic Botany* **18**, 205–221.

Vanhove, S., Vincx, M., van Gansbeke, D., Gijselinck, W. & Schram, D. 1992. The meiobenthos of five mangrove vegetation types in Gazi Bay, Kenya. *Hydrobiologia* **247**, 99–108.

Vann, J.H. 1980. Shoreline changes in mangrove areas (Brazil and Guiana). *Zeitschrift für Geomorphologie Supplementband* **34**, 255–261.

Van Santen, P., Augustinus, P.G.E.F., Janssen-Stelder, B.M., Quartel, S. & Tri, N.H. 2007. Sedimentation in an estuarine mangrove system. *Journal of Asian Earth Sciences* **29**, 566–575.

Van Wijnen, H.J. & Bakker, J.P. 2001. Long-term surface elevation change in salt marshes: a prediction of marsh response to future sea-level rise. *Estuarine Coastal and Shelf Science* **52**, 381–390.

Vaughan, T.W. 1909. Geologic work of mangroves in southern Florida. *Smithsonian Miscellaneous Collection* **52**, 461–464.

Wakamatsu, A. & Tomiyama, K. 2000. Seasonal changes in the distribution of Batillarid snails on a tidal flat near the most northern mangrove forest in Atago River Estuary, Kyushu, Japan. *Japanese Journal of Malacology* **59**, 225–243.

Walbert, K. 2002. *Investigations in the New Zealand mangrove, Avicennia marina var. australasica*. MSc thesis, University of Waikato, Hamilton, New Zealand.

Walling, D.E. 1999. Linking land use, erosion and sediment yields in river basins. *Hydrobiologia* **410**, 223–240.

Walsh, J.P. & Nittrouer, C.A. 2004. Mangrove-bank sedimentation in a mesotidal environment with large sediment supply, Gulf of Papua. *Marine Geology* **208**, 225–248.

Walton, M.E.M., Samonte-Tan, G.P.B., Primavera, J.H., Edwards-Jones, G. & Le Vay, L. 2006. Are mangroves worth replanting? The direct economic benefits of a community-based reforestation project. *Environmental Conservation* **33**, 335–343.

Ward, C.J. & Steinke, T.D. 1982. A note on the distribution and approximate areas of mangroves in South Africa. *South African Journal of Botany* **1**, 51–53.

Warren, J.H. 1987. *Behavioural ecology of crabs in temperate mangrove swamps*. PhD thesis, University of Sydney, Sydney, Australia.

Warren, J.H. 1990. The use of open burrows to estimate abundances of intertidal estuarine crabs. *Australian Journal of Ecology* **15**, 277–280.

Warren, J.H. & Underwood, A.J. 1986. Effects of burrowing crabs on the topography of mangrove swamps in New South Wales. *Journal of Experimental Marine Biology and Ecology* **102**, 2–3.

Wasson, R.J. 1994. Annual and decadal variation of sediment yield in Australia and some global comparisons. In *Variability in Stream Erosion and Sediment Transport (Proceedings of the Canberra Symposium, December 1994)*, L.J. Olive et al. (eds). Wallingford, U.K.: IAHS Press, 269–279.

Wasson, R.J. & Galloway, R.W. 1986. Sediment yield in the barrier range before and after European settlement. *Australian Rangeland Journal* **8**, 79–90.

Wells, A.G. 1983. Distribution of mangrove species in Australia. In *Biology and Ecology of Mangroves*, H.J. Teas (ed.). The Hague, The Netherlands: Dr. W. Junk, 57–76.

Wells, J.T. 1983. Dynamics of coastal fluid muds in low-, moderate-, and high-tide-range environments. *Canadian Journal of Fisheries and Aquatic Sciences* **40**, 130–142.

Wells, J.T. & Coleman, J.M. 1981. Physical processes and fine-grained sediment dynamics, coast of Surinam, South America. *Journal of Sedimentary Petrology* **51**, 1053–1068.

West, R.J., Thorogood, C.A., Walford, T.R. & Williams, R.J. 1985. *Estuarine Inventory for New South Wales, Australia*. Sydney: Division of Fisheries, Department of Agriculture New South Wales.

Whitfield, A.K. 1994. An estuary-association classification for the fishes of southern Africa. *South African Journal of Science* **90**, 411–417.

Wieser, W. 1953. Die Beziehung zwischen Mundhoelengestalt, Ernaehrungsweise und Vorkommen bei freilebenden marine Nematoden. *Archiv für Zoologie* **4**, 439–484.

Wildland Consultants. 2003. *Ecological restoration and enhancement of Waikaraka Estuary, Tauranga Harbour. Report prepared for the Waikaraka Estuary Management Group, Tauranga*. Tauranga, New Zealand: Waikaraka Estuary Management Group.

Wildland Consultants. 2005. Extent and effects of mangrove clearance in the Whangamata Harbour. Rotorua, New Zealand: Wildlands Consultants Ltd., Contract report No. 1274.

Wilkie, M.L. & Fortuna, S. 2003. Status and trends in mangrove area extent worldwide. Forest Resources Assessment Working Paper No. 63. Rome, Italy: Forestry Department, Food and Agriculture Organization of the United Nations.

Williams, R.J. & Meehan, A.J. 2004. Focusing management needs at the sub-catchment level via assessments of change in the cover of estuarine vegetation, Port Hacking, NSW, Australia. *Wetlands Ecology and Management* **12**, 499–518.

Williams, R.J. & Watford, F.A. 1997. Identification of structures restricting tidal flow in New South Wales, Australia. *Wetlands Ecology and Management* **5**, 87–97.

Wilmshurst, J.M. 1997. The impact of human settlement on vegetation and soil stability in Hawke's Bay, New Zealand. *New Zealand Journal of Botany* **35**, 97–111.

Wilton, K. 2001. Changes in coastal wetland habitats in Careel Bay, Pittwater, N.S.W. from 1940–1996. *Wetlands (Australia)* **19**, 72–86.

Wilton, K. 2002. *Mangrove and saltmarsh habitat dynamics in selected NSW estuaries*. PhD thesis, Australian Catholic University, Australia.

Wilton, K. & Saintilan, N. 2000. *Protocols for Mangrove and Saltmarsh Habitat Mapping*. Sydney, Australia: Australian Catholic University.

Wolanski, E. 1995. Transport of sediment in mangrove swamps. *Hydrobiologia* **295**, 31–42.

Wolanski, E., Jones, M. & Bunt, J.S. 1980. Hydrodynamics of a tidal creek-mangrove swamp system. *Australian Journal of Marine and Freshwater Research* **31**, 431–450.

Wolanski, E., Moore, K., Spagnol, S., D'Adamo, N. & Pattiaratchi, C. 2001. Rapid, human-induced siltation of the macro-tidal Ord River estuary, Western Australia. *Estuarine Coastal and Shelf Science* **53**, 717–732.

Wolanski, E., Spagnol, S. & Lim, E.B. 2002. Fine sediment dynamics in the mangrove-fringed, muddy coastal zone. In *Muddy Coasts of the World: Processes, Deposits and Function*, T. Healy et al. (eds). Amsterdam: Elsevier Science, 279–292.

Wong, Y.-S., Tam, N.F.Y., Chen, G.-Z. & Ma, H. 1997a. Response of *Aegiceras corniculatum* to synthetic sewage under simulated tidal conditions. *Hydrobiologia* **352**, 89–96.

Wong, Y.S., Tam, N.F.Y. & Lan, C.Y. 1997b. Mangrove wetlands as wastewater treatment facility: a field trial. *Hydrobiologia* **352**, 49–59.

Woodroffe, C.D. 1982. Litter production and decomposition in the New Zealand mangrove, *Avicennia marina* var. *resinifera*. *New Zealand Journal of Marine and Freshwater Research* **16**, 179–188.

Woodroffe, C.D. 1985a. Studies of a mangrove basin, Tuff Crater, New Zealand: I. Mangrove biomass and production of detritus. *Estuarine Coastal and Shelf Science* **20**, 265–280.

Woodroffe, C.D. 1985b. Studies of a mangrove basin, Tuff Crater, New Zealand: III. The flux of organic and inorganic particulate matter. *Estuarine Coastal and Shelf Science* **20**, 447–461.

Woodroffe, C.D. 1990. The impact of sea-level rise on mangrove shorelines. *Progress in Physical Geography* **14**, 483–520.

Woodroffe, C.D. 1992. Mangrove sediment and geomorphology. In *Tropical Mangrove Ecosystems*, A.I. Robertson & D.M. Alongi (eds). Washington, D.C., USA: American Geophysical Union, 7–41.

Woodroffe, C.D., Bardsley, K.N., Ward, P.J. & Hanley, J.R. 1988. Production of mangrove litter in a macrotidal embayment, Darwin Harbour, N.T., Australia. *Estuarine Coastal and Shelf Science* **26**, 581–598.

Woodroffe, C.D. & Davies, G. 2009. The morphology and development of tropical coastal wetlands. In *Coastal Wetlands—An Integrated Ecosystem Approach*, G.M.E. Perillo et al. (eds). Amsterdam: Elsevier, 65–88.

Wright, C.I., Lindsay, P. & Cooper, J.A.G. 1997. The effect of sedimentary processes on the ecology of the mangrove-fringed Kosi estuary/lake system, South Africa. *Mangroves and Salt Marshes* **1**, 79–94.

Wright, I.J. & Westoby, M. 2002. Leaves at low versus high rainfall: coordination of structure, lifespan and physiology. *New Phytologist* **155**, 403–416.

Wright, I.J., Westoby, M. & Reich, P.B. 2002. Convergence towards higher leaf mass per area in dry and nutrient-poor habitats has different consequences for leaf life span. *Journal of Ecology* **90**, 534–543.

Yan, Z., Wang, W. & Tang, D. 2007. Effect of different time of salt stress on growth and some physiological processes of *Avicennia marina* seedlings. *Marine Biology* **152**, 581–587.

Yates, E.J., Ashwath, N. & Midmore, D.J. 2004. Responses to nitrogen, phosphorus, potassium and sodium chloride by three mangrove species in pot culture. *Trees Structure and Function* **16**, 120–125.

Ye, Y., Tam, N.F.Y., Lu, C.Y. & Wong, Y.S. 2005. Effects of salinity on germination, seedling growth and physiology of three salt-secreting mangrove species. *Aquatic Botany* **83**, 193–205.

Young, B.M. & Harvey, L.E. 1996. A spatial analysis of the relationship between mangrove (*Avicennia marina* var. *australasica*) physiognomy and sediment accretion in the Hauraki Plains, New Zealand. *Estuarine Coastal and Shelf Science* **42**, 231–246.

Oceanography and Marine Biology: An Annual Review, 2010, **48**, 161-212
© R. N. Gibson, R. J. A. Atkinson, and J. D. M. Gordon, Editors
Taylor & Francis

THE EXPLOITATION AND CONSERVATION
OF PRECIOUS CORALS*

GEORGIOS TSOUNIS[1], SERGIO ROSSI[2], RICHARD GRIGG[3],
GIOVANNI SANTANGELO[4], LORENZO BRAMANTI[4] & JOSEP-MARIA GILI[1]

[1]*Departamento de Biología Marina, Institut de Ciències del Mar (CSIC),
Passeig Marítim de la Barceloneta, 37–49, 08003 Barcelona, Spain*
E-mail: gili@icm.csic.es, georgios@icm.csic.es (corresponding author)
[2]*Institut de CiènciaiTecnologia Ambientals (Universitat Autònoma de Barcelona),
Edifici Cn Campus UAB, Cerdanyola del Vallés (Barcelona) 08193, Spain*
E-mail: sergio.rossi@uab.cat
[3]*Department of Oceanography, University of Hawaii, Honolulu, Hawaii 96822, USA*
E-mail: rgrigg@soest.hawaii.edu
[4]*Departamento di Biologia, University of Pisa, Via A. Volta, 6 I-56126 Pisa, Italy*
E-mail: gsantangelo@biologia.unipi.it, lbramanti@biologia.unipi.it

Abstract   Precious corals have been commercially exploited for many centuries all over the world. Their skeletons have been used as amulets or jewellery since antiquity and are one of the most valuable living marine resources. Precious coral fisheries are generally characterized by the 'boom-and-bust' principle, quickly depleting a discovered stock and then moving on to the next one. Most known stocks are overexploited today, and populations are in decline. The unsustainable nature of most fisheries is clearly revealed by analyzing all available data. Precious corals belong to the functional group of deep corals and are important structure-forming organisms, so called ecosystem engineers, that provide shelter for other organisms, increasing biodiversity. Yet, their management is usually focused on single species rather than a holistic habitat management approach. This review compares the biology of precious corals as well as the historical ecology and the socioeconomy of their fisheries to improve precious coral management and conservation. The analysis demonstrates that a paradigm shift is necessary in precious coral exploitation, not only to conserve habitats of high biodiversity but also to achieve sustainable fisheries and stabilize a specialized jewellery industry.

Introduction

The exploitation of living marine resources for purposes other than for food and agricultural applications has a long tradition. Examples include coral, sponge and pearl fisheries, among others (Dall 1883, Russell-Bernard 1972, Pronzato 1999). Because they are in limited supply and coral jewellery can sell for high prices, precious corals are one of the most valuable marine resources. The use of precious corals as amulets or jewellery dates back millennia, and their industrial exploitation began many centuries ago (Tescione 1973). Their biology differs in many aspects from other commercially exploited marine organisms. Being sessile cnidarians, they are colonial organisms that depend on stable environmental conditions and are characterized by slow growth (Grigg 1984). The use

---

* Manuscript submitted 3 March 2009.

of scuba-diving, remotely operated vehicles and manned submersibles improved our understanding of the population dynamics and ecological significance of these key species (Grigg 1976, Rogers 1999). However, in a similar way, advancing technology also permitted an overefficient exploitation (Food and Agriculture Organization [FAO] 1988, U.S. Department of Commerce 1989, Santangelo & Abbiati 2001, Grigg 2004, Tsounis et al. 2007). Apart from the slow renewal rate of this resource, the reason that all known stocks are overexploited are likely due to the same dynamics that can be observed in other fisheries (Pauly et al. 2002). The conflict between sustainable management and short-term economical interests, as well as the tragedy of the commons, are universal factors that have led to overexploitation of many commercial species. However, since precious coral populations provide shelter to other marine organisms and thus increase biodiversity or act as a fish nursery, managers realized the need for holistic management plans (Western Pacific Regional Management Council [WPCouncil] 2007, Magnuson-Stevens Fishery Conservation and Management Act).

There are currently sufficient data to create complete management models for only a few species, so simple models have to be used (Grigg 2001, Tsounis et al. 2007). In the future, density-dependent full-population models may help predict exploitation limits that allow harvested populations to maintain their ability to fulfil their ecological function (Caswell 2001, Santangelo et al. 2007).

While ecologists develop better models to sustainably manage the marine environment, it is wiser to apply conservative exploitation limits (Sutherland 2000). In the case of precious corals, errors in ecosystem management take especially long to rectify and may have permanent consequences that likely affect their entire ecosystem. This is therefore a critical moment for managers and decision makers wishing to preserve not only these habitats of high biodiversity, but also a traditional industry that unites the craft of fishing with the talent of jewellery artists, together enriching our culture.

To contribute to improving precious coral fishery management, this review analyses historical data on the exploitation of precious corals, summarizes the current knowledge of their ecology, identifies gaps of knowledge and discusses their management and conservation.

Ecology of precious corals

Taxonomic classification and functional definition

Precious corals belong primarily to three orders of the class Anthozoa (phylum Cnidaria), which are the Gorgonacea, Zoanthidea (subclass Hexacorallia; see Cairns et al. 2002), and Antipatharia (subclass Ceriantipatharia; see Cairns et al. 2002). The most valuable species are red and pink corals of the genus *Corallium* in the order Gorgonacea, such as the legendary Mediterranean red coral (*Corallium rubrum,* Figure 1), as well as the Pacific species *Paracorallium japonicum, Corallium elatius, C. konojoi* and *C. secundum.* Another important group of precious corals are black corals in the order Antipatharia. At least 10 species (mostly from the genus *Antipathes*) from the 150 species known worldwide are used in the jewellery industry. Other gorgonians used for jewellery are gold corals from the families Gerardiidae and Primnoidae (*Primnoa resedaeformis, P. willeyi, Narella* spp., *Callogorgia* sp.), as well as bamboo corals in the family Isididae (*Acanella* spp., *Keratoisis* spp., *Lepidisis olapa*). Semiprecious species, mainly stylasterine corals, *Allopora* in the class Hydrozoa, blue corals (*Heliopora*), *Tubipora* and several gorgonians in the family Melitodiidae, are used for jewellery to some extent but are of low value due to their skeleton quality and thus are not dealt with here in detail. Reef-building stony corals (Madreporaria or Scleractinia) are also of commercial value, but generally not within the jewellery industry (due to their pores) and are thus not considered as precious corals.

As a result of their bathymetric range, habitat preference and ecological role, precious corals belong to the functional group called "structure-forming deep corals" (Rogers 1999, Lumsden et al. 2007). By increasing 3-dimensional structural complexity, they provide habitat for commercially

Figure 1   (See also Colour Figure 1 in the insert following page 212.) Unusually large red coral colony of nearly 20 cm (left) and an average sized colony of 3 cm height (right). (Courtesy of Georgios Tsounis (left) and Sergio Rossi (right).)

important fish species (Witherell & Coon 2000, Krieger & Wing 2002). Thus, the inclusion of precious corals in this group furthermore acknowledges their ecological significance. They have a complex branching morphology and sufficient size to act as refuge for other organisms. Some structure-forming deep corals are reef building, but precious corals are non-reef-building deep corals. Deep corals live mainly in continental shelves, slopes, canyons and seamounts, in depths of more than 50 m (although some species also extend into shallower water).

However, true deep-sea organisms are generally defined as those occurring deeper than the continental shelf (i.e., deeper than 200 m), whereas the term deep corals refers to corals that are neither shallow water corals nor true deep-sea organisms.

Deep corals are slow growing, long-lived organisms that are sensitive to disturbance such as dredging or extraction. Little is known about the gene flow between populations (Baco et al. 2006) and the need for their conservation results from the increasing anthropogenic impact on these organisms. Even though some deep precious corals have been exploited since antiquity, they have been known to science only since Carl von Linné (Linnaeus) published his *Systema Natura* in 1758. Today, they are the least-understood group of all corals (Rogers 1999, Roberts et al. 2006).

Deep coral communities are 'hot spots' of biodiversity and have been identified as a habitat for commercially important species of rockfishes, shrimp and crabs (Koenig 2001, Husebø et al. 2002, Krieger & Wing 2002, Freiwald et al. 2004, Martensen & Fosså 2006). Furthermore, their biodiversity may provide numerous targets for chemical and pharmaceutical research because several sponges and gorgonian species contain bioactive compounds that have medical potential (Ehrlich et al. 2006). Finally, due to their worldwide distribution and the fact that some gorgonian species can live for centuries, deep corals provide a source for geochemical and isotopic data that serve as a proxy for reconstructing past changes in ocean climate and oceanographic conditions (Smith et al. 1997, 2000, 2002, Adkins et al. 1998, Weinbauer et al. 2000, Risk et al. 2002, Frank et al. 2004, Thresher et al. 2004, Williams et al. 2007, 2009).

Primary threats to deep corals are damage by destructive fishing (i.e., bottom trawling; Witherell & Coon 2000, Krieger 2001, Hall-Spencer et al. 2002). Secondary threats include damage by entanglement in lost long lines and netting (Rogers 1999). The impact of pollution and siltation due to coastal development has not yet been studied. Secondary threats also include natural smothering by sediments, infection by parasites and invasion by alien species (Grigg 2004, Kahng & Grigg 2005, Kahng 2007). Global climate change and ocean acidification may alter the calcification of the skeleton of certain species and result in a weaker skeleton or lower rates (Hoegh-Guldberg et al.

2007). In the United States, the Magnus-Stevenson Act affirms the Regional Fishery Management Council's authority to protect deep-sea coral ecosystems as part of their fishery management plans without having to prove that corals constitute essential fish habitat.

A detailed review of deep corals can be found in the work of Rogers (1999), Roberts et al. (2006) and Lumsden et al. (2007). For the purpose of this review, the next section focuses on the biology of precious corals (Table 1).

General biological traits and ecological role of precious corals

All precious coral species are ahermatypic and non-reef building and occur in relatively deep water. Their skeletons consist of a complex of proteins or calcium carbonate. In general, they show a dendritic (arborescent) form, although some bamboo and black coral species grow in spiral or bushy form. Precious corals are long-lived organisms with low growth rates and low reproductive rates. Age at first reproduction is reached at a substantial age of more than a decade (except in *Corallium rubrum*, see p. 168). Many species can probably reach an age of more than a century (see Table 1).

Because precious corals lack zooxanthellae, they are benthic suspension-feeders, which means that they are sessile life forms on the seafloor that filter seston particles out of the water column (Cloern 1982, Officer et al. 1982). Their filtration rates are lower than those of sponges, bivalves and ascidians, but they do play an important role in pelago-benthic energy transfer processes (Fréchette et al. 1989, Kimmerer et al. 1994, Gili & Coma 1998, Riisgård et al. 1998, Arntz et al. 1999). They are usually gonochoric and brood their larvae, and some species synchronize their spawning (see next section). Fragmentation and reattachment (asexual reproduction; see Fautin 2002) appears to occur rarely in precious corals (Grigg 1994).

As in other corals, precious corals are important keystone species in various ecosystems because they provide 3-dimensional complexity to habitats, structuring and stabilizing the ecosystem (Hiscock & Mitchell 1980, Mitchell et al. 1993) and thus significantly increasing biodiversity (True 1970, Dayton et al. 1974, Jones et al. 1994, Jones & Lawton 1994, Gutiérrez et al., in press). An overview on the biology of the major precious coral groups is provided next.

Tropical pink and red corals (Corallium *and* Paracorallium *sp.)*

The genus *Corallium* (Cuvier 1798) along with the genus *Paracorallium* (Bayer & Cairns 2003) belong to the family Corallidae (Lamouroux 1812) and contain the most valuable precious corals due to their hard calcium-carbonate skeleton. Thirty-one *Corallium* species are known, of which seven species are currently used for jewellery (Cairns 2007): *Corallium secundum, C. regale, C. elatius, C. konojoi, C.* sp. nov., *C. rubrum,* and *Paracorallium japonicum.* Because it has been particularly well studied and is geographically separated from the other *Corallium* species, the Mediterranean *Corallium rubrum* is discussed separately (see next section).

Corallium species are found throughout the world in tropical, subtropical and temperate oceans (Grigg 1976, Bayer & Cairns 2003), including five species in the Atlantic, two from the Indian Ocean, three from the eastern Pacific Ocean, and 15 from the western Pacific Ocean (Grigg 1976, Cairns 2007). They have also been found on seamounts in the Gulf of Alaska (Baco & Shank 2005, Heifetz et al. 2005), the Davidson Seamount off the California coast (DeVogeleare et al. 2005) and the New England Seamounts in the Atlantic (Morgan et al. 2006). The depth range of this genus extends from about 100 to 2400 m (Bayer 1956), which is a broader range than found in *C. rubrum.* Dense tropical *Corallium* populations of commercial interest have been exploited at 200–500 and 1000–1500 m off Japan, the Island of Taiwan, Midway Island and the Emperor Seamounts. Near Japan, *Paracorallium japonicum* can be found at depths of 76 to 280 m on rocky bottoms in Sagami Bay on Japan's Pacific coast, as well as between the Ogasawara Islands and Taiwan, and near the Goto Islands. Pink coral (*Corallium elatius*) is found between 100 and 276 m at Wakayama at the

Table 1 Biology of precious coral species

Species	Common name	Zoogeographic distribution	Depth range (m)	Maximum height (cm)	Growth rate (height)	Growth rate (diameter) mm yr^{-1}	Maximum age (yr)	Reference
Corallium rubrum	Red coral	Mediterranean and neighbouring Atlantic shores	7–300	50	1.78 + 0.7 mm yr^{-1}	0.24 ± 0.05	ca. 100	Marchetti 1965, Tescione 1973, Zibrowius et al. 1984, Garrabou & Harmelin 2002
						0.34 ± 0.15		Marschal et al. 2004
						0.62 ± 0.15		Bramanti et al. 2007
Corallium secundum	Pink coral, angel skin, boké	Hawaiian Archipelago	340–475	75	0.9 cm yr^{-1}		45	Grigg 1974, 2002
Corallium sp. nov.	Midway deep-sea coral	Midway Island to Emperor Seamounts (W. Pacific)	700–1500	—	—	0.17	>90	Roark et al. 2005 Grigg 1984
Corallium konojoi	Shiro sango (white coral)	Japan to northern Philippine Islands	50–382	~80	—	0.58		Kithara 1902, Kishinouye 1903, Grigg 1984, N. Iwasaki & Suzuki in press
Corallium elatius	Momoiro sango (red coral)	Northern Philippines to Japan	150–330	110		0.19 ± 0.15		Kithara 1902, Grigg 1984, N. Iwasaki & Suzuki in press
						0.15		Hasegawa & Yamada in press
Corallium regale	Pink coral	Hawaii	390–500	—	0.58			Bayer 1956
Paracorallium japonicum	Aka sango (red coral)	Japan, Okinawa and Bonin Islands	100–300	~100		0.3 ± 0.14		Kithara 1902, Grigg 1984, N. Iwasaki personal communication
Antipathes griggi	Black coral	Major Hawaiian islands	30–100	250	6.4 cm yr^{-1}			Grigg 1976, 1984, Roark et al. 2005
Antipathes grandis	Black coral	Major Hawaiian islands	45–100	300	6.1 cm yr^{-1}			Grigg 1976, 1984
Antipathes salix	Black coral	Caribbean	190–330	250	4.5 cm yr^{-1}	0.18–1.149	50	Olsen & Wood 1980, Brook 1889

(continued on next page)

Table 1 (continued) Biology of precious coral species

Species	Common name	Zoogeographic distribution	Depth range (m)	Maximum height (cm)	Growth rate (height)	Growth rate (diameter) mm yr⁻¹	Maximum age (yr)	Reference
Gerardia	Gold coral	Hawaii, Caribbean	300–600	250	6.6 cm yr^{-1}		250 ± 70	Grigg 1984, Messing et al. 1990
							1800 ± 300	Druffel et al. 1995, Goodfriend 1997
						0.014–0.045	450 ± 2742	Roark et al. 2005
Primnoa resedaeformis	Gold coral	South-east Alaska to Amchitka, Aleutian Islands	50–80	—	4–5 m			Cimberg et al. 1981
		(NE Atlantic)	65–3200					Sherwood et al. 2005
			64–457		1.5–2.5 mm		320	Risk et al. 2002
Primnoa willeyi	Gold coral	South-east Alaska to Amchitka, Aleutian islands	50–80	—	—			Cimberg et al. 1981

Pacific coast, from the Ogasawara islands to the northern China Sea, and off the Goto Islands. One of the largest *C. elatius* specimens was harvested off Okinawa in 2006. It measured 1.1 m in height and 1.7 m in width and weighed 67 kg (Iwasaki & Suzuki in press). *Corallium konojoi* is distributed at all of these locations at a similar depth range. (N. Iwasaki personal communication, Kishinouye 1903, 1904, Eguchi 1968, Seki 1991, Nonaka et al. 2006, Nonaka 2010).

In the Hawaiian Islands, *C. secundum* has been found to grow on flat exposed substrata, whereas *C. regale* prefers encrusted uneven rocky bottom habitat (Grigg 1975). Both species are absent from shelf areas (<400 m depth), off populated islands, and substrata periodically covered with sand and silt (Grigg 1994). Similar to red coral (Tsounis et al. 2006c), other *Corallium* species exposed to the Kuroshio current have been found to dwell in shallower water than usual, which may be attributed to the high productivity of this turbid zone (Kuroshio means black current in Japanese). Because Mediterranean red coral thrives better under turbid conditions and has a varied diet, it may be that the diet spectrum among *Corallium* species is similarly broad, also including particulate organic matter (see p. 168).

Corallium species have a hard intense red or pink calcium carbonate skeleton, and Pacific species can grow to over 1 m in height (Iwasaki & Suzuki in press). The lifespan of most species has been estimated at about 45–100 yr. Growth rates are low (see Table 1). In *C. secundum*, studies using growth ring analysis resulted in an estimate of 9 mm height increase yr^{-1} (Grigg 1976), but radiometric studies indicated that growth rings in other species may not be annual and growth rates might be slower (Roark et al. 2006). However, even among modern radiometric techniques there are contradictions between the data (see p. 172).

Other species have been found to grow considerably slower: 0.3 ± 0.14 mm yr^{-1} (*Paracorallium japonicum*), 0.19 ± 0.04 mm yr^{-1} (*Corallium elatius*), and 0.58 mm yr^{-1} (*C. konojoi*) (N. Iwasaki personal communication). *Corallium konojoi* has been recorded to reach 30 cm, while *Paracorallium japonicum* and *C. elatius* reach a height of 1 m (Iwasaki & Suzuki in press). Few data on the reproductive biology are available at present for species other than *C. rubrum* and *C. secundum* (Kishinouye 1903, 1904). The latter reaches sexual maturity at 12 yr.

Corallium secundum has been found to coexist in the same depth zone and habitat with bamboo coral (*Lepidisis olapa*) and the parasitic gold coral (*Gerardia* spp.) at nearly 400 m depth off Makapu'u, Oahu (Grigg 1984). Populations of *Corallium secundum* in Hawaii (Makapu'u bed) were dominated by 15- to 20-yr-old colonies, and the oldest colonies found were 80 yr old (corresponding to a height of 80 cm). Natural mortality in the absence of fishing was estimated at 6% (Grigg 1984, 1994), three times higher than Mediterranean *C. rubrum* (Tsounis et al. 2007). Studies that used population trends to analyse fishing pressure found that coral abundance remained similar after a period of harvesting pressure that reduced the biomass by extracting older, larger corals (Grigg 2002). However, DNA microsatellite research suggested that the harvesting pressure on Hawaiian seamounts might have led to inbreeding suppression (Baco & Shank 2005).

Mediterranean red coral (Corallium rubrum)

The Mediterranean red coral (*Corallium rubrum* L. 1758, Gorgonacea, Octocorallia) is an arborescent gorgonian whose colonies can reach a height of 50 cm (Garrabou & Harmelin 2002). One of its main habitats is the hard substratum in the so-called coralligéne (Laubier 1966), where it is a characteristic species of high importance (Ballesteros 2006). The coralligenous zone extends from the lower photophilic algae to more than 100 m depth. Calcareous algae growing on coarse gravelly substrata coalesce the calcareous sediments to form a continuous, organogenic substratum on which a community develops that is comparable with tropical coral reefs in its diversity and complexity (Margalef 1985). The name 'coralligenous' may originate from findings of red coral branches and calcareous organisms in trawling hauls of semidark sublittoral bottoms with coarse gravel, which were thought to be "generators of coral" (Ros et al. 1984). According to other authors, the term coralligenous originated from the coralline algae (calcareous algae), which form the secondary substratum on which the whole coralligenous community settles (Sarà 1969).

Red coral is one of the most long-lived inhabitants of the coralligenous, possibly living for more than 100 yr (Riedl 1983, García-Rodríguez & Massó 1986a). It is a sciaphilous species that can be found in depths of 5–800 m, although more commonly at 30–200 m (Carpine & Grasshoff 1975, Rossi et al. 2008, Costantini et al. 2009), and is distributed throughout the Mediterranean and the neighbouring Atlantic coasts (Marchetti 1965, Tescione 1973, Zibrowius et al. 1984, Chintiroglou & Dounas-Koukouras 1989). Its commercial beds, however, are found in Sardinia, Corsica, Elba, southern Italy, Croatia, the Greek islands, Turkey, Mallorca, Alboran Sea, Costa Brava, southern France and the northern African coast (Liverino 1983). Other precious corals occur exclusively in deeper water, mainly below the euphotic zone (Grigg 1984).

Red coral is one of the most thoroughly studied gorgonians because it has been of interest to science since the controversies over whether it should be included in the plant or animal kingdom (Marsili 1707). Lacaze-Duthiers's pioneering (1864) study on the biology of red coral started a series of studies on its reproduction, growth and population dynamics.

Gorgonians usually have a protein (gorgonin) skeleton, but red coral grows a calcium carbonate skeleton that owes its red colour to a carotenoid pigment of unknown function (Cvejic et al. 2007). Furthermore, it is calcitic instead of aragonitic, as scleractinian skeletons usually are (Dauphin 2006). Long-term experimental data showed that it took 22 yr for red coral colonies growing within a cave to grow to colonies of 4- to 8-mm thick and 1.3–7 cm in height, with corresponding growth rates of 0.24 ± 0.05 mm yr^{-1} in base diameter and 1.78 ± 0.7 mm yr^{-1} in height (Garrabou & Harmelin 2002). The 22-yr-old corals developed only one to eight branches, further stressing how long it takes for these organisms to provide habitat structure. However, these corals grew in particular conditions (a cave) with relatively little water movement, and because suspension-feeders depend not only on seston concentration but also water movement to feed (Gili & Coma 1998), their energy input, and consequently growth rates, vary according to the habitat. Other *in situ* experimental observations therefore determined a diametric growth rate of red coral of 0.62 ± 19 mm yr^{-1} during its early life phase (Bramanti et al. 2005). Furthermore, a new sclerochronological method, which stains the organic skeleton matrix, found an average diametric growth rate of 0.35 ± 0.15 mm yr^{-1} in colonies from a variety of environments (Marschal et al. 2004).

Studies of the natural feeding of red coral have in fact found some variability between habitats, which may partly explain some of these findings. Red coral feeds mainly on organic matter and zooplankton (Tsounis et al. 2006c), although pico- and nanoplankton are also part of their diet (Picciano & Ferrier-Pages 2007). Due mainly to hydrodynamics, red coral colonies on the Costa Brava at 45 m depth captured more particles than shallower ones, and the seasonal fluctuation of capture rates was dampened by the constant availability of particulate organic matter (Tsounis et al. 2006c).

Probably the best-understood aspect of red coral biology is its reproductive cycle, thanks to a long series of studies that were motivated by its commercial importance. Most gorgonians seem to reproduce primarily sexually (Gili & Coma 1998), and because this method is more efficient in dispersing the population than asexual reproduction, most research has focused on sexual reproduction. *Corallium rubrum* is a gonochoric brooder with internal fertilization that releases lecitotrophic larvae during late summer (Vighi 1972). Female oocytes develop for more than a year, so that two generations can be found within the polyps, of which only the larger, mature ones are released (Santangelo et al. 2003, Tsounis et al. 2006a). Under certain environmental conditions, the polyps can reabsorb their oocytes and use the material to produce fewer, larger ones or not spawn at all (Santangelo et al. 2003, Pablo Lopez, University of Seville, personal communication). The male gonads develop in less than 12 months.

Red coral is characterized by a higher recruitment rate than most other octocoral species, which are generally considered organisms with relatively slow colonization and recovery rates (Grigg

1989, 2004, Garrabou & Harmelin 2002, Bramanti et al. 2007). *Corallium rubrum* reaches sexual maturity at an age of 3–10 yr (Santangelo et al. 2003, Torrents et al. 2005, Tsounis et al. 2006a), whereas even the relatively fast-growing black coral *Antipathes griggi* reproduces only after reaching 12–13 yr (Grigg 1976). Data showed that selective harvest in fish has lead to 'undesirable evolution' that results in commercial stocks growing to a smaller size (Conover et al. 2009). Similarly, it is possible that early reproduction and small size in *Corallium rubrum* is a response to the millennia of fishing pressure it has experienced.

Larvae are reported to show no pronounced phototaxis but do exhibit negative geotaxis and gregariousness (Weinberg 1979). They develop their negative geotaxis only after an initial period of positive geotaxis, and their free-swimming phase lasts only hours or days (Weinberg 1979, FAO 1983), which results in a limited dispersal distance of red coral larvae. It is therefore very likely that most populations are genetically isolated. In fact, recent studies demonstrated the occurrence of genetic differences among populations even over short distances of less than 3 km (del Gaudio et al. 2004, Calderon et al. 2006, Costantini et al. 2007a,b, 2009). As a result of slow growth and short larval dispersal distance, populations in some overharvested locations have been quickly reduced to few young populations that may not be connected with each other (Weinberg 1978, Plujà 1999, Rossi et al. 2008). This vulnerability to isolation has implications for the management of this species because it is often assumed that deep populations act as a refuge by providing larvae to the shallower ones. Furthermore, since red coral shows a uniquely broad depth range and remarkable morphological variation, the question whether these populations actually belong to the same species has been raised (Y. Benayahu personal communication). Future research may confirm whether deep populations are able to contribute larvae to shallow-water habitats.

Further implications of the reproductive biology of the species for the management of the fishery become apparent when taking population structure into account. To our knowledge, the only extensive study on the abundance of *C. rubrum* has been conducted on the Costa Brava and indicated an extreme patchiness, with a total colony abundance on coralligenous hard substratum (20–50 m depth) of 3.42 ± 4.39 colonies m^{-2} (Tsounis et al. 2006b). Due to geographic variation, this study is certainly not representative for the whole Mediterranean. The only comparable data available are on the abundance within patches, which varies considerably among geographic locations as well as between depths and habitats and is not indicative of total abundance. Values recorded range among 1300 colonies m^{-2} in Calafuria, Italy, 130 colonies m^{-2} at the Costa Brava (Spain), and 400–600 colonies m^{-2} in France (Santangelo et al. 1993a, 1999, Garrabou et al. 2001, Tsounis et al. 2006b, Rossi et al. 2008). For other regions, presence or absence may be recorded, but in general there is a lack of data on abundance.

Red coral abundance on the Costa Brava is inversely proportional to depth; deep habitats are characterized by a more scattered distribution, in contrast to the dense patches observed in shallow water (Tsounis et al. 2006b, Rossi et al. 2008). This distribution pattern can have implications for the harvest of deep populations because a low abundance may make exploitation unfeasible. In general, the extreme patchiness results in such a high standard deviation that abundance data are of no particular use for abundance comparisons and identification of population decline. In the case of *C. rubrum*, size and age structures give a better indication of population decline because the fishery is size selective, and a decline in numbers will inevitably coincide with a decline of large colonies. Data on population structure are also more useful in identifying a trend in proportion of mature and immature colonies, which is more useful as a basis for management decisions that need to ensure recruitment is sufficient for the population to survive.

The presently observed size–frequency distribution in shallow water shows that unprotected populations consist mainly of small corals under 8 cm height, with an average height of 3.1 ± 0.16 cm, and an average basal diameter of 5.1 ± 2.0 mm yr^{-1} (Tsounis et al. 2006b), which corresponds to an age of 8–15 yr (Marschal et al. 2004, Bramanti et al. 2007). In comparison, a population protected for 15 yr had an average height of 4.2 ± 2.5 cm, and an average basal diameter of 6.9–2.4 mm (Tsounis et al. 2006b), that corresponds to an age of 11–20 yr (Marschal et al. 2004, Bramanti et al. 2007). In the same area, palm-sized colonies could be found in the 1960s at 35 m depth (J.G. Harmelin personal communication).

Results from two biological studies on the stocks off Morocco revealed a similar situation and confirmed the observed pattern; average height of colonies in two harvested stocks was 7.25 cm (Topo–Cala Iris) and 6.4 cm (Sidi Hsein), with an overall size range of 3–13 cm (Abdelmajid 2009, Zoubi 2009). These data originate from subsampling the harvesters' catches, so the actual average size may be even lower.

The only two published demographic studies on red coral populations below 50 m show that harvesting pressure is proportional to the degree of accessibility of the sites to divers (Rossi et al. 2008, Angiolillo et al. 2009). These populations were heavily dredged until 1994 and appear to consist of 10- to 16-cm tall colonies that show a well-developed branching pattern but an average basal diameter of only 9 mm (Rossi et al. 2008, Angiolillo et al. 2009).

Causes of the young population structure other than fishing impact can be excluded, or at least regarded as insignificant compared with fishing mortality, as natural disturbances are likely to affect young colonies more than older ones. Therefore, a dense, young population structure may indicate an early recovery stage after a disturbance. The study of natural *C. rubrum* populations certainly has to deal with the lack of a baseline; however, video observations by coral divers harvesting at 90–110 m in Sardinia confirmed the frequent occurrence of 50- to 60-cm tall colonies in these populations (M. Scarpati personal communication). Natural mortality in red coral is relatively low, a trait shared with other precious corals (Harmelin 1984). Natural mortality may be increased by the parasitic boring sponges *Spiroxya heteroclite* and *Cliona sarai,* which affect the corals by perforating their bases until they are perfused with holes and lose structural stability. Except on granite, it appears that the sponges enter the corals via the substratum (G. Bavestrello personal communication). Older corals are affected at a higher rate, with a noticeable increase after 4 yr (Corriero et al. 1997). Interestingly, the apical parts are never infected by the sponges, so it can be deduced that they cannot penetrate the live coenosarcs. Apart from the fact that colonies above a certain size may no longer be supported by the perforated stem, these corals are also of less value for the jewellery industry (Bavestrello et al. 1992, Corriero et al. 1997). Other parasites may potentially increase mortality (Abbiati & Santangelo 1989), and interspecific competition may further influence the population structure of red coral (FAO 1983, Giannini et al. 2003). However, both have an insignificant effect compared with harvesting.

Another cause for increased mortality in *Corallium rubrum* is mass mortalities (Cerrano et al. 2000, Garrabou et al. 2001, Bramanti et al. 2005, 2007), which have been reported as early as or earlier than 1983 in the French maritime province (FAO 1983), but very little is still known about the cause of these episodes. The fact that they manifest themselves during abnormally warm summers suggests that temperature tolerance may be a contributing factor, as may pathogenic agents or pollutants (Garrabou et al. 2001). *Corallium rubrum* is reported to be tolerant of high temperatures and even sudden temperature peaks (FAO 1983). However, because Mediterranean suspension-feeders suffer a trophic crisis due to low plankton abundance and water movement in summer (Coma et al. 2000, Rossi & Tsounis 2007), physiological stress (starvation) could be a further contributing factor. Not surprisingly, the affected red coral populations seem to be the shallower ones because the partial or total mortality decreases with depth and is insignificant below 40 m (Linares et al. 2005). In general, though, natural mortality appears to be low, a conclusion based on a single experiment on growth and survival in a cave (Garrabou & Harmelin 2002).

Consequently, the extremely young population structure is a direct result of overharvesting (Santangelo et al. 1993b, Santangelo & Abbiati 2001, Tsounis et al. 2007). The history of intensive fishing (see p. 176) also has its impact on scientific studies because even the oldest marine pro-tected areas are too young to serve as a baseline for the study of the red coral population structure. Specimens in museums and private collections demonstrate that this species can reach a size of more than 50 cm (Bauer 1909, Barletta et al. 1968, Cicogna & Cattaneo-Vietti 1993, Garrabou & Harmelin 2002), but the percentage of these large colonies in a natural population is not known.

Black coral (Antipatharia)

Seven families and about 150 black coral species are known today, most of which are antipathar-ians (hexacorals with branched or unbranched skeletons). Their protein skeleton is dark brown to black and of a consistency resembling hardwood, but their tissue and polyps make them appear rust, yellow, green or white. The genus can be found in all oceans, from New Zealand fjords in the temperate Pacific (Grange 1997), through the tropical Pacific (Grigg 1965), the Caribbean (Olsen & Wood 1980), to the Canary Islands and Mediterranean submarine canyons (Davidoff 1908). They mainly prefer deep-water habitats of tropical and subtropical oceans (Lumsden et al. 2007). Only a few species are of commercial interest. Many species, even harvested ones, are little known, thus the following information relates especially to the main commercial species.

The best-studied black coral species are the two most important commercial species, *Antipathes grandis* and *A. griggi* (a redescription of *A. dichotoma*, Opresko 2009), which are distributed over the tropical Pacific between 30 and 100 m depth (Grigg 1976). Together with red coral and the semi-precious bamboo coral, these black corals are the only precious corals that dwell at depths shallow enough to be harvested by scuba divers.

Populations in the Hawaiian Islands show the highest densities on hard, sloping substratum in areas exposed to 0.5- to 2-knot currents (Grigg 1965). *Antipathes* spp. preferably settle in depres-sions, cracks or other rugged features along steep ledges, with few colonies found on smooth basal-tic substratum (Grigg 1965). The lower depth limit of the distribution coincides with the top of the thermocline in the Hawaiian Islands (ca. 100 m; Lumsden et al. 2007). Some other black coral species occur in shallow water, underneath ledges and in caves (*Cirrhipathes anguina* can occur at 4 m depth). Depth appears to influence the distribution of various coral taxa, whereas substratum and environmental conditions (flow, sedimentation) seem to influence patchiness of precious corals (Grigg 1976).

Several black coral species, especially *A. grandis* and *A. griggi*, are large branching species and therefore important structure-forming corals (Parrish et al. 2002). Their erect branching struc-ture creates substratum for attachment of sponges, tubeworms, barnacles, molluscs, anemones and echinoderms and provides shelter from predators for small fishes, as well as a sleeping perch for large fishes and rock lobsters (Warner 1981, Grange 1985). A study in the Hawaiian Islands revealed that populations of the endemic and highly endangered Hawaiian monk seal (*Monachus schauin-slandi*) forage on fish species thriving among black coral communities (Goodman-Lowe 1998, Parrish et al. 2002, Boland & Parrish 2005, Longenecker et al. 2006). Species that associate with *Antipathes* include 17 different pontoniine shrimp from the Indo-Pacific (Australia, Madagascar, Kenya, Maldives, Indonesia, Zansibar, New Caledonia, Borneo and Hawaii) and the Caribbean (Spotte et al. 1994). Many invertebrate species have been found only among antipatharians (Love et al. 2007).

Black coral is known to grow to heights exceeding 4 m. Growth rates are low, as *in situ* obser-vations and growth ring analysis demonstrated (Brook 1889, Grigg 1976, Olsen & Wood 1980; see Table 1)—6.12 cm yr^{-1} (*A. grandis*) and 6.42 cm yr^{-1} (*A. griggi*)—although not as low as temperate water black corals (1.6–3 cm yr^{-1}, *A. fjordensis*; Grange 1997) or *Corallium* species. The oldest specimens of *Antipathes grandis* and *A. griggi* observed in Hawaii are reported to be 75 yr old (WPCouncil 2007), which was supported by radiometric data (Roark et al. 2005).

Other black coral species have been estimated to live longer than a century (Love et al. 2007, Williams et al. 2007), while Roark et al. (2005) estimated a lifespan of 2377 yr for *Leiopathes glaberrima.* The diet of *Antipathes grandis* and *A. griggi* is largely unknown (Grigg 1965, Lewis 1978, Warner 1981).

Sexual maturity in *A. grandis* and *A. griggi* is reached at a size of 64–80 cm, corresponding to an age of 10–12.5 yr (Grigg 1976). This is typical for the K-selection generally observed in deep corals. Observations of natural recruitment seem to be scarce, however, indicating recruitment limitation in Hawaiian *Antipathes* populations. The temperate species *A. fjordensis,* a gonochoric broadcast spawner with a seasonal reproductive pattern (Parker et al. 1997), reaches maturity at a similar size range (70–105 cm) but at a notably higher age of 31 yr (Parker et al. 1997).

In *A. grandis* and *A. griggi,* fertilization takes place externally in the water column (broadcast spawning), and their larvae are negatively phototactic. Data on deep-water genera of black corals are still scarce. Their morphology can be branched, feathered (*Myriopathes, Bathypathes* spp., *Stauropathes* and *Leiopathes*) or whip-like (*Stichopathes* spp.).

The age structure of an unfished *Antipathes griggi* population showed that juvenile corals under 20 yr old dominated the populations, and that 40-yr-old corals made up 2% of the population (Grigg 1976).

The invasive octocoral *Carijoa riisei* (snowflake coral) was first observed in Pearl Harbour in 1972 and has spread to the eight main Hawaiian Islands (Kahng & Grigg 2005). It overgrew large areas of substratum and adult colonies of black coral in 70- to 100-m depth in the Au'Au channel off Maui. The invasion may have contributed to a decrease in recruitment, diminishing a deep-water refuge that was assumed to contribute to recruitment of the intensively harvested shallower stocks. Results suggest that the invasion may have slightly abated (Grigg 2004, Kahng 2007), although further data are needed for confirmation (WPCouncil 2006, 2007). Therefore, the observed decline in biomass appears to be mainly due to harvesting of the stocks shallower than 70 m (Grigg 2004) rather than the invasion of *C. riisei.*

Little is known about black coral populations outside Hawaii.

Gold corals (mainly Gerardia spp.)

The Gerardiidae is a family about which very little is known, and its taxonomy is not well defined (Lumsden et al. 2007). Gold corals of the family Gerardiidae are found on hard substrata such as basalt and carbonate hard grounds on seamounts in the north and equatorial Pacific and Atlantic Oceans. In Hawaii, their depth range is 350–600 m and in the Straights of Florida at around 600 m (Messing et al. 1990). Gold corals are found in commercial quantities in the Hawaiian Islands (Grigg 1984, 2002). A *Gerardia* sp. fishery in Turkey was abandoned after learning that it served as substratum for several shark species to lay their eggs (B. Öztürk personal communication).

Gold corals in the order Zoanthidae are known to broadcast spawn during mass spawning events, but nothing is known about the reproductive strategies of *Gerardia* species except that their larval stages settle out on other coral species (particularly bamboo corals) and eventually overgrow those colonies (Lumsden et al. 2007). Gold corals of the genus *Primnoa* have skeletons of protein that are abundantly infused with calcite ($CaCO_3$) spicules (Grigg 1984).

Gold corals undoubtedly are quite long lived. However, there is considerable variation between data on the lifespan of *Gerardia* species among the various studies, probably resulting from different methods. Estimates by Grigg (2002) using growth ring counts indicated a maximum longevity of less than 100 yr for Hawaiian *Gerardia* species. A study using radiocarbon dating on *Primnoa resedaeformis* confirmed a similar maximum age of 78 yr in the north-western Atlantic Channel off Nova Scotia and confirmed that growth rings are annual for this species (Sherwood et al. 2005). Sherwood & Edinger (2009) found a slightly older age of 200 ± 30 yr for the same species and area. Risk et al. (2002) reported a lifespan of more than 300 yr for another *P. resedaeformis* colony collected off Nova Scotia. Goodfriend (1997) calculated an amino acid racemization age of 250 ± 70 yr on an Atlantic *Gerardia* specimen.

In contrast to these studies, Druffel et al. (1995) dated the same specimen to 1800 ± 300 yr. Slower growth rates and older lifespan were also reported by Roark et al. (2006), who used ^{14}C dating to measure growth rates that corresponded to ages of 450–2742 yr. The differences were explained by arguing that radiocarbon dating revealed that growth rings were not annual in these species (Roark et al. 2006). On the other hand, carbon dating accuracy can be affected by corals feeding on ^{14}C sources that have a different origin (and thus age) than assumed (Druffel & Williams 1990, Druffel et al. 1992). While this type of error is usually accounted for (Roark et al. 2006, Sherwood & Edinger 2009), there remain differences in growth rate measurements, even between radiocarbon studies. For example, Roark et al. (2006) reported a lifespan of 2377 yr for the black coral *Leiopathes glaberrima,* whereas Williams et al. (2009) found this and other black corals to have shorter lifespans of a few decades or several centuries.

Clearly, further research is necessary to resolve and understand these contradicting data (Parrish & Roark 2010). Therefore, the WPCouncil has issued a 5-yr moratorium on the fishing of gold coral until the lifespan is confirmed (WPCouncil 2006). However, the consensus of all studies at this point is that precious corals, including gold corals, are particularly long-lived species and are therefore vulnerable to overexploitation.

Other precious corals

Other important although less-valuable precious corals are the gold corals *Primnoa resedaeformis* and *P. willeyi* within the family Primnoidae. Primnoidae is a large family containing more than 200 species that are among the most common of the large gorgonians (Etnoyer & Morgan 2003, 2005). The family exhibits a particularly broad depth range of 25–2600 m, while most species occur shallower than 400 m. The commercially important *P. resedaeformis* dwells between 91 and 548 m (Cairns & Bayer 2005).

They grow branching colonies with a skeleton of a horn-like protein called gorgonin and can reach 4–5 m in height (Krieger 2001). Linear growth rates in the Gulf of Alaska have been found to be low, with just 1.6–2.32 cm yr^{-1}, and diametrical growth was estimated at 0.36 mm yr^{-1} (Andrews et al. 2002). Linear growth in *P. resedaeformis* is higher in the first 30 yr of its life (1.8–2.2 cm yr^{-1}) than after that (0.3–0.7 cm yr^{-1}). The oldest corals sampled in the Canadian Atlantic were 60 yr old (Buhl-Mortensen & Mortensen 2005), but large colonies in Nova Scotia were found to be hundreds of years old (Risk et al. 1998, 2002). Very little is known about their reproduction, but it is likely that they are gonochoric broadcast spawners similar to other octocorals (Fabricius & Alderslade 2001).

Another group of precious corals that deserves mention is bamboo corals in the family Isisidae (*Acanella* spp., *Lepidisis, Olapa*). This family consists of over 150 species that are primarily deep-water species. The most common deep-water genera are *Isidella, Keratoisis* and *Acanella*. Bamboo corals consist of skeletons with alternating sections of calcium carbonate and proteins (Lumsden et al. 2007). They are important structure-forming species in the Gulf of Mexico, Hawaii, south-eastern U.S. waters, the north-eastern Pacific and Indopacific (Fabricius and Alderslade 2001, Etnoyer & Morgan 2003). Although some genera have been reported from 10 to 120 m (Fabricius and Alderslade 2001), most species occur below 800 m (Etnoyer & Morgan 2005), with the deepest recorded at 4851 m (Bayer & Stefani 1987). Some species are bioluminescent (Etnoyer 2008).

Their morphology can be whip-like but is usually branched, bushy or fan-like, with sizes ranging from tens of centimetres to over 1 m (Verrill 1883). Their skeleton consists of heavily calcified internodes and gorgonin nodes, giving it a bamboo-like appearance. Diametric growth of *Lepidisis* sp. in New Zealand waters ranges between 0.05 and 0.117 mm yr^{-1}, with maximum ages between 43 and 150 yr (Tracey et al. 2007).

Culture and history of precious coral exploitation

Cultural significance of precious corals

Few other natural resources have fascinated humankind more than precious corals. The findings of perforated red coral beads with Paleolithic human remains demonstrates that corals have been treasured by humans for at least 25,000 yr (Tescione 1965, Skeates 1993). The history of precious corals starts with the Mediterranean red coral, the precious coral *par excellence*. In addition to its ornamental use in the Neolithic (3000–5000 yr ago), red coral also developed a tremendous cultural importance, as its appearance in decorative arts of the Minoan and Mycenean civilization documents (Tescione 1965).

Greek mythology originally elevated red coral to magical status (see Ovid's *Metamorphoses*). The ancient myth of Perseus killing the Gorgon monster Medusa, feared for her petrifying stare, states that on laying down Medusa's severed head, her blood seeped into seaweeds that were subsequently petrified and stained red. Red coral was born, and coral is therefore a symbol of rebirth. The legend of Medusa further states that Perseus gave Medusa's head to the goddess Athena, who used it as a shield against her enemies, which may explain why coral talismans are used for protection. It is also said that in petrifying the algae, some of Medusa's magic was conferred to the corals. The Romans ingested and applied coral powder as an antidote to poison, a cure against stings, comfort for fainting spirits, to counteract fascinations, to protect humans against sorcery, to purify the blood and to cure imbecility of the soul, melancholy, mania and other maladies. Protection against the shade of Satan is attributed to red coral in Christian cultures; in various other religions, it is a protector against the evil eye and any misfortune (Tescione 1973, Wells 1983). In Iranian beliefs, red coral protects against lightning and storms. In Buddhism, precious corals are revered as treasure from paradise and are used to decorate statues of Buddha (Kosuge 1993). The attractive blood red colour made it a symbol for the blood of Christ. Its hardness and the tree-like shape of its skeleton made it a symbol for immortality (hence the name 'tree of life'). Coral necklaces or coral branches are an element in many works of spiritual or religious art, such as the works of the fifteenth century Renaissance artists Piero della Francesca and Andrea Mantegna. Entire churches have been built thanks to economic wealth that coral fishing created, such as the church of San Giorgio built in 1154 in Portofino (Italy), which contains a chapel dedicated to coral fishermen (Mazzarelli 1915), the church of San Giovanni at Cervo, Italy (Liverino 1983), and the Sant Esteve in Begur (Spain).

As Marco Polo reported, red coral has been found as far away from the Mediterranean as ancient Tibetan temples. It has also been used as a decoration in Chinese clothes dating back several millennia (Knuth 1999). In the 1950s, coral beads were still widely used as ornaments on kimono belts in Japan (Liverino 1983). Similar protective powers were also attributed to black coral. According to Indonesian folklore, a black coral bracelet worn on the right arm increases virility, whereas on the left it cures rheumatism (Wells 1983). The name *Antipathes* translated from Latin means 'against disease', and Albertus Magnus mentioned corals used as cures. In addition to its ornamental use and its importance in religious rituals, precious corals were one of the most sought after exchange products on the oriental markets of the Phoenicians and Egyptians about 15 centuries before Christ. Pliny mentioned trade of red coral with India. Today, its use as an aphrodisiac or a cure is limited compared with its use as a raw material in the manufacture of jewellery and artistic sculptures (see p. 183). The high cultural importance of precious coral still lives; black coral is the official state gem of Hawaii, and red coral art has never lost its sacred fascination. When considering various management options (listed in that section), the cultural importance of the fishery and art should be taken into account.

History of coral exploitation

The origin of the name 'coral' is generally attributed to the Greek word for pebble ('korallion'), but Hebrew ('goral', meaning small pebble) and Arabic ('garal', small stone) origins are also possible, maybe due to the names given to them by early traders travelling between Europe and the Middle East (Hickson 1924). In ancient times, Mediterranean red coral was collected when washed up on beaches after heavy storms had broken off branches in shallow water. During these times, traders listed it as spice (S.J. Torntore personal communication).

Intentional precious coral exploitation, however, did not start until about 5000 yr ago in the Mediterranean, when iron hooks were used to harvest red coral (Grigg 1984). Greek islanders called these tools 'kouralió' (Tescione 1973). It can be assumed that this was done while free diving. At a later stage, the free divers were equipped with Japanese goggles that must have made the harvesting quite efficient, considering the warm summer water in the Mediterranean and the fact that red coral can still be found as shallow as 7 m in shadowed overhangs and crevices (G. Tsounis personal observation). Ancient artworks show swimmers emerging with large coral branches, indicating an abundance of large specimens. The efficiency of free diving should not be underestimated, as other examples of free-diving fisheries show. Greek sponge divers descended to considerable depths, even down to 80 m, using a marble stone as a weight (Mayol 2000), and the legendary Japanese Ama pearl divers made, and still make, a living by diving for pearls and shellfish in the Okinawa Islands and the Izu Peninsula (Hong et al. 1991, Hlebica 2000). The last example furthermore suggests that the early coral free divers might have harvested the sea for food at the same time as looking for red coral. Regarding the efficiency of early coral diving, it is significant to note that a recent study found that children of sea nomads in Thailand, who play and fish in the sea without goggles, possess better underwater vision than other children (an improvement that can be achieved by training; Gislén et al. 2003, 2006). We can therefore assume that harvesting pressure on precious corals in shallow water was considerable. Fishing by free diving was probably practiced in various Mediterranean Islands. The first coral-harvesting tools that could be employed from boats are dated to the fourth to third centuries BC, that is, during the Hellenistic, Roman-Carthaginian and late Etruscan ages (Galili & Rosen 2008).

Precious coral 'fishing' became more efficient during the fourth century BC when the Greeks or Arabs developed a dredging device known as 'ingegno' or 'Saint Andrews cross' (Galasso 1998, 2001). It consisted of a wooden cross with nets attached that was dragged along the bottom, entangling red coral in the Mediterranean Sea. Political struggles over existing coral beds, and the discovery of new ones, dictated the production at this early phase. Industrial-scale exploitation for red coral began during the early 1800s when the Kingdom of Naples probably employed more than 1000 boats dredging for coral. By 1870, this number had decreased to 612 (Tescione 1973).

Throughout history, there were various centres of coral jewellery manufacture. In the tenth and eleventh centuries, Marsa'el Karez on the northern African coast was the largest coral port and trade centre. In the fourteenth century, Barcelona was famous for its coral art (Lleonart & Camarassa 1987); later, the main activity shifted to Lisbon and in the seventeenth century to Marseille. Parallels between these shifts have been attributed to Jewish migrations (G. De Simone personal communication). From 1100 to 1600, Genoa was already one of the important centres, and in that century nearby Torre del Greco finally established a firm position in the coral-fishing business and remains the main centre of *Corallium rubrum* jewellery manufacture today (Tescione 1973, FAO 1988).

Many coral fisheries in the world are characterised by depleted stocks (Grigg 1989, Santangelo et al. 1993b, Santangelo & Abbiati 2001). This boom-and-bust exploitation, more similar to coal mining than to a fishery, made for unstable yields, with various peaks and troughs. As early as the 1800s, the coral-fishing fleets made substantial trips to foreign shores in search of new stocks (Tescione 1973). The following figures from G. Tescione's extensive compilation of historic data demonstrate the size of the industry. In 1862, there were 347 boats fishing for corals, which in 1864 rose to 1200 vessels fishing, with 24 factories in Torre del Greco (Italy) and 17,000 persons employed in total (Tescione 1973, FAO 1983). Political regime changes and wars were also responsible for unstable fishing, and coral fishing stopped completely during the WWI in 1914–1918. At about the same time, imports of Japanese coral started to reduce demand, and thus fishing, dramatically. The Italian coral industry adapted to market demand and crafted large coral pieces of the popular pale pink coral, but it struggled to survive. Demand for Mediterranean coral increased again, and apart from another pause during WWII (in 1941, there were only 5 boats active, which increased to 31 in 1947), larger-scale coral fishing was resumed. In 1982, there were 200 boats, 50 divers, and 150 factories, employing 4000 workers and 1600 fishermen.

A discovery of large beds of dead, subfossil red coral between Sicily and Tunis in the 1880s led to a 'coral rush' of 2000 vessels into the small area and quickly depleted those grounds, while lowering prices and reducing fishing in other areas (Tescione 1973). The area where the so-called Sciacca coral was found is subject to subsidence, volcanic activities and earthquakes, making it likely that undetermined geological processes had a fatal impact on the populations (Cicogna & Cattaneo-Vietti 1993). The coral trade therefore depended not only on the state of the resource but also on market forces and the political situation. Ironically, the Sciacca coral that drove prices down in 1880 is the most valuable *C. rubrum* variety today.

During the Industrial Age, the Saint Andrews cross was abandoned in favour of a modified metal version called 'barra italiana'. It was made of a heavy iron bar with nets attached along its length. Even with the most efficient coral dredges, it is estimated that only about 40% of the corals broken off the substratum are entangled and retrieved (FAO 1983). In 1876, dredging was forbidden temporarily by royal decree in Italy, allowing only diving in helmeted suits. In the light of the immense ecological damage that dredging inflicts on coral habitats (Thrush & Dayton 2002), coral dredging in European Union waters was banned in 1994, and scuba-diving using advanced technology remains the dominant exploitation method today. After the invention of the Cousteau/Gagnan Aqualung in 1943, scuba-diving quickly found its application in coral harvesting because it allowed divers to selectively pick large corals in protected crevices that were inaccessible to dredging. The first 5 kg of red coral harvested by scuba were brought to Torre del Greco in 1954 by the sport diver Leonardo Fusco, who became a professional scuba coral harvester after discovering coral while free diving in Palinuro near Naples (Liverino 1983). In his first season, he harvested 250 kg, which he sold for 15,000 Italian lire kg^{-1}. Quickly, the small circle of scuba pioneers in Italy (Guido Garibaldi, Alberto Novelli and Ennio Falco) became coral fishermen and discovered further banks in Sardinia, Elba and Corsica (Roghi 1966). Accidents also started to occur that same year. One of the most spectacular and valuable source of *C. rubrum* was discovered at that time in the legendary Capo Caccia Cavern in Sardinia at a depth of 37 m (Liverino 1983). Liverino reported that in 1956 divers worked at 30–35 m, but in 1958 at 40–45 m. By 1964, an ever-growing group of divers was working at depths of 72 m, and inevitably a long list of accidents was the result of the spreading 'coral fever' among the young divers (Liverino 1983). Leonardo Fusco reported that in 1955 he harvested at 60 m in the Gulf of Naples, but in 1964 he had to descend to 90 m. Similarly, the pioneer Fausto Zoboli is reported to have said that he worked as one of the first at 60 m in 1964 (near Rome), while in 1971 he was forced to work at 100 m in Alghero, Sardinia (Liverino 1983). Others similarly documented that by the late 1950s divers in France and Italy already had to descend to depths of 80 m, and at times to even more than 100 m, to find coral (Galasso 2000). In 1974, helium-based mixed-gas diving techniques developed by the French ocean engineering company COMEX started to spread among coral divers, permitting them to work at 120 m for 20 min without the dangers of nitrogen narcosis (Liverino 1983).

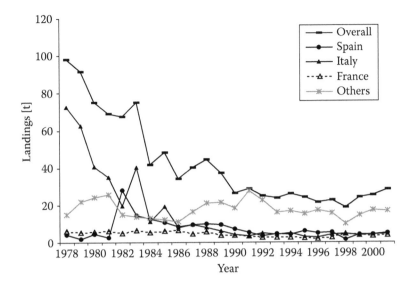

Figure 2 Red coral harvest in the Mediterranean. Note that dredging was phased out from the early 1990s until it was banned EU wide in 1994. (Data from FAO 1984, http://www.fao.org)

The 14-mile long Scherchi Channel from Sicily to Tunisia was regarded as a coral 'el Dorado', with 80 divers from Italy, France and Spain harvesting 70–120 t in 1978, starting at 60 m and gradually working their way down to 130 m. This submarine canyon was described as an oasis for marine fauna and flora, uncontaminated and exceptionally rich in rare species. In 1979, there were 366 boats at work (283 of them were registered in Italy) and 150 divers (Liverino 1983). Algeria and Morocco developed a local management programme with the aid of Italian coral divers (Liverino 1983). Then, in 1977, following Jacques Cousteau's recommendations, Algeria placed a temporary ban on coral fishing. Tunisia allowed fishing by Italian dredging boats ('corallinas') and French divers until it developed its own fishery in 1974. The last major peak recorded lists 100 t of *C. rubrum* in 1978, but from then on, the reported landings remained below 30 t (Figure 2). In the 1980s, France reported a production of about 5 t yr^{-1} by 40 licensed divers and 1 t by dredging.

In contrast to the development in the Mediterranean, the Hawaiian black coral fishery relied early on scuba diving since the 1960s. In the United States, selective harvesting methods (manned submersibles, remotely operated vehicles [ROVs]) have been required since 1980 (J. Demello, WPCouncil, personal communication). However, dredges are not completely banned in other parts of the world.

Unlike Mediterranean red coral and Hawaiian black coral, Japanese coral inhabits depths beyond 100 m and was not washed up on beaches. *Corallium rubrum* was imported into Japan, was regarded as one of the most valuable presents to state officials, and is still highly revered (Kosuge 1993, S. Kosuge personal communication). In the early 1800s, other *Corallium* precious coral species were discovered in the Pacific Ocean. Off Japan, coral dredging started soon after (1804) but was initially prohibited. The shoganates did not, however, enforce this law strictly and published a document in 1838 that stated that harvesters were prohibited from selling freely and should offer the corals to the government (Kosuge 1993). Thus, coral fishing in Japan flourished only after the Meiji Reform abolished the feudal system in 1968. Konojoi Ebisuya is credited with developing the Japanese fisheries, including the invention of a more efficient coral-gathering net (Kosuge 1993).

Coral was exported directly or through the Japanese bank Mitsui (especially to the Netherlands). The Pacific coral export increased steadily because its pale pink colour and large size enjoyed great popularity (Tescione 1973). *Corallium* species have been harvested in the western Pacific islands,

Table 2 Changes in the fishing grounds of the Japanese precious coral (*Corallium* and *Paracorallium*) fishery illustrating ever more extensive fishing trips

Year	Species	Location
1950/51	*P. japonicum* (Aka), *C. elatius* (Momo)	Hachijo (150 miles off Tokyo)
1952/53	*P. japonicum, C. elatius*	Amani Island
1952/53	*C. elatius, C. konojoi* (White)	Amani and Goto islands
1960	*C. elatius*	Okinawa
1961/63	*P. japonicum, C. elatius*	Sumizu off Okinawa and Hachijo
1965	*C. elatius, C. secundum* (Boké)	South China Sea
1965	*C.* sp. nov. (Midway coral)	Midway

Source: Data from Liverino 1983.

Note: Japanese common name in parentheses.

including Japan, Taiwan and the Philippines, for 200 yr (Grigg 1994). The dredging devices used in the Pacific were coral mops (tangle nets, weighed down with natural stones) that were pulled along the bottom at 100–330 m (Grigg 1984). The 'rascle' is a similar device, consisting of a V-shaped metal frame supporting a net (FAO 1983).

The Pacific fishery extracted *Corallium* species off Japan, around Okinawa, off the Bonin Islands and off Taiwan, at either 200–500 m depth or 1000–1500 m. The golden era is said to have been from 1954 to 1970 (Liverino 1983). In 1950 to 1955, the harvest concentrated on the islands of Shikoku and Kyushu. Tosa Bay and the islands off Goto were noted as particularly rich in *C. elatius* and *Paracorallium japonicum* (Liverino 1983). Subsequently, after discovering banks at the Ryu-Kiu Islands south of Kagoshima (Okinawa, Amani, Miyado) and those south-east of Osaka and Yokohama (Ogosawara, Hachijo, Sumisu), the fleet started making ever-more-distant fishing trips (Table 2). Inshore vessels of more than 5-t displacement were used, while long-range boats displaced 150–180 t and were manned by a 25- to 30-man crew, which did 2- to 5-mo trips. The boats could employ up to 18 minidredges simultaneously. The dredges consisted of a round 10-kg stone that had five nets attached to it (Liverino 1983). Working this way, a boat could harvest up to 80 kg day^{-1}. The fishery and trade were locally managed by the three fishery associations: the All Japan Sango Fishing Association, the Sukomo Kyodo Kumiai and the Goto Sango Kogei Kyodo Kumiai.

Eventually, the fishing trips by the Japanese fishermen were so far reaching that in 1965 they discovered a very large bed of pink coral (*C. secundum*) in 400 m at the Milwaukee Banks, Mellish Bank and surrounding seamounts in the Emperor Seamount chain north of Midway Island, near the Hawaiian Archipelago. Subsequently, most of the world's tropical *Corallium* landings came from that bed (Grigg 1994). About 200 vessels from Taiwan and Japan made up to 7 trips yr^{-1} (Convention on International Trade in Endangered Species of Wild Fauna and Flora [CITES] 2007). The fishery reached a peak of 150 t in 1969 but had finally depleted the stocks, and yield remained low for the next 5 yr until a previously undescribed deep-water species (Midway coral, *Corallium* sp. nov.) was discovered at a depth of 900–1500 m. The colour of this species is spotty (sometimes called 'scotch'), varying between pink and white, and it was extremely abundant (Grigg 1994). This discovery led to another coral rush, with over 100 boats from Japan and Taiwan involved in the harvesting. In 1973, four boats from Suao were guided by the research vessel Jungmei 6 and harvested coral worth 90,000 Italian lire in just 1 wk (Liverino 1983). In total, the 100 boats working the entire season made over 3 billion Italian lire. Production peaked at 300 t in 1981 (Liverino 1983, Grigg 1994). Over the whole Pacific, the production totalled about 400 t at that time (FAO 1983). The principal extraction method was dredging, but manned submersibles and ROVs were also employed to a small extent in the 1980s for exploration purposes (Grigg 1989, CITES 2007).

In 1981–82, the market was flooded with Midway coral, and prices fell so low that many fishermen suffered losses or even went out of business (only 21 vessels in 1981). Buyers in Italy paid

2000 lire kg^{-1} instead of the initial 80,000 (Liverino 1983). Yield consequently sank (84 t), but demand rose again, and the fishing continued for many years, until the stocks were depleted and yield dropped a few years later to about 10% (20–30 t) of the maximum and ceased with less than 3 t in 1988 (FAO 1983). In 1991, an all-time low of less than 3 t of precious coral production was recorded throughout the Pacific (Grigg 1994), with prices for raw material at unprecedented heights (Grigg 1994), indicating the depletion of all known stocks. Coral fishing has not been resumed in those areas; one reason cited is that the corals found in by-catch of fishing trawlers is of low quality. Precious coral yields worldwide (all species) reached a peak of 450 t yr^{-1} in 1984 but fell to about 28–54 t during the last 18 yr (http://www.fao.org).

In 1923, coral banks were discovered in Taiwan, and in 1934 the Penghu government sent the research vessel Kai Peng Maru to study the stocks and set up a limitation of 20–40 vessels, which contributed to the longevity of the fishery (Liverino 1983). In 1962, coral was harvested by the same fishermen off Hong Kong at a depth of 70–100 m. In 1968, new banks off the Pratas Islands were discovered (Liverino 1983).

In 1958, large black coral populations consisting primarily of *Antipathes griggi* and *A. grandis* were discovered in the Hawaiian Islands between 30 and 90 m deep (the presence of the species was first reported during the Challenger Expedition; Brook 1889). Traditionally, these stocks were harvested by scuba divers. The black coral jewellery industry grew steadily between 1960 and 1970. In 1966, pink coral was discovered in the Makapu'u bed off Oahu, and a small group of fishermen dredged the bed using tangle nets (Grigg 1994). Prior to 1974, harvest was low, and the sales generated only around US$70,000, but the industry doubled within 6 mo during that year. In the 1980s, technological advances in coral processing led to a dramatic decrease of the amount of coral needed to produce the same value of finished product. Furthermore, import of cut and polished black coral from Taiwan affected the demand for coral by the Hawaiian industry, which consumed less than 2 t yr^{-1} on average (Oishi 1990). At that time, about 70 t of raw material were processed in Taiwan, most of which originated in the Philippine Islands, primarily *Cirrhipathes anguina* (whip coral) (Carleton 1987). About 90% of black coral sold in Hawaii are beads, rings, bracelets and necklaces made of *C. anguina* that has been worked in Taiwan. *Antipathes* species maintained their popularity due to their denser and higher-quality skeleton and their designation as the official state gem of Hawaii. However, import of inexpensive black coral from the Philippines and Tonga (Harper 1988) and more efficient manufacturing combined to keep the demand within sustainable limits. The Hawaiian black coral fishery was the first fishery that was managed on the basis of an extensive fishery research programme, which was conducted at the University of Hawaii in the early 1970s (Grigg 1976, Grigg 2001).

The same programme also discovered a small bed of pink coral (*Corallium secundum*) at 400 m off Makapu'u, Oahu, and developed a selective harvesting system employing a manned submersible. Maui Divers of Hawaii incorporated this system and harvested 0.5–2 t pink coral, gold coral (*Gerardia* sp.) and bamboo coral (*Lepidisis olapa*) per year from the same area. Non-selective fishing gear was banned in favour of non-destructive harvest. However, the operation was discontinued in 1978 because of high operating costs (a diving accident resulting in two deaths during the launch of the submersible increased insurance costs). It is estimated that the fishery extracted 32% of the standing stock in the time from 1966 to 1978 (Grigg 1994). Since then, the industry in Hawaii has relied on stockpiles of gold corals and imports of pink and red corals, mostly from Taiwan and Japan (Grigg 1994). An attempt to harvest pink corals domestically within the Exclusive Economic Zone (EEZ) used tangle nets at Hancock Seamount (Emperor Seamounts chain) but was cancelled after 450 kg of dead, low-quality *Corallium secundum* were brought to the surface (Grigg 1994). The Hawaiian pink and gold coral fishery was revived by American Deepwater Engineering (ADE), which used two 1-person submersibles to exploit an established bed and an exploratory area in 1999–2000. In 2000, the company harvested 1.2 t of *C. secundum*, 150 kg of *Gerardia* sp. and 61 kg of *Corallium regale* (Grigg 2002). However, the declaration of the north-western

Hawaiian Islands as a U.S. Coral Reef Ecosystem Reserve per Executive Order 13196 in the year 2000 excluded two-thirds of the precious coral deep-water habitats from exploitation. Given this reduction of potential supply, combined with marginal investment returns due to high operational costs, ADE suspended operations in 2001 (Grigg 2002). Foreign poaching has been a problem in the past because during the 1980s Japanese and Taiwanese coral vessels continuously violated the EEZ near the Hancock Seamounts. In 1980, about 20 Taiwanese coral draggers reportedly poached about 100 t of *Corallium* from seamounts within the EEZ north of Gardner Pinnacles and Laysan Island (Grigg 1994). However, it appears that since the 1980s poaching within the EEZ by foreign coral fishing has been negligible, in part due to the general fishing activity in the area. Fishing has now been terminated with the declaration of the area as a national monument, so it is not clear if current enforcement will prevent the reoccurrence of poaching (J. DeMello, WPCouncil, personal communication).

Socioeconomy

Modern exploitation methods

Harvesting methods

There is a variety of methods used to harvest precious corals today. At depths that can be accessed by scuba-diving, it is used as the exclusive harvesting method. Traditional compressed air scuba-diving is used to harvest populations between about 30 and 80 m deep, while mixed-gas scuba (see p. 181), is used between 80 and 150 m deep. Traditional air diving gear is cheap and readily available, whereas only a small number of divers invest in more expensive mixed-gas diving.

In contrast to dredging, scuba harvesting inflicts little direct damage to non-target species in the same habitat. However, coral exploitation by diving allows absolute selectivity only in theory. In practice, it has been reported that divers sometimes make a 'clean sweep' of an entire precious coral population at one site (FAO 1988). Since the market value of small coral branches has risen to US$240 kg^{-1}, the alternative of doing business with immature coral taken in shallow water appeals even to licensed fishermen and poachers, although this practice appears to affect only shallow-water populations in certain regions (e.g., Costa Brava, Spain). Poachers using air scuba have been convicted of harvesting up to 30 kg of young coral from one shallow dive site in 1 day (Fisheries Department, Government of Catalonia, personal communication). The increased amount of raw material that poaching makes available to the market may reduce the price, thereby causing further damage to licensed divers. In some cases, poachers reportedly sell to licensed divers, who resell the corals, or sell directly to Taiwanese buyers. In Spain, fishing licenses are issued without requiring special fees, but the selection of the few licensees is based on a consideration of personal and family history in coral fishing.

Towards the eastern Mediterranean, coralligenous habitats occur in deeper water, and coral fishermen rely on mixed-gas diving for harvesting, sometimes aided by ROVs for prior surveying of the target sites (personal observation). These divers selectively harvest large colonies once they have located them using the ROV. No data on poaching of these deep populations are available, so it is difficult to estimate poaching intensity. Another consideration is unconfirmed information that license holders may occasionally harvest areas outside the designated stocks, as opportunist amateurs do.

The method of removing coral is assumed to influence its recovery potential. Ideally, responsible divers cut the red coral base instead of extracting the whole colony. Leaving the base in place leaves a chance that this colony might regrow. This regrowth has been observed on a few occasions (Rossi et al. 2008) but is not well studied. Furthermore, the considerable time pressure and the difficulty of working underwater and at depth is quite incapacitating, so that red coral divers may not be able to consistently perform a precise size selection or partial harvest of corals. In fact, some studies demonstrated that up to 60–70% of confiscated poachers' catches were entire corals with the substratum still attached to their base (Hereu et al. 2002, Linares et al. 2003).

In the Mediterranean, harvesting concentrates on the warmer summer months (May–October), removing most of the corals at a time they produce their larvae. This practice should be of little impact (Kwit et al. 2004) but is an important fact to take into consideration in a socioeconomic analysis because it means that most of the divers are effectively working part time and have often invested into other businesses as well (e.g., hotels, aquaculture). On the Costa Brava, all divers are natives and older than 45 yr. In Hawaii and Spain, teams of three or more divers often work from one vessel. In Sardinia, 80% were said to be non-natives. Divers usually dive alone using mixed gas and from boats that are crewed by one coxswain and one diver (Andaloro & Cicogna 1993). In Hawaii and the Costa Brava (Spain), current coral fishermen are in their 50s and will leave the fishery in the near future due to the hazards of the occupation (WPCouncil 2007). The youngest licensed diver on the Costa Brava is 45 yr old.

In Morocco, 50-t boats with an onboard dry decompression chamber are mandatory. Two divers work from this boat and complete their decompression onboard instead of underwater. In Hawaii, the water temperature and visibility make for better diving conditions, and coral can be harvested year round, although currents, swell and sharks do present a challenge for Hawaiian black coral divers. In both cases, ocean conditions allow access only on relatively calm days, even though the Mediterranean fishery often operates close to shore.

The tools used to detach the coral and containers to store them vary between divers because they have traditionally been individual solutions. Some sort of pick is frequently used in the Mediterranean, perhaps with modifications such as a chisel incorporated into its shaft. The detached coral is usually put into a basket made of netting. These nets are often clipped to a rope during decompression and lifted to the boat (lift bags seem to be rarely used to lift a dive's red coral catch because the dimensions and weight are manageable). Also noteworthy are modern dive lights that consist of powerful rechargeable batteries that are attached to the diving tanks, and the small but bright light head itself is fixed to the forearm or head. This allows the diver to quickly find corals in crevices and work with two hands. Fishermen remark that diving safety issues are rarely considered in management plans and that divers are not invited to scientific consultation meetings (M. Scarpati, personal communication).

The Hawaiian black coral grows on exposed substrata and is harvested when at least 1.2 m tall and more than 2.54-cm thick. It is therefore harvested differently from Mediterranean red coral because the divers need to cut or break the thick keratinous coral stem with an axe or sledge. Selecting corals that exceed the minimum size is far easier in the case of Hawaiian black coral than it is with Mediterranean red coral (where millimetres of base diameter need to be distinguished). The coral harvest used to be tied to the boat anchor and buoyed to the surface using an inflating lift bag after the dive. Today, it is more common for the divers to use lift bags to transport coral to the surface, while the boat follows the buoys. Mixed gas, rebreather diving and ROVs are not yet used in Hawaii (Bruckner et al. 2008).

Modern scuba technology

Modern technology has made scuba-diving, and thus coral fishing, much more efficient. This includes better thermal protection suits that reduce the risks of hypothermia. Hypothermia affects cognitive capacity and vital aspects of diving physiology and can reduce the efficiency of decompression, which is mandatory after the deep and long dives that coral diving requires.

In the late 1990s, the so-called technical diving industry brought mixed-gas diving techniques from offshore commercial diving and rebreather technology from the military to recreational diving, making these techniques widely available and giving divers greater access to deeper water (Pyle 2000). Few coral divers use rebreathers at the moment, but this can be expected to change. It can be assumed that especially younger divers will use modern techniques to access corals in excess of 100 m. On the Costa Brava, divers are reported to mainly dive to 30–50 m deep using air because red coral grows as shallow as 20 m in this area. In any case, the recommended maximum depth for air diving of 40 m (Bove & Davis 1997) has always been, and still is, routinely exceeded. The preference for traditional scuba is due to decades of experience with the simple and rugged equipment as well as lower cost of operations. This also implies that divers will only take the risk to dive deeper than 50 m if no corals can be found in shallower water. In other Mediterranean areas, such as Italy, coral fishermen dive much deeper because of depletion of the shallower stocks (FAO 1988).

ROVs and manned submersibles

Because robotic extraction is not practical and not permitted in many fisheries, ROVs are increasingly being employed to scout a potential coral bed, improving the yield per dive. Basic ROVs are available today for as little as US$5000 and consist of a motorized real-time video camera that is controlled from the boat via a cable that also transmits the video signal to a topside monitor and recorder. ROVs can also be equipped with a robotic arm that permits remote-controlled harvesting, although this option raises the acquisition cost considerably. ROVs allow harvesting at greater depths and with fewer time restraints than scuba-diving but at lower cost than using manned submersibles.

ROVs have been used for the exploration of new beds since 1983 in Japan (Grigg 1994, CITES 2007). In general, however, remote harvesting is considered impractical compared with direct methods (scuba, manned submersibles): Currents, nets and the topography of coral habitats make it difficult to manoeuvre the tethered machines, and without dedicated technicians, a minor malfunction may easily render an ROV unusable for an entire expedition. Also, the ROV tether may damage precious corals if not carefully used (WPCouncil 2007).

The methodology for harvesting with manned submersibles originates from the commercial application of an exploration and sampling protocol that was developed in the course of a long-term Sea Grant research programme at the University of Hawaii in the early 1970s. It was used in Hawaii in 1972–1978 and experienced a short-lived renaissance in 1999–2001 (see p. 179). Apart from Hawaii, submersibles have been used in the Mediterranean, Japan and Taiwan to support exploration efforts of the fishery (Grigg 1989). In the state of today's market, the operational costs are so high that the profit margins do not justify their use. An exception is their current use in Japan to access deeper precious coral beds (N. Iwasaki personal communication).

Operational costs using manned submersibles are high due to the necessity of a large mother ship equipped with a heavy-duty crane. The aforementioned protocol applied in the 1970s in Hawaii reduced these costs considerably by designing a launch-and-recovery system operating from a platform that could be submerged to 20 m to release the submarine and be made buoyant again for towing the sub back into port.

Progress on the marine technological front during the 1990s allowed the construction of small, lightweight, low-cost submersibles, such as the Deepworker 2000 made by the Canadian company Nuytco Research. This class of submersibles represents one-person vehicles that weigh less than 2 t and can operate at a depth of 610 m. Their low cost and small size make it possible to use them in tandem (Grigg 2001), which increases safety. However, the overall operating costs in relation to precious coral prices are presently still too high, considering furthermore that only about 20% of the time each year are there adequate weather conditions (especially wind and swell) to launch a submersible.

Use of precious corals

The raw material for the precious coral jewellery industry is the skeleton of the corals. In general, it is easier to work with the species with harder skeletons. In some species, such as *P. japonicum,* there is considerable waste of material due to imperfections in the skeleton structure (S.J. Torntore personal communication).

Precious corals are used primarily for jewellery pieces like rings, pendants, amulets, necklaces, earrings and carved art objects such as statues. The market can be divided into the ethnic market that sells mainly rough coral beads, the tourist market and the high-end luxury fashion market that is the realm of jewellers' shops (Torntore 2002).

Despite extensive legends, present medical uses of precious corals appear to be negligible. Red coral powder is still being sold as a cure against various maladies or as an aphrodisiac. However, similar to the fraud with worked coral pieces that are made of plastic or low-quality species, it is often powder made from sponge coral (*Melithea* sp.) that is sold as *Corallium* powder (S.J. Torntore personal communication). In general, it is reef-building coral species that are frequently used as bone prostheses because their pores are quickly filled with capillaries (Pechenik 2005). Both stony corals and precious corals are sold as curios or decorations and for aquaria.

The genus *Corallium* is the most coveted group of precious corals. Their skeletons are appreciated for their hardness, purity and colour, and the lower abundance of these corals further increases their value. Depending on colour, striations (colour patterns also called 'anima') and consistency, whether collected alive or as fossils and whether they are infected by boring sponges, the material is divided into quality categories when priced. Colour varies according to species and locality, and popularity of colours on the market follows fashion. The dark red colour of *C. rubrum* has become the gold standard for the industry. Thin branches are of lower quality, and red coral skeletons from colonies infected by boring sponges are mainly of interest to the ethnic market because their skeleton is not solid but contains holes, which do not allow the piece to be given a polished surface (FAO 1983).

Coral branches of less than 7 mm in diameter used to be of negligible use, but this changed with the introduction of composite coral manufacture, a hardened mix of coral powder and a plastic such as epoxy (FAO 1988). This allows small branches and fragments to be ground to powder and formed into larger blocks, such as beads. Chemical analysis or inspection of growth rings under the microscope can identify coral pieces made from reconstituted coral (Smith et al. 2007). There are no data about the species that are used preferably to manufacture this type of jewellery, although some wholesalers state that they can sell coral powder only as a medical potion to Asian markets. It is unlikely that high-quality manufacturers in the fashion market produce reconstituted coral, and although FAO reports confirmed its existence, some industry insiders consider it a myth (G. Tsounis personal observation). In any case, reconstituted coral and small coral fragments are more likely to be sold in the tourism and ethnic marketplaces.

Precious corals are supplied to jewellery manufacturers as whole dried colonies, unworked branches or polished beads. In the last state, it is not straightforward to identify the species. Bamboo coral is often dyed red or black and sold as *Corallium* or *Antipathes*, whereas sponge coral (*Melithea* sp.) is often impregnated with acrylic polymers and sold as *Corallium*. Another inexpensive species that is sold as precious coral is rose coral (Scleractinia). Finally, there are bracelets from Bakelite or celluloid on the market that are also sold as precious coral jewellery.

Economy and trade

Prices for unworked precious corals have varied throughout time, depending on demand. The trade of precious coral has been increasing in recent years, judging from *Corallium* spp. import data into the United States (the largest importer of precious corals; CITES 2007) and a recent 50% increase in sales volume recorded in Hawaii (Grigg 2004).

Prices of *C. rubrum* have risen steadily over the last decades because demand always exceeds the diminishing supply (FAO 1983, 1988). *Corallium rubrum* is sold for relatively high prices; high-quality raw colonies are sold for US$1500 kg^{-1}, worked beads sell for US$30–50 g^{-1}, and necklaces cost up to US$25,000 (Torntore 2002). Manufacturing 1 kg of beads takes 115.5 h or 10.5–14.5 days. Prices for *C. rubrum* have risen from US$100–900 kg^{-1} to US$230–2900 kg^{-1} (FAO 1983, Moberg & Folke 1999). Today, even thin juvenile branches are bought for US$230–300 kg^{-1}, whereas they were practically worthless some decades ago (FAO 1988, Tsounis et al. 2007). Single, large *C. rubrum* colonies with a base diameter greater than 4 cm are reportedly sold for as much as EUR 45,000 per colony.

In Europe, the specialized red coral jewellery industry situated in Torre del Greco, near Naples (Italy), is estimated to generate more than US$230 million yr^{-1} (Assocoral personal communication). Around 270 companies participate in the labour-intensive manual manufacture of jewellery. The majority are small family businesses consisting typically of a father and sons. A substantial number of those specialize in outsourced processes for the larger companies (Ciro Condito, Assocoral, personal communication). The three oldest companies (Liverino, Antonio De Simone and Ascione) deserve special mention because they are significantly larger than the rest. More than 90% of all *C. rubrum* extracted has been processed in Torre del Greco since the 1800s (see p. 175). However, only 30% of the processed corals are *C. rubrum* now because 70–80% are imported from Japan and Taipei (Castiligliano & Liverino 2004). In fact, Italy has become a major importer of precious corals (CITES 2007). Industry insiders say that the black market is significant, maybe totalling 50% of the trade.

Tropical *Corallium* species are also of high value. In 2001, *C. secundum* was priced at US$187 kg^{-1}, *C. regale* was worth US$880 kg^{-1}, and *Gerardia* was sold for US$400 kg^{-1} (Grigg 1984). Recently, a large *Corallium elatius* colony 1.1 m high and weighing 67 kg was reported to have been sold for about US$100,000–300,000 (N. Iwasaki personal communication). It is not clear, however, if colonies of this size occur in sufficient quantities to make fishing trips commercially worthwhile. Large jewellery pieces of *C. elatius* that were sold to tribal groups in Nigeria during the 1960s are now being bought back by the industry to be resold to the luxury market, indicating a shortage of large tropical *Corallium* colonies.

Black coral sold for a relatively low price of about US$15 kg^{-1} (Grigg 2001) during the 1980s when Taiwan and the Philippines started to export large numbers (70 t to Hawaii alone in 1987; Carleton 1987) but rose to about $70 kg^{-1} today. The Hawaiian precious coral industry imports pink coral from Taiwan and is estimated to have generated US$30 million in 2004 with over 100 retailers (Grigg 2004) and in recent years US$70 million yr^{-1} (WPCouncil 2006). Recent sales volume has risen 50% because in 2001 the value was US$15 million (Grigg 2001), already much higher than the US$2 million reported in 1969 (Grigg 1994).

The United States is the largest importer of precious corals, including unworked coral from China and Italy. While the United States does not export coral or coral products, a large part is sold to tourists (especially in Hawaii). Taiwan exported 90% of the worked black coral on the market, previously importing much of the raw material from the Philippines (CITES 2007). Japan and China are also important manufacturing centres, with an annual value in 1982 of US$50 million (CITES 2007), although a large proportion of the exports are semifinished products (such as beads; S. J. Torntore personal communication).

In 2002, the data show a massive export of *Corallium* sp. from Italy into the United States that was about five times higher than the previous and following years. This situation may reflect the discharge of stockpiling rather than the discovery of a new bed. Italy and China contributed about 90% of all precious coral imports into the United States, while Italy's contribution decreased from 50% in 2002 to less than 4% in 2006. Thailand's contribution to the overall import volume rose from 0.2% in 2001 to 5% in 2006. China and Taipei (pooled here to reveal their impact on the South Pacific) were responsible for 84% of 1,807,357 precious coral products imported into the United States in 2006.

Recent yield data

The overall landings of red coral from the Mediterranean have been relatively stable at 25–30 t over the last 15–20 yr. The annual harvest is about 25% (25 t) of what it was in the early 1980s (Figure 2). However, as the more detailed discussion that follows will show, a stable yield should not be mistaken for a sustainable fishery because coast guard controls indicate that fishermen have been forced to harvest ever-smaller colonies in recent decades (Linares et al. 2003, Fisheries Department–Government of Catalonia personal communication), and a progression towards deeper beds and more remote stocks may maintain the yield. Today's main stocks are located on the Costa Brava (Spain), Corsica, Sardinia, and the northern African coast, although harvesting also occurs in Sicily, Mallorca and some other locations.

In addition to the general FAO landings data, which are pooled per country, there are a few detailed case studies. In Morocco, harvesting was done by foreign fleets and divers until the 1980s. Two boats with two divers each have been active since 1984 (10 boats in 2004) and are issued 500-kg individual annual quotas for specific areas that are managed using 10-yr rotation systems (Zoubi 2009). Al Hoceima was the most prosperous zone, with 8.6 t landings in 1985, but yield decreased after 3 yr and continued to fall during the 1990s. After closing the region for 12 yr (the fleet diverted to other areas and towards the Atlantic), the region was opened, only to be depleted within a year, and the fishery shifted to Tofino, where 145 kg were harvested in 2004 and 510 kg in 2005. Exploration for new stocks began in 2007 and identified Sidi Hsein as a stock with potential for continuing the fishery. In Sardinia there are 20–30 licenses (16–17 boats) harvesting ca. 1.1–1.4 t per year in 2008–2009 (Cannas et al. 2009, Doneddu 2009).

Thirteen of the total of 16 licensed divers in Spain are authorized to participate in the Costa Brava fishery (as a comparison, there are 25 active licenses on the island of Sardinia). One of those licenses is restricted to so-called interior waters (i.e., areas within transects connecting capes), whereas three licenses are issued only for the area outside. Nine licenses are for all of the Costa Brava, and any license is in general valid for more than one region of Spain (Fisheries Department, Government of Catalonia, personal communication).

The Costa Brava fishery concentrates on the Cap de Creus area, a 190-km^2 sparsely inhabited peninsula that is exposed to frequent northern gales. The length of the coastline measures about 42 km. In contrast to the easily definable coral beds in Hawaii, red coral habitat is very heterogeneously scattered, forming small patches in microhabitats (Tsounis et al. 2006b). Recruitment and habitat preference are not yet well understood, but competition with other benthic species and a short dispersal distance seem to be influential factors. A bionomic study in a nearby area (Ros et al. 1984) showed that 1 km of shoreline corresponds to about 0.01 km^2 of coral habitat. The annual yield for inshore waters on the Costa Brava ranges from 0.8 to 1.7 t, whereas the overall yield of inshore and offshore waters in Spain was about 4–5 t (Fisheries Department of Catalonia, personal communication). These numbers do not, however, take into account the severe problem of intensive poaching in the Mediterranean (see p. 180). An additional problem is that current management does not record the size of the harvested corals.

In the Pacific, a similar stabilization of yield following a dramatic decline can be observed (Figure 3). Current production in the Pacific is about 5% (20 t) of the yield in 1982 (400 t; FAO 1983). After a peak of 400 t in the mid-1980s, the yield of *Paracorallium japonicum, Corallium regale, C.* sp. nov. and *C. konojoi* remained at around 10 t after the 1990s (data from FAO).

Pink and gold coral were not harvested in Hawaii after 1979 because the cost of selective harvesting of deep beds was too high, and apart from a brief operation in 1999–2000 (1.2 t), these species are not currently harvested in the EEZ.

Black coral landings in Hawaii have increased considerably in recent years (WPCouncil 2006). Landings in the last 7 yr make up 58% of the total catch since 1985. From 900 kg yr^{-1} between

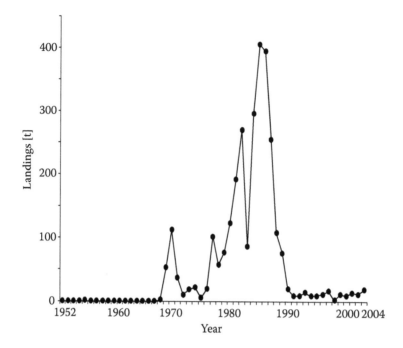

Figure 3 *Corallium* spp. harvest in the Pacific. The graph shows peaks that mark the discovery-exploitation-depletion cycle of several stocks located in seamounts (except the decline due to overproduction in 1982), and after 1991 a more constant yield from harvesting coastal stocks. (Data from FAO Fisheries Statistics Programme, http://www.fao.org.)

1985 and 1991, the yield rose to about 3.5 t yr^{-1} in 1999–2005 in the same coral bed (the Au'Au channel), and the maximum sustainable yield (MSY) of 5 t (Grigg 1976) was exceeded in at least 1 yr (WPCouncil 2006). New beds were not discovered for this diving-based fishery. Since 1980, virtually all black coral harvested in the Hawaiian Islands has been taken from the Au'Au Channel bed (with an areal coverage of 1.7 km^2). The Kauai Bed with an area of 0.4 km^2 has also been harvested, but to a lesser extent. Currently, only about five coral divers are active in this region, which is a surprisingly low number considering the amount of biomass removed and the impact on the population structure.

In the past, the hope of any precious coral fishery has depended on the discovery of new beds. Nearly two decades ago, southern oceans had been identified as the most promising areas for the future of precious coral exploitation, and in fact this is where most precious corals are harvested today. Potential new areas include channel waters around Madagascar (Grigg & Brown 1991). The Committee for Coordination of Joint Prospecting for Mineral Resources in South Pacific Offshore Areas (CCOP-SOPAC) reported *Corallium* sp. in Tasmania, the Cook Islands, Fiji, Kiribati, Solomon Islands, Tonga, Vanuatu and Western Samoa but not in commercial quality or abundance (Harper 1988). Within the U.S. EEZ, the following areas are still open for exploration: around the Hawaiian Islands, Guam and the Commonwealth of the Northern Marianas, American Samoa and all EEZs of the U.S. Pacific Island possessions. Commercial harvesting of precious corals has not been reported in the United States outside the Hawaiian Archipelago except for potions (Lumsden et al. 2007).

Bamboo coral harvest in Bone Bay, Sulawesi (Indonesia), appears to have increased significantly in recent years (S. Ferse, Centre for Tropical Ecology, Bremen, personal communication) because exports of more than 100 t were reported in 2005 (Department of Fishery and Marine Affairs [Dinas Perikanan dan Kelautan Bone sudah tercatat]). This development may represent the phenomenon called 'fishing down the price list', that is, shifting to the next-available resource after depleting the most valuable ones.

Management and conservation

This section attempts to give an overview of the management and conservation of precious coral fisheries in various countries. Gold and pink corals are no longer fished in the U.S. EEZ for economic reasons (see p. 179), and because the Asian fisheries are only now being thoroughly studied, it is necessary to focus on a comparison of the two best-documented case studies: the Mediterranean red coral fishery and the Hawaiian black coral fishery. The differences in socioeconomy and in the ecology of the species help to highlight causes and effects and to better understand precious coral exploitation in general.

Coral harvesting in territorial waters is controlled by a licensing system permitting a controlled number of divers to harvest coral. Harvest quota are used to further control the yield, and usually minimum size limits are used as well. However, management based only on harvest quota is not effective in preventing overharvesting (WPCouncil 2007) because immature corals cannot be protected in this way.

Management

Precious coral fisheries in Asia

Most fisheries in Asia have not yet been studied in detail, and there are no published data on some species. Some countries, such as Thailand and the Philippines, have banned coral fishing. The majority of black coral landings in the past were reported to have been harvested in the Philippines (Ross 2008). However, coral trade in the Philippines was banned in 1977, and any stockpiles had to be sold 3 yr after the closing date. In July 2008, China listed the four *Corallium* species *C. elatius*, *C. konojoi*, *C. japonicum* and *C. secundum* in Appendix III of CITES, thereby controlling their export (CITES 2008).

Coral fisheries in Japan are authorized by prefectural governors and therefore differ from each other. Red, pink and white corals are harvested by traditional stone-weighted non-selective tangle nets in Kochi. Since 1983, in waters extending from Kagoshima to Okinawa, harvesters have used manned and unmanned underwater vehicles and follow self-imposed size limits (Iwasaki & Suzuki in press). There is no official quota for these fisheries because the research needed to manage the stocks has only recently been initiated (N. Iwasaki personal communication). According to unpublished information by Sadao Kosuge (Institute of Malacology, Tokyo), yields have been stable over the last decade.

Of 24 areas, only three are open to fishing. One area is fished by submersible, while strong currents in the other two allow only tangle nets to be used. Each boat deploys one dredge for 4 h per day, and is active for about two weeks per year. Yield per boat and year is about 12 kg, and 120–230 vessels are active. One of these areas (Kochi prefecture) consists of soft bottoms with deposits of dead coral, which is targeted using dedicated gear (resulting in 80% of the Kochi area's catch). Total catch in Japan is 3.6–4.8 t in 2003–2008 (S. Kosuge unpublished data).

The Taiwanese precious coral fishery began in 1929 and in 1983 was limited to 150 vessels. Currently, there are 53 vessels harvesting *Corallium* sp. in five regions, each vessel with an annual quota of 200 kg vessel^{-1} over a 220-day activity limit for each year (Chih-Shin Chen, Institute of Marine Affairs and Resource Management, National Taiwan Ocean University, personal communication). The fishermen employ traditional non-selective gear consisting of the tangle nets typical for Asia deployed at a slow speed of 1.5 knots. Only 2% of the harvested coral in Taiwan is live coral; 83% is dead coral, and a further 15% is dead coral that has been on the seafloor long enough to decay. Penalties for poaching are severe in Taiwan because they include a 3-yr imprisonment and a fine of US$20,000 (C.-S. Chen personal communication).

Hawaiian black coral

In Hawaii, commercial black coral beds are located in state *and* federal waters. State waters include areas within 3 mi of islands as well as interisland waters, where black coral is predominantly found, and harvest is regulated by the Department of Land and Natural Resources (DLNR), Division of Aquatic Resources (DAR). The area outside the state of Hawaii falls under federal jurisdiction and is referred to as the U.S. EEZ (Grigg 1994, National Oceanic and Atmospheric Administration [NOAA] 2006). Precious coral exploitation in this zone has been managed since 1983 by the National Marine Fisheries Service of the NOAA through the Precious Coral Fisheries Management Plan of the WPCouncil.

The Federal Fishery Management Plan by the WPCouncil classifies precious coral beds as established beds, conditional beds, refugia beds and exploratory permit areas (Grigg 1994). Selective harvesting gear is mandatory, although until 1999 conditional and exploratory beds could be harvested with non-selective gear (Bruckner et al. 2008). The Makapu'u bed (the densest and most productive bed, measuring 1.7 m^2) and Au'Au channel bed (0.4 m^2) are currently the only established beds. Conditional beds, for which yield has been estimated relative to their size assuming identical conditions to known beds, are Kea-hole Point, Kaena Point, Brooks Banks and 180 Fathom Bank (Grigg 1994). The WP Council bed between Nihoa and Necker Island is the only designated refugium (Bruckner et al. 2008).

The need for *sustainable* management of precious coral fishery was first recognized when the precious coral jewellery industry in Hawaii began to grow steadily. In response, the University of Hawaii set up a research programme to study the ecology and fisheries management of precious corals in 1970 (Grigg 1976). The state of Hawaii management programme is thus the first precious coral fishery management based on the ecological characteristics of the species.

The study revealed that the minimum size limit at which divers were harvesting black coral voluntarily was above the age at first reproduction. The fishermen voluntarily refrained from harvesting black coral colonies smaller than 1.22 m in height and 2.54 cm in base diameter (reproductive maturity in *Antipathes griggi* is reached at about 64–80 cm; Grigg 1976). Smaller colonies were of little value to the curio industry, and it made economic sense to spend the dive time harvesting larger coral. Small black coral trees were already of some value to the curio or display market, which bought them at US$50 kg^{-1} instead of US$10 kg^{-1}, but apparently as long as large coral trees can be harvested and sold for more to the jewellery industry, the small corals are not harvested intensively by these fishermen.

The UH-Seagrant study applied the Beverton-Holt model to estimate an MSY of about 5 t for the Au'Au Channel stock (Makapu'u bed) and 1.25 t for the Kauai bed (Grigg 1976). This model defines maximum production in a fishery as the point at which natural mortality losses are balanced by population growth (Beverton & Holt 1957). Despite being a simple model that has been surpassed by more advanced models in various fisheries, it has been used successfully to assess various coral populations (Grigg 1976, García-Rodríguez & Massò 1986b, Tsounis et al. 2007, Goffredo & Lasker 2008, Knittweis et al. 2009).

The age at maximum yield per recruit for *A. griggi* was estimated to be 22–40 yr, corresponding to corals that measure 1.7 and 3.2 m in height (the oldest black corals can reach 3.5 m across and more than 4 m in height). However, corals continued to be harvested when they had reached 1.22 m, instead of 1.7 m. The reason for the discrepancy is shown by an analysis of *optimum* yield. Harvesting all corals exceeding the height limit of MSY would, in theory, provide a 100% efficiency of the fishery. Lower efficiency than that, however, may result in more profit if catch per unit effort and optimum yield are considered. Therefore, the most economic and yet sustainable strategy often is to fish at low intensity and catch the coral at an earlier age than at MSY (Grigg 1976). As

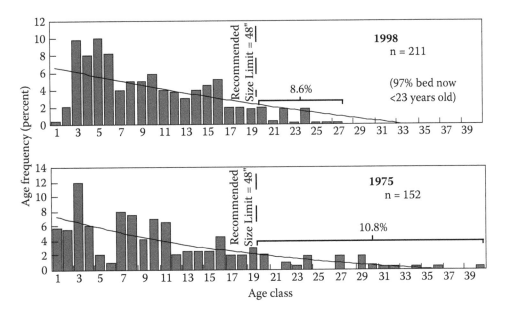

Figure 4 Age frequency distribution of black coral off Maui, Hawaii, in 1975 and 1998. (Data from Grigg 2001, with kind permission from *Pacific Science*, University of Hawai'i Press.)

long as the harvested corals are significantly older than the age at first reproduction, a reproductive cushion is ensured (i.e., the colonies are allowed to contribute to recruitment for several generations before they are harvested, ensuring sufficient recruitment and thus survival of the population). On the other hand, this policy means that care must be taken to control fishing intensity by using monitoring programmes. This practice does not produce maximum yield but allows for maximum profit (thus called *optimum yield*) because yield per fishing effort is maximized (achieving maximum yield may in some cases result in less profit if it requires disproportionally higher fishing effort).

The new management rules did not conflict with the fishermen's interest of harvesting large colonies and were voluntarily accepted. The Hawaiian black coral fishery may therefore be the only sustainable precious coral fishery in history (Grigg 2001). The comparison of the population structure between 1975 and 1998 showed that the oldest and largest colonies were no longer present, but that the structure of the population at the younger spectrum remained unchanged (Figure 4).

However, in 1998 demand in black coral started to increase (sales increased by 50%; see 'Economy and Trade' section), and the state of Hawaii lowered the size limit, introducing a grandfathering scheme that allowed veteran divers (those who had reported black coral harvest in the preceding 5 yr) to collect corals of 0.9 m in height to meet the rising demand. Another reason that the fishery has become more efficient is the availability of detailed bathymetric maps and the adoption of GPS (global positioning system) navigation (Bruckner et al. 2008). Use of this equipment has led to a decline in black coral biomass of 25% (Grigg 2004), possibly posing a threat to the population. In 1998, after 23 yr of harvesting, no colonies older than 27 yr remained (Grigg 2001), but enough mature colonies still remained in the population (maturity is reached at 10–12.5 yr; see p. 172). Three yr later, no colonies older than 24 yr were left (Grigg 2004), illustrating a biomass loss due to increased intensity of fishing. A further negative impact on the black coral populations is the aforementioned invasion by the octocoral *Carijoa riisei*. The affected populations are beyond the depth range and were previously thought to be important contributors to larval recruitment of the overharvested populations but were found to be non-reproductive (R. Grigg unpublished data).

Responding to this situation, the state of Hawaii has returned to the minimum size limit of 1.2 m. Recent surveys furthermore suggest that MSY be adjusted downward by approximately 25% (Grigg

2004, Parrish 2006). Also, the state of Hawaii is in the process of setting up an inventory-tracking system using bar codes to gather data on the harvest and trade of black coral (T. Montgomery, DAR, personal communication). Adaptive management (Walters 1986) is likely to improve this fishery because two factors favouring the recovery of the stocks are (1) the ecology of the species is fairly well understood and (2) the conclusions have been successfully applied for more than 20 yr and a steady state has been achieved. The reasons that lead to biomass decrease are known, and the precious coral management system can be described as better organized than other precious coral fisheries. Furthermore, demand is not excessively high because the industry does not depend solely on production from the Hawaiian Islands, and there is a low number of fishermen (5 licenses), with no indications of poaching. For these reasons, and due to relatively fast growth, Hawaiian black coral beds have the best potential for recovery and sustainable harvesting among precious coral fisheries.

Mediterranean red coral

The management of the Mediterranean red coral fishery was traditionally based on social, market and political considerations, but ecology was taken into account as early as 1882, when Professor C. Parona from the University of Cagliari was asked to help improve the efficiency of the fishery. In the 1870s, the naturalists Cavolini, Milne-Edwards, Marsili, Lacaze-Duthiers, Issel and Canestrini, as well as the fishermen themselves, studied the distribution and reproduction of red coral to create a basis for management decisions (although their recommendations did not have much influence; see Tescione 1973). The growth rate was only studied for the first time half a century later (Dantan 1928). Fishery statistics before the 1980s were documented by noblemen and various governmental organizations and were summarized by Tescione (1973). In Italy, the fishery was for a long time essentially unregulated, giving absolute freedom to dredging and divers (Arena et al. 1965, Liverino 1983).

Responding to the dramatic decrease of yields in the late 1970s and early 1980s (Chouba & Tritar 1988, Cattaneo-Vietti et al. 1998, Santangelo & Abbiati 2001), the FAO (1983, 1988) hosted technical consultation meetings; however, these resulted in few management changes other than banning coral dredging. Intrusions of foreign poachers into national waters off the island of Alboran in the 1980s initiated efforts to ban dredging as well as to list *C. rubrum* in CITES Appendix II.

With the advent of selective scuba harvesting, the voluntary minimum harvesting size of 7 mm was eventually established as a rule, and for a long time smaller corals continued to be of little value to the industry (FAO 1983). Similar to the Hawaiian fishery, later studies found red coral to reproduce at an earlier age, so that the protection of immature juveniles was ensured. Determining MSY was not possible at that time. Although the biometry of red coral had been studied (Marín & Reynald 1981), the population structure of red coral was first studied in 1986 (García-Rodríguez & Massó 1986c). A subsequent study was then able to determine the MSY at 80 yr (García-Rodríguez & Massó 1986b), which has since been confirmed by Tsounis et al. (2007), who obtained a very similar value (although slightly higher due to using lower growth rates) of 98 yr. In contrast, the corals that are legally fished are about 14 yr old. García-Rodríguez and Massó (1986a) recommended a minimum size of 8.6 mm, but to date this has not been adopted, and 7 mm remains a widespread minimum size of harvest. Exceptions are Algeria, where 8 mm is applied (CITES 2007), and Sardinia, where a minimum size of 10 mm does not interfere with the intent of the divers to harvest only large colonies in very deep water. Morocco has not set up any size limit and manages its stocks through quotas (Abdelmajid 2009, Zoubi 2009).

Although red coral becomes sexually mature at an age of 3–4 yr, fertility does not reach 100% before 6–9 yr (Santangelo et al. 2003, Tsounis et al. 2006a). The legal minimum of 7 mm in diameter corresponds to about 11-yr-old colonies, depending on the growth rate, which varies according to habitat and geographic region (see p. 168). Thus, fishing does not allow the majority of the population to reach its full reproductive potential (Figure 5). Three yr of reproductive buffer may not

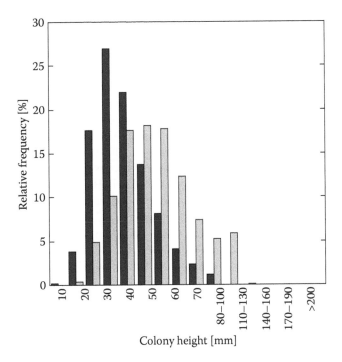

Figure 5 *Corallium rubrum* colony height distribution on the Costa Brava (black) and the Medas Islands Marine Park (grey) (both in Spain). The slight but significant size difference is the result of about 14 years of protection. (Data from Tsounis et al. 2006b, with kind permission of Reports for Polar and Marine Research.)

be long enough to ensure high recruitment if compared with black coral in Hawaii, which reaches maturity at 10–12.5 yr but is not fished until it is 20 yr old. This is significant in modular, highly branched organisms such as corals because only a small fraction of the older colonies contributes the majority of the recruits (Miller 1996). In some species, up to 98% of the recruits are produced by the older half of the population (Babcock 1991, Sakai 1998, Beiring & Lasker 2000).

Again, similar to the Hawaiian fishery, red coral harvesting below the age limit corresponding to MSY was continued, increasing the catch per unit effort. In both fisheries, the main factor that has led to overharvesting and decreasing biomass and yields is the problem of balancing fishing effort against age at first capture. When harvesting at an age below MSY to achieve optimum/economic yield, fishing effort needs to be sufficiently low. If the fishery is too efficient for the assumed effort or the effort misjudged (e.g., due to poaching or forged statistics), the stocks can be harvested down to the targeted age at first capture. In fact, this appears to be the cause for the documented overharvesting.

Harvesting down a stock to the minimum size limit is not desirable because it is unsustainable and depletes the stocks. More sophisticated models are likely to allow better predictions of harvesting effects based on a certain fishing effort. Leslie-Lewis transition matrices, for example, are age-structured, nonlinear models based on demographic data (Caswell 2001) and can take density dependence into account as well. The advantage lies in linking reproduction, growth, mortality and demographic structure into one model. These models have been used to simulate the trends of a red coral population over time (Santangelo et al. 2007).

The results of the transition matrix modelling show that the present, extremely young red coral populations may not be able to recover from the combination of overfishing and frequent mass mortalities (Santangelo et al. 2007). If natural catastrophes of any kind occur with an elevated frequency, the affected population can only survive if it has a strong recruitment. Recruitment potential is directly linked to the number of polyps; thus, heavily fished populations with only young colonies are driven to local extinction. Strategies to avoid this so-called Allee effect of populations decreasing below critical size (Allee et al. 1949, Stephens et al. 1999, Dulvy et al. 2000) are setting adequate quotas, controlling poaching and the introduction of a minimum size limit higher than 7 mm. Conservative management plans such as this one usually conflict with short-term socioeconomic interests, and decision makers therefore hesitate to implement them (Bearzi 2007). However, in case of the red coral fishery, the additional dilemma lies in the fact that the fishery has left only a small fraction of the populations with colonies that surpass this limit. The fishery in Sardinia, however, decided to ban coral harvest in waters shallower than 80 m. The management was strong enough to allow it to shut down the fishery completely in 2007 to study the stocks and revise management (Cannas et al. 2009, Doneddu 2009). Furthermore, considerable no-take areas are in place, although their size may not reach 80% of the total stock, as in Hawaii. These measures represent a favourable development, and when the data are analysed and published, the Sardinia situation may well turn out to be a case study of sustainable management of *C. rubrum*.

In general, however, all known *C. rubrum* stocks have been recognized as overexploited since 1989 (FAO 1988, Cognetti 1989, Santangelo & Abbiati 2001, Tsounis et al. 2007), but the full extent of the situation in the Mediterranean is only now being understood through a recent accumulation of data. No discoveries of new coral beds have followed the overexploitation of known stocks. The fishery progressed towards deeper depths and smaller colonies and, as the data showed, is approaching limits in both aspects. The fishery has not extended its reach into depths beyond 130 m because dredges are prohibited, and submersibles are not economic. ROVs are occasionally used for scouting, but remotely controlled extraction does not appear to be feasible. Ironically, there is hope within the industry of exploiting deeper strata, but they were depleted in the 1980s, and scuba-diving has been able to access coral in crevices and cave entrances only because this species is cryptic in shallow water. It appears that a vast part of the deep populations is presently not of high commercial viability because red coral grows at low densities at these depths and at more exposed positions, so that crevices in deep habitats contain fewer large corals than shallow water crevices. The most important aspect is that these populations have not yet recovered from centuries of heavy dredging, and part of their habitat may be irreversibly occupied by fast-growing *Lophelia pertusa* deep corals (Rossi et al. 2008, Orejas et al. 2008). Exploiting the deeper populations of Mediterranean red coral can be problematic because they may more vulnerable to harvesting than shallow-water colonies. Consequently, the future of Mediterranean red coral fishery is likely to see a further reduction of yield in one way or another.

The only way to meet demand is to import Pacific species, whose stocks are not yet studied, and the sustainability of their fisheries therefore remains questionable. The question of how to achieve sustainability of the Mediterranean fishery is an enormous socioeconomical challenge. Optimum yield for an overharvested stock is focused on rebuilding its full capacity (Magnuson-Stevens Fishery Conservation and Management Act, U.S. Department of Commerce 2007), but this requires a reduction of yield.

Conservation

Habitat conservation considerations

The importance of non-reef-building deep-water coral species for ocean biodiversity creates a conflict between habitat conservation and traditional precious coral fishery management. Whereas fishery management allows an intensive exploitation of a population, accepting a significant modification of its age structure, habitat conservation intends to maintain a population structure that guarantees a minimum functionality. Overfishing is in some cases defined as a biomass reduction of the target species population to 20–30% of the baseline (CITES 2007, WPCouncil 2007). Today, however, precious coral fishery managers realize the need for an ecosystem approach in fishery management rather than single-species-oriented management (WPCouncil 2007), and recent management research has expanded into connectivity of coral taxa in the Hawaiian Archipelago (Baco & Shank 2005, Baco et al. 2006) and associations with other species. This ecosystem-based fishery management (EBFM) represents a holistic approach that considers the connectedness of different species and links between species and environment, rather than managing species as if they were in isolation (Bruckner et al. 2008).

In the case of deep corals, studies highlight with increasing tendency the importance of deepwater corals as ecosystem engineer species for improving biodiversity and productivity and stability of their ecosystem (see p. 171). An intensively exploited population, even if sustainably managed, is surely less efficient in providing these 'ecological services' (Moberg & Folke 1999) than a population with a balanced age structure (containing both recruits and very old individuals). Future precious coral management is therefore challenged with determining, and maintaining, the level of modification of population structure by fishing pressure that still allows an acceptable ecosystem engineering effect.

The implementation of marine protected areas may be an important measure to ensure the biodiversity of nearby commercial areas if refugia are large enough and interconnected. In contrast, traditional rotation systems (in use since the Middle Ages) that harvest one stock and then work on others while the exhausted one is left to recover are problematic. The severe depletion of one stock is not in line with habitat management because it does not account for the ecological services of the corals. If the spatial scale chosen is too broad, overharvesting of one area may interrupt the gene flow between coral populations, as well as of invertebrate and fish populations seeking shelter in their branches. Therefore, management should instead monitor the fishery ensuring that no age class is eliminated completely. The present precious coral fisheries are far from this ideal, but lessons could be learned by examining the plenter principle that is applied to terrestrial forestry (O'Hara et al. 2007). As a starting point, given the lack of models, the maximum size of colonies in a population could be determined by ROV transects and a percentage of them prohibited from being harvested to maintain habitat structure. In a way, this approach would add maximum size limits (protecting habitat structure and thus biodiversity) to the already common minimum size limits and protect the reproductive potential.

International conventions versus local management

As illustrated, local precious coral fishery management has in many cases proven inadequate, so there is no doubt that management strategies need to be revised. One reason is the 'tragedy of the commons' that is typical for most fisheries, that is, the overexploitation of common resources by competing fishermen who exceed quotas for fear their competitors will soon exhaust the commonly exploited resource anyway. However, the fact that precious corals are sessile, non-migrating species may offer a possibility of avoiding this problem by granting indigenous exploitation rights or forms of comanagement and stewardship. Overfishing is, however, in part due to poaching and the location of several precious coral stocks in international waters. Therefore, the question arises regarding whether international conventions might help to improve precious coral management as it does with other fisheries.

At present, it is up to each country to manage their stocks because, unlike many migratory fish species, precious corals have not been included in regional fisheries management organizations as is the case in other fisheries, such as the International Commission for the Conservation of Atlantic Tunas (ICCAT), the Western and Central Pacific Fisheries Commission (WCPFC), the Indian Ocean Tuna Commission (IOTC) or the General Fisheries Commission of the Mediterranean (GFCM). The approach taken by the last organisation has been suggested as a means of managing *C. rubrum* fisheries, but it is unclear if the GFCM recommendations would be enforced. The European Union placed *C. rubrum* in Annex V of the European Union Habitat Directive (2007), and it is also listed in Appendix III of the 1979 Bern Convention on the Conservation of European Wildlife and Natural

Habitat Protection (Council of Europe 1979). Consequently, the currently observed overexploitation is in large part due to lack of enforcement.

The management and conservation of precious coral populations in international waters has not been addressed by international treaties, although fishermen operating out of ports in the United States do need a permit for high-seas fishing. In other countries, seamounts in international waters can currently be harvested without licenses. It is likely that these unstudied habitats contain the last natural populations, treasuring a significant biodiversity so that they should be protected by declaring them, for example, as UNESCO (United Nations Educational, Scientific and Cultural Organisation) World Heritage Sites in Danger. Some organizations, such as the Northwest Atlantic Fisheries Organization (NAFO) are designing international protection treaties for the UN, vulnerable marine ecosystems (VMEs), which might serve as role models for other precious coral habitats. The declaration of the north-western Hawaiian Islands as a marine national monument served a similar purpose. However, in all cases of declaring precious coral communities in international areas as protected, there is the problem of enforcement. Trade control may thus be an important further measure (see next section). International trade control can be an efficient tool in discouraging poaching and illegal trade as well as fostering research into better management, but strong local management and enforcement will certainly remain the basis of precious coral management.

Control of the precious coral trade through CITES

The ecological importance and vulnerability of precious corals has led to several proposals to include these species in Appendix II of CITES. CITES is an agreement between 175 governments that was created in 1975 to ensure that international trade of wild animals does not threaten their survival. Unlike the Red List of the International Union for Conservation of Nature (IUCN), which identifies and documents species in danger of extinction, CITES actively controls trade. Both organisations distinguish various levels of threat, but the most relevant category of protection for precious corals is presently Appendix II (which contains the bulk of CITES listed species). It includes species not necessarily threatened with extinction but in which trade must be regulated to protect declining populations and ensure sustainable exploitation. More drastic measures include the entire prevention of precious coral import into the United States via the Lacey Act, closing down the largest segment of the precious coral market. CITES Appendix II, however, does not ban exploitation but instead regulates trade.

Several orders and families containing precious coral species have been included in Appendix II, such as Antipatharia (black corals), Coenothecalia (blue corals), Tubiporidae (organ-pipe corals) and Stylasteridae (lace coral). Some of them, like blue corals, are not traded at all, and little is known about their biology.

Since 1986, proposals to include *Corallium* species were rejected because of lack of data that could confirm whether these species were threatened. The existence of refuge populations in deep water were assumed, and as species with a broad depth range and scattered distribution, they are less likely to suffer from extinction. On the other hand, their high value, the slow recovery of overexploited stocks and the short larval dispersal do increase the risk of genetic isolation and extinction, as recent data indicate. In general, it is thought that commercial species go *economically* extinct before the species is threatened by *ecological* extinction. But, multispecies exploitation sustains the industry, while part-time poaching may lead to local extinction of some isolated populations. Furthermore, factors such as manufacture of reconstituted coral from small branches and mass mortality epidemics make the risk of extinction difficult to predict.

What is beyond doubt, however, is the unsustainable nature of most precious coral fisheries outside Hawaii. For this reason, and because errors regarding precious coral management may take decades to rectify, the United States proposed the inclusion of the genus *Corallium* in Appendix II of CITES in 2007.

Several issues arose that led to some controversy regarding this proposal. First, identification was seen as a hindrance to enforcing the implied trade control because *Corallium* species are not easily identified. In fact, once worked and polished, it becomes unfeasible for an enforcement officer to distinguish *Corallium* species (even fake coral can be hard to identify, and DNA identification is not economical). However, the need to identify *Corallium* species already exists because China listed four *Corallium* species in Appendix III in 2008. Also, identification of *Corallium* species is easier than *Antipathes* species, which is already listed (E. Cooper, WWF Canada, personal communication). An effective solution would be to list the family Coralliidae instead of the genus. This solution is certainly desirable because of the family's ecological significance and vulnerability and it would reduce administrative effort. However, it is not clear whether it is necessary to identify *Corallium* shipments to species level. On the one hand, if the whole family is listed, identification to that level is feasible (E. Cooper personal communication). On the other hand, conservation efforts are local and on the species level, so that an accurate identification may help identify the origin of harvested coral.

Other concerns are that international trade control might be seen as a substitute for local management. In the case of black coral, effective local management was in place before the listing, and the CITES listing affects the industry because each permit requires 2 h of employee time and a $129 processing fee, increasing the costs of the business (R. Grigg unpublished manuscript). Processing time per permit takes 1 month. Problems occur when jewellery under warranty is sent back to the manufacturer for repair and is held in customs due to missing permits (T. Montgomery, DAR, personal communication). In response to these administrative burdens, some companies no longer sell outside the United States. Overall, less black coral is sold, and in this sense, black coral resources are conserved because less raw material is purchased (R. Grigg unpublished manuscript). Regarding the significance of a Coralliidae listing for the Italian coral jewellery industry, this type of burden would be especially severe on small family businesses that are typical in Torre del Greco (C. Condito, Assocoral, personal communication). The industry is also concerned about losing sales due to the resulting stigma of a CITES listing (G. De Simone personal communication).

Perhaps the most difficult subject of discussion is the question of how to deal with stockpiles of coral that were harvested before the convention. These are, by definition, exempt from the treaty, but existing standard procedures such as delayed implementation may not deal effectively with the potentially large quantities. However, all these considerations are also of a general nature and not coral specific.

Further discussion during the initial evaluation of the 2007 proposal (including a preparatory FAO consultation) arose over uncertainties whether the criterion of biomass decline was met for all species. A lack of data was identified for several species, but further species can be proposed for listing based on 'lookalike criteria' in comparison with similar species. Landings do exist for several species but were criticised as not reflecting biomass trends, as when fishing intensity decreases independently (e.g., if market forces render fishing economically infeasible), catch declines may erroneously indicate a biomass decline. However, the change of prices does easily identify if the reason for the declining yields (Figures 2 and 5) is dependent on the availability of corals. Two cases are known when an overabundance of corals drove prices down and forced the industry to stop fishing (other than during the world wars): (1) the 1880 crisis that was caused by the discovery of large beds of fossilised Sciacca coral (see p. 176) and (2) in 1982 the consequences of harvesting large amounts of Midway coral. In other known cases of reduced fishing activity, prices increased and have continued to do so in recent years.

Recent catch declines do therefore reflect biomass declines, if changes in harvesting methods and fishing effort are accounted for. Furthermore, the only two case studies that evaluated population structure/biomass, or standing stock, are of the precious corals in Hawaiian waters and a few populations in the Mediterranean. Identifying population biomass decline requires a time series comparison or comparison with unfished populations. In any case, population decline in colonial animals must be identified by analyzing polyp numbers, not colony numbers, because the polyps are the reproductive modules (Bruckner 2009). Population structure data serve well for that purpose, in contrast to abundance data, which are associated with high standard deviations due to patchy distribution (see p. 169).

In any case, with the exception of black coral in Hawaii, there is no baseline because no other species has been studied before intensive harvesting began. Considering that today's populations lack the very old colonies that contained a number of polyps several magnitudes higher than today's small/young colonies, it is certain that populations of most species in question did decline dramatically, bringing with it a reduction of recruitment.

While local management remains the basis of conservation (Pani & Berney 2007, Pani 2009), supporters of the CITES listing argue that a listing in Appendix II would stimulate this kind of research and management because CITES obliges each member country to assess its fishery to be able to issue non-detrimental findings to harvesters who apply for an export permit. (However, the listings of black coral in Appendix II have not resulted in further local management programmes.) CITES is not meant to be a substitute or a replacement for local or regional management but rather a means of ensuring that harvest and trade are legal. It is difficult to assess and foresee management options in all countries. In Spain, for example, a CITES listing may lead to more severe penalties, including jail terms rather than fines, and might therefore strengthen local management.

These complexities are reflected in the initial adoption of the 2007 proposal in CITES Committee One and its subsequent overturning during deliberations in Plenary on the final day of the listing due to concerns regarding its implementation (Morell 2007). Between the CITES Conference of Parties 14 in 2007 and the time of publication of this review, two *ad hoc* workshops were held to clarify these issues and discuss whether and how the listing can be implemented.

The consensus was that the FAO GFCM (General Fisheries Commission of the Mediterranean) may be the appropriate organisation to effectively manage *Corallium rubrum*, and that shallow water populations should be protected (University Parthenope, in press). The United States and European Union submitted an updated proposal to be evaluated at the Conference of Parties 15 in 2010. In a consulting function to CITES Conference of Parties (CoP), the FAO found that the data do not meet the decline criteria and recommends strong local management to prevent unsustainable harvesting, while highlighting implementation issues and the administrative burden of a CITES listing (FAO 2007; FAO 2009). The IUCN (International Union for Conservation of Nature) and TRAFFIC (the Wildlife Trade Monitoring Network) also advise the CoP and came to the conclusion that the data do not meet the CITES criteria (IUCN & TRAFFIC. 2007).

Active restoration

Rearing red coral to provide the jewellery industry with unlimited resources has been tested with no notable success (Cicogna & Cattaneo-Vietti 1993). The slow growth rates imply that the risks of the operation are high and the investment return low. However, due to the long recovery time of devastated precious coral populations, it is feasible to design programmes that combine *ex situ* or *in situ* rearing with transplantation techniques. This procedure could enable ecosystem managers to actively restore habitats that have suffered local extinction and create a network of stepping stones that can ensure sufficient gene flow and recruitment. The recovery of complex coral ecosystems, which takes decades or more (Dayton 2003), might be accelerated in this way. Active restoration is seen as the future of conservation and has evolved notably during the last decade (Young 2000, Rinkevich 2005). Restoration methods such as transplantation of coral fragments, branches and whole colonies have been pioneered since the 1970s through pilot studies, yet restoration of coral reefs is still in its infancy (Edwards & Clarke 1998) and has until recently not been a widely applied management option (Lindahl 1998).

The most significant advance in coral reef restoration over the last decade has probably been the introduction of the concept of 'coral gardening', which is a two-step protocol for the mariculture of coral recruits (spats, nubbins, coral fragments and small coral colonies) in nurseries (Rinkevich 1995). *In situ* pools, installed in sheltered coastal zones (Epstein et al. 2001), and *ex situ* tanks (Becker & Mueller 2001) are used to mariculture coral recruits to an adequate size, thus providing ecosystem managers with an unlimited source of colonies for transplantation (Rinkevich 2005). This concept represents the most efficient and advanced approach to ecosystem restoration.

However, efforts have focused on tropical reef-building corals, and there are few data on temperate and cold-water coral restoration (Grigg 1984, Cicogna & Cattaneo-Vietti 1993, Montgomery 2002). The settlement and early-life-phase biology of *Corallium rubrum* have been studied *in situ* using marble tiles (Bramanti et al. 2005, 2007), providing the basis for the subsequent transplantation of the tiles. Future research may further refine transplantation protocols, including the evaluation of measures that increase growth rate (Rinkevich 2005).

Nevertheless, considerable deep coral and temperate coral restoration research will be necessary to prepare conservation scientists and managers to actively support the natural recovery of these habitats. This research needs to include precious coral communities devastated by harvesting as well as other deep coral communities that were destroyed as a result of dredging or natural events.

Of course, prevention is to be preferred for many reasons (including financial considerations) over active restoration. It should be the last resort because the only feasible outcome that can realistically be expected is an increase in the natural recovery rate (Edwards & Clarke 1998). Future research may test appropriate methods and estimate the rate of increase of recovery. In some cases, recovery would not occur without active restoration (habitats geographically isolated from the gene pool), but it would still be a slow process and assumes the concept would prove feasible in deep corals.

Summary and conclusions

The analysis of historical data revealed that today's precious coral populations are heavily modified by commercial extraction. These are the slowest-growing organisms of any fishery known, past or present. In some cases, such as the *Corallium* sp. harvest at the Emperor Seamounts, the populations have been devastated, and later sampling found only dead and broken coral; in the case of *Antipathes* sp., there are few exploitable populations left in diving range. *Corallium rubrum* populations in the Mediterranean Sea resisted exploitation due to their uniquely high reproductive capacity, but their population structure in shallow water has changed from providing a forest-like habitat to one resembling a grass plain. The Mediterranean red coral was dredged for centuries until the populations declined, then divers harvested corals inaccessible to dredges until depths reachable with traditional air scuba became largely depleted. Since the industrial exploitation of *C. rubrum* began centuries before marine science was born, the baseline of the population structure under undisturbed conditions is not known. Recent research indicates that, despite the hopes of the industry, populations below scuba limits have not yet recovered from dredging significantly enough to expose them to the pressure of a deep coral fishery. Poaching appears to be a severe problem for the shallow-water populations, and some areas need urgently improved protection.

The single example in Hawaii of a fishery where local management plans have been based on a prior study of growth rate, population structure and other ecological parameters shows that sustainable management is possible, at least for faster-growing species. The analysis of all available data showed that any exploited precious coral populations that have been studied have been harvested down to the target age or size.

Nevertheless, the U.S. and European jewellery industries have been able to grow despite limited or declining stocks because they have shifted to imports from Asian stocks and species. This shift is worrying because 80% of the coral harvested in Taiwan is dead coral, which means that these beds will not renew themselves significantly. The size of landed coral may already be decreasing in tropical corals because jewellers have started buying back large coral pieces that can be cut to high-fashion jewellery. Furthermore, there are first signs that even economically less-valuable species such as bamboo coral have become exposed to heavy harvesting pressure.

Thus, although there remain controversies about the age of some species, about the existence of undiscovered stocks and a lack of biological data for many species, there is little doubt that most all known precious coral stocks have been overexploited. In fact, given the nature of the exploitation, the terms 'harvesting' and 'fishery' inaccurately imply a renewal of the resource, which in reality rarely occurs. In management terms, the majority of fisheries can more precisely be characterized as 'coral mining'.

The recent studies and the proposal to list the genus *Corallium* in CITES Appendix II mark 2007 as a watershed year for the precious coral fishery. There remains little doubt that deep and precious coral habitats are in urgent need of better protection, even though management and conservation are made more difficult by the lack of scientific data on the population structure and basic biology of many species.

Precious corals stand out from many other known fished species due to their significance as structure-forming organisms. However, the discussion about sustainable fisheries and species preservation is only starting to acknowledge the need to ensure complex and diverse precious coral habitats. Recent management plans have emerged that have the goal of considering habitat in a holistic approach, but until now, the specific measures of how to achieve this goal in precious coral stocks have yet to be pioneered. At present, there are only few refugia, and not all of them had been put in place before harvesting began, so effectively few virgin populations are known. In the case of shallow populations, there are probably none left.

The need for a paradigm shift in precious coral resource management is therefore apparent. Implementing sustainable management based on best-available knowledge and maintaining high habitat complexity will not only provide a high biodiversity and productivity of deep coral habitats but will also ensure the survival of the coral jewellery industry. These unique traditional industries have survived political and market force-induced hardships, but they depend on the fishery and ecosystem managers to advance the state of the art of precious coral management and ensure a future for this craft.

Recommendations

In the light of the data presented in this review, we make the following recommendations for the future protection, management and conservation of precious corals:

1. Legally binding transnational management, as typical in many other fisheries. This recommendation is more likely to unite expertise and resources in an effort to adapt local management plans and enforce them. At present, however, CITES is the only international organisation with legal authority that has been proposed. Because it controls trade, rather than management, further support is necessary.
2. Identification of potential unfished virgin populations. If such populations exist, a fraction of them should be protected, for example, as UNESCO World Heritage Sites. Overharvested sites, which most likely means all shallow-water populations in air-diving range (about 0–70 m depth), should be exempt from fishing. All other sites that show a moderate impact, or new sites, may be continued to be harvested under new guidelines and a reduced number of licenses. This will require further research, and care must be taken that once new

populations are identified, they are protected before a coral rush begins. Further models for international precious coral habitat protection treaties may be the example of UN VMEs, which are designed by NAFO.

3. Improved monitoring, as is common in other fisheries. Usually, the only published landings data list the weight and location of coral harvested. Some other data are available from authorities, but to our knowledge there is usually no information on harvested colony size. Authorities need to collect data on the size of the landed coral, possibly through observers on fishing boats or at harbours. Data on the minimum and, more important, the maximum size of the harvested coral will give an additional indication of when the stocks are over-harvested. Should fishermen consistently fail to harvest large corals, it is a reliable indication that a stock is depleted.

In any case, regular fishery-independent monitoring of the stocks is advisable. In the absence of landings data, price increase (e.g., for previously little harvested species such as bamboo corals) could provide indications for further investigation. In populations for which recruitment is thought to be limited, this variable should also be monitored. Early warning programmes for invasive species could be designed by educating coral divers and providing contacts to whom observations could be reported. In cases where doubts about severe overharvesting exist, moratoria might be put in place until the data mentioned are obtained and analysed. This proposed procedure should not be confused with rotation harvesting, which appears to have more disadvantages than advantages.

4. Improved local enforcement against poaching where necessary. In Hawaii, Japan and the Mediterranean, it is reported that licensed fishermen long ago developed voluntary guidelines for harvesting to sustain their livelihood. The failure to stop poaching is hurting these wise practitioners. In countries where only fines are given, penalties for poaching need to be increased. Penalties for unlicensed fishing are lighter than those for poaching wildlife on land in some countries. Protection of red coral through CITES would lead to stricter penalty measures, at least in Spain, and would therefore strengthen local management.

5. Revision of yield quotas and revision of minimum size limits. In the Mediterranean, the minimum allowable base diameter limit should be at least 10 mm (following the example of Sardinia), and the colonies must be branched to a certain degree to preserve a minimum number of polyps that provide larvae. This minimum size is far smaller than size at MSY, so the fishery must operate at considerably lower efficiency than 100%. If fishing effort cannot be controlled, then the minimum size should be much higher. Morphology is subject to geographic variation, so the minimum branching allowed for harvesting must be determined locally. In Hawaii, experience shows that the grandfathering scheme is detrimental, so adherence to the traditional 1.22-m size limit is recommended.

6. Ban the trade of precious corals for display purposes in the curio and aquarium trade as well as jewellery made of composite coral powder because both practices provide incentive for the harvest of immature colonies.

7. In traditional fisheries using tangle nets, government incentives should be given to develop a selective fishery (e.g., using ROVs). The ROV footage will furthermore provide fishery-dependent data for better management.

8. Research on how cold-water, deep and precious corals act as ecosystem engineers and increase biodiversity and productivity of their habitat. Furthermore, the causes of mass mortality, including the recovery of populations from mortality and the growth of partially harvested colonies should be studied.

9. Research on the population structure and distribution of little-known species (or deep populations in the case of *C. rubrum*) and study of the larval biology and dispersion limits

between shallow and deep populations. In many species, there are no data on growth rate and population structure. Species for which growth rate estimates span a wide range would benefit from further research (i.e., *Gerardia* sp.).

Ecological research should be accompanied by a socioeconomic component that looks at how to implement the newfound knowledge. In the Mediterranean, for example, implementation issues rather than lack of knowledge need to be addressed.

10. Creating a network of microreserves. The main objective would be to ensure gene flow between deep and shallow populations and between populations along the coast. Microreserves would have the secondary function of protecting all those organisms associated with the corals. It is important that commercial stocks also contain microreserves, which should be monitored. The goal is to maintain a minimum proportion of large colonies within the stocks, at a distance at which they can provide larvae to the habitat, and provide shelter to associated organisms. For the same reason, rotation harvesting should be abandoned in favour of truly sustainable harvesting of each population.

The slower the growth rate of a species, the more conservative the yield quotas should be and the smaller should the fraction of the harvestable stock be in relation to the total population.

11. Humane reduction of licenses where necessary. The number of licenses is often adequate if poaching is reduced and there is adherence to quotas. However, where necessary, license numbers should be reduced by not transferring licenses of retiring divers.

Acknowledgements

Sincere thanks to Andy Bruckner, Ida Fiorillo, Jean-Georges Harmelin, Nozomu Iwasaki, Sadao Kosuge, Tony Montgomery, Frank Parrish, Robin Gibson, Susan Torntore, Illaria Vielmini and an anonymous referee for helpful comments and information.

References

Abbiati, M. & Santangelo, G. 1989. A record of *Corallium rubrum* associated fauna: *Balssia gasti* (Balls 1921) (Crustacea: Decapoda). *Atti della Società Toscana dei Scienze Naturali Memorie* **96**, 237–241.

Abdelmajid, D. 2009. *Red Coral* Corallium rubrum, *Linné. 1758*. Royaume du Maroc. Casablanca, Morocco: Institut National de Recherche Halieutique.

Adkins, J.F., Cheng, H., Boyle, E.A., Druffel, E.R.M. & Edwards, R.L. 1998. Deep-sea coral evidence for rapid change in ventilation of the deep North Atlantic 15,400 years ago. *Science* **280**, 725–728.

Allee, W.C., Emerson, A.E., Park, O., Park, T. & Schmidt, K.P. 1949. *Principles of Animal Ecology*. Philadelphia: Saunders.

Andaloro, F. & Cicogna, F. 1993. Fishing red coral: problems and management. In *Red Coral in the Mediterranean Sea, Art, History and Science*, F. Cicogna & R. Cattaneo-Vietti (eds). Rome: Ministero delle Risorse di Agricole e Alimentari e Forestali, 131–157.

Andrews, A.H., Cordes, E.E., Mahoney, M.M., Munk, K., Coale, K.H., Cailliet, G.M. & Heifetz, J. 2002. Age, growth and radiometric age validation of a deep-sea, habitat-forming gorgonian (*Primnoa resedaeformis*) from the Gulf of Alaska. *Hydrobiologia* **471**, 101–110.

Angiolillo, M., Canese S., Giusti, M., Cardinali, A., Bo, M., Salvati, E. & Greco, S. 2009. Presence of *Corallium rubrum* on coralligenous assemblages below 50 m. In *Proceedings of the 1st Mediterranean Symposium on the Conservation of the Coralligenous and other Calcareous Bio-Concretions*, P. Pergent-Martini & M. Brichet (eds). Tunis: Regional Activity Center for Specially Protected Areas, 46–51.

Arena, P., Sará, R. & Bombace, G., 1965. *Campagna Preliminare di Ricerche sul Corallo nei Mari Della Sicilia*. Messina, Italy: Centro Sperimentale Per L' Industria dlla Pesca E dei Prodotti del Mare.

Arntz, W.E., Gili, J.M. & Reise, K. 1999. Unjustifiably ignored: reflections on the role of benthos in marine ecosystems. In *Biogeochemical Cycling and Sediment Ecology*, J.S. Gray et al. (eds). Dordrecht: Kluwer Academic, 105–124.

Babcock, R.C. 1991. Comparative demography of three species of scleractinian corals using age- and size-dependent classification. *Ecology* **61**, 255–244.

Baco, A., Clark, A.M. & Shank, T.M. 2006. Six microsatellite loci from the deep-sea coral *Corallium lauuense* (Octocorallia: Coralliidae) from the islands and seamounts of the Hawaiian Archipelago. *Molecular Ecology Notes* **6**, 147–149.

Baco, A.R. & Shank, T.M. 2005. Population genetic structure of the Hawaiian precious coral *Corallium lauuense* using microsatellites. In *Cold-water Corals and Ecosystems,* A. Freiwald & J.M. Roberts (eds). Heidelberg: Springer, 663–678.

Ballesteros, E. 2006. Mediterranean coralligenous assemblages: a synthesis of present knowledge. *Oceanography and Marine Biology An Annual Review* **44**, 123–195.

Barletta, G., Marchetti, R. & Vighi, M. 1968. Ricerche sul corallo rosso, Part IV: Ulteriori osservazioni sulla distribuzione del corallo rosso nel Tirreno. *Istituto Lombardo* **102**, 119–144.

Bauer, M. 1909. *Precious Stones*. Leipzig: Tauchnitz, 2nd edition.

Bavestrello, G., Cattaneo-Vietti, R. & Senes, L. 1992. Micro and macro differences in Mediterranean red coral colonies in and outside a cave. *Bollettino dei Musei e degli Istituti Biologici dell'Università di Genoa* **58**, 117–123.

Bayer, F.M. 1956. Descriptions and redescriptions of the Hawaiian octocorals collected by the U.S. Fish Commission steamer "Albatross" (2. Gorgonacea: Scleraxonia). *Pacific Science* **10**, 67–95.

Bayer, F.M. & Cairns, S.D. 2003. A new genus of the scleraxonian family Coralliidae (Octocorallia: Gorgonacea). *Proceedings of the Biological Society of Washington* **116**, 222–228.

Bayer, F.M. & Stefani, J. 1987. New and previously known taxa of isidid octocorals (Coelenterata: Gorgonacea), partly from Antarctic waters, with descriptions of new taxa. *Proceedings of the Biological Society of Washington* **100**, 937–991.

Bearzi, G. 2007. Marine conservation on paper. *Conservation Biology* **21**, 1–3.

Becker, L.C. & Mueller, E. 2001. The culture, transplantation and storage of *Montastraea faveolata, Acropora cervicornis* and *Acropora palmata*: what we have learned so far. *Bulletin of Marine Science* **69**, 881–889.

Beiring, E.A. & Lasker, H.R. 2000. Egg production by colonies of a gorgonian coral. *Marine Ecology Progress Series* **196**, 169–177.

Beverton, R.J.H. & Holt, S. 1957. On the dynamics of exploited fish populations. *Fishery Investigations Ministry of Agriculture, Fisheries and Food (Great Britain) Series 2,* **19**, 1–533.

Boland, R.C. & Parrish, F.A. 2005. Description of fish assemblages in the black coral beds off Lahaina, Maui, Hawaii. *Pacific Science* **59**, 411–420.

Bove, A.A. & Davis, J.C. (eds). 1997. *Diving Medicine*. Philadelphia: Saunders, 3rd edition.

Bramanti, L., Magagnini, G., De Maio, L. & Santangelo, G. 2005. Recruitment, early survival and growth of the Mediterranean red coral *Corallium rubrum* (L 1758), a 4-year study. *Journal of Experimental Marine Biology and Ecology* **314**, 69–78.

Bramanti, L., Rossi, S., Tsounis, G., Gili, J.M. & Santangelo, G. 2007. Settlement and early survival of red coral on artificial substrates in different geographic areas: some clues for demography and restoration. *Hydrobiologia* **580**, 219–224.

Brook, G. 1889. Report on the Antipatharia. *Report of the Scientific Results of the Voyage of the Challenger Zoology* **32**, 5–222.

Bruckner, A., De Angelis, P. & Montgomery, T. 2008. Case study for black coral for Hawaii. In *International Expert Workshop on CITES Non-detriment Findings*, Cancun, November 17th–22nd 2008. Geneva: CITES. Online. Available HTTP: http://www.conabio.gob.mx/institucion/cooperacion_internacional/TallerNDF/Links-Documentos/WG-CS/WG9-AquaticInvertebrates/WG9-CS1%20BlackCoral/WG9-CS1.pdf (accessed July 2009).

Bruckner, A. & Roberts, G. (eds). 2009. Proceedings of the First International Workshop on Corallium Science, Management and Trade. NOAA Technical Memorandum. Washington, DC: National Oceanic and Atmospheric Administration, National Marine Fisheries Service, Silver Spring, MD.

Bruckner, A.W. 2009. Rate and extent of decline in *Corallium* (pink and red coral) populations: are existing data adequate to justify a CITES Appendix II listing? *Marine Ecology Progress Series* **397**, 319–332.

Buhl-Mortensen, L. & Mortensen, P.B. 2005. Distribution and diversity of species associated with deep-sea gorgonian corals off Atlantic Canada. In *Cold-Water Corals and Ecosystems*, A. Freiwald & J.M. Roberts (eds). Berlin: Springer, 849–879.

Cairns, S. 2007. Deep-water corals: an overview with special reference to diversity and distribution of deep-water scleractinian corals. *Bulletin of Marine Science* **81**, 311–322.

Cairns, S.D. & Bayer, F.M. 2005. A review of the genus *Primnoa* (Octocorallia: Gorgonacea: Primnoidae), with the description of two new species. *Bulletin of Marine Science* **77**, 225–256.

Cairns, S.D., Calder, D.R., Brinckmann-Voss, A., Castro, C.B., Fautin, D.G., Pugh, P.R., Mills, C.E., Jaap, W.C., Arai, M.N., Haddock, S.H.D. & Opresko, D.M. 2002. *Common and scientific names of aquatic invertebrates from the United States and Canada: Cnidaria and Ctenophora*. 2nd edition. Bethesda, Maryland: American Fisheries Society Special Publication **28**, 1–115.

Calderon, I., Garrabou, J. & Aurelle. D. 2006. Evaluation of the utility of COI and ITS markers as tools for population genetic studies of temperate gorgonians. *Journal of Experimental Marine Biology and Ecology* **336**, 184–197.

Cannas, R., Caocci, F., Follesa, M.C., Grazioli, E., Pedoni, C., Pesci, P., Sacco, F., Cau., A. 2009. Multi-disciplinary data on the status of red coral (*Corallium rubrum*) resource in Sardinian seas (Central Western Mediterranean. Presentation in: International Workshop "Red Coral Science, Management, and Trade: Lessons from the Mediterranean" University of Naples "Parthenope" Naples, September 23rd–26th 2009. Online. Available HTTP: http://dsa.uniparthenope.it/rcsmt09/ (accessed: December 2009).

Carpine, C. & Grasshoff, M. 1975. Les Gorgonaires de la Méditerranée. *Bulletin de l'Institut Océanographique* **71**, 1–140.

Castigliano, A. & Liverino S. 2004. *Il Corallo: Aspetto Storico—Geografico de una Tradizione Millenaria.* Napoli: Loffredo editore.

Caswell, H. 2001. *Matrix Population Models: Construction, Analysis and Interpretation,* Sunderland, MA: Sinauer, 2nd edition.

Cattaneo-Vietti, R., Bavestrello, G. & Cerrano, C. 1998. *Corallium rubrum* management: state of the art. *Bollettino dei musei e degli Istituti Biologici dell'Università di Genoa* **64**, 61–71.

Carleton, C. 1987. Report on study of the marketing and processing of precious coral products in Taiwan, Japan and Hawaii. *South Pacific Forum Fisheries Report*, 87/13.

Cerrano, C., Bavestrello, G., Nike-Bianchi, C., Cattaneo-Vietti, R., Bava, S., Morganti, C., Morri, C., Picco, P., Sara, G., Sciaparelli, S., Siccardi, A. & Spogna, F. 2000. A catastrophic mass-mortality episode of gorgonians and other organisms in the Ligurian Sea (northwestern Mediterranean), summer 1999. *Ecology Letters* **3**, 284–293.

Chintiroglou, H. & Dounas-Koukouras, C. 1989. The presence of *Corallium rubrum* (Linnaeus, 1758) in the eastern Mediterranean Sea. *Mitteilungen aus dem Zoologischen Museum Berlin* **65**, 145–149.

Chouba, L. & Tritar, B. 1988. The exploitation level of the stock of red coral (*Corallium rubrum*) in Tunisian waters. *Bulletin du Musée d'Histoire Naturelle de Marseille* **56**, 29–35.

Cicogna, F. & Cattaneo-Vietti, R. (eds). 1993. *Red Coral in the Mediterranean Sea, Art, History and Science.* Rome: Ministero delle Risorse di Agricole e Alimentari e Forestali.

Cimberg, R.L., Gerrodette, T. & Muzik, K. 1981. Habitat requirements and expected distribution of Alaska coral. Final Report to NOAA for contract 27-80. Boulder, CO: NOAA.

Cloern, J.E. 1982. Does the benthos control phytoplankton biomass in south San Francisco Bay? *Marine Ecology Progress Series* **9**, 191–202.

Cognetti, G. 1989. FAO Congress on *Corallium rubrum. Marine Pollution Bulletin* **20**, 95 only.

Coma, R., Ribes, M., Gili, J.M. & Zabala, M. 2000. Seasonality in coastal ecosystems. *Trends in Ecology and Evolution* **15**, 448–453.

Conover, D.O., Munch, S.B. & Arnott, S.A. 2009. Reversal of evolutionary downsizing caused by selective harvest of large fish. *Proceedings of the Royal Society Series B*, in press. doi:10.1098/rspb.2009.0003. Also published online, at: http://rspb.royalsocietypublishing.org/content/early/2009/02/27/rspb.2009.0003.full.pdf+html

Convention on International Trade in Endangered Species of Wild Fauna and Flora (CITES). 2007. Consideration of proposals for amendment of Appendices I and II. Conference of Parties 14, Proposal 21. Geneva.

Convention on International Trade in Endangered Species of Wild Fauna and Flora (CITES). 2008, April 2. Notification to parties. No 2008/027. Geneva.

Corriero, G., Abbiati, M. & Santangelo, G. 1997. Sponges inhabiting a Mediterranean red coral population. *Marine Ecology* **18**, 147–155.

Costantini, F., Fauvelot, C. & Abbiati, M. 2007a. Genetic structuring of the temperate gorgonian coral (*Corallium rubrum*) across the western Mediterranean Sea revealed by microsatellites and nuclear sequences. *Molecular Ecology* **16**, 5168–5182.

Costantini, F., Fauvelot, C., Abbiati, M. 2007b. Fine-scale genetic structuring in *Corallium rubrum* (L): evidences of inbreeding and limited effective larval dispersal. *Marine Ecology Progress Series* **340**, 109–119.

Costantini, F., Taviani, M., Remia, A., Pintus, E., Schembri, P.J., Abbiati, M. 2009. Deep-water *Corallium rubrum* (L., 1758) from the Mediterranean Sea: genetic characterisation. *Marine Ecology*, in press. doi: 10.1111/j.1439-0485.2009.00333.x

Council of Europe 1979. Convention on the conservation of European wildlife and natural habitats. Online. Available HTTP: http://www.conventions.coe.int/Treaty/en/Treaties/Word/104.doc (accessed December 2009).

Cvejic, J., Tambutté, S., Lotto, S., Mikov, M., Slacanin, I. & Allemand, D. 2007. Determination of canthaxanthin in the red coral (*Corallium rubrum*) from Marseille by HPLC combined with UV and MS detection. *Marine Biology* **152**, 855–862.

Dall, W.H. 1883. Pearls and pearl fisheries. *American Naturalist* **17**, 731–745.

Dantan, J.L. 1928. Recherches sur la croissance du corail rouge *Corallium rubrum* Lamarck. *Bulletin de la Societé Zoologique de France* **53**, 42–46.

Dauphin, Y. 2006. Mineralizing matrices in the skeletal axes of two *Corallium* species (Alcyonacea). *Comparative Biochemistry and Physiology. Part A, Physiology* **145**, 54–64.

Davidoff, M.V. 1908. Russische Zoologische Station in Villefranche (Riviera). *Internationale Revue der Gesamten Hydrobiologie und Hydrographie* **1**, 295–297.

Dayton, P.K. 2003. The importance of the natural sciences to conservation. *American Naturalist* **162**, 1–13.

Dayton, P.K., Robilliard, G.A., Paine, R.T. & Dayton, L.B. 1974. Biological accommodation in the benthic community at McMurdo Sound, Antarctica. *Ecological Monographs* **44**, 105–128.

Del Gaudio, D., Fortunato, G., Borriello, M., Gili, J.M., Buono, P., Calcagno, G., Salvatore, F. & Sacchetti, L. 2004. Genetic typing of *Corallium rubrum*. *Marine Biotechnology* **6**, 511–515.

DeVogeleare, A., Burton, E.J., Trejo, T., King, C.E., Clague, D.A., Tamburri, M.N., Cailliet, G.M., Kochevar, R.E. & Douros, W.J. 2005. Deep-sea corals and resource protection at the Davidson Seamount, California, USA. In *Cold-water Corals and Ecosystems*, A. Freiwald, & J.M. Roberts (eds). Berlin: Springer, 1189–1198.

Doneddu, R. 2009. Red coral management strategies in Sardinian waters: a successful experience. In: International Workshop "Red Coral Science, Management, and Trade: Lessons from the Mediterranean" University of Naples "Parthenope" Naples, September 23rd-26th 2009. Online. Available HTTP: http://dsa.uniparthenope.it/rcsmt09/ (accessed: December 2009).

Druffel, E.R.M., Griffin, S., Witter, A., Nelson, E., Southon, J., Kasgarian, M. & Vogel, J. 1995. *Gerardia*: bristlecone pine of the deep-sea? *Geochimica et Cosmochimica Acta* **59**, 5031–5036.

Druffel, E.R.M. & Williams, P.M. 1990. Identification of a deep marine source of particulate organic carbon using bomb ^{14}C. *Nature* **347**, 172–174.

Druffel, E.R.M., Williams, P.M., Bauer, J.E. & Ertel, J.R. 1992. Cycling of dissolved and particulate organic matter in the open ocean. *Journal of Geophysical Research C: Oceans and Atmospheres* **97**, 15639–15659.

Dulvy, N.K., Metcalfe, J.K., Glanville, J., Pawson, M.G. & Reynolds, J.D. 2000. Fishery stability, local extinction and shifts in community structure in skates. *Conservation Biology*, **14**, 283–293.

Edwards, A.J. & Clarke, S. 1998. Coral transplantation: a useful management tool or misguided meddling? *Marine Pollution Bulletin* **37**, 474–478.

Eguchi, M. 1968. The red coral of Sagami Bay. In *The Hydrocorals and Scleractinian Corals of Sagami Bay*. Biological Laboratory of the Imperial Household (ed.). Tokyo: Maruzen Co. Ltd. **15**, 53, C1–C80.

Ehrlich, H., Etnoyer, P., Litvinov, S.D., Olennikova, M., Domaschke, H., Hanke, T., Born, R., Meissner, H. & Worch, H. 2006. Biomaterial structure in deep-sea bamboo coral (Gorgonacea: Isididae): perspectives for the development of bone implants. *Materialwissenschaft und Werkstofftechnik* **37**, 553–557.

Epstein, N., Bak, R.P.M. & Rinkevich, B. 2001. Strategies for gardening denuded coral reef areas: the applicability of using different types of coral material for reef restoration. *Restoration Ecology* **9**, 432–442.

Etnoyer, P. 2008. A new species of *Isidella* bamboo coral (Octocorallia: Alcyonacea: Isididae) from northeast Pacific seamounts. *Proceedings of the Biological Society of Washington* **121**, 541–553.

Etnoyer, P. & Morgan, L. 2003. Occurrences of habitat-forming deep sea corals in the northeast Pacific Ocean. A report to NOAA's Office of Habitat Conservation. Redmond, WA: Marine Conservation Biology Institute.

Etnoyer, P. & Morgan, L.E. 2005. Habitat-forming deep-sea corals in the Northeast Pacific Ocean. In *Cold-water Corals and Ecosystems*, A. Freiwald & J.M. Roberts (eds). Berlin: Springer, 331–343.

European Union Habitat Directive 2007. Council Directive 92/43/EEC on the Conservation of natural habitats and of wild fauna and flora, Appendix V. Online. Available HTTP: http://www.ec.europa.eu/environment/nature/index_en.htm (accessed December 2009).

Fabricius, K. & Alderslade, P. 2001. *Soft Corals and Sea Fans*. Townsville, Australia: Australian Institute of Marine Science.

Fautin, D.G. 2002. Reproduction of Cnidaria. *Canadian Journal of Zoology* **80**, 1735–1754.

Food and Agriculture Organization (FAO). 1983. Technical consultation on red coral resources of the Western Mediterranean. *FAO Fisheries Report* **306**, 1–142.

Food and Agriculture Organization (FAO). 1988. GFCM technical consultation on red coral of the Mediterranean. *FAO Fisheries Report* **413**, 1–159.

Food and Agriculture Organization (FAO) 2007. Report of the second FAO ad hoc expert advisory panel for the assessment of proposals to amend appendices I and II of CITES concerning commercially-exploited aquatic species, Rome, 26–30 March 2007. FAO Fisheries Report No. 833. Rome, FAO.

Food and Agriculture Organization (FAO) 2009. Fisheries advisory panel offers recommendations on CITES proposals. Food and Agricultural Organization. Online. Available HTTP: http://www.fao.org/news/story/en/item/38195/icode/ (accessed December 2009).

Frank, N., Paterne, M., Ayliffe, L., van Weering, T., Henriet, J.P. & Blamart, D. 2004. Eastern North Atlantic deep-sea corals: tracing upper intermediate water $\Delta^{14}C$ during the Holocene. *Earth and Planet Science Letters* **219**, 297–309.

Fréchette, M., Butman, C.A. & Geyer, W.R. 1989. The importance of boundary-layer flow in supplying phytoplankton to the benthic suspension feeder, *Mytilus edulis* L. *Limnology and Oceanography* **34**, 19–36.

Freiwald, A., Fosså, J.H., Grehan, A., Koslow, T. & Roberts, J.M. 2004. Cold-water coral reefs. Cambridge, UK: United Nations Environment Programme, World Conservation Monitoring Centre.

Galasso, M. 1998. Unterwasserfunde in West-Sardinien. *Zeitschrift für Unterwasserarchaeologie* **1**, 18–31.

Galasso, M. 2000. Pesca del *Corallium rubrum* in Sardegna nell'antichità attraverso l'indagine archeologica, cartografica e I rilevamenti in mare. In *L'Africa Romana. Atti del XIV Convegno di Studi*, M. Khanoussi et al. (eds). Rome: Carocci editore, 1159–1200.

Galasso M. 2001. *La Pesca del Corallo in Sardegna: Evoluzione, Persistenze e innovazioni Tecniche*. Sassari, Italy: Centro Studi sul Corallo (*Corallium rubrum* L.).

Galili, B. & Rosen, E. 2008. Ancient remotely-operated instruments recovered under water off the Israeli Coast. *International Journal of Nautical Archaeology* **37**, 283–294.

Garrabou, J., Perez, T., Sartoretto, S. & Harmelin, J.G. 2001. Mass mortality event in red coral (*Corallium rubrum*, Cnidaria, Anthozoa, Octocorallia) populations in the Provence region (France NW Mediterranean). *Marine Ecology Progress Series* **207**, 263–272.

Garrabou, J.M. & Harmelin, J.G. 2002. A 20-year study on life-history traits of a harvested long-lived temperate coral in the NW Mediterranean: insights into conservation and management needs. *Journal of Animal Ecology* **71**, 966–978.

García-Rodríguez, M. & Massó, C. 1986a. Algunas bases para la determinación de la edad del coral rojo (*Corallium rubrum* L.). *Boletin Instituto Español de Oceanografia* **3**, 61–64.

García-Rodríguez, M. & Massó. C. 1986b. Modelo de explotación por buceo del coral rojo (*Corallium rubrum* L.) del Mediterráneo. *Boletin Instituto Español de Oceanografia* **3**, 75–82.

García-Rodríguez, M. & Massó, C. 1986c. Estudio biométrico de poblaciones de coral rojo (*Corallium rubrum* L.) del litoral de Gerona (NE de España). *Boletin Instituto Español de Oceanografia* **3**, 61–64.

Giannini, F., Gili, J.M. & Santangelo, G. 2003, Relationship between the spatial distribution of red coral (*Corallium rubrum*) and co-existing suspension feeders at Medas Island Marine Protected Area (Spain). *Italian Journal of Zoology* **70**, 233–239.

Gili, J.M. & Coma, R. 1998. Benthic suspension feeders: their paramount role in littoral marine food webs. *Trends in Ecology and Evolution* **13**, 316–321.

Gislén, A., Dacke, M., Kröger, R.H.H., Abrahamsson, M., Nilsson D.E. & Warrant, E.J. 2003. Superior underwater vision in a human population of sea gypsies. *Current Biology* **13**, 833–836.

Gislén, A., Warrant, E.J., Dacke, M. & Kröger R.H.H. 2006. Visual training improves underwater vision in children. *Vision Research* **46**, 3443–3450.

Goffredo, S. & Lasker, H.R. 2008. An adaptive management approach to an octocoral fishery based on the Beverton-Holt model. *Coral Reefs* **27**, 751–761.

Goodfriend, G.A. 1997. Aspartic acid racemization and amino acid composition of the organic endoskeleton of the deep-water colonial anemone *Gerardia*: determination of longevity from kinetic experiments. *Geochimica et Cosmochimica Acta* **61**, 1931–1939.

Goodman-Lowe, G. 1998. Diet of Hawaiian monk seal (*Monachus schauinslandi*) from the Northwestern Hawaiian Islands during 1991–1994. *Marine Biology* **132**, 535–546.

Grange, K.R., 1985. Distribution, standing crop, population structure and growth rates of an unexploited resource of black coral in the southern fjords of New Zealand. In *Proceedings of the Fifth International Coral Reef Congress*, Tahiti, **6**, 217–222.

Grange, K.R. 1997. Demography of black coral populations in Doubtful Sound, New Zealand: results from a 7-year experiment.. In *Proceedings of the 6th International Conference on Coelenterate Biology*. J.C. den Hartog (ed.). Leiden, The Netherlands: National Museum of Natural History, 185–193.

Grigg, R.W. 1965. Ecological studies of black coral in Hawaii. *Pacific Science* **19**, 244–260.

Grigg, R.W. 1974. Growth rings: annual periodicity in two gorgonian corals. *Ecology* **55**, 876–881.

Grigg, R.W. 1975. Distribution and abundance of precious corals in Hawaii. In *Proceedings of the 2nd International Coral Reef Symposium, 2. Great Barrier Reef Communications*, B.M. Cameron et al. (eds). Brisbane: Great Barrier Reef Committee, 235–240.

Grigg, R.W. 1976. Fisheries management of precious and stony corals in Hawaii. UNIHI-SEAGRANT, TR-77-03. Honolulu: University of Hawaii Sea Grant Program.

Grigg, R.W. 1984. Resource management of precious corals: a review and application to shallow water reef building corals. *Marine Ecology* **5**, 57–74.

Grigg, R.W. 1989. Precious coral fisheries of the Pacific and Mediterranean. In *Marine Invertebrate Fisheries: Their Assessment and Management*, J.F. Caddy (ed.). New York: Wiley, 636–645.

Grigg, R.W. 1994. History of the precious coral fishery in Hawaii. *Precious Corals & Octocoral Research* **3**, 1–18.

Grigg, R.W. 2001. Black coral: history of a sustainable fishery in Hawaii. *Pacific Science* **55**, 291–299.

Grigg, R.W. 2002. Precious corals in Hawaii: discovery of a new bed and revised management measures for existing beds. *Marine Fisheries Review* **64**, 13–20.

Grigg, R.W. 2004. Harvesting impacts and invasion by an alien species decrease estimates of black coral yield off Maui, Hawai'i. *Pacific Science* **58**, 1–6.

Grigg, R.W. & Brown, G. 1991. Tasmanian gem corals. *The Australian Gemnologist* **17**, 399–404.

Gutiérrez, J.L., Jones, C.G., Byers, J.E., Arkema, K.K., Berkenbusch, K., Committo, J.A., Duarte, C.M., Hacker, S.D., Hendriks, I.E., Hogarth, P.J., Lambrinos, J.G., Palomo, M.G. & Wild, C. (in press). Physical ecosystem engineers and the functioning of estuaries and coasts. In *Treatise of Estuaries and Coastal Ecosystems*, E. Wolanski & D. McLusky (series eds). Volume 7, Functioning of Estuaries and Coastal Ecosystems, C.H.R. Heip et al. (eds), chapter 5. Amsterdam: Elsevier.

Hall-Spencer, J., Allain, V. & Fossa, J.H. 2002. Trawling damage to northeast Atlantic ancient coral reefs. *Philosophical Transactions of the Royal Society B: Biological Sciences* **269**, 507–511.

Harmelin, J.G. 1984. Biologie du corail rouge. Paràmetres de populations, croissance et mortalité naturelle. Etat de connaissances en France. *FAO Fisheries Report* **306**, 99–103.

Harper, J.R. 1988. Precious coral prospecting strategies for the South Pacific region. CCOP/SOPAC Technical Report 84. Suva, Fiji Islands: Committee for Co-ordination of Joint Prospecting for Mineral Resources in South Pacific Offshore Areas (CCOP/SOPAC).

Hasegawa, H. & Yamada, M. (in press). Chemical analysis of carbonate skeletons in precious corals. In *Biohistory of Precious Corals,* N. Iwasaki (ed.). Kanagawa, Japan: Tokai University Press.

Heifetz, J., Wing, B.L., Stone, R.P., Malecha, P.W. & Courtney, D.L. 2005. Corals of the Aleutian Islands. *Fisheries Oceanography* **14**, 131–138.

Hereu, B., Linares, C., Diaz, D. & Zabala, M. 2002. Avaluació de l´episodi d´espoli de corall vermell (*Corallium rubrum*) de la zona de la Pedrosa (Costa de Montgrí) i de les mostres incautades els dies 21 I 22 de Desembre de 2002. Barcelona: Departament de Medi ambient, Generalitat de Catalunya.

Hickson, S.J. 1924. *An Introduction to the Study of Recent Corals*. London: Manchester University Press.

Hiscock, K. & Mitchell, R. 1980. The description and classification of sublittoral epibenthic ecosystems. In *The Shore Environment, 2: Ecosystems*, J.H. Price et al. (eds). London: Academic Press, 323–370.

Hlebica, J. 2000. The enduring Ama: Japan's venerated breath-hold divers still ply their trade. *Historical Diver* **8**, 18–22.

Hong, S., Henderson, J., Olszowka, A., Hurford, W.E., Falke, K.J., Qvist, J., Radermacher, P., Shiraki, K., Mohri, M. & Takeuchi, H. 1991. Daily diving pattern of Korean and Japanese breath-hold divers (Ama). *Undersea Biomedical Research* **18**, 433–443.

Hoegh-Guldberg, O., Mumby, P.J., Hooten, A.J., Steneck, R.S., Greenfield, P., Gomez, E., Harvell, C.D., Sale, P.F., Edwards, A.J., Caldeira, K., Knowlton, N., Eakin, C.M., Iglesias-Prieto, R., Muthiga N., Bradbury, R.H., Dubi, A. & Hatziolos, M.E. 2007. Coral reefs under rapid climate change and ocean acidification. *Science* **318**, 1737–1742.

Husebø, A., Nøttestad, L., Fosså, J.H., Furevik, D.M. & Jørgensen, S.B. 2002. Distribution and abundance of fish in deep-sea coral habitats. *Hydrobiologia* **471**, 91–99.

IUCN & TRAFFIC. 2007. Summaries of the IUCN/TRAFFIC Analyses of the Proposals to Amend the CITES Appendices at the 14th Meeting of the Conference of the Parties. The Hague, Netherlands 3–15 June 2007. IUCN Species Programme and Species Survival Commission and TRAFFIC. Online. Available HTTP: http://www.traffic.org/cites-cop-papers/traffic_pub_cop14_5.pdf (accessed December 2009).

Iwasaki, N. & Suzuki, T. (in press). Biology of precious corals. In *Biohistory of Precious Corals*, N. Iwasaki (ed.). Kanagawa, Japan: Tokai University Press.

Jones, C.J. & Lawton, J.H. 1994. *Linking Species and Ecosystems*. Berlin: Springer.

Jones, C.J., Lawton, J.H. & Shachak, M. 1994. Organisms as ecosystem engineers. *Oikos* **69**, 373–386.

Kahng, S. 2007. Ecological impacts of *Carijoa riisei* on black coral habitat. Black Coral Science Management Workshop April 18–19. Honolulu: Western Pacific Fisheries Management Council, 27–30. Online. Available HTTP: http://www.wpcouncil.org/precious/Documents/2006%20Black%20Coral%20Scence%20and%20Management%20Workshop%20ReportSCANNED%20VERSION.pdf (accessed December 2009).

Kahng, S. & Grigg, R.W. 2005. Impact of an alien octocoral, *Carijoa riisei*, on black corals in Hawaii. *Coral Reefs* **24**, 556–562.

Kimmerer, W.J., Gartside, E. & Orsi, J.J. 1994. Predation by an introduced clam as the likely cause of substantial declines in zooplankton of San Francisco Bay. *Marine Ecology Progress Series* **113**, 81–93.

Kishinouye, K.J. 1903. Preliminary note on the Coralliidae of Japan. *Zoologische Anzeiger* **26**, 623–626.

Kishinouye, K.J. 1904. Notes on the natural history of corals. *Journal of Fishery Bureau* **14**, 1–32.

Kithara, T. 1902. On the coral fishery of Japan. *Journal of Fishery Bureau* **13**, 1–13.

Knittweis, L., Jompa, J., Richter, C. & Wolff, M. 2009. Population dynamics of the mushroom coral *Heliofungia actiniformis* in the Spermonde Archipelago South Sulawesi, Indonesia. *Coral Reefs* **28**, 793–804.

Knuth, B.C. 1999. *Gems in Myth, Legend and Lore*. Thornton, CO: Jewelers Press.

Koenig, C.C. 2001. *Oculina* banks: habitat, fish populations, restoration and enforcement. Report to the South Atlantic Fishery Management Council. North Charleston, SC: South Atlantic Fishery Management Council.

Kosuge, S. 1993. History of the precious coral fisheries in Japan. *Precious Corals and Octocorals Research* **1**, 30–38.

Krieger, K.J. 2001. Coral (*Primnoa*) impacted by fishing gear in the Gulf of Alaska. In *First International Symposium on Deep-Sea Corals*, J.H.M. Willison et al. (eds). Halifax: Ecology Action Center, Dalhousie University and Nova Scotia Museum, 106–116.

Krieger, K.J. & Wing, B.L. 2002. Megafaunal associations with deepwater corals (*Primnoa* spp.) in the Gulf of Alaska. *Hydrobiologia* **471**, 83–90.

Kwit, G., Horvits, C. & Platt, W.J. 2004. Conservation of slow-growing, long-lived tree species: input from the demography of rare understory conifer *Taxus floridiana*. *Conservation Biology* **18**, 432–443.

Lacaze-Duthiers, H. 1864. *Histoire Naturelle du Corail*. Paris: Baillère.

Laubier, L. 1966. Le coralligéne des Albéres. *Annales de l'Institut Océanographique de Monaco. Monographie Biocénotique* **43**, 139–316.

Lewis, J.B. 1978. Feeding mechanisms in black corals (Antipatharia). *Journal of Zoology* **186**, 393–396.

Linares, C., Coma, R., Diaz, D., Zabala, M. & Hereu, B. 2005. Immediate and delayed effects of a mass mortality event on gorgonian population dynamics and benthic community structure in the NW Mediterranean Sea. *Marine Ecology Progress Series* **305**, 127–137.

Linares, C., Diaz, D. & Zabala, M. 2003. Avaluació dels danys ocasionats per un espóli de corall vermell a la cala fredosa (Reserva Natural de Cap de Creus, parc natural del cap de Creus) l´Abril de 2003. Barcelona: Departament de Medi Ambient, Generalitat de Catalunya.

Lindahl, U. 1998. Low-tech rehabilitation of degraded coral reefs through transplantation of staghorn corals. *Ambio* **27**, 645–650.

Liverino, B. 1983. Il Corallo—Esperienze e ricordi di un corallaro. Banca di credito populare Torre del Greco. Torre del Greco, Italy: Li Causi Editore.

Lleonart, J. & Camarassa, J.M. 1987. *La Pesca a Catalunya El 1722 Segons Un Manuscrit De Joan Salvador I Riera*. Barcelona: Museu Maritim, Diputacio de Barcelona.

Longenecker, K., Dollar, R.A. & Cahoon, M.K. 2006. Increasing taxonomic resolution in dietary analysis of the Hawaiian monk seal. *Atoll Research Bulletin* **543**, 101.

Love, M.S., Yoklavich, M.M., Black, B.A. & Andrews, A.H. 2007. Age of black coral (*Antipathes dendrochristos* Opresko, 2005) colonies, with notes on associated invertebrate species. *Bulletin of Marine Science* **80**, 391–400.

Lumsden, S.E., Hourigan, T.F., Bruckner, A.W. & Dorr, G. (eds) 2007. The state of deep coral ecosystems of the United States. NOAA Technical Memorandum CRCP-3. Silver Spring, MD: NOAA, National Marine Fisheries Service.

Marchetti, R. 1965. Ricerche sul corallo rosso della costa ligure e toscana. Distribuzione geographica. *Istituto Lombardo, Accademia di Scienze e Lettere* **99**, 255–278.

Margalef, R. 1985. *Western Mediterranean (Key Environments)*. Oxford, UK: Pergamon Press.

Marín, L. & Reynald, L. 1981. Premiers données biometriques sur le corail rouge, *Corallium rubrum* LMCK, de Corse. *Rapport du Congrès de la Commission Internationale pour l'Exploration Scientifique de la Mer Méditerranée* **27**, 171–172.

Marschal, C., Garrabou, J. & Harmelin, J.G. 2004. A new method for measuring growth and age in the precious red coral *Corallium rubrum* (L.). *Coral Reefs* **23**, 423–432.

Marsili, Count Lois Ferdinand. 1707. Extrait d´une lettre écrite de Cassis, prés de Marseille, le 18 décember 1706 à M. Abbé Bignon, touchon queques branches de corail qui ont fleuri. *Le Journal des Scavans* **35**, 346–359.

Mayol, J. 2000. *Homo Delphinus*. New York: Idelson-Gnocchi.

Mazzarelli, G. 1915. Banchi di corallo esplorati dalla R. Nave "Volta" nell'estate del 1913. Rome: Ministerio di Agricoltura, Industria e Commercio, Annali dell'Industria.

Messing, C.G., Neuman, A.C. & Lang, J.C. 1990. Biozonation of deep-water lithotherms and associated hardgrounds in the northeastern Straits of Florida. *Palaios* **5**, 15–33.

Miller, K.J. 1996. Piecing together the reproductive habitats of New Zealands endemic black corals. *Water and Atmosphere* **4**, 18–19.

Mitchell, N., Dardeu, M.R. & Schroeder, W.W. 1993. Colony morphology, age structure and relative growth of two gorgonian corals, *Leptogorgia bebes* (Verrill) and *Leptogorgia virgulata* (Lamarck) from the northern Gulf of Mexico. *Coral Reefs* **12**, 6–70.

Moberg, F. & Folke, C. 1999. Ecological goods and services of coral reef ecosystems. *Ecological Economics* **29**, 215–233.

Montgomery, A.D. 2002. The feasibility of transplanting black coral (Order Antipatharia). *Hydrobiologia* **471**, 157–164.

Morell, V. 2007. Corals: Suffering from whiplash. *Science* **316**, 5832.

Morgan, L.E., Tsao, C. & Guinotte, J.M. 2006. Status of deep sea corals in U.S. waters with recommendations for their conservation and management. Bellevue, WA: Marine Conservation Biology Institute.

Mortensen, P.B. & Fosså, J.H. 2006. Species diversity and spatial distribution of invertebrates on *Lophelia* reefs in Norway. In *Proceedings of the 10th International Coral Reef Symposium, Okinawa*. Yoshimi Suzuki et al. (eds). Tokyo: Japanese Coral Reef Society, 1849–1868.

National Oceanic and Atmospheric Administration (NOAA). 2006. *Precious Coral Fishery*. Honolulu, Hawaii: Pacific Islands Fisheries Science Center of the National Marine Fisheries Service (NMFS). Online. Available HTTP: http://www.pifsc.noaa.gov/wpacfin/hi/dar/Pages/hi_fish_6.php (accessed 18 March 2009).

Nonaka, M., Katherine, M.M. & Uchida, S. 2006. Capture, study and display of precious corals. In *Proceedings of the 10th International Coral Reef Symposium*, Okinawa. Yoshimi Suzuki et al. (eds). Tokyo: Japanese Coral Reef Society, 1821–1831.

Nonaka, M. & Muzik, K. 2010. Recent harvest records of commercially valuable precious corals in the Ryukyu Archipelago. *Marine Ecology Progress Series* **397,** 269–278.

Officer, C.B., Smayda, T.J. & Mann, R. 1982. Benthic filter feeding: a natural eutrophication control. *Marine Ecology Progress Series* **9**, 203–210.

O'Hara, K.L., Hasenauer, H. & Kindermann, G. 2007. Sustainability in multi aged stands: an analysis of long-term plenter systems. *Forestry* **80**, 163–181.

Oishi, F. 1990. *Black Coral Harvesting and Marketing Activities in Hawaii.* Honolulu: State of Hawaii Division of Aquatic Resources,

Olsen, D.A. & Wood, R.S. 1980. Investigations on black coral in Salt River Submarine Canyon St. Croix, U.S. Virgin Islands. St. Croix, U.S. Virgin Islands: Division of Fish and Wildlife. Available HTTP: http:// www. aoml.noaa.gov/general/lib/CREWS/Cleo/St.%20Croix/salt_river189.pdf (accessed December 2009).

Opresko, D.M. 2009. A new name for the Hawaiian Antipatharian coral formerly known as *Antipathes dichotoma* (Cnidaria: Anthozoa: Antipatharia). *Pacific Science* **63**, 277-291.

Orejas, C., Gori, A. & Gili, J.M. 2008. Growth rates of live *Lophelia pertusa* and *Madrepora oculata* from the Mediterranean Sea maintained in aquaria. *Coral Reefs* **27**, 255 only.

Pani, M. 2009. Is CITES the right tool for the management and conservation of precious coral? In: *International forum on precious coral. Report of the first meeting in Hong Kong, March 7, 2009,* Sadao Kosuge (ed.). Kochi (Japan): The international association on promoting of precious coral study, 21-24.

Pani, M. & Berney, J. 2007. A review of the proposal to include the genus *Corallium* in Appendix II of CITES. IWMC World Conservation Trust. Rome and Lausanne. Online. Available HTTP: http://www.iwmc. org/IWMC-Forum/MarcoPani-JaquesBerney/D1242_IWMC_ENG_LR.pdf (accessed December 2009), 12pp.

Parker, N.R., Mladenov, P.V. & Grange, K.R. 1997. Reproductive biology of the antipatharian black coral *Antipathes fordensis* in Doubtful Sound, Fjordland, New Zealand. *Marine Biology* **130**, 11–22.

Parrish, F.A. 2006. Precious corals, and subphotic fish assemblages. *Atoll Research Bulletin* **543**, 425–438.

Parrish, F.A., Abernathy, K., Marshal, G.J. & Buhleier, B.M. 2002. Hawaiian monk seals (*Monachus schauin-slandi*) foraging in deep-water coral beds. *Marine Mammal Science* **18**, 244–258.

Parrish, F.A. & Roark, B.E. 2010. Growth validation of gold coral *Gerardia* sp. in the Hawaiian Archipelago. *Marine Ecology Progress Series* **397,** 163–172.

Pauly, D., Christensen, V., Guénette, S., Pitcher, T.J., Sumaila, U.R., Walters, C.J., Watson, R. & Zeller, D. 2002. Towards sustainability in world fisheries. *Nature* **418**, 689–695.

Pechenik, J.A. 2005. *Biology of the Invertebrates.* Boston: McGraw Hill, 5th edition.

Picciano, M. & Ferrier-Pages, C. 2007. Ingestion of pico- and nanoplankton by the Mediterranean red coral *Corallium rubrum. Marine Biology* **150**, 773–782.

Plujà, A. 1999. El corall vermell. *Farella, Revista de Llançà* **8**, 23–27.

Pronzato, R. 1999. Sponge-fishing, disease and farming in the Mediterranean Sea. *Aquatic Conservation Marine and Freshwater Ecosystems* **9**, 485–493.

Pyle, R. (2000). Assessing undiscovered fish biodiversity of deep coral reefs using advanced self-contained diving technology. *Marine Technology Journal* **34**, 82–91.

Riisgård, H.U., Jensen, A.S. & Jürgensen, C. 1998. Hydrography, near-bottom currents, and grazing impact of the filter-feeding ascidian *Ciona intestinalis* in a Danish fjord. *Ophelia* **49**, 1–16.

Rinkevich, B. 1995. Restoration strategies for coral reefs damaged by recreational activities: the use of sexual and asexual recruits. *Restoration Ecology* **3**, 241–251.

Rinkevich, B. 2005. Conservation of coral reefs through active restoration measures: recent approaches and last decade progress. *Environmental Science and Technology* **39**, 4333–4342.

Riedl, R. 1983. *Fauna und Flora des Mittelmeers.* Hamburg: Parey, 3rd edition.

Risk, M.J., Heikoop, J.M., Snow, M.G. & Beukens, R. 2002. Lifespans and growth patterns of two deep-sea corals: *Primnoa resedaeformis* and *Desmophyllum cristagalli. Hydrobiologia* **471**, 125–131.

Risk, M.J., MacAllister, D.E. & Behnken, L. 1998. Conservation of cold water and warm water seafans: threatened ancient gorgonian groves. *Sea Wind* **12**, 2–21.

Roark, E.B., Guilderson, T.P., Dunbar, R.B. & Ingram, B.L. 2006. Radiocarbon-based ages and growth rates of Hawaiian deep-sea corals. *Marine Ecology Progress Series* **327**, 1–14.

Roark, E.B., Guilderson, T.P., Flood-Page, S.R., Dunbar, R.B., Ingram, B.L., Fallon, S.J. & McCulloch, M.T. 2005. Radiocarbon-based ages and growth rates for bamboo corals from the Gulf of Alaska. *Geophysical Research Letters* **32**, L04606.

Roberts, J.M., Wheeler, A.J. & Freiwald, A. 2006. Reefs of the deep: the biology and geology of cold-water coral ecosystems. *Science* **312**, 543–547.

Rogers, A.D. 1999. The biology of *Lophelia pertusa* (Linnaeus, 1758) and other deep-water reef-forming corals and impacts from human activities. *International Revue of Hydrobiology* **84**, 315–406.

Roghi, G. 1966. La febre rossa. La strada del corallo: una storia italiana. *Mondo Sommerso* **6**. Online. Available HTTP: http://www.gianniroghi.it/ (accessed 7 October 2009).

Ros, J., Olivella, I. & Gili, J.M. 1984. *Els Sistemes Naturals de les Illes Medes*. Barcelona: Institut d´Estudis Catalans.

Ross, M.A. 2008. A quantitative study of the stony coral fishery in Cebu, Philippines. *Marine Ecology* **5**, 75–91.

Rossi, S. & Tsounis, G. 2007. Temporal and spatial variation in protein, carbohydrate, and lipid levels in *Corallium rubrum* (Anthozoa, Octocorallia). *Marine Biology* **152**, 429–439.

Rossi, S., Tsounis, G., Padrón, T., Orejas, C., Gili, J.M., Bramanti, L., Teixidor, N. & Gutt, J. 2008. Survey of deep-dwelling red coral (*Corallium rubrum*) populations at Cap de Creus (NW Mediterranean). *Marine Biology* **154**, 533–545.

Russell-Bernard, H. 1972. Kalymnos, island of sponge fishermen. In *Technology and Social Change*, P.J. Pelto & H. Russell-Bernard (eds). London: Macmillan, 278–316.

Sakai, K. 1998. Delayed maturation in the colonial coral *Gonasteria aspera* (Scleractinia): whole-colony mortality, colony growth and polyp egg production. *Research on Population Ecology* **40**, 287–292.

Santangelo, G. & Abbiati, M. 2001. Red coral: conservation and management of an over-exploited Mediterranean species. *Aquatic Conservation Marine Freshwater Ecosystems* **11**, 253–259.

Santangelo, G., Abbiati, M. & Caforio, G. 1993a. Age structure and population dynamics in *Corallium rubrum*. In *Red Coral in the Mediterranean Sea: Art, History and Science*, F. Cicogna & R. Cattaneo-Vietti (eds). Rome: Ministero delle Risorse Agricole, Alimentari e Forestali, 145–155.

Santangelo, G., Abbiati, M., Giannini, F. & Cicogna, F. 1993b. Red coral fishing trends in the western Mediterranean Sea during the period 1981–1991. *Sciencia Marina* **57**, 139–143.

Santangelo, G., Bongiorni, L., Giannini, F., Abbiati, M. & Buffoni, G. 1999. Structural analysis of two red coral populations dwelling in different habitats. In *Red Coral and Other Mediterranean Octocorals: Biology and Protection*, F. Cicogna et al. (eds). Rome: Ministero delle Risorse di Agricole e Alimentari e Forestali, 23–43.

Santangelo, G., Bramanti, L. & Iannelli, M. 2007. Population dynamics and conservation biology of the over-exploited Mediterranean red coral. *Journal of Theoretical Biology* **244**, 416–423.

Santangelo, G., Carletti, E., Maggi, E. & Bramanti, L. 2003. Reproduction and population sexual structure of the overexploited Mediterranean red coral *Corallium rubrum*. *Marine Ecology Progress Series* **248**, 99–108.

Sarà, M. 1969. Research on coralligenous formation; problems and perspectives. *Publicazzioni della Stazione Zoologica di Napoli* **37**, 124–134.

Seki, K. 1991. Study of precious coral diver survey with open circuit air SCUBA diving at 108 m. *Annals of Physiological Anthropology* **10**, 189–199.

Sherwood, O.A. & Edinger, E.N. 2009. Ages and growth rates of some deep-sea gorgonian and antipatharian corals of Newfoundland and Labrador. *Canadian Journal of Fisheries and Aquatic Sciences* **66**, 142–152.

Sherwood, O.A., Scott, D.B., Risk, M.J. & Guilderson, T.P. 2005. Radiocarbon evidence for annual growth rings in a deep sea octocoral (*Primnoa resedaeformis*). *Marine Ecology Progress Series* **301**, 129–134.

Skeates, R. 1993. Mediterranean coral: its use and exchange in and around the alpine region during the later Neolithic and copper age. *Oxford Journal of Archaeology* **12**, 281–292.

Smith, C.P., McClure S.F., Eaton-Magana, S. & Kondo, D.M. 2007. Pink to red coral: a guide to determining origin of color. *Gems and Gemology* **43**, 4–15.

Smith, J.E., Risk, M.J., Schwarcz, H.P. & McConnaughey, T.A. 1997. Rapid climate change in the North Atlantic during the Younger Dryas recorded by deep-sea corals. *Nature* **386**, 818–820.

Smith, J.E., Schwarcz, H.P. & Risk, M.J. 2002. Patterns of isotopic disequilibria in azooxanthellate coral skeletons. *Hydrobiologia* **471**, 111–115.

Smith, J.E., Schwarcz, H.P., Risk, M.J., McConnaughey, T.A. & Keller, N. 2000. Paleotemperatures from deep-sea corals: overcoming 'vital effects'. *Palaios* **15**, 25–32.

Spotte, S., Heard, R.W., Bubucis, P.M., Manstan, R.R. & McLelland, J.A. 1994. Pattern and coloration of *Periclimenes rathbunae* from the Turks and Caicos Islands, with comments on host associations in other anemone shrimps of the West Indies and Bermuda. *Gulf Research Reports* **8**, 301–311.

Stephens, P.A., Sutherland, W.J. & Freckleton, R.P. 1999. "What is the Allee effect?" *Oikos* 87, 185–190.

Sutherland, W.J. 2000. *The Conservation Handbook: Research, Management, and Policy.* Oxford, UK: Blackwell Science.

Tescione, G. 1965. *Il corallo nella storia e nell'arte.* Napoli, Italy: Montanino Editore.

Tescione, G. 1973. *The Italians and Their Coral Fishing.* Naples: Fausto Fiorentino.

Thresher, R., Rintoul, S.R., Koslow, J.A., Weidman, C., Adkins, J.F. & Proctor, C. 2004. Oceanic evidence of climate change in southern Australia over the last three centuries. *Geophysical Research Letters* 31, L07212.

Thrush, S.F. & Dayton, P.K. 2002. Disturbance to marine benthic habitats by trawling and dredging: implications for biodiversity. *Annual Review of Ecology and Systematics* 33, 449–473.

Torrents, O., Garrabou, J., Marschal, C. & Harmelin, J.G. 2005. Age and size at first reproduction in the commercially exploited red coral *Corallium rubrum* (L.) in the Marseilles area (France, NW Mediterranean). *Biological Conservation* 121, 391–397.

Torntore, S.J. 2002. *Italian coral beads: characterizing their value and role in global trade and cross-cultural exchange.* PhD thesis, University of Minnesota.

Tracey, D.M., Neil, H., Marriott, P.A., Allen, H., Cailliet, G.M. & Sánchez, J.A. 2007. Age and growth of two genera of deep-sea bamboo corals (family Isididae) in New Zealand waters. *Bulletin of Marine Science* 81, 393–408.

True, M.A. 1970. Étude quantitative de quatre peuplements sciaphiles sur substrat rocheux dans la région marseillaise. *Bulletin Institute Océanographique Monaco* 69, 1–48.

Tsounis, G., Rossi, S., Aranguren, M., Gili, J.M. & Arntz, W.E. 2006a. Effects of spatial variability and colony size on the reproductive output and gonadal development cycle of the Mediterranean red coral (*Corallium rubrum* L.). *Marine Biology* 148, 513–527.

Tsounis, G., Rossi, S., Gili, J.M. & Arntz, W.E. 2006b. Population structure of an exploited benthic cnidarian: the case study of red coral (*Corallium rubrum* L.). *Marine Biology* 149, 1059–1070.

Tsounis, G., Rossi, S., Gili, J.M. & Arntz, W.E. 2007. Red coral fishery at the Costa Brava (NW Mediterranean): case study of an overharvested precious coral. *Ecosystems* 10, 975–986.

Tsounis, G., Rossi, S., Laudien, J., Bramanti, L., Fernández, N., Gili, J.M. & Arntz, W.E. 2006c. Diet and seasonal prey capture rates in the Mediterranean red coral (*Corallium rubrum* L.). *Marine Biology* 149, 313–325.

University Parthenope, in press. Red coral science, management and trading in the Mediterranean. Conclusions from science and management workshop (26th Sept). University of Parthenope, Naples, Italy. Online. Available HTTP: http://dsa.uniparthenope.it/rcsmt09/ (accessed December 2009).

U.S. Department of Commerce. 1989. The coral fishery and trade of Japan—foreign fishery developments. *Marine Fisheries Review* 50, 665. Available HTTP: http://www. findarticles.com/p/articles/mi_m3089/is_n4_v51/ai_9346396/?tag=content;col1 (accessed December 2009).

U.S. Department of Commerce. 2007. *Magnuson-Stevens Fishery Conservation and Management Act, as amended through January 12, 2007.* NOAA National Marine Fisheries Service. Online. Available HTTP: http://www.nmfs.noaa.gov/msa2005/docs/MSA_amended_msa%20_20070112_FINAL.pdf (accessed: December 2009).

Verrill, A.E. 1883. Report on the Anthozoa, and on some additional material dredged by the Blake in 1877–1879, and by the U.S. Fish Commission Steamer "Fish Hawk", in 1880–82. *Bulletin of the Museum of Comparative Zoology Harvard* 11, 1–72.

Vighi, M. 1972. Ètude sur la reproduction du *Corallium rubrum* (L.). *Vie Milieu* 23, 21–32.

Walters, C.J. 1986. *Adaptive Management of Renewable Resources.* New York: Macmillan.

Warner, G.F. 1981. Species descriptions and ecological observations of black corals (Antipatharia) from Trinidad. *Bulletin of Marine Science* 31, 147–163.

Weinbauer, M.G., Brandstatter, F. & Velimirov, B. 2000. On the potential use of magnesium and strontium concentrations as ecological indicators in the calcite skeleton of the red coral (*Corallium rubrum*). *Marine Biology* 13, 801–809.

Weinberg, S. 1978. Mediterranean octocoral communities and the abiotic environment. *Marine Biology* 49, 41–57.

Weinberg, S. 1979. The light dependent behaviour of planula larvae of *Eunicella singularis* and *Corallium rubrum* and its implication for octocorallian ecology. *Bijdragen tot de Dierkunde* 49, 145–151.

Wells, S. 1983. Precious corals commercially threatened. *IUCN Invertebrate Red Data Book.* Gland, Switzerland: IUCN, 35–42.

Western Pacific Regional Management Council (WPCouncil). 2006. Black coral science management workshop, April 18–19. Honolulu. Online. Available HTTP: http://www.wpcouncil.org/precious/Documents/2006%20 Black%20Coral%20Science%20and%20Management%20Workshop%20Report-SCANNED%20 VERSION.pdf (accessed December 2009).

Western Pacific Regional Management Council (WPCouncil). 2007. Fishery management plan for the precious coral fisheries of the western Pacific region. Honolulu: WP Council. Online. Available HTTP: http:// www.wpcouncil.org/hawaii/PreciousCorals.htm (accessed December 2009).

Williams, B., Risk, M.J., Ross, S.W. & Sulak, K.J. 2007. Stable isotope records from deepwater antipatharians: 400-year records from the south-eastern coast of the United States of America. *Bulletin of Marine Science* **81**, 437–447.

Williams, B., Risk, M.J., Ross, S.W. & Sulak, K.J. 2009. Deep-water antipatharians: proxies of environmental change. *Geology* **34**, 773–776.

Witherell, D. & Coon, C. 2000. Protecting gorgonian corals off Alaska from fishing impacts. In *First International Symposium on Deep-Sea Corals*, J.H.M. Willison et al. (eds). Halifax: Ecology Action Center and Nova Scotia Museum, 117–125.

Young, T.P. 2000. Restoration ecology and conservation biology. *Biological Conservation* **92**, 73–83.

Zibrowius, H., Montero, M. & Grasshoff, M. 1984. La ripartition du *Corallium rubrum* dans l'Atlantique. *Thetis* **11**, 163–170.

Zoubi, A. 2009. *An Overview on the Main Marine Resources (Commercial And Non-commercial Groups) at the Moroccan Mediterranean*. Casablanca, Morocco: National Institute for Fisheries Research.

Oceanography and Marine Biology: An Annual Review, 2010, **48**, 213-266
© R. N. Gibson, R. J. A. Atkinson, and J. D. M. Gordon, Editors
Taylor & Francis

THE BIOLOGY OF VESTIMENTIFERAN TUBEWORMS

MONIKA BRIGHT[1] & FRANÇOIS H. LALLIER[2,3]

[1]Department of Marine Biology, University of Vienna, Althanstrasse 14, A-1090 Vienna, Austria
E-mail: monika.bright@univie.ac.at
[2]UPMC—Paris 6, Laboratoire Adaptation et Diversité en Milieu Marin,
Station Biologique de Roscoff, 29680 Roscoff, France
[3]CNRS UMR 7144, Station Biologique de Roscoff, 29680 Roscoff, France
E-mail: lallier@sb-roscoff.fr

Abstract   Vestimentiferan tubeworms, once erected at a phylum level, are now known to comprise a part of the specialised deep-sea polychaete family Siboglinidae. Their widespread and abundant occurrence at hydrothermal vents and hydrocarbon seeps has fostered numerous studies of their evolution and biogeography, ecology and physiology. Harbouring autotrophic, sulphide-oxidising, intracellular bacterial symbionts, they form large populations of 'primary' producers with contrasting characteristics, from fast-growing, short-living species at vents, to slow-growing, long-living species at seeps. These different life strategies and the ways they modify the biogeochemistry of their respective environments have consequences on the macro- and meiofaunal assemblages that develop within vestimentiferan bushes. New findings indicate that postlarval recruits get infected through the skin by free-living bacteria for which growth is rapidly and specifically limited by the host to mesoderm cells around the gut that further transform into the characteristic trophosome. The resulting internal location of symbionts prompts specific adaptations of the hosts to fulfil their metabolic requirements, including unusual sulphide and carbon dioxide assimilation and transport mechanisms. Symbiont genome sequencing has improved our knowledge of potential bacterial metabolism and should rapidly open the way for new research approaches to resolve the intricate physiological relationships between a eukaryotic host and its chemoautotrophic bacterial symbionts.

Introduction

Since the discovery of *Riftia pachyptila* around deep-sea hydrothermal vents on the Galapagos Spreading Centre (GSC) in 1977 (Corliss et al. 1979), a considerable amount of work has been devoted to the biology of vestimentiferan tubeworms, a group of animals with rather low diversity and less than 30 recognised species mostly living in the deep sea. Yet, the conspicuous aggregations they form around deep-sea hot vents or cold seeps and their strict reliance on endosymbiont chemo-autotrophic metabolism have granted vestimentiferans, and among them *R. pachyptila*, the rank of model organisms. However, since the reviews by Tunnicliffe (1991) and Childress & Fisher (1992), there has not been a comprehensive review of the biology of these tubeworms despite the publication of more than 500 papers on the topic to date (444 record count with "Riftia" in topic on the Web of Science as of July 2009; 358 with "vestimentifer*" and 628 with "Riftia OR vestimentifer*"). Some recent reviews have included vestimentiferan tubeworms in a broader coverage of symbiosis in annelids (Bright & Giere 2005) or chemosynthetic symbiosis (Dubilier et al. 2008, Vrijenhoek 2010) or have dealt with phylogeny and biogeography (McMullin et al. 2003, Halanych 2005), palaeontology (Little & Vrijenhoek 2003, Campbell 2006), morphology and anatomy (Southward et al. 2005), development (Southward 1999, Southward et al. 2005), reproduction and dispersal (Tyler & Young 1999), physiology and biochemistry (Minic & Herve 2004), while others have focused on the bacterial symbionts (Stewart et al. 2005, Vrijenhoek 2010).

The aim of the present review is to summarise recent work on the host biology in a broad sense and on host–symbiont relationships. The review briefly considers the extant species known to date and their phylogeny, mostly inferred from recent molecular studies, and introduces the present hypothesis regarding the phylogeography of Vestimentifera. The two differing habitats occupied by vestimentiferan tubeworms are described (i.e., hydrothermal vents and cold seeps), and the different life strategies associated with them are considered, including the abiotic conditions to which they are exposed, the biotic interactions they develop and the role of tubeworms as foundation species and their associated fauna. An important part of the review is devoted to the description of the life cycle of vestimentiferans, focusing on *R. pachyptila*, from the aposymbiotic larval stage to the symbiotic adult stage, embracing the brief but crucial symbiont transmission phase. The last part of the review focuses on the progress made in understanding the physiology and biochemistry of the chemoautotrophic host–symbiont association. The general use of the term *tubeworm* in this review refers to vestimentiferans and not to other taxa.

Systematics, phylogeny and biogeography

What are vestimentiferan tubeworms?

Vestimentifera is a taxon of marine deep-sea worm-like animals living in chitinous tubes and lacking a digestive tract. Instead, they harbour chemoautotrophic bacteria in an internal organ, the trophosome, and derive their metabolic needs from these bacteria. They share these characteristics with some other animals, namely, the Monilifera (*Sclerolinum*) and Frenulata (Figure 1), forming together the taxon Siboglinidae (Caullery 1914), previously referred to as Pogonophora, and nested within the Annelida. From a cladistic point of view, as stated by Rouse (2001), "Vestimentifera can be defined as the first siboglinid and all its descendants to have a vestimentum as seen in the

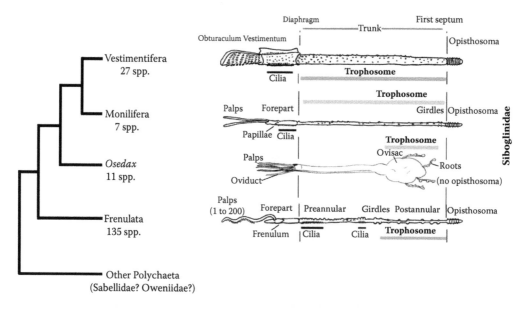

Figure 1 Schematic tree depicting the position of Vestimentifera within the polychaete family Siboglinidae and comparing the main morphological features of the different taxa of the group. (Drawings for Vestimentifera, Monilifera and Frenulata were modified from Southward et al. 2005.) See text for details.

holotype of *Riftia pachyptila*", the vestimentum being a "waistcoat-like" body region "with lateral flaps that enfold the anterior part of the body, behind the plume" (Southward 2000).

Historically, it took almost a century and a rather complicated path to reach this present status. A more detailed account may be found in Rouse (2001) and Pleijel et al. (2009), but the story may be summarised as follows. The first *Siboglinum* species described by Caullery (1914) was not assigned to any phylum, but Uschakov (1933) described *Lamellisabella* as a polychaete. Revision by Johansson (1939) resulted in the erection of the class Pogonophora in 1939. Pogonophora reached phylum status in 1944, and Ivanov (1951) united *Siboglinum* and *Lamellisabella* within this phylum. Pogonophora were often considered as deuterostomes at that time with a tripartite body plan. The description of the segmented, chaetae-bearing terminal part of *Siboglinum fiordicum* (Webb 1964) progressively brought the Pogonophora back to the protostomes. The description of *Lamellibrachia barhami* by Webb (1969) introduced a new class, Vestimentifera, in the phylum Pogonophora. The discovery of *Riftia pachyptila* and other vent species, however, prompted Jones (1985) to place Vestimentifera at phylum rank besides Pogonophora. This view was challenged by many authors, and finally Rouse & Fauchald (1997) reunited the two phyla within the annelids as the family Siboglinidae Caullery, 1914.

Siboglinids are now firmly anchored within the polychaete annelids, even though their sister taxon within polychaetes is still debated, with some favouring Sabellida (Schulze & Halanych 2003) and others Oweniidae (Rousset et al. 2004, Struck et al. 2007). As can be seen in Figure 1, Frenulata are basal to the Siboglinidae, and Vestimentifera form the crown group with Monilifera as a sister taxon. This schematic phylogenetic tree is a summary based on molecular and morphological sources (Boore & Brown 2000, Halanych et al. 2001, Schulze 2003, Halanych 2005, Jennings & Halanych 2005). Frenulata typically inhabit anoxic sediments, Monilifera live on decaying organic matter or reduced sediments, and Vestimentifera are mostly found at hydrocarbon seeps and hydrothermal vents. Other siboglinids can be found on whale falls; these belong to the recently discovered genus *Osedax* with two species described initially (Rouse et al. 2004), and now the number of species described has increased to nine (Glover et al. 2005, Fujikura et al. 2006, Goffredi et al. 2007, Rouse et al. 2008) with several other species awaiting description (R. C. Vrijenhoek, F. Pradillon, personal communications). According to molecular data, these species fall between frenulates and moniliferan/vestimentiferan tubeworms (Rouse et al. 2004), and although they are undoubtedly siboglinid polychaetes, they have aposymbiotic dwarf males. The females lack an opisthosome, and their trophosome harbours heterotrophic symbionts and is located in the most posterior body regions, the ovisac and the so-called roots, which penetrate deeply into whale bones (Goffredi et al. 2007).

Morphology and anatomy of the adult vestimentiferan tubeworm

This part briefly introduces the external morphology and general anatomy of adult Vestimentifera. More detailed descriptions may be found elsewhere (e.g., Gardiner & Jones 1993, Southward 2000, Southward et al. 2005), with a noteworthy list of useful characters provided in Appendix 1 of Schulze (2003). Further details are given if appropriate in the context of this review.

The body of adult tubeworms is enclosed in a chitinous tube that is closed at the posterior end, a character in contrast to the open-ended tube of Monilifera and Frenulata. In *Riftia pachyptila*, the tube is cylindrical, virtually straight, and rather flexible; it can reach a length of more than 2 m and a diameter of 5 cm at the apex, with a tube wall thickness of 2–3 mm (Gaill & Hunt 1986, Gardiner & Jones 1993). The constant diameter of the tube of *R. pachyptila* and the existence of basal partitions, or clumps, not occupied by the worm (Gaill et al. 1997) contrast with the tubes of other vent-endemic species (e.g., within genera *Tevnia*, *Oasisia*, *Ridgeia*), which are conical, tapering at the basal end, more or less twisted, harder and often reinforced with annular extensions (collar). Seep genera (e.g., *Lamellibrachia*, *Escarpia*) have long tapering tubes, generally hard and thick at their anterior, water-immersed part, but much thinner and fragile at their posterior, sediment-buried part (see, e.g., Andersen et al. 2004); the functional rationale for this is explained in the section 'Acquisition of inorganic substrata from the environment', p. 247.

The anterior part of the body, the obturacular region, consists of a branchial plume that can be extended from the tube and is composed of filaments, partly fused into lamellae and supported by a rigid, collagenous extension, the obturaculum (Figure 1) (Andersen et al. 2001). When the animal withdraws into its tube, the terminal part of the obturaculum blocks the entrance of the tube; it may be flat or may bear saucer-like or rod-like collagenous extensions in some species or even among some individuals within the same species (Southward et al. 1995, Andersen et al. 2004). Insertion of branchial lamellae is basal relative to the obturaculum in most species (the "Basibranchia" in Jones 1988) with the exception of *Riftia pachyptila* (the only "Axonobranchia" in Jones 1988), for which lamellae are inserted along the whole length of the obturaculum. In lamellibrachids and alaysiids the outer lamellae fuse to form a sheath surrounding the other lamellae.

Following the obturacular region, the vestimentum is a muscular region with two ventral wings that fold dorsally. When the animal extends its branchial plume from the tube, the vestimentum comes up to the rim of the tube and contracts, blocking the animal in this position. A ciliated area is located ventrally on the vestimentum, and gonopores open dorsally, extended by external paired ciliated grooves in males only. Conspicuous pores on the epidermis are connected to pyriform, chitin-producing glands (Gaill et al. 1992, Shillito et al. 1993) involved in tube biosynthesis. The vestimentum also encloses the heart, which is an extended and muscular portion of the dorsal vessel in the anterior part of the vestimentum; the brain, which is anteroventral; and the excretory organ, which is posterodorsal and adjacent to the brain (Gardiner & Jones 1993, Schulze 2001a, Gardiner & Hourdez 2003).

The trunk forms the most extended part of the adult body, although its relative length varies greatly during growth and between individuals in most species (see, e.g., Fisher et al. 1988, Andersen et al. 2004). Externally, the epidermis is relatively smooth and covered by a collagenous cuticle, with scattered pores connected to chitin-producing glands similar to those of the vestimentum. Internally, a muscular layer of circular and longitudinal muscles surrounds a vast coelomic cavity mostly filled with coelomic fluid (CF) and the conspicuous trophosome, a soft and highly vascularised tissue harbouring the symbiotic bacteria. The ontogeny and anatomy of the trophosome is described in detail in the life-cycle section of this review. Gonads and blood vessels run along the length of the trunk; their description can be found in the life-cycle and physiology sections of the review.

The opisthosome forms the posterior end of the body; it is a short, multisegmented region bearing rows of uncini that function to anchor the animal within its tube. The uncini of Vestimentifera, and Siboglinidae in general, are considered homologous to those of Terebellida and Sabellida, implying a monophyletic origin of these three taxa (Bartolomaeus 1995, 1997), although the phylogenetic value of the uncini has been questioned (Schulze 2001b) as their similar design may be the result of functional convergence.

Present state of vestimentiferan tubeworm systematics and phylogeny

Table 1 presents a comprehensive list of the 19 species of vestimentiferan tubeworms described to date, beginning with *Lamellibrachia barhami* Webb, 1969 and ending with *L. juni* Miura & Kojima, 2006 and *Oasisia fujikurai* Miura & Kojima, 2006. It also includes eight unnamed species recognised through their CO1 gene sequence (Kojima et al. 2001, Kojima et al. 2002, Kojima et al. 2003) but not yet described. A number of additional unnamed species have been reported in various cruise reports but without formal description or even gene sequence comparison, therefore preventing a confident assignation at the generic level. For example, species of vestimentiferan have been found recently in Mediterranean seep sites and North Sea mud volcanoes and await formal description.

Table 1 Taxonomic list of described[a] species of Vestimentifera

Species	Habitat[c]	Distribution[b]	References
Alaysia spiralis	HV	Lau	Southward 1991
Alaysia sp. *A1*	CS	Nankai	Kojima et al. 2003
Alaysia sp. *A2*	HV	Okinawa	Kojima et al. 2003
Alaysia sp. *A3*	HV	Manus	Kojima et al. 2003
Alaysia sp. *A4*	HV	Okinawa	Kojima et al. 2003
Arcovestia ivanovi	HV	Lau, Manus	Southward & Galkin 1997
Escarpia laminata	CS	GoM	Jones 1985
Escarpia southwardae	CS	GoG	Andersen et al. 2004
Escarpia spicata	SV, CS, WF	GoC, SCB	Jones 1985
Escarpia sp. *E1*	CS	Nankai	Kojima et al. 2002
Lamellibrachia barhami	SV, CS	GoC, JFR	Webb 1969
Lamellibrachia columna	SV	Lau	Southward 1991
Lamellibrachia juni	SV	Kermadec	Miura & Kojima 2006
Lamellibrachia luymesi	CS	GoM, Amazon	van der Land & Nørrevang 1975
Lamellibrachia satsuma	SV, CS	Kago, Nankai, Nikko	Miura et al. 1997
Lamellibrachia victori	CS	Uruguay	Mañé-Garzón & Montero 1985
Lamellibrachia sp. *L1*	SV	Okinawa	Kojima et al. 2001
Lamellibrachia sp. *L2*	CS	Nankai	Kojima et al. 2001
Lamellibrachia sp. *L4*	SV	Manus	Kojima et al. 2001
Oasisia alvinae	HV	EPR 21°N–18°S	Jones 1985
Oasisia fujikurai	HV	Kermadec	Miura & Kojima 2006
Paraescarpia echinospica	CS	PNG, Java	Southward et al. 2002
Ridgeia piscesae	HV	Explorer, JFR, GoC?	Jones 1985
Riftia pachyptila	HV	EPR 21°N–31°S, Galap, GoC	Jones 1981
Seepiophila jonesi	CS	GoM	Gardiner et al. 2001
Siphonobrachia lauensis	SV	Lau	Southward 1991
Tevnia jerichonana	HV	EPR 13°N–21°S	Jones 1985

[a] Kojima et al. (2001, 2002, 2003) mentioned unnamed species of *Lamellibrachia* (L1 to L7), *Escarpia* (E1, E2) and *Alaysia* (A1 to A4) in the western Pacific HV and CS sites and grouped them in several phylotypes based on CO1 molecular phylogeny.

[b] Distribution: Amazon = Amazon passive margin; EPR = East Pacific Rise; Explorer = Explorer Ridge; GSC = Galapagos Spreading Center; GoC = Gulf of California; GoG = Gulf of Guinea; GoM = Gulf of Mexico; JFR = Juan de Fuca Ridge; Kago = Kagoshima Bay; Kermadec = Kermadec back arc basin; Lau = Lau back arc basin; Manus = Manus back arc basin; Nankai = Nankai Trough; Nikko = Nikko Seamount; SCB = Santa Catalina Basin; Uruguay = Uruguay passive margin.

[c] Habitat: CS = cold seeps; HV = hydrothermal vents; SV = sedimented vents; WF = whale falls.

Species diversity is maximum in *Lamellibrachia* (nine spp.), followed with *Alaysia* (five spp.) and *Escarpia* (four spp.), whereas vent-endemic species are mostly represented by monospecific genera. However, recent molecular data point to several morphologically indistinguishable *Oasisia* species (Hurtado et al. 2002) and signal the need for a reexamination of monospecificity in vent species. This low species diversity within vent genera has been interpreted as evidence of a recent radiation of vent-endemic species from sediment-dwelling ancestors (e.g., Schulze & Halanych 2003).

Two alternative phylogenies are presented in Figure 2, summarising present ideas on the evolution within Vestimentifera. Broadly, all studies point to three major groups within Vestimentifera: the lamellibrachids, escarpids, and vent-endemic species as a crown group. Several molecular studies supported *Lamellibrachia* as the most basal extant vestimentiferan using 28S ribosomal DNA (rDNA; Williams et al. 1993), 18S rDNA (Halanych et al. 2001), or the CO1 gene (Black et al. 1997).

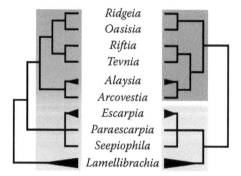

Ridgeia
Oasisia
Riftia
Tevnia
Alaysia
Arcovestia
Escarpia
Paraescarpia
Seepiophila
Lamellibrachia

Figure 2 Alternative hypotheses for Vestimentifera phylogeny at the genus level. On the left side there is a continuum of species from seeps (light-grey shade) to vents (dark-grey shade) in contrast to the situation depicted on the right, showing two sister groups, one with seep species, the other with vent species. Clarification of the phylogenetic position of *Alaysia* and *Arcovestia* species would help resolve current uncertainty.

Using the same marker, CO1, Kojima et al. (2003) placed escarpids at the phylogenetic base but with no significant support. Cladistic approaches based on morphological characters are also in favour of a basal lamellibrachid clade (Schulze 2003) or, alternatively, a basal clade formed by lamellibrachids and escarpids (Rouse 2001). The existence of a vent-endemic species clade as the crown group of Vestimentifera has been repeatedly postulated from some molecular (Black et al. 1997) or morphological (Schulze 2003) studies, although this interpretation was weakly supported by other analyses (Williams et al. 1993, Halanych et al. 2001). The position of a clade uniting *Alaysia*-like and *Arcovestia* vestimentiferans (Kojima et al. 2003) is more problematic; it is seldom mentioned in the literature since these two taxa are generally not included (Halanych 2005). However, when they are included, they form a weakly supported clade with other vent-endemic species (Kojima et al. 2003), although morphological studies separated them, with *Arcovestia* closer to vent species and *Alaysia* closer to escarpids (Schulze 2003). The question remains regarding whether there is a clear-cut distinction between seep (and sedimented vents) and vent species (Figure 2, right), the latter including *Alaysia* and *Arcovestia* species, or if there is a continuous gradation from seep to vent species (Figure 2, left). To further resolve within-Vestimentifera phylogeny, other molecular markers or a combination of them is needed to increase the resolution of future analysis, which should include *Alaysia*-like species. Other arguments may come from a more comprehensive approach, namely, phylogeography, uniting the phylogenetic signal of molecular and morphological markers with our understanding of the biogeography and geological history of the oceans.

Where do vestimentiferan tubeworms live?

Vestimentiferans inhabit mainly hydrothermal vents and cold-seep environments in the deep sea, with the exception of *Lamellibrachia satsuma* living as shallow as 80 m in Kagoshima Bay (Hashimoto et al. 1993, Miura et al. 1997). Most commonly, species are restricted to one or the other habitat type. Some *Lamellibrachia* species, however, thrive in both environments (Jones 1985, Black et al. 1997, Southward & Galkin 1997, Tunnicliffe et al. 1998). *Escarpia spicata* exhibits the broadest habitat range, living at vents and seeps and on whale falls (Black et al. 1997).

Figure 3 shows the geographic and habitat distribution of known species on vent and seep sites around the world, based on the species and references list in Table 1. Cold-seep Vestimentifera have been found in all oceans to date except in the Indian Ocean (although *Paraescarpia echinospica* has been reported at a seep site of the Java Trench; Southward et al. 2002). In contrast, vent vestimentiferan tubeworms are only known from Pacific hydrothermal vents, with the crown

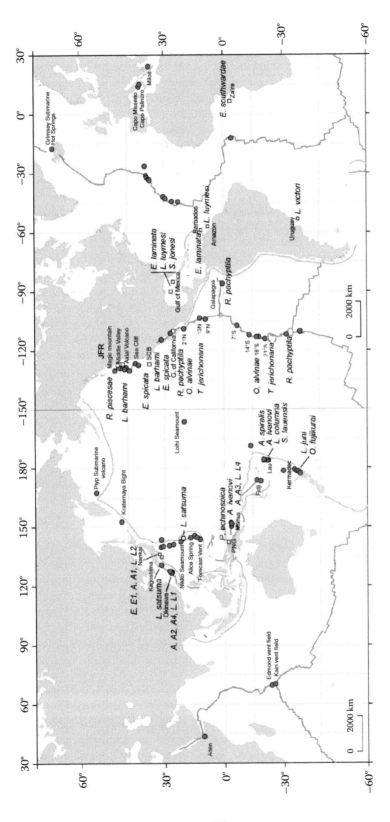

Figure 3 (See also Colour Figure 3 in the insert following page 212.) Geographic localisation of hydrothermal vent and cold seep sites explored to date (2008) with indication of identified Vestimentifera species. Refer to Table 1 for complete species names and references. (Map modified from Desbruyères et al. 2006.)

group species limited to the eastern Pacific. An interesting parallel exists between the western and eastern Atlantic: *Escarpia southwardae* found in the Gulf of Guinea is close to *E. laminata* (western Atlantic) and to *E. spicata* (eastern Pacific), suggesting an eastward route from the Pacific to the Atlantic (Van Dover et al. 2002, Schulze 2003, Andersen et al. 2004). This may be tentatively explained by larval dispersal and patterns of deep oceanic currents, which at present flow mainly from the Atlantic to the Pacific due to physical and geographical constraints (Van Dover et al. 2002). However, large areas are still poorly explored; the subduction zones of the eastern Pacific, the South Atlantic and Indian ridges and the circum-Antarctic ridges deserve more exploration. If vestimentiferan tubeworms are found in these areas, these data should yield important information regarding phylogeography of extant vent species. So far, mapping biogeographical data onto phylogeny (Schulze 2003) leads to ambiguous conclusions, except for the recent radiation of vent species on the East Pacific Rise (EPR).

A mention should be made about whale falls: only one vestimentiferan occurrence (*E. spicata*) has been documented off southern California (Feldman et al. 1998) despite considerable exploration and experimentation efforts in recent years (reviewed in Smith & Baco 2003 and Fujiwara et al. 2007). In addition, two occurrences of an unknown species of *Lamellibrachia* have been reported on rotting organic cargo from shipwrecks (Dando et al. 1992, Hughes & Crawford 2006).

Understanding the evolution and biogeographic patterns of Vestimentifera should also take fossil data into account. These have been reviewed (Campbell 2006) following reports about the apparent discrepancy between age estimates based on fossils and molecular phylogenies (Little & Vrijenhoek 2003). Molecular data indicated divergence estimates no older than mid-Mesozoic (about 100 million yr), whereas fossil data gave a Silurian or Devonian origin (over 400 million yr). The fossil remains consist mainly of preserved tubes that have been calcified (Little et al. 1997, Peckmann et al. 2005), and the comparison with chitinous tubes from living species is therefore difficult (Little & Vrijenhoek 2003, Campbell 2006).

From the various data reported here, it appears that Vestimentifera have recently radiated from seep species distributed worldwide to vent species restricted to the Pacific. The differences between these two habitats (illustrated in Figure 4) may prove to be an ecological constraint sufficient to drive the recent speciation events observed so far (Halanych et al. 2001, Schulze & Halanych 2003); the next section of the review details the ecological setting in which vestimentiferan tubeworms flourish.

Habitat and ecology

Contrasting habitat characteristics and environmental conditions distinguish the vent and seep habitats in which vestimentiferan tubeworms thrive. While vents are relatively short-lived, subject to major disturbances and waxing and waning of fluid flow, seeps are relatively long-lived and stable. Nevertheless, in both sulphide (i.e., ΣH_2S total concentration of labile species of sulphide, and especially H_2S, the most toxic form for aerobic organisms; see Le Bris et al. 2003) is present and utilised by the endosymbionts of tubeworms. At vents, it is produced geothermally and emerges from cracks and crevices in the earth's crust (see Van Dover 2000). At seeps, the majority of sulphide is produced biologically via sulphate reduction utilising methane of other hydrocarbons as electron donors (Arvidson et al. 2004, Joye et al. 2004).

Most of our ecological knowledge in vestimentiferans is based on a few species, *Riftia pachyptila* and *Ridgeia piscesae* from vents and *Lamellibrachia luymesi* and *Seepiophila jonesi* from seeps. It has become apparent that some species from vents and seeps have evolved different life strategies in accordance with their different habitat types and characteristics, but some species can bridge the gap between these habitats and thrive under very broad environmental conditions. On the one hand, *Riftia pachyptila* grows extremely fast, but only in areas with relatively vigorous diffuse vent flow, and is relatively short-lived (Fisher et al. 1988, Hessler et al. 1988, Shank et al. 1998). In

Figure 4 (See also Colour Figure 4 in the insert.) The two contrasted ecological settings of Vestimentifera are illustrated by the most-studied species: a bush of *Riftia pachyptila* from the EPR at 12°50′N (© Ifremer-Hope 1999) and one of *Lamellibrachia luymesi* from the Gulf of Mexico (© C.R. Fisher). Diagrams on the right show (top) how vent species such as *Riftia pachyptila*, fixed to the hard rock substratum, get both oxygen and sulphide through their branchial plume from the mixed fluid and deliver them through circulation to the internally located but environmentally acquired bacteria (black triangle) and (bottom) in seep species such as *Lamellibrachia luymesi*, sulphide is acquired from the sediment through the tapering, buried tube and trunk 'roots'. BR = branchial plume; TR = trophosome.

contrast, the seep tubeworms grow extremely slowly in areas with low seep flow and are extremely long-lived (Fisher et al. 1997, Julian et al. 1999, Bergquist et al. 2000, Cordes et al. 2005a, Cordes et al. 2007a). *Ridgeia piscesae*, however, can grow very rapidly when exposed to higher vent flow but is also capable of living in areas with lower vent flow, where it then grows much more slowly (Urcuyo et al. 2007).

Often, hydrothermal vents and seeps are visually easily recognised through their conspicuous populations of megafauna. Among these are Vestimentifera (Figure 4), which build bush-like aggregations and act as foundation species by creating this physical structure that provides living space for other species (Corliss et al. 1979, Paull et al. 1984, Kennicutt et al. 1985, Tunnicliffe 1991, Sarrazin & Juniper 1999, Tsurumi & Tunnicliffe 2003).

Hydrothermal vent tubeworms

Deep-sea hydrothermal vents are found at midocean ridges and back-arc basins. Catastrophic volcanic eruptions, tectonic disturbances and hydrothermal vent fluid circulation form a transient, relatively short-lived environment (see Van Dover 2000). At the fast-spreading EPR, large-scale disturbances that kill existing communities can occur on a decadal scale (Haymon et al. 1993, Shank et al. 1998, Tolstoy et al. 2006). Small-scale disturbances due to fluid flow alternation and changes in fluid composition occur over shorter time intervals (Fustec et al. 1987, Jollivet 1993, Shank et al. 1998). Vent fluid composition, flow rates and mixing with ambient seawater are variable in dramatic and unpredictable ways (see Childress & Fisher 1992, Fornari et al. 1998) and create unstable physicochemical conditions regarding temperature, sulphide, oxygen and pH gradients. Striking spatial patterns of typical macrofauna assemblages along a gradient of hydrothermal fluid flux can nevertheless be distinguished.

At the EPR and the Galapagos Spreading Center (GSC), the vestimentiferans *Riftia pachyptila*, *Tevnia jerichonana* and *Oasisia alvinae* live in conditions of vigorous diffuse flow (Hessler & Smithey 1983, Hessler et al. 1985, Haymon et al. 1991, Sarrazin et al. 1997, Shank et al. 1998). The fluids are generally enriched in sulphide, methane, hydrogen, carbon dioxide, silicate and in some cases ferrous iron. However, their temperature and chemical composition are highly variable in space over scales of centimetres and in time over scales of seconds (see Le Bris et al. 2006a). Most often, temperature was used as a proxy of the hydrothermal vent fluid contribution and related chemical parameters in diffuse flow habitats (Johnson et al. 1988a,b, Shank et al. 1998, Mullineaux et al. 2000, Urcuyo et al. 2003, Hunt et al. 2004). In addition, various techniques for *in situ* chemical measurements have considerably increased understanding of the conditions under which tubeworms live (Johnson et al. 1986, 1988a,b, Childress et al. 1993, Sarradin et al. 1998, Luther et al. 2001, 2008, Le Bris et al. 2003, 2006a,b).

In aggregations of *Riftia pachyptila*, temperature ranges from ambient deep-sea temperatures of about 2°C to a maximum of about 30°C, pH can be as low as 4.4, and sulphide can be as high as 330 μM (Luther et al. 2001, Le Bris et al. 2003, 2006a,b). Within a site, the overall sulphide–temperature relationship varies only slightly. There is high temperature and chemical variability on small spatial and temporal scales. For example, at organism level, the branchial plume experiences higher pH and lower temperature conditions than the tube (Le Bris et al. 2003). Further, aggregations from different sites also showed considerable differences in correlations between temperature and sulphide or pH, not necessarily related to the age of a site (Le Bris et al. 2006a).

The density of *R. pachyptila* per surface area of basalt on which they grow has been reported as about 2000 ind. (individuals) m^{-2} (Shank et al. 1998) and between about 550 and 3500 ind. m^{-2} (calculated from data in Govenar et al. 2005). Individuals reached tube lengths of 3 m with individual biomasses up to 650 g wet weight (Grassle 1986, Fisher et al. 1988). The surface area of tubes of *R. pachyptila* was at least a magnitude higher than the surface area of the basalt (Govenar et al. 2005). The biomass correlated positively with the tube surface area. In most aggregations, some individuals of *Tevnia jerichonana* or *Oasisia alvinae* were found (Govenar et al. 2005); mostly, they grow on the basalt underneath large *Riftia pachyptila* bushes and are difficult to sample (M.B. personal observations).

Colonisation of *R. pachyptila* was found to be rapid and growth fast, and this species may be rapidly replaced by successional species (Hessler et al. 1988, Lutz et al. 1994, Shank et al. 1998).

The first time series study monitoring tubeworm aggregations was reported by Hessler et al. (1988) at Rose Garden, GSC, where fast tubeworm colonisation was followed by mussel overgrowth. Long-term photographic and video-recording observations along a 1.37-km long transect at the 9°50′N EPR region for a period of 5 yr after the 1991 eruption revealed that the most common pattern of sequential colonisation was from *Tevnia jerichonana* (11 mo posteruption), to *Riftia pachyptila* (32 mo posteruption), to the mussel *Bathymodiolus thermophilus,* which appears last and has sometimes been known to replace *Riftia pachyptila*. However, development of *R. pachyptila* aggregations without prior colonisation by *Tevnia jerichonana* (at least in the size visible in photographs and videos) was also observed (Shank et al. 1998).

Ridgeia piscesae lives at varying diffuse flow habitats in the Northeast Pacific. This species exhibits a greater tolerance to varying physicochemical conditions than any other known Vestimentifera. It inhabits sulphide edifices (smokers or chimeys) with a more dynamic, relatively high-temperature diffusive flow of hydrothermal fluid and basalt with relatively stable, low-temperature diffuse flow (Southward et al. 1995, Sarrazin et al. 1997, 1999, Urcuyo et al. 1998). Several growth forms or morphotypes can be distinguished in this species (Black 1991, Southward et al. 1995, Tunnicliffe et al. 1997, Tsurumi & Tunnicliffe 2003). Their occurrence is correlated with specific physicochemical characteristics (Sarrazin et al. 1997, Urcuyo et al. 1998). Environmental conditions at sulphide edifices are known to change on the scale of months (Sarrazin et al. 1997), while diffuse flow from basalt is much more stable in time. Also *R. piscesae* bushes exhibit a range of architectural types based on the degree of branching complexity (Tsurumi & Tunnicliffe 2003).

Using accepted terminology, the long-skinny morphotype of *R. piscesae* was exclusively found at low-temperature basalt. It is reported to grow over a metre in length (Urcuyo et al. 1998). The surface of tubes increases 200-fold compared with the surface of the basalt on which the aggregation grows (Urcuyo et al. 2003). There was a spatial gradient of temperature from the base of the tube to the plume, ranging from consistently elevated temperatures compared with ambient temperature (+0.07–22°C anomaly) to only slightly elevated temperatures (Urcuyo et al. 2003, 2007). Long-term temperature recording showed that over 90% of time temperature at the plume level was less than 1°C above ambient (Urcuyo et al. 2003). Plume-level sulphide concentration was less than 0.1 μM, while measurements at the base were about 100 μM (Urcuyo et al. 2003). Sulphide and temperature were correlated within each site, and this relationship was different between sites (Urcuyo et al. 2007). At one site, no sulphide was detected at plume level, while at the other site low levels of sulphide were measured in 60% of the measurements (Urcuyo et al. 2007).

Other morphotypes of *R. piscesae*, including the short-fat morphotype, occur on active sulphide structures exposed to high temperatures lower than 45°C (Martineu et al. 1997, Sarrazin et al. 1997, 1999).

A model of community succession describing six stages of succession was proposed by Sarrazin et al. (1997), later extended and supported by additional studies (Sarrazin & Juniper 1999, Sarrazin et al. 1999, Govenar et al. 2002). Following stages I and II lacking tubeworms, colonisation of *R. piscesae* can be found occasionally during stage III with low abundances (212 ind. m^{-2}), while at stage IV dense populations of small tubeworms are present. By growth of these tubeworms, the community turns into stage V LF (low flow). Alternatively, a stage V HF (high flow) community can directly develop from stage I or II with higher temperatures. In later successional stages IV,V-HF, and V-LF, tubeworm abundances range from 10,310 to 66,903 ind. m^{-2} (Sarrazin & Juniper 1999, Govenar et al. 2002). Surface areas provided by tubeworms were calculated for several collections and the data showed that in assemblages at stage IV the small tubeworms increased the surface area 1.5 times, in V-LF with large tubeworms by about 22 times and in V-HF, also with large tubeworms, by about 27 times (Sarrazin & Juniper 1999). Stage VI is characterised by senescent tubeworms. Mean temperatures were quite similar in all stages except the stage VI with senescent tubeworms (Sarrazin et al. 2002).

Seep tubeworms

At active and passive continental margins, cold seeps with seepage of higher hydrocarbons and gas are found throughout the world's oceans. Mostly, the expelled fluids are similar in temperature to the surrounding deep-sea water, but differ chemically. They lack oxygen and may contain methane or sulphide or high salt concentrations, but they lack the high levels of heavy metal concentrations typical of hydrothermal vents (see Sibuet & Olu 1998, Sibuet & Olu-Le Roy 2002, Levin 2005). Such seeps can persist for centuries; for example, in the Gulf of Mexico (GoM) estimates for stable seepage of individual sites are in excess of 10,000 yr (Roberts & Aharon 1994).

The vestimentiferan aggregations composed of *Lamellibrachia luymesi* and *Seepiophila jonesi* were studied in detail at the upper Louisiana slope, GoM (Bergquist et al. 2003, Cordes et al. 2005c). Most of these tubeworm bushes were dominated by *Lamellibrachia luymesi,* with an average of 74.3% (Bergquist et al. 2003) and 72.4% (Cordes et al. 2005b) of the species present in the bushes, but in some of the aggregations both species were almost equally distributed or highly dominated by *L. luymesi*. Abundance was from about 200 to over 9500 ind. m^{-2} surface area of sediment (Bergquist et al. 2003). The tube surface area increased the sediment surface from which the tubeworms grew between 2.6- to 26-fold (Bergquist et al. 2003). Such aggregations can extend 2 m above the sediment surface (MacDonald et al. 1989, 1990). Population sizes of individual aggregations varied between 150 and 1500 individuals, but aggregations covering tens to hundreds of square metres are common (see Cordes et al. 2005b). With their posterior extensions, termed *roots,* the animals penetrate deeply into the sediment. Such large animals were found to experience nearly ambient deep-sea conditions at the plume level resulting from little mixing of seep fluids with overlaying bottom water (MacDonald et al. 1989, Scott & Fisher 1995, Julian et al. 1999, Freytag et al. 2001). Sulphide concentrations rarely exceed 0.1 µM at the plume level and usually are well below 0.1 µM; sulphide concentrations are higher at the roots (Bergquist et al. 2003, Cordes et al. 2005b), which were found to acquire sulphide from the porewater of the sediments (Julian et al. 1999, Freytag et al. 2001). In general, a decrease in sulphide concentration with the age of the aggregations was found (Cordes et al. 2005b).

A model of community succession was proposed by Bergquist et al. (2003). Concerning tubeworms, active recruitment occurs when sulphide concentrations are high and authigenic carbonate precipitation allows larval settlement onto hard substrata. This recruitment phase lasts as long as sulphide is present and may last for several decades. When growth of aggregations depletes sulphide in bottom waters, animals use their roots to take up sulphide from the porewaters of the sediment. This phase may last for several centuries (Bergquist et al. 2003). In a diagenetic model, Cordes et al. (2005a) estimated that hypothetical release of sulphate from the roots fuelling sulphate reduction in the sediment could augment exogenous sulphide production and support moderate-sized aggregations for hundreds of years.

Role of tubeworms as foundation species

Despite the low number of tubeworm-associated community studies, which are restricted to a few foundation species, and difficulties in direct comparisons of data (differences in sampling methods, extraction of fauna, standardisation of abundance, classification and distinction between macro- and meiofauna size classes), it appears that the abundance of associated macrofauna is lower but the species richness is higher at seeps compared with vents. Further, as Tsurumi & Tunnicliffe (2003) pointed out and Gollner et al. (2007) later showed, the meiofauna size class considerably contributes to the total number of species colonising tubeworm aggregations. However, meiofaunal species occur in low abundances, in contrast to the highly abundant macrofauna.

Quantitative or semiquantitative studies of the associated communities are limited to a few species: *Ridgeia piscesae* (Sarrazin et al. 1999, Govenar et al. 2002, Tsurumi & Tunnicliffe 2003), *Riftia pachyptila* (co-occurring with the smaller species *Tevnia jerichonana* and *Oasisia alvinae*) (Govenar et al. 2004, 2005, Gollner et al. 2007) and the mixed aggregations of *Lamellibrachia luymesi* and *Seepiophila jonesi* (Bergquist et al. 2003, Cordes et al. 2005b). The so-called Bushmaster junior/ senior or Chimneymaster are hydraulically actuated collection devices of different sizes that are placed over the tubeworm aggregations and closed at the surface of the substratum to collect whole communities. In contrast, there is a higher risk of potential loss of parts of the community by sampling with grabs (Sarrazin et al. 1999, Tsurumi & Tunnicliffe 2003).

Separation of fauna from sediment and distinction between different faunal size classes also varied considerably between studies and limits a direct comparison of community structure. Sarrazin & Juniper (1999) distinguished, without sieving the samples, between the class of macroscopic mega- and macrofauna and submacroscopic macro- and meiofauna. Other researchers used nets of different sizes to separate the macrofauna from the meiofauna, while animals contained on a net with a mesh aperture of either 250 µm (Govenar et al. 2002), 1 mm (Bergquist et al. 2003, Govenar et al. 2005, Gollner et al. 2007, Tsurumi & Tunnicliffe 2003) or 2 mm (Cordes et al. 2005b) were considered to be macrofauna. So far, the entire community of mega-, macro-, and meiofauna at species level has been studied only in *Riftia pachyptila* aggregations (Govenar et al. 2005, Gollner et al. 2007).

The two different measures of abundance further limit the possibilities of direct comparisons. Either the abundance of associated fauna was calculated per tube surface area of the foundation species (Sarrazin et al. 1999, Tsurumi & Tunnicliffe 2003, Cordes et al. 2005c, Govenar et al. 2005) or abundances of foundation species and associated animals were standardised to the surface area of the substratum (basalt, sulphide chimneys or sediment) (Govenar et al. 2002, Bergquist et al. 2003, Gollner et al. 2007).

Macrofauna associated with *R. pachyptila* aggregations were studied at the hydrothermal vent 9°50′N EPR region (Govenar et al. 2004, 2005). A total of 46 associated macrofauna species were collected from eight aggregations (Govenar et al. 2005) and a single aggregation (Govenar et al. 2004). An aggregation of *R. pachyptila* that grew within a year after clearing the area artificially revealed the co-occurrence of *Tevnia jerichonana* (Govenar et al. 2004); however, other samples taken from the same location and another location revealed the presence of *T. jerichonana* and *Oasisia alvinae*. The abundance of associated macrofauna was between 1723 and 8216 ind. m^{-2} tube surface area (Govenar et al. 2005). The tubes increased the surface available for associated animals between about 7- and 144-fold. Species richness ranged from 19 to 35 species per aggregation and was positively correlated with the tube surface area. Despite differences in age and physicochemical characteristics of the two sites studied, the structure and composition of associated macrofauna communities were remarkably similar, and one of the univariate measures of diversity, the Shannon-Wiener index, was low and between H′ log e 1.23 to 2.14.

In six of these eight aggregations, the meiofauna community was studied in detail (Gollner et al. 2007). Abundances were highly variable (<1 to 976 ind. 10-cm^{-2} surface area of basalt) and in the majority of samples were well below the average abundance values of meiofauna in deep-sea samples. Considering that the tube surface area increased the overall surface considerably, meiofauna can be considered comparatively rare in the habitat. Abundance was positively correlated with tube surface area and with the volume of sediment found within the bushes of tubeworms. A total of 33 species were found, increasing the overall number of macro- and meiofauna species in *Riftia pachyptila* aggregations to 79, with a ratio between macro- and meiofauna species of about 1.4:1. While the macrofauna communities were very similar at both sites, the meiofauna communities differed considerably, with H′ significantly higher at the older Riftia Field site (1.75 to 2.00) than at the younger Tica site (0.44 to 1.35). The average Bray-Curtis dissimilarity was almost 70%.

The associated fauna of *Ridgeia piscesae* aggregations from various locations of the Juan de Fuca Ridge was studied. Following the criteria of the community succession model of Sarrazin et al. (1997), detailed data are available for assemblages III and IV (one collection each Sarrazin &

Juniper 1999, Govenar et al. 2002), V-HF (one collection Sarrazin & Juniper 1999, three collections Govenar et al. 2002) and V-LF (one collection Sarrazin & Juniper 1999, one collection Bergquist et al. 2007). In addition, Tsurumi & Tunnicliffe (2003) based their analyses on 51 collections from a wide range of geographic sites and substratum types but did not use this classification. The abundance of associated macrofauna calculated per tubeworm surface area ranged from 200 to 50,000 ind. m^{-2} (Tsurumi & Tunnicliffe 2003), from 30,000 to about 250,000 ind. m^{-2} (Sarrazin & Juniper 1999), and from 12,000 to over 100,000 ind. m^{-2} (Govenar et al. 2002). Overall, the number of associated species in each aggregation had a large range (4–28), with a total species richness of 37 for the 51 samples studied (Tsurumi & Tunnicliffe 2003). The Shannon–Wiener index of species diversity H' reported in earlier studies falls within the range 1.34–2.19 (Sarrazin & Juniper 1999) and 0.90–1.31 (Govenar et al. 2002) and corroborates a general trend of lower abundance and lower diversity on relatively short-lived sulphide edifices than on more stable basalt (Tsurumi & Tunnicliffe 2003, Bergquist et al. 2007).

At cold-seep sites on the upper Louisiana slope of the GoM, the macrofauna associated with tubeworm aggregations of mixed *Lamellibrachia luymesi* and *Seepiophila jonesi* was studied (Bergquist et al. 2003, Cordes et al. 2005b). The tube surface area increased the sediment surface from which the tubeworms grew 2.6- to 26-fold (Bergquist et al. 2003). After studying seven aggregations, the total number of associated species found was 65 (Bergquist et al. 2003). Later, by studying 13 additional aggregations, this number increased to at least 90 species (Cordes et al. 2005b). A single aggregation contained 22–44 (Bergquist et al. 2003) and 14–47 species (Cordes et al. 2005b). Species richness was correlated with habitat size (measured as tube surface area) (Cordes et al. 2005b). The Shannon–Wiener diversity index H' ranged from 1.39 to 2.86 (Bergquist et al. 2003) and 1.37 to 3.10 (Cordes et al. 2005b). Applying a population growth model, Cordes et al. (2005b) estimated the age of the 13 studied aggregations to be between 8 and 157 yr. Faunal abundances decreased with increasing age of aggregations, species considered to be seep endemics dominated communities of young aggregations, while generalist species also known from the surrounding deep sea dominated older aggregations. Diversity was not linearly correlated with aggregation age; the lowest diversity was found in the youngest and oldest aggregations. These studies on the associated community together with laboratory experiments and modelling approaches suggest that *Lamellibrachia luymesi* not only provides habitat for an associated fauna but also can alter the biogeochemistry of the seep sites and reduce sulphide levels in the water around their tubes (Julian et al. 1999, Freytag et al. 2001, Bergquist et al. 2003, Cordes et al. 2003, 2005a,c).

Nutritional links between tubeworms and associated fauna

Lethal predation by killing and consuming whole large tubeworms has not been documented so far. However, considerable numbers of gametes and larvae released into the water column might be directly consumed in the pelagic environment (Bergquist et al. 2003); settled larvae and small juveniles might fall prey to mobile grazers (Micheli et al. 2002).

Partial predation by 'nipping' plume parts was observed directly by many scientists in the vent species *Riftia pachyptila* and *Ridgeia piscesae* (see Micheli et al. 2002, M.B. personal observations). In contrast, there is no evidence for predation in seep species (Bergquist et al. 2003, Cordes et al. 2007a). Feeding preference experiments showed that the brachyuran crab *Bythograea thermydron* and the galatheid crab *Munidopsis subsquamosa* were attracted by *Riftia pachyptila* as food and not as a biogenic structure (Micheli et al. 2002). Galatheid crabs and polynoid polychaetes (and possibly also the spider crab *Macroregonia macrochira*) forage on plumes of *Ridgeia piscesae* (Tunnicliffe et al. 1990, Juniper et al. 1992). Stable isotope ratios of crabs were consistent with a diet including *Riftia pachyptila* (Fisher et al. 1994). Similar studies identified *Ridgeia piscesae* as part of the diet of several polynoid polychaetes (Bergquist et al. 2007), whereas no predator was found to feed directly on *Lamellibrachia luymesi* and *Seepiophila jonesi* (MacAvoy et al. 2005).

Several parasitic siphonostomatoid copepods with typical mouth structures suitable for cutting round holes into host tissue were found associated with, and were thought to feed on, tubeworms (see Ivanenko & Defaye 2006). For example, several *Ceuthocetes* species were identified from aggregations of *Riftia pachyptila* (Gollner et al. 2006); *Dirivultus dentaneus* and *D. spinigulatus* were found associated with *Lamellibrachia barhami* and *Paraescarpia echinospica*, respectively (Humes & Dojiri 1980, Humes 1988, Southward et al. 2002). Other species that may feed on tubeworms are the phyllodocid polychaetes *Galapagomystides aristata* from vents and *Protomystides* sp. from seeps found associated with *Riftia pachyptila* and *Escarpia laminata*, respectively. These phyllodocids contained blood in their digestive systems, which was thought to stem from their tubeworm hosts (Jenkins et al. 2002, Cordes et al. 2007b).

Life cycle and symbiont transmission

In Vestimentifera, all evidence points to a biphasic life cycle with a pelagic larva and a benthic adult such as is common in many marine invertebrates. However, this pelago-benthic cycle is complicated by the uptake of the specific symbiont and the subsequent transformation from an aposymbiotic larva to a symbiotic entity comprised of a host with a specific endosymbiotic phylotype belonging to the subdivisison of *Gammaproteobacteria* (but also see 'Symbionts', p. 233). To take these overlapping but not identical life-cycle phases and the key events of settlement and symbiont transmission into account, tubeworms are considered in this review to have an aposymbiotic phase (fertilised egg within mother, embryonic and pelagic larval development, settlement of larva and metamorphosis) and a symbiotic phase (symbiont transmission in metamorphosing larva, juvenile and adult phase ending with the death of the individual) (Figure 5).

Further, at least four phases of nourishment can be distinguished: (1) lecithotrophy in the pelagic environment (Marsh et al. 2001), (2) microphagy after settlement (Southward 1988, Jones & Gardiner 1989, Nussbaumer et al. 2006) and maybe even in the late pelagic phase, (3) both microphagy and symbiotic nutrition and (4) only symbiotic nutrition via translocation of metabolites and direct digestion of symbionts (Bosch & Grassé 1984a, Felbeck & Jarchow 1998a, Bright et al. 2000).

Aposymbiotic phase

Spermatozoa and oocytes

Vestimentiferans are gonochoric with paired gonads (see Gardiner & Jones 1993, Southward 1999, Southward et al. 2005). The sex ratio was estimated to be 1:1 in several species (Young et al. 1996, Thiebaut et al. 2002). No evidence for periodicity in reproductive output was reported (see Tyler & Young 1999). All attempts to detect symbionts in eggs and ovaries of *Riftia pachyptila* and *Ridgeia piscesae* using *in situ* hybridisation techniques have failed (Cary et al. 1993, M.B. unpublished data). Also, testes of *Riftia pachyptila* did not contain symbionts (M.B. unpublished data).

According to Franzén (1956), the spermatozoa of all vestimentiferan species investigated so far are classified as 'modified sperm type'; following the more recent classification of Rouse & Jamieson (1987), such spermatozoa are defined as entaqua-sperm (van der Land & Nørrevang 1977, Gardiner & Jones 1985, 1993, Jones & Gardiner 1985, Cary et al. 1989, Southward 1993, Marotta et al. 2005). For this sperm type, release into the surrounding water is suggested, and from there they reach the female in some way (see Rouse 2005).

Mature oocytes are quite small: *Riftia pachyptila* eggs in the gonad have a diameter of 78 μm (Jones 1981) but are 105 μm after spontaneous release onboard various research vessels at ambient pressure (Cary et al. 1989), *Ridgeia piscesae* eggs are 90–100 μm (Southward et al. 1995), *Seepiophila jonesi* (as *Escarpia* sp.) are 115 μm, *Lamellibrachia luymesi* (as *Lamellibrachia* sp.) are 105 μm (Young et al. 1996) and *L. satsuma* fertilised eggs are 100 μm (Miyake et al. 2006).

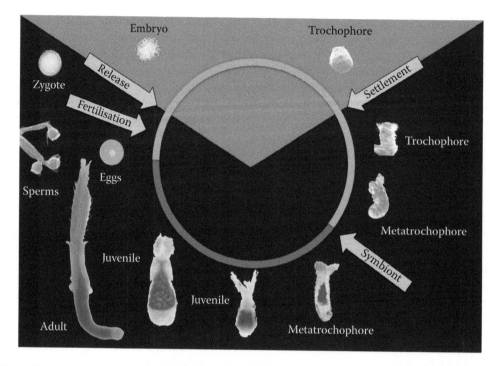

Figure 5 (See also Colour Figure 5 in the insert.) Schematic life cycle of Vestimentifera. Adults with separate sexes produce sperm and eggs. Sperm bundles are released and taken up by females, in which internal fertilisation takes place, and aposymbiotic zygotes are released into the water column and disperse. Embryonic and larval development into a trochophore takes place in the pelagic environment. On settlement and further larval development into a metatrochophore, metamorphosis and uptake of symbionts from the environment are initiated. The symbiotic metatrochophore with a trophosome develops into a small juvenile in which the trophosome is present as a one-lobule stage. On growth, the trophosome expands into a multilobule stage, and animals become mature. Green = aposymbiotic life stages; red = symbiotic stages (photographs of sperm bundles from Marotta et al. 2005; zygote, embryo, and pelagic trochophore from Marsh et al. 2001; sessile trochophore from Gardiner & Jones 1994).

Fertilisation and spawning

Internal fertilisation was first suggested by Gardiner & Jones (1985) and later by other authors (Malakhov et al. 1996a, Hilario et al. 2005). All studies were based on the finding of sperm in the female genital tract of *Riftia pachyptila* (Gardiner & Jones 1985), *Ridgeia piscesae* (Malakhov et al. 1996a), *Lamellibrachia satsuma* (Miyake et al. 2006) and *Riftia pachyptila*, *Ridgeia piscesae*, *Tevnia jerichonana*, *Lamellibrachia luymesi* and *Seepiophila jonesi* (Hilario et al. 2005). In *Ridgeia piscesae*, Southward & Coates (1989) had speculated that fertilisation should be either internal or just external to the female gonopores.

The findings of some sperm and eggs outside the body of vestimentiferans (Southward & Coates 1989, Southward 1999, MacDonald et al. 2002) as well as occasional *in situ* observations of expulsion of some products into the water column (Van Dover 1994, Hilario et al. 2005, M.B. personal observations) are in accordance with an internal fertilisation mode. As pointed out by Hilario et al. (2005), such exudates could be spermatozoa, sperm bundles, unfertilised eggs, zygotes, developing embryos or even larvae as they never have been collected directly.

The mode of release, however, apparently differs between tubeworm species. 'Spawning' events in the form of forceful expulsions have been observed in *Riftia pachyptila* (Van Dover 1994, M.B. personal observations). As described by Van Dover (1994) and captured by coincidence in a video

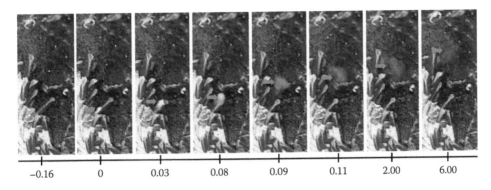

Figure 6 (See also Colour Figure 6 in the insert.) Video sequence of the expulsion of a white cloud from one individual of *Riftia pachyptila* at the East Pacific Rise 9°50′N region, location Tica, in December 2003, Alvin dive 3948. Red arrows mark the location of the released product. Time is given in seconds as ± from the time of release (0).

sequence (Figure 6), small, white clouds of 10- to 15-cm diameter were released forcefully. When originating from females, the presumed eggs were slightly negatively buoyant and sank between tubeworms, while such clouds of sperm released from males were neutrally buoyant and dispersed within 10 s in the water column. In *Ridgeia piscesae* (Southward & Coates 1989, MacDonald et al. 2002) and *Tevnia jerichonana* (Southward 1999), however, the transfer of spermatozoa appeared to be as clumps. These thread-like white objects obviously do not dissolve immediately after release and sink between other tubeworms so that they can be collected. They were identified as sperm masses (Southward & Coates 1989) found on plumes, the dorsal side of the obturaculum and vestimentum and sometimes even on the female gonopores or in the ovisacs. In addition, released eggs were found on plumes (MacDonald et al. 2002).

Embryonic development

Based on an *in situ* experiment, embryogenesis is suggested not to begin until inseminated oocytes are released into the water column (Hilario et al. 2005). Using artificial insemination techniques, larvae of *Lamellibrachia luymesi, Seepiophila jonesi* (Young et al. 1996), *Lamellibrachia satsuma* (Miura et al. 1997, Miyake et al. 2006) and *Riftia pachyptila* (Marsh et al. 2001) were reared and studied with light microscopy (LM) and scanning electron microscopy (SEM) techniques. In addition, some information based on transmission electron microscopy (TEM) sections is available for *Lamellibrachia satsuma* (Miyake et al. 2006). *Lamellibrachia luymesi* and *Seepiophila jonesi* were found to develop similarly at 1, 50 and 100 bar, 9°C. *Riftia pachyptila* was reared at 250 bar, 2°C. The shallow-water species *Lamellibrachia satsuma* was maintained at ambient pressure, 16°C.

In all species, the cleavage pattern was spiral. The development into ciliated larvae took only 3 days in *L. luymesi* and *Seepiophila jonesi* (Young et al. 1996), 5 days in *Lamellibrachia satsuma* (Miyake et al. 2006) and 34 days in *Riftia pachyptila* (Marsh et al. 2001).

Aposymbiotic, pelagic trochophore

The larvae of all these artificially reared species had a trochus in the anterior body region, interpreted as a prototroch. A second ciliary band, located posteriorly to the prototroch, was described in *Lamellibrachia luymesi, Seepiophila jonesi* (Young et al. 1996) and *Riftia pachyptila* (Marsh et al. 2001) but was lacking in *Lamellibrachia satsuma* (Miura et al. 1997, Young 2002). Following the progress of development in detail, 15-day-old larvae of *Lamellibrachia* sp. exhibited a wide prototroch (Young 2002). After 21 days, two ciliary bands had developed (C.M. Young personal communication). Whether these bands represented the prototroch and the metatroch or a wide prototroch that later separated into two is not yet clear. No apical plate, mouth opening or anus was detected in any species, although Young (2002) suggested that in the 21-day-old larva of *Lamellibrachia* sp. a mouth might be present. A telotroch was also described in both *Lamellibrachia* species (Miura et al. 1997, Young 2002, Miyake et al. 2006). According to Rouse (1999), these larvae (with a prototroch, protonephridia and an apical plate, which is lost in some) are definitely of the trochophore type.

Dispersal and settlement

No larvae such as those described have ever been found in the water column despite intensive searches (Mullineaux et al. 2005). However, Harmer et al. (2008) detected a tubeworm genetic sequence in a pelagic water sample taken in the vicinity of vents at the EPR 9°50′N region. *In situ* hybridisation of 18S rRNA techniques to identify larvae have been developed (Pradillon et al. 2007), but appropriate material has yet to be made available for testing.

Artificially reared *Riftia pachyptila* larvae were estimated to disperse more than 100 km along the axial summit collapse trough over a period of 38 days, passively for about the first 3 wk and then actively using their cilia for the next 2 wk (Marsh et al. 2001). This calculation was based on the assumption that these larvae are exclusively lecithotrophic during their entire pelagic phase, based on measurements of protein and lipid content, respiration rates and a model of the flow regime along the ridge axis of the 9°50′N EPR region. Modelling the flow regime in two different regions of the EPR using current records showed that the dispersal potential of *R. pachyptila* can be quite different even when the average daily current speed is similar. Maximal dispersal distance at 13°N was 241 km but only 103 km at 9°50′N (Mullineaux et al. 2002).

Artificially reared *Lamellibrachia satsuma* larvae could be maintained in the laboratory for 45 days (Miyake et al. 2006). They swam in Petri dishes for about a month when no suitable substratum was offered. They attached rapidly to any substratum offered 10 days after fertilisation; however, there was no further development. With a current speed model, Miyake et al. (2006) estimated that larvae can disperse about 2000 km in the Kuroshio Subgyre area.

Knowledge is scarce on habitat selection, spatial and temporal recruitment and settlement processes during which vestimentiferan larvae move to the substratum, explore, attach and begin their benthic life (Qian 1999). Regardless of inhabiting soft sediments (seeps or vents) or hard substrata (basalt, sulphide chimneys at vents or whale carcasses), vestimentiferans appear to need a hard substratum on which to settle. Vestimentiferan tubeworms also recruit to artificial substrata as scientists have often noted using many experimental devices made of plastic, glass or ropes deployed at locations with appropriate environmental conditions. At sedimented seeps, larvae are suggested to settle on carbonate rocks, the by-products of hydrocarbon degradation (Fisher et al. 1997). Only *Paraescarpia echinospica* was also collected from mud without obvious hard substratum being present (Southward et al. 2002). At vents, small specimens are found on basalt but also very often on tubes of larger specimens. This shows that recruitment to those places obviously is possible, but not that larvae are capable of selectively choosing this location. Discontinuous and synchronised recruitment within a vent site was reported for *Riftia pachyptila* based on size–frequency histograms (Thiebaut et al. 2002). However, a different population structure study using size–frequency histograms came to the conclusion that recruitment occurred throughout the year (Govenar et al. 2004).

The co-occurring vent species *Tevnia jerichonana*, *Riftia pachyptila* and *Oasisia alvinae* were the subject of several studies. Deploying artificial substrata at sites with different environmental conditions at the 9°50′N EPR vent region during a period of 3 yr, two of five substrata were colonised by small tubeworms located in 'warm' vent areas (Mullineaux et al. 1998). Subsequently, basalt blocks were deployed in this region in three different tubeworm aggregations. Identifying the settled tubeworms by molecular methods, it was found that all three species colonised the experimental blocks regardless of the dominance of tubeworm species in the initial assemblage and independent of temperature measured on the blocks. *Riftia pachyptila* and *Oasisia alvinae* were never found on blocks without *Tevnia jerichonana*, but *T. jerichonana* colonised blocks also without the other two species, even when these blocks were deployed in *Riftia*-dominated clumps. This pattern strongly supported the facilitation/competition hypothesis suggesting that *Tevnia jerichonana* facilitates colonisation of *Riftia pachyptila* and *Oasisia alvinae* (Mullineaux et al. 2000). An earlier long-term photo- and videographic documentation of community succession in this area showed that at the macroscopic level, most often *Tevnia jerichonana* clumps preceded the establishment of *Riftia pachyptila*; however, *R. pachyptila* clumps without pre-existing *Tevnia jerichonana* were also documented (Shank et al. 1998). Also, the findings of many small individuals of *Riftia pachyptila* settled on tubes of larger conspecific but not on *Tevnia jerichonana* tubes are in contrast to the hypothesis (Thiebaut et al. 2002).

In the following years, another experiment was conducted to test the facilitation/competition hypothesis at a later successional community stage (Hunt et al. 2004). When using tubes of *T. jerichonana* or *Riftia pachyptila* or artificial plastic tubes glued to blocks and deployed for several months, the species pattern of colonisation differed from previous studies. Colonists of *R. pachyptila* and *Oasisia alvinae* were found on blocks without *Tevnia jerichonana*. Further, no difference in the number of colonists between blocks with natural and artificial tubes was found, although it is important to note that tubes of live *Riftia pachyptila* touched all blocks due to the specific deployment in *R. pachyptila* aggregations. The authors suggested that *Tevnia jerichonana* might be important for colonisation of new vents and young developing communities, while *Riftia pachyptila* might act as settlement cue in later successional community stages. Predator exclusion experiments using the same blocks, caged to exclude predators, uncaged or controls with cages having one side open, revealed that tubeworms colonised not only cages in the vestimentiferan zone, as expected, but also the lower-flow bivalve zone. Colonisation in the vestimentiferan zone was statistically similar regardless of treatment after 5 and 8 mo (Micheli et al. 2002).

Aposymbiotic, sessile trochophore

The smallest sessile larval stage collected from washing of large *Ridgeia piscesae* aggregations was only 58 μm long. Based on the prototroch and two types of larval chaetae (ciliary and hooked uncini), it was identified as a trochophore. No metatroch, neurotroch, telotroch or apical organ was detected. Also, neither mouth opening nor anus was apparent when examining this specimen using SEM techniques (Jones & Gardiner 1989).

Aposymbiotic, sessile metatrochophore

The next developmental stage can be classified as a segmented larva (metatrochophore) according to Heimler (1988). This stage is characterised by an elongated hyposphere, larval segments and adult organs such as head appendages being formed. The smallest of these metatrochophore larvae were aposymbiotic. Two specimens 150 and 270 μm long (Southward 1988) and two specimens 200 and 250 μm long (Nussbaumer et al. 2006) were studied in detail (Figures 7A,B, 8A–C). Although named 'juvenile' by Southward (1988) and Jones and Gardiner (1989), such specimens of *Ridgeia piscesae* (Southward 1988, Jones & Gardiner 1989), *Riftia pachyptila* (Jones & Gardiner 1989) and unidentified vestimentiferan tubeworms from the 9°50′N EPR region (co-occurring *R. pachyptila*, *Tevnia jerichonana* or *Oasisia alvinae*) (Nussbaumer et al. 2006) exhibited a prototroch. They had a transient digestive system with mouth and anus, cilia at the position of a neurotroch, two palps also termed tentacles (Southward 1988, Nussbaumer et al. 2006) or branchial filaments (Jones & Gardiner 1989) and larval chaetae termed setae by Jones & Gardiner (1989). No trophosome was present. Feeding on bacteria (Southward 1988) or bacteria and diatoms (Nussbaumer et al. 2006) was evident from remnants found in endodermal midgut cells.

Adopting the terminology of Rouse & Fauchald (1997), the body is divided into (1) a presegmental region (termed cephalic region in Southward 1988) with a prostomium containing the brain and a peristomium, the area surrounding the mouth also containing a corresponding unpaired, minute peristomial coelomic cavity and the prototroch, and (2) a segmental region with either two or three

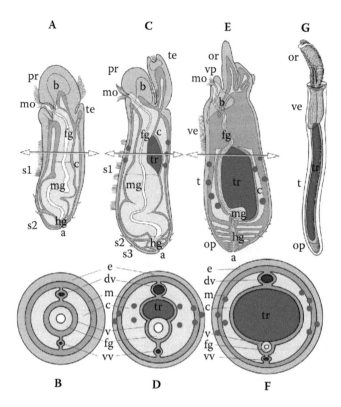

Figure 7   (See also Colour Figure 7 in the insert.) Schematic drawing of life-history stages: (A) aposymbiotic metatrochophore with (B) corresponding cross section; (C) symbiotic metatrochophore with symbionts invading and (D) corresponding cross section; (E) symbiotic juvenile with trophosome in one-lobule stage and (F) corresponding cross section; (G) adult with trophosome in multilobule stage. Pink, symbiont-housing trophosome or symbionts (tr); blue, digestive system, mouth opening (mo), ventral process (vp), foregut (fg), midgut (mg), hindgut (hg) and anus (a); purple, blood vascular system with dorsal blood vessel (dv) and ventral blood vessel (vv). Body regions of larvae: pr = prototroch; s1–3 = chaetigers 1 to 3; and te = palps. Body regions of juvenile and adult: or = obturacular region; ve = vestimentum; t = trunk; op = opisthosome. Tissues: e = epidermis; m = muscles; c = coelom; v = visceral mesoderm. (From Nussbaumer et al. 2006.)

segments and, correspondingly, paired coelomic cavities and larval chaetae. Only the first two segments were chaetigers having four sets of larval chaetae each. The postsegmental region, the pygidium, including the growth zone at the posterior end of the body, was minute and quite indistinct.

While Nussbaumer et al. (2006) clearly described the palps developing from the first segment, confirming Rouse's (2001) interpretation based on the fact that palps are located behind the prototroch, Southward (1988) speculated that they either originate from the presegmental region (first segment in Southward 1988) or the first segment (second segment in Southward 1988). Both Southward (1988) and Nussbaumer et al. (2006), however, clearly saw the elongated region, which on symbiont transmission will develop into the trunk containing the trophosome, as a single, highly elongated segment.

Symbiont transmission

Symbionts

In investigations to date, the symbionts of vestimentiferan tubeworms have been detected by molecular methods free living in the benthic and pelagic environments and found associated with the developing tube of small individuals (but not large ones), in the skin of the metatrochophore larvae and juveniles and in the trophosome of metatrochophore larvae, juveniles and adults.

The majority of molecular studies of the identity of the symbionts revealed that each host species houses a single, specific endosymbiont (but see next paragraph). Based on 16S rRNA sequences, two phylotypes of *Gammaproteobacteria* with 4.3% sequence divergence are known (Feldman et al. 1997, Di Meo et al. 2000, Nelson & Fisher 2000, McMullin et al. 2003, Thornhill et al. 2008, Vrijenhoek 2010). No information is yet published for the symbionts of the host genera *Arcovestia*, *Alaysia* and *Paraescarpia*. Phylotype 1 has a larger geographic distribution (Atlantic, Pacific, GoM), diversity of host species (species of *Lamellibrachia*, *Escarpia*, *Seepiophila*, *Arcovestia* and *Alaysia*; S. Johnson unpublished observations) and habitat type (seeps, vents, one example whale fall) than phylotype 2. Three groups of symbionts can be distinguished in phylotype 1. Group 1 contains the symbionts of *Lamellibrachia barhami* from the Oregon Slope and Middle Valley, *Escarpia laminata* from Florida Escarpment and *Lamellibrachia columna* from Lau Basin. The symbionts of group 2 are present in *Lamellibrachia* sp. and *Escarpia laminata* from Alaminos Canyon, *Lamellibrachia barhami* from Vancouver Margin, Monterey Bay, and *Escarpia spicata* from a whale fall in Santa Catalina, California. Group 3 symbionts are found in *Seepiophila jonesi*, *Lamellibrachia* cf. *luymesi* and a new escarpid species from the Louisiana slope. Phylotype 2 has been found so far only in the East Pacific vent species *Ridgeia piscesae* and the co-occurring *Riftia pachyptila*, *Tevnia jerichonana* and *Oasisia alvinae*. The genome of this symbiont, Candidatus *Endoriftia persephone* from *Riftia pachyptila*, has been published (Robidart et al. 2008).

Two studies claimed that several symbiotic phylotypes were extracted from tubeworm trophosomes. Two sequences of *Epsilonproteobacteria* were obtained from *Lamellibrachia* sp. from a cold seep in the Sagami Bay and visualised in the trophosome by *in situ* hybridisation (Naganuma et al. 1997a,b). Later, a picture was published (Figure 3 in Naganuma et al. 2005) in which the different types of these presumed bacteria were, in the authors' opinion, sperm of the testes. Using terminal-restriction fragment length polymorphism (t-RFLP), *Ridgeia piscesae* collected from Juan de Fuca Ridge, Axial Caldera, Explorer Ridge and Magic Mountain were found to contain either only the previously known γ-proteobacterial endosymbiont or up to four additional operational transcribed units belonging to other γ-proteobacterial clusters, *Alphaproteobacteria*, and Cytophaga-Flavobacterium-Bacteroides. However, no fluorescent *in situ* hybridisation (FISH) was applied in this study, and in our opinion a conclusive proof for multiple symbionts remains to be established.

Recently, the symbiont of the vent tubeworms was detected free living in the environment at the EPR 9°50′N sites. Natural collections of basalt, artificially deployed rocks and glass slides from various locations within and next to tubeworms, but also about 100 m away from the axial summit collapse trough on bare basalt, and water samples taken about 1 m away from tubeworm aggregations with a pelagic pump revealed the presence of the free-living stage (Harmer et al. 2008). Various growth types, such as rods and chains of rods, were stained with a symbiont-specific probe applying *in situ* hybridisation (Harmer et al. 2008, Nussbaumer personal communication) (Figure 8A).

A diverse bacterial community of *Epsilon-*, *Alpha-*, *Delta-*, and *Gammaproteobacteria* inhabits the tube of adult *Riftia pachyptila*. The specific symbiont was not part of this community (Lopez-Garcia et al. 2002, Nussbaumer et al. 2006). The developing tube of small specimens, however, also harbours the symbiont (Nussbaumer et al. 2006) (Figure 8B). This finding suggests that prior to symbiont uptake, the free-living symbionts aggregate in the developing tube and remain there only for a certain time.

For a horizontally transmitted bacterium, symbiotic life is facultative. Such bacteria exhibit a population in the environment as well as a population associated with the host. The free-living population serves as an inoculum for the symbiosis. Symbiont release mechanisms from the host into the environment are quite common in symbiosis with horizontal transmission; however, in vestimentiferan tubeworms they have not been found yet. The free-living counterpart of the symbionts

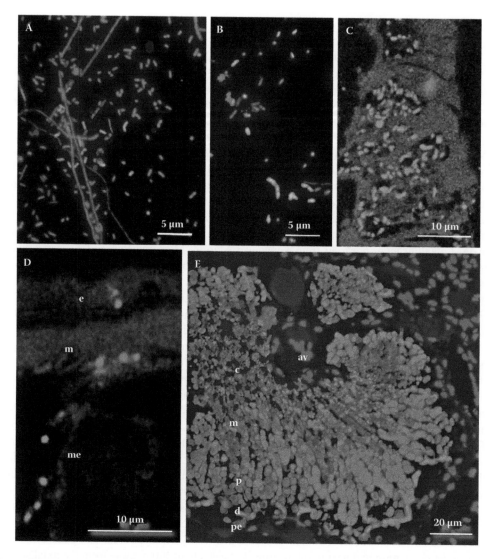

Figure 8 (See also Colour Figure 8 in the insert.) Fluorescent *in situ* hybridisation with symbiont-specific (pink) and eubacterial (blue) probes. (A) Free-living bacterial community containing symbionts (pink) colonising on glass slide during 1-yr deployment (courtesy of A.D. Nussbaumer). (B) Free-living bacterial community containing symbionts (pink) on and in developing tube of metatrochophore (courtesy of A.D. Nussbaumer). (C) Symbionts (pink) in developing trophosome in metatrochophore. (D) Symbionts (pink) in epidermis (e), muscles (m), and undifferentiated mesoblastem (me). (E) Juvenile one-lobule stage with symbionts (pink) and host nuclei (blue): av = axial blood vessel; distinct zonation of central (c), median (m), peripheral (p) and few degrading bacteriocytes (d), peritoneum (pe) (C, D and E from Nussbaumer et al. 2006).

is part of the free-living microbial community, and selection pressure relies on successful competition under present environmental conditions. At the same time, such a versatile bacterium exhibits the necessary repertoire for associating with their hosts. The metagenome of Candidatus *Endoriftia persephone* of *Riftia pachyptila* displays remarkable motility (Hughes et al. 1998) and chemoreception, suggesting that the free-living counterparts of these endosymbionts can use chemotaxis to reach their respective hosts (Robidart et al. 2008). The horizontally acquired microbes do not experience a significant selective pressure for genome reduction, as opposed to strictly vertically transmitted symbionts, which do not exist in a free-living state. The genome of Candidatus *Endoriftia perse-phone* is 3.3 Mb, and its guanosine plus cytosine (GC) content is 60% (Robidart et al. 2008).

Transmission

Evidence for horizontal transmission came from series of sessile developmental stages and was also supported by the finding of the free-living symbiont in the environment (Harmer et al. 2008). The locations of entry for bacteria and the tissue developing into the trophosome were hypothesised to be either the mouth and the endodermal midgut (Southward 1988, Jones & Gardiner 1989) or the skin and the visceral mesoderm (Nussbaumer et al. 2006). Consequently, the evolution of symbiosis is pictured quite differently. One interpretation is that a feeding larva ingests bacteria and encloses them in food vacuoles in the midgut. One bacterial species, however, evades digestion and starts proliferating in the endodermal cells, which as a consequence transform into the trophosome (Southward 1988, Jones & Gardiner 1989). Another explanation is that one bacterial species manages to invade the skin of the larva in a way similar to a pathogen infection process and migrates through several tissues (epidermis, musculature, undifferentiated mesoblastem) until it reaches the visceral mesoderm between dorsal blood vessel and gut, which initiates proliferating into the trophosome (Nussbaumer et al. 2006). While the former hypothesis is based solely on microscopy, identifying the occurrence and location of bacteria, the latter hypothesis also relies on FISH using symbiont-specific and general γ-proteobacterial and eubacterial probes.

Another line of evidence for horizontal transmission is the lack of phylogenetic congruence and cospeciation between symbiont and host, which would reflect cotransmission over evolutionary timescales. Furthermore, the time depth of diversification of host and symbiont phylogenies is often dissimilar. Vestimentiferans reveal no evidence of cospeciation when comparing symbiont 16S rRNA gene- and host COI-based trees (Feldman et al. 1997, Nelson & Fisher 2000, McMullin et al. 2003). The sequence divergence of the two symbiotic phylotypes suggests a separation over 200 million years ago (mya), whereas host radiation was possibly as recent as 60 mya (Vrijenhoek 2010).

Symbiotic phase

Symbiotic metatrochophore

The overall morphology of the only specimen studied (unidentified vestimentiferan from the 9°50′N EPR region) by reconstruction from serial TEM/LM sections revealed how the development from the aposymbiotic to the symbiotic metatrochophore proceeded. The symbiotic metatrochophore developed the first two palps, the peristomal coelomic cavity was closed, and a set of midventrally located patches of cilia in the first chaetiger developed (Nussbaumer et al. 2006) (Figures 7C,D, 8D, 9C,D). Also, the trophosome started to develop and was composed only of peritoneal cells and bacteriocytes containing rod-shaped symbionts. The digestive system was transient, composed of a slit-like mouth opening, a fore-, mid-, and hindgut and an anus. Apparently, the gut was still functioning, as evidenced by degrading bacteria and diatoms. A few symbionts, identified by FISH using a set of symbiont-specific, γ-proteobacterial and eubacterial probes, were found in the epidermis, the muscles and the undifferentiated mesoblastem in the trunk region containing part of the foregut (Figure 9D). In contrast to the symbionts in the prospective trophosome being contained in vacuoles, in the other tissues the symbionts are either intracellular but free in the cytoplasm or intercellular between various tissues (Nussbaumer et al. 2006). These findings suggest that the symbionts first aggregate in the developing tube, invade the host larva specifically in the trunk region and migrate through several tissues by entering and exiting host cells until they reach the prospective trophosome area where they establish and proliferate. Interpartner recognition processes, known to play a pivotal role in horizontally transmitted symbionts, have not been studied in this symbiosis yet. Further, symbionts invading and travelling within a host are usually subjected to a vigorous host

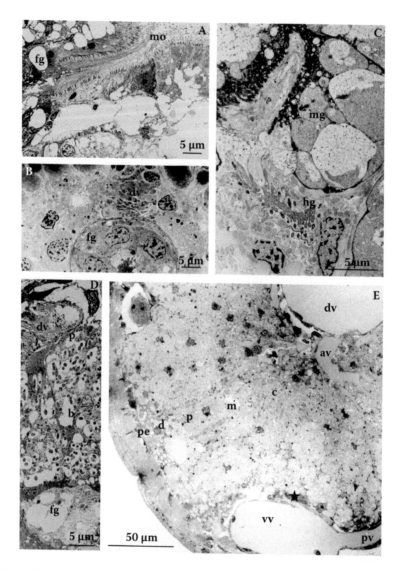

Figure 9 Transmission electron micrographs of aposymbiotic metatrochophore. (A) Mouth opening (mo) and foregut (fg). (B) Foregut (fg) and dorsal blood vessel (dv). (C) Midgut (mg) and hindgut (hg). (D) Symbiotic metatrochophore with dorsal blood vessel (dv), foregut (fg), and developing trophosome: bacteriocytes (b) containing rods and peritoneum (p). (E) Symbiotic juvenile with trophosome in one-lobule stage exhibiting axial blood vessel (av), distinct zonation of central (c), median (m), peripheral (p), and few degrading bacteriocytes (d), peritoneum (pe), and peripheral blood vessels (pv); note dorsal blood vessel extends to axial blood vessel and ventral blood vessel (vv) extends to peripheral blood vessels; star indicates foregut. (From Nussbaumer et al. 2006.)

immune response. Interestingly, Candidatus *E. persephone* possesses a wide repertoire of defence-associated genes (Robidart et al. 2008).

Metamorphosis

The onset of metamorphosis, the process during which larvae go through morphological and physiological changes to complete transition from the pelagic larva to the benthic juvenile, is known to commence before, concurrently or immediately after settlement (Qian 1999). The termination of metamorphosis is marked by the loss of the prototroch (Heimler 1988) and is accomplished in vent vestimentiferans (*Ridgeia piscesae* and unidentified specimens from the 9°50′N EPR) during the symbiotic phase. The exact onset of metamorphosis is not known as there is not enough information on the morphology of the planktonic and benthic trochophore available. Thus, the development of a functioning digestive system, for example, could either happen on settlement or already be present during the planktonic phase. The latter scenario would imply a switch from lecithotrophic to planktotrophic lifestyle and thus potentially a prolongation of the possible time period to reside and disperse in the water column. Further, delayed settlement has been suggested by Jones & Gardiner (1989) as the examination of small sessile specimens with SEM revealed quite a plasticity in occurrence of mouth, anus and other features related to the size (and possibly age?).

Symbiotic early juveniles

Previously unpublished observations by one of us (M.B.) have shown that, at the early juvenile stage of unidentified specimens from the 9°50′N EPR), a small, functioning trophosome with all cell cycle stages of bacteriocytes and associated aposymbiotic tissues is present (one-lobule stage) concurrently with a mouth opening on the tip of the so-called ventral process and a transient, still-functioning digestive system. The palps increase in number during development, and the vestimentum develops (Figures 7E,F, 8E, 9E).

This so-called one-lobule stage of the trophosome in such small juveniles resembles a single lobule of the typical multilobed organ of larger juveniles and adults. The entire one-lobule organ takes up most of the inner trunk region. The single central axial blood vessel is a direct extension of the dorsal blood vessel with a myoepithelium composed of myocytes, so-called non-bacteriocytes (epithelial cells) and central bacteriocytes. Most of the bacteriocytes are apolar and show a distinct zonation of central bacteriocytes containing rods, median ones with small cocci and peripheral ones with large cocci and a zone of degradation, similar to the multilobule stage, described by Bright & Sorgo (2003) (Figure 10A). The bacteriocytes are enclosed by a peritoneum. Numerous small blood vessels extend from the ventral blood vessel to surround and connect with the peritoneum,

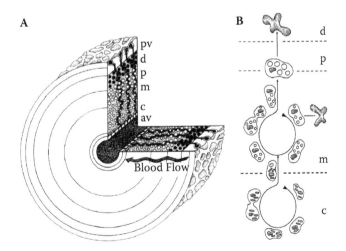

Figure 10 (A) Schematic drawing of one lobule of trophosome in larger juveniles and adult (from Bright & Sorgo 2003 with permission from John Wiley & Sons). (B) Schematic drawing of bacteriocyte cell cycle with terminal differentiation (from Pflugfelder et al. 2009 with permission from Springer Science): av = axial blood vessel, distinct zonation of central (c), median (m), peripheral (p), degrading bacteriocytes (d), peritoneum basal to peripheral blood vessels (pv); note blood flow is from periphery to the central region.

thus allowing vascular blood (VB) to circulate between these peritoneal cells but also into the inter-spaces between the bacteriocytes.

Termination of transmission was evidenced by massive apoptosis in the epidermis, somatic musculature and undifferentiated mesoblastem in the trunk region. Further symbiont uptake was evidently stopped by this apoptosis since in larger juveniles symbionts were never found in these tissues. It remains to be studied in detail if and how the host regulates this process.

Symbiotic late juveniles and adults

In large juveniles and adults, the digestive system ceases to function. Remnants of the gut, however, can still be found in juveniles of *Riftia pachyptila* with body lengths of at least a few millimetres (M.B., unpublished data). The trophosome was studied in detail in several species: *Lamellibrachia luymesi* (van der Land & Nørrevang 1975, 1977), *Riftia pachyptila* (Cavanaugh et al. 1981, Bosch & Grassé 1984a,b, Hand 1987, Gardiner & Jones 1993, Bright & Sorgo 2003), *Ridgeia piscesae* (Jones 1985, De Burgh 1986, De Burgh et al. 1989, Malakhov et al. 1996b) and several *Lamellibrachia* species (Kim & Ohta 2000).

The trophosome is multilobed in larger juveniles and adults instead of being composed only of a single lobule as in small juveniles. However, the cellular organisation is identical. Several ultra-structural findings led to the hypothesis of a bacteriocyte cell cycle with terminal differentiation, initially formulated by Bosch & Grassé (1984a,b) and later supported by Gardiner & Jones (1993) and modified by Bright & Sorgo (2003). Finally, using immunohistochemistry to identify the location and abundance of proliferating and mitotically active host cells, as well as cells undergoing apoptosis in *Riftia pachyptila* and *Lamellibrachia luymesi*, Pflugfelder et al. (2009) could verify the existence of such a cell cycle (Figure 10B). Bacteriocytes undergoing DNA synthesis and mitosis are restricted to central and median bacteriocytes but are completely lacking in the periphery; apoptosis is mainly in the periphery. Thus, the authors proposed that tissue-specific unipotent bacteriocyte stem cells containing rod-shaped symbionts undergo a cell cycle with division in the central lobule region. Semidifferentiated daughter cells containing small coccoid symbionts then migrate into the median region, where they either undergo another round of cell division or initiate programmed cell death. Migration of these differentiated cells containing large coccoid symbionts to the periphery is followed by the fully differentiated cells entering apoptosis, including digestion of the symbiont. As the diameter of each lobule is constant (resulting from cell proliferation in the centre being in balance with cell death in the periphery), net growth of the organ is accomplished by lobules growing in length or by ramification of lobules, both from proliferating cells in the centre and median regions (Pflugfelder et al. 2009; see next section, 'Growth and lifespan').

Growth and lifespan

Various morphological characters were used to measure tubeworm size. The tube length has proved to be a valid character in several species (e.g., *Lamellibrachia luymesi, Seepiophila jonesi, Tevnia jerichonana, Ridgeia piscesae*) since the worms consistently occupy the entire tube (Fisher et al. 1997, Shank et al. 1998, Bergquist et al. 2000, 2002, Govenar et al. 2002, Cordes et al. 2005b, 2007a).

In *Riftia pachyptila* tube length has been shown not to be a reliable measure of animal size. This species is capable of moving within its tube and usually occupies only a fraction of its tube (Gaill et al. 1997, Ravaux et al. 1998, 2000). Fisher et al. (1988), for example, found a tube 2 m long with an animal only 80 cm in length. Within the tube, sometimes chitinous septa are built (Gaill et al. 1997, Ravaux et al. 2000). Moreover, *R. pachyptila* is capable of contracting its trunk region considerably and thus decreases its overall length, so that especially in fixed material, the total length is an expression of the state in which the animal died (M.B. personal observation). Further, fixation in formalin or ethanol also leads to overall shrinkage of the tissue. Fisher et al. (1988) found that the only reliable size measure in *R. pachyptila* was the relation between the volume of the CF and the total wet weight. Thiebaut et al. (2002), in contrast, found that the length of the plume and that of the vestimentum as well as the width of the plume, vestimentum and opisthosome are unaffected by formalin fixation, although it affects total length. The width of the vestimentum was then used by these authors as a biometric index representative of the individual size. Moreover, they found that the tube length and vestimentum width were correlated. Govenar et al. (2004) found a strong negative correlation between the length an animal occupies in its tube and the tube length in one aggregation of *Riftia pachyptila* but pointed out that this relationship was unlikely to hold true for animals from other microhabitats.

Despite these limitations, growth rates have been estimated from the length of the tube, even in *R. pachyptila*. In time series studies taking *in situ* photographs, Lutz et al. (1994) estimated the increase in tube length at 85 cm yr^{-1}, thus making *R. pachyptila* the fastest-growing invertebrate. Monitoring tubeworm aggregations at the 9°50′N EPR region for several years posteruption, tubes of *R. pachyptila* exceeding 2 m in length were found 3.5 yr after their appearance. In contrast, the smaller *Tevnia jerichonana* was found to grow about 30 cm yr^{-1} (Shank et al. 1998). Another study to estimate growth was carried out using vestimentum width of three populations collected at the same vent site (Riftia Field, 9°50′N EPR) over a period of 22 days. Although changes in environmental conditions, known to influence the growth rate, cannot be neglected as pointed out by the authors, they proposed that the differences in size–frequency histograms revealing several cohorts are due to growth and aging of the population since the time period considered in this study was short and the samples came from the same vent field. They estimated a mean increase in vestimentum width of 0.193 mm d^{-1}. To compare these growth rates with the one using time series photographs (Lutz et al. 1994), Thiebaut et al. (2002) used the relation between vestimentum width and tube length, which was found to be correlated, and estimated a tube growth rate of as much as 160 cm yr^{-1}.

Direct measurement of growth using cable ties as a banding device for individual tubeworms as well as staining the tubes of entire tubeworm aggregations have been extremely useful. The increase of growth was monitored with cable ties and video analyses in two cold-seep species over a period of 4 yr. The majority of *Lamellibrachia luymesi* grew very little in at least one of these 4 yr, on average only 0.77 cm yr^{-1}, while *Seepiophila jonesi* did not grow at all (Fisher et al. 1997). *In situ* staining and measurements of tube-length increase over time confirmed earlier studies. Large individuals with tube lengths over 2 m of *Lamellibrachia luymesi* are estimated to be older than 200 yr (Bergquist et al. 2000, 2002). Growth in this species was best described by a model of declining growth rates with individual size (Cordes et al. 2005a). *Seepiophila jonesi* also grows extremely slowly and reaches ages comparable to *Lamellibrachia luymesi*, as established with a growth model including a size-specific probability of growth and an average growth rate that does not vary with individual size (Cordes et al. 2007a). In accordance with the wide range of environmental conditions under which *Ridgeia piscesae* can thrive and develop different growth forms, it is not surprising that growth rates can also vary from as much as 95 cm yr^{-1} (Tunnicliffe et al. 1997) to as little as 3 mm yr^{-1} in the long-skinny morphotype (Urcuyo et al. 2003, 2007).

A different approach was used by estimating cell proliferation and cell death in the epidermis applying immunohistochemical techniques as a measure of animal growth in *Riftia pachyptila* and *Lamellibrachia luymesi* (Pflugfelder et al. 2009). Interestingly, proliferation activities in both species are higher than in any other invertebrate studied so far and are comparable only to wound-healing processes or tumour growth. While in *Riftia pachyptila* cell death is downregulated, leading to fast growth, in *Lamellibrachia luymesi* it is upregulated, leading to a fast turnover of tissues and rejuvenation, which might be important for this extremely long-lived species.

Physiology and host–symbiont relationships

The originality of the Vestimentifera as far as functional physiology of the symbiosis is concerned relies on the following paradoxical situation:

- Symbionts are acquired from the environment, implying they are able to live freely in the vent or seep environments and suggesting their association with the vestimentiferan is facultative, although they are obviously growing well within their host.
- But for the hosts, it is an obligatory symbiosis, at least at the juvenile and adult stages (see 'Symbiotic phase', pp. 235–239), with bacteria remotely located within an internal organ, thus prompting sophisticated adaptations of the host systems to provide the bacteria with the inorganic, and potentially toxic, molecules they need.
- Or, as stated by Fisher et al. (1989): "The host tube-worms reap the benefits of an auto-trophic life style, while providing their symbionts with an environment which free-living sulfide-oxidising bacteria can only regard with envy".

To our knowledge, nobody has been able to cultivate these bacteria so far, strongly limiting physiological studies and suggesting a profound transformation on the onset of symbiosis. Yet, the trophosome of vestimentiferan tubeworms is actually an extremely efficient organ to cultivate these bacteria, with tremendous yields reaching 10^9 to 10^{11} bacterial cells per gram of tissue (Powell & Somero 1986). In this section, what is known about the metabolic requirements and performance of the bacterial symbionts, including recent insights from genomic studies, is first reviewed. The host adaptations to fulfil these requirements and take advantage of the symbiosis are then examined, and responses to additional environmental constraints within the hydrothermal vents and cold-seep ecosystems are considered. Over the last decade, and starting from the state-of-the-art situation depicted in the review by Childress and Fisher (1992), the processes by which metabolites (carbon, sulphide, oxygen, nitrogen) and waste products (protons, sulphate) cycle among environment, host and symbionts and the corresponding enzymes involved in both host and symbiont metabolism have been progressively, yet incompletely, unravelled. Figures 11 and 12 summarise these findings for vent (*Riftia pachyptila*) and seep (*Lamellibrachia luymesi*) model tubeworms and should be referred to throughout this section.

Metabolic requirements and performance of the symbionts

The initial description of symbionts of *Riftia pachyptila* inferred their chemoautotrophic potential from microscopy (Cavanaugh et al. 1981), biochemistry (Felbeck 1981) and stable isotope ratios (Rau 1981). The symbionts looked like thiotrophic free-living gram-negative bacteria with sulphur inclusions (Cavanaugh et al. 1981), the trophosome tissue demonstrated highly significant adenosine triphosphate (ATP) sulphurylase and RuBisCO (ribulose 1,5-bisphosphate carboxylase/oxygenase) activities (Felbeck 1981), with peculiar stable isotope signatures for ^{13}C (Rau 1981), *R. pachyptila* tissues being less depleted in ^{13}C than other thiotrophic symbioses. Mussels and clams have $\partial^{13}C$ values of about −30 per mil (‰), in the range that is expected for carbon derived from chemoauto-trophic bacteria, but the $\partial^{13}C$ value of *R. pachyptila* was much higher (~−15‰).

Live symbiont metabolism

Since the symbionts could not, and still cannot, be cultivated, subsequent studies of these bacteria have been limited to those performed on freshly isolated or preserved material. Isolation and puri-fication of symbiotic bacteria from live tubeworms is possible through dissection of trophosomal tissue and size fractionation on a Percoll gradient (Distel & Felbeck 1988). The bacterial suspen-sion obtained is relatively pure, and bacteria stay metabolically active for a few hours, depending on the pressure, temperature and composition of the incubation medium (Distel & Felbeck 1988, Wilmot & Vetter 1990, Scott et al. 1994). These experiments showed that the symbionts of *R. pac-hyptila* were very specific in their needs, metabolising only sulphide, and not thiosulphate as other

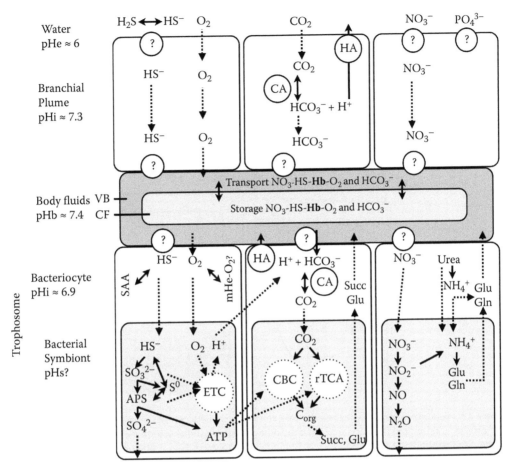

Figure 11 Schematic physiological model for the symbiosis between the vent tubeworm *Riftia pachyptila* and its thiotrophic bacteria regarding the principal metabolites: oxygen, sulphide, carbon dioxide and nitrate. For clarity, the various metabolic pathways (O_2 and H_2S, CO_2, NO_3^-) are shown on separate cells but co-occur in plume epithelial or trophosome bacteriocyte cells. See text for details. APS = adenosine phosphosulphate; CA = carbonic anhydrase; CBC = Calvin-Benson cycle; CF = coelomic fluid; ETC = electron transport chain; HA = proton-ATPase; mHe = myohemerythrin; rTCA = reverse tricarboxylic acid cycle; SAA = sulphur-amino acids; VB = vascular blood.

free-living or symbiotic thioautotrophic bacteria (Wilmot & Vetter 1990), and fixing carbon only from molecular carbon dioxide and not bicarbonate (Scott et al. 1998, 1999).

By measuring oxygen consumption rates of isolated symbiont suspension from *R. pachyptila*, it was possible to demonstrate that they oxidise sulphide but not thiosulphate or sulphite (Wilmot & Vetter 1990). Heterotrophy was also tested using malate, succinate, pyruvate or fumarate as substrata, all of which were not oxidised. The bacteria are able to oxidise even small amounts of sulphide (5 μM) and have maximal respiration rates at concentrations greater than 1 mM sulphide, showing no inhibition in oxygen consumption up to sulphide concentrations of 2 mM. The products of sulphide oxidation are mostly insoluble elemental sulphur and, to a lesser extent, soluble sulphate and polysulphide (Wilmot & Vetter 1990). Polysulphide increase is only transient, and the final soluble product of sulphide oxidation is sulphate. Similar results were obtained by measuring carbon fixation in a trophosome incubation experiment (Fisher et al. 1989); carbon fixation occurred

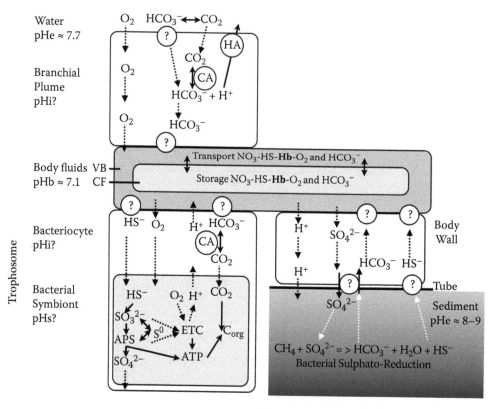

Figure 12 Schematic physiological model for the symbiosis between the seep tubeworm *Lamellibrachia luymesi* and its thiotrophic bacteria regarding the principal metabolites: oxygen, sulphide, carbon dioxide and nitrate. See text for details. Abbreviations as in Figure 11. (Adapted from Dattagupta et al. 2006.)

only with sulphide, but it was obviously inhibited by increasing oxygen concentrations, suggesting that sulphide oxidation and carbon fixation are not tightly coupled.

Carbon fixation by isolated symbionts was also measured by ^{14}C-inorganic carbon incubation and subsequent analysis of ^{14}C incorporation rates and apparent half-saturation constant ($K_{1/2}$) (Scott et al. 1998, 1999). By setting various pH and total inorganic carbon levels in the suspension medium, it is possible to discriminate which form of carbon is preferred: carbon dioxide or bicarbonate. Reliance on CO_2 with relatively low affinity is characteristic of cells adapted to growth with abundant total inorganic carbon. Alternatively, higher affinity and reliance on both CO_2 and HCO_3^- are typical of organisms adapted to growth with low total inorganic carbon. The three vestimentiferan symbionts tested (from *R. pachyptila*, *Ridgeia piscesae* and *Seepiophila jonesi*) all incorporated carbon from CO_2 and not HCO_3^-. This is related to the absence of carboxysome in these symbiotic bacteria, in contrast to the free-living, sulphide-oxidising *Thiomicrospira crunogena*, which is able to use both CO_2 and HCO_3^- (Scott et al. 1998). However, symbionts from *Ridgeia piscesae* had a higher affinity for CO_2 than those from *Riftia pachyptila* ($K_{1/2}$ of 7.6 μM vs. 49 μM; Scott et al. 1999). Since both symbionts belong to the same phylotype, this difference has been attributed to substratum availability, in relation to the higher inorganic carbon content in the blood of *R. pachyptila* (often above 30 mM; Childress et al. 1993, Toulmond et al. 1994) compared with *Ridgeia piscesae* (<10 mM; Scott et al. 1998).

Symbiont metabolic capabilities

A number of discrete biochemical studies have improved understanding of vestimentiferan symbiont capabilities for CO_2 fixation (RuBisCO; Robinson et al. 1998, 2003), sulphide oxidation (ATP sulphurylase; Renosto et al. 1991, Laue & Nelson 1994, Beynon et al. 2001) or nitrogen metabolism (Lee & Childress 1994, Lee et al. 1999, Minic et al. 2001, 2002, Minic & Herve 2003).

Until recently (see 'Genomics of the symbionts', p. 244), and since the initial characterisation of RuBisCO activity in vestimentiferan symbionts (Felbeck 1981), it has been postulated that fixation of CO_2 occurred through the Calvin-Benson cycle. Further investigations showed that vent vestimentiferan symbionts (from *Riftia pachyptila* and *Tevnia jerichonana*) did not express the most common form I RuBisCO (*cbbL* gene) but instead the form II RuBisCO (*cbbM* gene) in contrast with bivalve symbionts (Robinson & Cavanaugh 1995). A form II RuBisCO was also expressed in *Lamellibrachia* sp. (probably *Lamellibrachia* sp. L1; Table 1) (Elsaied et al. 2002), but *in situ* hybridisation analyses were not convincing due to confusion between sperm and bacteria on trophosome sections (see 'Symbionts', p. 233). Form II RuBisCO, a dimer of large subunits, has a lower specificity for CO_2 and thus is more effective under CO_2-rich conditions (Tabita et al. 2007). The kinetic properties of form II RuBisCO from *Riftia pachyptila* symbionts demonstrate a typical low affinity (K_{CO2} = 240 µmol L^{-1}) and a low CO_2/O_2 specificity factor (Ω = 8.6) compared with form I (Robinson et al. 2003). The degree of isotopic discrimination by the enzyme was determined to be 19.5‰, lower than that for form I (Robinson et al. 2003). Together, these properties may help explain the particular $\partial^{13}C$ signature in vestimentiferans: the isotopic discrimination by form II RuBisCO combines with a large CO_2 gradient between the host and the symbiont that drives an enrichment in ^{13}C of the CO_2 pool available to the symbionts (Scott 2003).

Felbeck (1981) initially reported high activities of adenosine phosphosulphate (APS) reductase and ATP sulphurylase in trophosome extracts from *R. pachyptila*, indicating that symbiont production of ATP could occur through the APS pathway. However, this pathway may also function in the reverse direction, as in dissimilatory sulphate reduction and assimilatory biosynthetic pathways (Madigan et al. 2009). High activity and kinetic properties of ATP sulphurylase of *R. pachyptila* (Renosto et al. 1991), specifically a high K_{cat} (296 s^{-1}) for ATP synthesis that may be related to the specific nucleotide sequence of the *sopT* gene (Laue & Nelson 1994) and structure of the active site (Beynon et al. 2001), strongly support the idea that this pathway derives sufficient energy from sulphide oxidation to sustain the chemoautotrophic metabolism of the symbiont. Elemental sulphur (S^0) accumulates in symbiont-containing tissues of thiotrophs (Vetter & Fry 1998) and has been localised to membrane-bound vesicles in the symbionts (Pflugfelder et al. 2005), where it may serve as an internal store of partially oxidised sulphur. This store may be further oxidised (i.e., re-enter the APS pathway) if oxidants are present (mainly O_2 and eventually NO_3^-; Hentschel & Felbeck 1993) but may also generate sulphide in case of anoxia (Arndt et al. 2001). The complete oxidation of sulphide to sulphate produces electrons shuttled through an electron transport chain (ETC) system, and the resulting proton gradient generates ATP through oxidative phosphorylation.

Nitrogen metabolism has also been the subject of several studies (Lee & Childress 1994, Lee et al. 1999, Minic et al. 2001, 2002, Minic & Herve 2003). It appears that the main source of nitrogen for the symbiont may be nitrate (NO_3^-) that can be reduced to nitrite (NO_2^-) with the enzyme nitrate reductase (Lee et al. 1999), eventually converted to ammonia (NH_3) and ultimately be incorporated into amino acids due to the activities of glutamine synthetase and glutamate dehydrogenase (Lee et al. 1999). As well as amino acids, nucleotide synthesis needs nitrogen. Interestingly, the symbiont possesses the enzymatic equipment for the biosynthesis of pyrimidine nucleotides through the *de novo* pathway but lacks the enzymes of the salvage and catabolic pathways (Minic et al. 2001, 2002).

Numerous other nutrients (e.g., phosphorus, vitamins, metals) are obviously necessary for the complete metabolism of the symbionts, but the assimilation and conversion pathways involved have not been studied so far.

Genomics of the symbionts

A long-awaited major breakthrough is the publication of the metagenome of the symbiont of *R. pachyptila* (Robidart et al. 2008) and incidentally its denomination as Candidatus *Endoriftia persephone*. The main findings reported by Robidart et al. relate to the metabolic possibilities that could be expressed in the free-living form of *E. persephone* (e.g., heterotrophic capacity, flagellum synthesis) and, in addition to the confirmation of the different pathways described, the existence of an alternative carbon fixation pathway, the reductive tricarboxylic acid (reverse TCA or rTCA) cycle. Energy requirements to fix CO_2 differ among these two pathways: the rTCA cycle requires only one molecule of ATP for one of CO_2, whereas the Calvin-Benson cycle requires three molecules of ATP (reviewed by Nakagawa & Takai 2008). Moreover, as mentioned by Markert et al. (2007), the stable carbon isotopic fractionation between inorganic CO_2 and biomass by the rTCA cycle is less than that by the Calvin-Benson cycle, providing an attractive alternative explanation for the anomalously high carbon isotope values found in Vestimentifera. It should be noted that oxygen-sensitive enzymes are involved in the rTCA cycle (Nakagawa & Takai 2008), prompting a reexamination of the experiments conducted on isolated symbionts regarding the oxygenation characteristics of the suspension medium.

Genome knowledge allows for expression (transcriptome and proteome) studies with a reliable identification of major proteins produced under various conditions. In the first study of this kind, Markert et al. (2007) confirmed the quantitative importance of the sulphide oxidation pathway and the reality of the rTCA cycle for carbon fixation. The three major enzymes DsrA (dissimilatory sulphite reductase), AprA/AprB (APS reductase) and SopT (ATP-sulphurylase), catalysing the oxidation of H_2S to SO_4^{2-} and yielding ATP, constitute more than 12% of the total cytosolic symbiont proteome (Markert et al. 2007). Several genes encoding sulphur transferases are also present (Robidart et al. 2008) and would allow direct use of thiosulphate by the symbionts, in contrast to findings from earlier physiological studies (Wilmot & Vetter 1990). The use of elemental sulphur as an electron sink when oxygen is absent, postulated by Arndt et al. (2001), may be possible via a sulphur reductase (Robidart et al. 2008). The four enzymes allowing the rTCA cycle are abundant and show a high activity level, providing strong evidence that symbionts of *R. pachyptila* effectively use the rTCA cycle for autotrophic carbon fixation as well as the Calvin-Benson cycle (Markert et al. 2007). Future physiological studies are now needed to determine the relative importance of each carbon fixation pathway as well as numerous other metabolic pathways revealed by genomic sequence annotation (Robidart et al. 2008).

Host adaptations to fulfil symbiont metabolic needs

From this information, the requirements of symbionts for chemoautotrophy may be summarised as follows: they need carbon, in the form of molecular carbon dioxide, sulphide and oxygen for energy, and nitrogen, mainly in the form of nitrate and ammonium. The acquisition from the environment and subsequent transport of these molecules by the host worm to the internal location of the symbionts has been puzzling researchers ever since the chemotrophic nature of the symbiosis was postulated. Evidence for host adaptations to perform these functions has mainly been brought by whole-organism experiments carried out in pressurised aquaria, discrete biochemical analyses and, more recently, broad-range gene and protein expression studies.

Autotrophic balance of intact animals

The first whole-organism experiments that gave evidence for autotrophy of the *R. pachyptila* symbiosis, by analysing gas fluxes in and out of flow-through pressurised aquaria, were those of Fisher et al. (1988) and Childress et al. (1991). These experiments established that, given appropriate (i.e., environmentally realistic) amounts of oxygen and sulphide, the *R. pachyptila* symbiosis could eventually consume carbon dioxide instead of producing it as in other animals. Without sulphide, O_2 consumption was around 2.7 μmol g^{-1} h^{-1} with a corresponding ΣCO_2 production of 2.2 μmol g^{-1} h^{-1}, whereas consumption of 2.5 μmol g^{-1} h^{-1} sulphide (and 5.5 μmol g^{-1} h^{-1} O_2) neutralised net ΣCO_2 fluxes, and maximum ΣCO_2 uptake reached 2.7 μmol g^{-1} h^{-1} for 5.5 μmol g^{-1} h^{-1} sulphide and 9 μmol g^{-1} h^{-1} O_2 consumptions (Childress et al. 1991).

These initial experiments have been repeated several times since this work was reported, and the last refinements, using semicontinuous mass spectrometry measurements in addition to gas chromatography calibration, were performed and published by Girguis & Childress (2006). Their results showed that substrata concentrations and environmental temperature can strongly govern oxygen and sulphide uptake rates of *R. pachyptila* as well as net carbon uptake: symbiont carbon fixation was observed to be highest after sufficient oxygen and sulphide had been acquired by *R. pachyptila* and when temperatures were relatively high (Girguis & Childress 2006). Under similar conditions to those used by Childress et al. (1991) (conditions termed "better" in Girguis & Childress 2006), they obtained somewhat higher yet comparable flux values considering temperature effects (8–13°C in Childress et al. 1991; cf. 15°C in Girguis & Childress 2006) with a maximum ΣCO_2 uptake of 3.7 μmol g^{-1} h^{-1} for 7.8 μmol g^{-1} h^{-1} sulphide uptake and 12.8 μmol g^{-1} h^{-1} O_2 uptake. Pushing the symbiosis in *R. pachyptila* to its limits ("best" conditions in Girguis & Childress 2006) with external concentrations of 256 μmol L^{-1} ΣH_2S, 197 μmol L^{-1} O_2 and 10.8 μmol L^{-1} ΣCO_2 at 15°C), they measured fluxes of 12.7 μmol g^{-1} h^{-1} ΣCO_2 for 11.9 μmol g^{-1} h^{-1} ΣH_2S and 25.1 μmol g^{-1} h^{-1} O_2. Optimal temperature for maximal CO_2 uptake was around 26°C, close to the maximal thermal tolerance of *R. pachyptila,* which is above 30°C (Girguis & Childress 2006); at the optimal temperature, O_2 and sulphide consumptions of 35 and 16 μmol g^{-1} h^{-1} yielded a ΣCO_2 uptake of nearly 16 μmol g^{-1} h^{-1}.

In cold-seep species, as stated, there is not enough sulphide in the water column around the plumes of vestimentiferan tubeworms to sustain a sulphide-driven autotrophic mode of life. However, sulphide present in the sediment has been proposed as the source of sulphide for adult *Lamellibrachia luymesi* taken up through the buried part of their tubes and body (Julian et al. 1999). In a modified setup taking this into account with a split-vessel, flow-through respirometry chamber (Freytag et al. 2001), it was possible to show net CO_2 (and O_2) uptake at the plume level when sulphide was passed over the roots. At 5°C, consumption rates were lower than in *Riftia pachyptila*; using sulphide concentrations around the roots in the range 200–400 μmol L^{-1}, Freytag et al. measured 1–3 μmol g^{-1} h^{-1} CO_2 uptake for 3–5 μmol g^{-1} h^{-1} sulphide and 7–10 μmol g^{-1} h^{-1} O_2 consumptions.

Acquisition of inorganic substrata from the environment

As shown, two organs of vestimentiferan hosts may serve as an entry point for the necessary metabolites to be delivered to the symbionts: the branchial plume and, in the case of cold-seep tubeworms, the posterior end of the trunk, or root. The mechanisms by which these substances cross the epithelium and enter the body fluids have been the subject of several studies using direct, physiological or indirect, biochemical approaches.

As for any exchange surface and for uncharged molecules such as O_2, H_2S or CO_2, the first uptake mechanism is diffusion, which is proportional to the surface area exposed to the external medium and inversely proportional to the diffusion distance (Fick's law). Although a respiratory role for the branchial plume of Vestimentifera is obvious based on anatomical considerations (e.g., Gardiner & Jones 1993), actual surface area estimates of complicated surfaces such as respiratory organs are extremely tedious and, as a result, have seldom been performed. For Vestimentifera, one study dealt with *R. pachyptila* (Andersen et al. 2002) and another with the two morphotypes of *Ridgeia piscesae* (Andersen et al. 2006). *Riftia pachyptila* has a mean specific branchial surface area (SBSA) of 22 cm² g⁻¹, second highest among all aquatic animals. Mean diffusion distance for the branchial filament wall was 15 μm, with a minimal diffusion distance of 2 μm in the pinnules that cover the free tip of the filaments (Andersen et al. 2002). In *Ridgeia piscesae*, two morphotypes occur: a short-fat, chimney-dwelling and a long-skinny, basalt-dwelling morphotype (see 'Hydrothermal vent tubeworms', p. 223 and Southward et al. 1995). Mean SBSA (24 cm² g⁻¹) is similar in both morphotypes and comparable to the values obtained for *Riftia pachyptila* as well as the diffusion distances for the chimney-dwelling morphotype, whereas there is a 20% increase in diffusion distance for the basalt-dwelling morphotype (Andersen et al. 2006). To date, the branchial plume of seep species and the root part of the trunk in seep species or basalt-dwelling *Ridgeia piscesae* have not been subjected to detailed studies of their surface area or diffusion distances, preventing substantiated comparisons. The only estimate based on average geometric considerations yields a specific root surface area of 8 cm² g⁻¹ in *Lamellibrachia luymesi* (Dattagupta et al. 2006), about a third of the SBSA in vent species.

Diffusion is the only uptake mechanism known for uncharged molecules such as oxygen (but see discussion of the role of haemoglobin [Hb] in facilitated diffusion and transport), but alternative pathways may exist for chemicals that also occur in ionic form. In water, this is the case for carbon dioxide ($CO_2 \Leftrightarrow HCO_3^-$; $pK \approx 6.2$ at 10°C), sulphide ($H_2S \Leftrightarrow HS^-$, $pK \approx 6.8$ at 10°C) and the various forms of nitrogen (NH_4^+, NO_2^-, NO_3^-), the dissociation of which mainly depends on the environmental pH and temperature around the exchange organ. Using whole-organism experiments as described and by varying the external conditions while measuring VB and CF substrata concentrations, it has been possible to trace the most plausible acquisition pathways for carbon dioxide and sulphide.

Carbon dioxide total concentration in the body fluids of *Riftia pachyptila* can reach extremely high values, up to 50 mmol L⁻¹ (Childress et al. 1993), and CO_2 uptake is correlated to environmental CO_2 and not HCO_3^- concentrations (Goffredi et al. 1997b, Girguis & Childress 2006). CO_2 diffusion is facilitated in the vent environment, where pH values are lower and ΣCO_2 higher than in normal seawater due to mixing with acidic and CO_2-enriched hydrothermal vent fluid, thus providing a high CO_2 partial pressure P_{CO_2} gradient across the branchial epithelium. Diffusion alone can account for the high ΣCO_2 values measured in CF and VB because the P_{CO_2} gradient is maintained by rapid internal transformation of CO_2 into HCO_3^- through carbonic anhydrase (CA) activity and by the active elimination of protons through H^+-ATPase (adenosine triphosphatase) activity (Goffredi et al. 1997b, 1999, Goffredi & Childress 2001), thus maintaining a higher pH in the body fluids relative to the external pH. These findings have been substantiated by similar studies on *R. pachyptila* and other species (Girguis & Childress 1998, 2006, Girguis et al. 2002, Dattagupta et al. 2009) and by more detailed examination of ATPase and CA in branchial plume tissue. *Riftia pachyptila* has unusually high ATPase activity (646 μmolP$_i$ g⁻¹ h⁻¹), with a large proportion of its ATPases (38%) being P- and V-H⁺ATPases, the forms devoted to proton transport (Goffredi & Childress 2001). In *R. pachyptila*, branchial epithelium V-H⁺ATPase is mostly colocalised with CA on the apical side, and both enzymes probably allow CO_2 entry while regulating intracellular pH (De Cian et al. 2003a). In comparison, seep species have much lower ATPase activities (48–57 μmolP$_i$ g⁻¹ h⁻¹ including plume and root), suggesting that other routes may be active for proton elimination and carbon uptake (Dattagupta et al. 2009), such as proton channels and HCO_3^- transport across the body wall epithelium. The hypothesis is supported by the presence of vascular connections between the trophosome and the body wall in *Lamellibrachia luymesi* that are absent in *Riftia pachyptila* (van der Land & Nørrevang 1977, Gardiner & Jones 1993).

High levels of CA activity (254 µmol CO_2 g^{-1} min^{-1}) have been reported for branchial plume tissue of *Riftia pachyptila* (Kochevar et al. 1993, Kochevar & Childress 1996, Goffredi et al. 1999). Similar activity levels occur in *Tevnia jerichonana* branchial plume tissue, but much lower activities have been measured in seep-species plume tissue (Kochevar & Childress 1996). A protein sequence for *Riftia pachyptila* plume CA clustered with cytosolic CA isoforms (De Cian et al. 2003c), and immunolocalisation showed it is mostly present at the apical side of the branchial epithelium (De Cian et al. 2003a). Confirmation of high transcription levels of CA messenger RNA have been reported in transcriptomic approaches in *R. pachyptila* (Sanchez et al. 2007b) and *Ridgeia piscesae* (Ruan et al. 2008), and two different CA genes are expressed in the branchial plume of *Riftia pachyptila*, with one form specifically expressed in this tissue, the other common with the CA gene expressed in the trophosome (Sanchez et al. 2007a). Further functional characterisation and physiological studies are needed to clarify the precise role of each of these isoforms, with some of these specialised for CO_2 conversion or better poised for acid–base regulation (Henry 1996).

Because sulphide is toxic to aerobic organisms (Grieshaber & Volkel 1998), thiotrophic symbioses constitute very unusual models by which sulphide becomes an essential nutriment. Initially, experiments with *R. pachyptila* exposed to various external sulphide concentrations and pH showed a correlation of body fluid total sulphide (ΣH_2S) with the ionic bisulphide ion HS^- and not with the diffusible H_2S form (Goffredi et al. 1997a). However, additional measurements and refinements in analytical procedures revealed that both H_2S and HS^- could be acquired (Girguis & Childress 2006) depending on the external pH and the corresponding main form of sulphide in the water (H_2S below pK_S, HS^- above pK_S). Within an environmentally realistic concentration range (10–400 µmol L^{-1}), ΣH_2S and O_2 uptake are highly correlated, which is consistent with the sulphide-oxidising metabolism of the symbionts (Girguis & Childress 2006).

In seep species, there is hardly any sulphide in the water surrounding the branchial plume high above the sediment. Sulphide is sometimes detectable at the base of tubes or at the surface of sediment and increases sharply with sediment depth, reaching 1.5 mmol L^{-1} at 20 cm (Julian et al. 1999). The thin (70-µm) tube of the root is permeable to H_2S (Julian et al. 1999), and sulphide uptake does occur under these conditions, allowing inorganic carbon uptake (Freytag et al. 2001). Based on these results and additional measurements of sulphate and proton concentrations, Dattagupta et al. (2006) proposed a model (see also Figure 12) in which sulphate excretion fuels microbial consortia in the sediment (Boetius et al. 2000); the resulting sulphide production may diffuse through the thin tube and enter the tubeworm through the body wall epithelium. This sulphate/sulphide cycle would be sufficient to support the long life of seep tubeworms according to growth models (Cordes et al. 2005a). In vent tubeworms, and even more so in seep tubeworms, further studies that examine membrane transport are needed to confirm and detail these pathways in plume and body wall epithelia, both at the apical (cell–environment interface) and basal (cell–body fluid interface) membranes (as illustrated by the numerous question marks in Figures 11 and 12).

Nitrogen uptake in *R. pachyptila* takes place in the form of nitrate, not ammonium or dinitrogen (Lee & Childress 1994, Hentschel et al. 1998, Girguis et al. 2000). With environmental nitrate concentrations around 40 µmol L^{-1} around vents (Lee & Childress 1994), nitrate uptake amounts to 3.5 µmol g^{-1} h^{-1} (Girguis et al. 2000). However, the mechanism of nitrate uptake, a highly unusual function in animals, is yet unknown. *Riftia pachyptila* can also take up free amino acids (Childress et al. 1984), but this is not believed to be important in the normal physiology of the worm. To our knowledge, there are no data on the uptake of nitrogen compounds in seep species, whereas both ammonium and nitrate might be present in the environment (Lee & Childress 1994).

Transport of inorganic metabolites in the body fluids

The body fluids of *R. pachyptila* accumulate high amounts of oxygen, sulphide, carbon dioxide and nitrate, well above the concentration of these molecules in the surrounding environment. The central role of Hb in the transport of oxygen and sulphide was suspected very early, which is not surprising considering the large amount (4% blood and 26% CF, W/W; Childress et al. 1984) and bright red colour of these fluids in *R. pachyptila*. The role of body fluids in sulphide accumulation and transport (Arp & Childress 1981, 1983), and subsequently its reversible binding to Hb, independently of oxygen binding was established early (Arp et al. 1987). Vestimentifera have extracellular Hbs in both the VB and CF. VB contains two forms, HbV1 and HbV2, whereas CF contains one form, HbC1 (Zal et al. 1996b). HbV1 is similar to the high molecular weight (~3500 kDa) hexagonal bilayer haemoglobin (HBLHb), which is specific of annelids, while HbV2 and HbC1 are extracellular Hb of lower molecular weight (~400 kDa) (Zal et al. 1996b). Analysis of their subunit composition demonstrated that these two forms differ slightly, one globin chain being absent from HbC1 (Zal et al. 1996a). The Hb concentration is also higher in VB ([haem] $\approx$ 3 mmol L^{-1}) than in CF (0.5 mmol L^{-1}) (F.L. personal observations). The involvement of cysteins in the mechanism of sulphide binding, initially proposed (Arp et al. 1987), was substantiated by the finding and localisation of free cysteins on some of the globin chains constituting both VB and CF Hbs (Zal et al. 1996a, 1997) and convincing equilibration studies (Zal et al. 1998). An evolutionary scenario was proposed (Bailly et al. 2002, 2003) by which sulphide binding could have appeared in annelid Hbs to avoid its toxicity and evolved in reversible binding for sulphide transport in thiotrophic Siboglinidae. However, crystallographic data showed that these cystein residues are located beneath the surface of the molecule, questioning their efficient role in sulphide binding (Flores et al. 2005). Instead, these authors have proposed that zinc ions bound to HbC1 may chelate sulphide, but their experiments explained only 57% of sulphide binding with this Hb and only 18% with HbV1. Increasing the zinc chelator concentration in a subsequent analysis, these authors obtained a complete inhibition of sulphide binding in HbC1 (Flores & Hourdez 2006). At the same time, the structure of the 400-kDa Hb from the Frenulata *Oligobrachia mashikoi* was also obtained (Numoto et al. 2005), and mercury binding to free cystein residues showed that these sites were readily accessible to mercury and that sulphide binding could be stabilised by close-by phenylalanine residues. Additional work is clearly needed to fully understand the possibly multiple mechanisms of sulphide binding to Vestimentifera Hb and their precise functional characteristics under physiologically relevant conditions to get the whole picture of combined oxygen and sulphide transport, which is so important for the symbiosis to operate.

Carbon dioxide also accumulates in high amounts in body fluids of *Riftia pachyptila* (Childress et al. 1993), and most of it is not bound to Hb as carbamate but is present in the ionic form bicarbonate, the most preponderant form at body fluid pH values (7.1–7.4) (Toulmond et al. 1994). The two body fluids, VB and CF, have different behaviour in relation to CO_2; gel filtration analysis of the fluids revealed that the protein-free fractions retained 64% of the CO_2 in VB and 80% in CF, in relation with their respective Hb content (Toulmond et al. 1994). An alternative hypothesis for CO_2 transport as HCO_3^- was proposed early (Felbeck 1985), involving branchial carboxylation of pyruvate into C4 compounds (Felbeck & Turner 1995) malate and succinate, which are readily labelled in the effluent following perfusion experiments with dissected plumes and vestimentums under atmospheric pressure. Further studies with catheterised worms maintained under pressure have shown that CO_2 from the environment is apparently transported initially as inorganic carbon in the blood of *R. pachyptila* (Felbeck et al. 2004). Nevertheless, the carbon stored in the large quantities of organic acids found in the body fluids, and which may originate either from the branchial plume or the trophosome based on enzymatic capacities (Felbeck 1985), could function as a store to maintain carbon fixation in the trophosome during periods of low CO_2 availability (Felbeck et al. 2004).

Nitrate transport must also occur if nitrate is utilised as the main nitrogen source for bacterial biosynthetic metabolism (Lee et al. 1999) or occasional respiration (Hentschel & Felbeck 1993), and nitrate is found at elevated concentrations in the body fluids of *R. pachyptila*: up to 0.2 mmol L^{-1} in CF and in the range 1–4.5 mmol L^{-1} in VB (Hentschel et al. 1998, Pospesel et al. 1998). Hb, or a blood-borne protein in sufficient amount to accommodate the high nitrate concentrations measured, is a serious candidate to bind nitrate (Hahlbeck et al. 2005), but the mechanism is as yet unknown.

Utilisation of inorganic metabolites and bacteriocyte–symbiont relationships

In the trophosome, bacteriocytes must take up available molecules in the blood to provide them to the symbionts localised within the vacuoles. In addition, organic molecules produced by the symbionts must transit through the bacteriocytes and be delivered to the body fluids to benefit the host metabolism. The cellular physiology of bacteriocytes is thus central to the functioning of the symbiosis in Vestimentifera. Although bacteriocyte isolation has proved feasible in *R. pachyptila* (De Cian et al. 2003b), few studies have exploited this approach, and most of these have been done on trophosome tissue incubations. The rest of current knowledge on bacteriocyte–symbiont relationships is based on indirect evidence, mostly from biochemical and molecular studies.

Regarding oxygen and considering the microaerophilic nature of the bacteria (see, e.g., Fisher et al. 1989, Hentschel & Felbeck 1993, Girguis & Childress 2006), the free oxygen concentration should be maintained very low within the bacteriocytes, but no direct measurements are yet available. The high affinity of extracellular Hbs in VB and CF may help to limit the amount of free oxygen in the bacteriocytes. Interestingly, the expression level of myohemerythrin (mHe), a non-haem, di-iron oxygen-binding protein present in *R. pachyptila* (Bailly et al. 2008) and *Ridgeia piscesae* (Ruan et al. 2008), appears especially high in the trophosome (Sanchez et al. 2007b) and could play a role in oxygenation regulation in the vicinity of bacteria. Further studies on the functional properties of mHe relative to those of VB and CF Hbs are needed.

Intracellular storage of sulphide in the bacteriocytes could be mediated by sulphur-amino acids (SAAs). Early on, high thiotaurine levels were reported in symbiotic tissues of Vestimentifera (Alberic 1986, Pruski et al. 2000, Yin et al. 2000). Thiotaurine synthesis from hypotaurine appears to be a specific adaptation to the thiotrophic mode of life (Pruski et al. 2001), and its rate is related to sulphide exposure, either experimentally (*in vitro* tissue incubation; Pruski & Fiala-Medioni 2003) or naturally *in situ* (Brand et al. 2007). While the precise function (i.e., transport or storage of sulphide) of hypotaurine and thiotaurine has yet to be established, it seems to be a host attribute: free-living sulphide-oxidising bacteria (*Thiobacillus hydrothermalis*) do not metabolise thiotaurine unless trophosome extract is added to the medium (Pruski et al. 2001). However, recent evidence showed that hypotaurine and thiotaurine also accumulate in non-symbiotic annelids exposed to high-sulphide environments, redefining their role in sulphide detoxication (Ortega et al. 2008, Yancey et al. 2009). In the symbionts, sulphide may also be stored as elemental sulphur (S^0) granules, as stated, and be further oxidised to sulphate with oxygen or nitrate as final electron acceptors. However, in case of anoxia, it seems that elemental sulphur may be reduced back to H_2S that accumulates in the blood (Arndt et al. 2001). The extent of anaerobic sulphide production relates to the colour of the trophosome (Pflugfelder et al. 2005) and hence on the elemental sulphur content of symbionts (Arndt et al. 2001).

To provide molecular CO_2 to the symbionts, bacteriocytes have to draw from the large pool of HCO_3^- in the body fluids and convert it with the help of CA. Metabolically produced CO_2 may also participate in inorganic carbon supply to the bacteria. CA is highly active in trophosome extracts from *Riftia pachyptila* with a level similar to that observed in the plume (Kochevar et al. 1993, Kochevar & Childress 1996, Goffredi et al. 1999). In *Tevnia jerichonana*, CA activity in the trophosome is lower than that in the plume, but in seep species (*Lamellibrachia luymesi* cited as *Lamellibrachia* sp. and *Seepiophila jonesi* cited as Escarpid, undescribed), CA activity is higher in the trophosome compared with the plume (Kochevar & Childress 1996).

Immunolocalisation of cytosolic CA and *in situ* hybridisation with CA-cDNA probe in *Riftia pachyptila* trophosome lobules demonstrated that CA is expressed and found in bacteriocytes and peritoneal cells (De Cian et al. 2003a). A physiological investigation using various inhibitors on isolated bacteriocyte suspensions from the trophosome tissue of *R. pachyptila* revealed a complex interaction of CA with two important ion-transporting enzymes: the vacuolar-type V-H⁺ATPase and the Na⁺K⁺-ATPase (De Cian et al. 2003b). These enzymes seem to be involved in the transepithelial transport processes for electrolytes and CO_2, and the authors suggested that in the trophosome, the proton efflux might generate a local acidification of the outer layer of the membrane, facilitating the influx of CO_2. There is one cytosolic isoform of CA (RpCAtr) highly expressed in the trophosome (De Cian et al. 2003c, Sanchez et al. 2007a), in contrast with the branchial plume where this isoform is much less expressed than another, more specific branchial isoform (RpCAbr) (Sanchez et al. 2007a). In addition, immunoblots of membrane-bound protein fractions from the trophosome of *R. pachyptila* revealed proteins with CA activity, but the sequence of this putative membrane-bound CA is yet unknown (De Cian et al. 2003c). Similar characterisation and localisation studies on CA in seep species are lacking.

Based on purified symbiont incubations, the transfer of organic carbon from the symbionts to the host appears to be mainly in the form of succinate and glutamate (Felbeck & Jarchow 1998a,b). Succinate plays a pivotal role in the *R. pachyptila* symbiosis since it is also the main end product of anaerobic metabolism during the first 40 h of anoxic exposure (Arndt et al. 1998a,b). The phosphagen creatine phosphate, not analysed in these studies, should also play a role on the onset of anaerobiosis, as suggested by large amounts of creatinine (1128 µmol L^{-1}) and high level of creatine kinase (176 µmol.g^{-1}.min^{-1}) in the trophosome (De Cian et al. 2000). The trophosome also has high levels of glycogen, up to 100 µmol glycosyl units g^{-1} (Arndt et al. 1998b). Glycogen reserves are localised in both host cells and symbionts, which contribute equally to the total glycogen content of the trophosome (Sorgo et al. 2002). However, utilisation of glycogen only appeared after long-term (48-h) anoxic exposure (Arndt et al. 1998b).

Regarding nitrogen compounds, glutamate and glutamine may be the main molecules derived from the ammonia issued from nitrate reduction (Lee et al. 1999) and that could be excreted to the host. However, the complex sharing of enzymatic activities between host and symbionts for many N-compound pathways has neen noted (see 'Symbiont metabolic capabilities', p. 243 and Minic & Herve 2004). In addition, unusually high levels of uric acid (1–2 mmol L^{-1}) and urea (0.9 mmol L^{-1}) have been reported in the trophosome of *R. pachyptila* (De Cian et al. 2000). While the role of uric acid is still elusive, urea transporters have been found in the genome of the symbiont (Robidart et al. 2008), and urease activity is present in the trophosome (De Cian et al. 2000), allowing for a complex shuttle of urea and ammonia between the two partners.

Conclusion

For a group that was little known some 30 yr ago, and remains difficult to access due to its mostly deep-sea habitat, much understanding has been gained in all aspects of the biology of Vestimentifera. Of course, knowledge of their biodiversity, as well as that of many other deep-sea-dwelling organisms, is still far from complete. Exploration of unknown areas of the ocean bottom, especially in the Southern Hemisphere, is necessary to identify new species and supplement knowledge of biogeographic traits and the evolutionary history in this taxon. This effort is considered by the international community in programmes such as ChESS (Census of Marine Life program on Biogeography of Deep-Water Chemosynthetic Ecosystems) (Tyler et al. 2002).

To develop a global approach of symbiosis-related gene expression (transcriptome) and protein production (proteome) in vestimentiferan tubeworms, there is an urgent need for host genome sequencing. First attempts using suppression subtractive hybridisation (SSH) libraries in *R. pachyptila* (Sanchez et al. 2007b) or expressed sequence tag (EST) libraries in *Ridgeia piscesae* (Nyholm et al. 2008) have been confronted with too many unknown sequences due to the lack of phylogenetically close model organisms. The recent availability of some annelid genomes (e.g., *Capitella capitata*) should help in this respect. However, with the present development of new high-throughput techniques, genome or transcriptome sequencing of the host in a model vestimentiferan species should be accessible in the near future. Choosing model vestimentiferan tubeworms for this purpose should rely on ecological (vent and seep species), evolutionary (crown and basal groups) and physiological (acquired knowledge) bases, and for this purpose there are two obvious species, *Riftia pachyptila* and *Lamellibrachia luymesi*.

The problem of the control of symbiont genome expression by the host, necessary for a stable and efficient mutualistic interaction, is one of general biological significance. Vestimentiferan representatives certainly stand out as model systems to study such problems as the infectious mode of symbiont transmission, the effective elimination of bacteria from cells outside trophosome tissue and the very efficient growth and metabolic control of bacteria. Controlling bacterial growth within a single cell type may indeed prove useful in designing novel ways of dealing with infectious bacteria, including those eliciting disease in humans (Hentschel et al. 2000).

Acknowledgements

This work was supported by the Austrian Science Fund (FWF) grant P20282-B17 (M.B.) and the French National Research Agency (ANR) Deep Oases grant 06-BDIV-005-02 (F.H.L.). We would like to thank Andrea Nussbaumer for providing yet unpublished FISH micrographs, Greg Young and Steve Gardiner for helpful discussions and pictures and Heidemarie Grillitsch for the drawing of Figure 5. Warm thanks are extended to Chuck Fisher, Stéphane Hourdez, Sabine Gollner and Ann Andersen for their comments on an earlier version of this chapter.

References

Alberic, P. 1986. Occurrence of thiotaurine and hypotaurine in the tissues of *Riftia pachyptila* (Pogonophora, Vestimentifera). *Comptes Rendus de l'Académie des Sciences, Série III* **302**, 503–508.

Andersen, A.C., Flores, J.F. & Hourdez, S. 2006. Comparative branchial plume biometry between two extreme ecotypes of the hydrothermal vent tubeworm *Ridgeia piscesae*. *Canadian Journal of Zoology—Revue Canadienne de Zoologie* **84**, 1810–1822.

Andersen, A.C., Hamraoui, L. & Zaoui, D. 2001. The obturaculum of *Riftia pachyptila* (Annelida, Vestimentifera): ultrastructure and function of the obturacular muscles and extracellular matrix. *Cahiers de Biologie Marine* **42**, 219–237.

Andersen, A.C., Hourdez, S., Marie, B., Jollivet, D., Lallier, F.H. & Sibuet, M. 2004. *Escarpia southwardae* sp nov., a new species of vestimentiferan tubeworm (Annelida, Siboglinidae) from West African cold seeps. *Canadian Journal of Zoology—Revue Canadienne de Zoologie* **82**, 980–999.

Andersen, A.C., Jolivet, S., Claudinot, S. & Lallier, F.H. 2002. Biometry of the branchial plume in the hydrothermal vent tubeworm *Riftia pachyptila* (Vestimentifera; Annelida). *Canadian Journal of Zoology—Revue Canadienne de Zoologie* **80**, 320–332.

Arndt, C., Gaill, F. & Felbeck, H. 2001. Anaerobic sulfur metabolism in thiotrophic symbioses. *Journal of Experimental Biology* **204**, 741–750.

Arndt, C., Schiedek, D. & Felbeck, H. 1998a. Anaerobiosis in the hydrothermal vent tube-worm *Riftia pachyptila*. *Cahiers de Biologie Marine* **39**, 271–273.

Arndt, C., Schiedek, D. & Felbeck, H. 1998b. Metabolic responses of the hydrothermal vent tube worm *Riftia pachyptila* to severe hypoxia. *Marine Ecology Progress Series* **174**, 151–158.

Arp, A.J. & Childress, J.J. 1981. Blood function in the hydrothermal vent vestimentiferan tube worm. *Science* **213**, 342–344.

Arp, A.J. & Childress, J.J. 1983. Sulfide binding by the blood of the hydrothermal vent tube worm *Riftia pachyptila*. *Science* **219**, 295–297.

Arp, A.J., Childress, J.J. & Vetter, R.D. 1987. The sulfide-binding protein in the blood of the vestimentiferan tube worm, *Riftia pachyptila*, is the extracellular haemoglobin. *Journal of Experimental Biology* **128**, 139–158.

Arvidson, R.S., Morse, J.W. & Joye, S.B. 2004. The sulfur biogeochemistry of chemosynthetic cold seep communities, Gulf of Mexico, USA. *Marine Chemistry* **87**, 97–119.

Bailly, X., Jollivet, D., Vanin, S., Deutsch, J., Zal, F., Lallier, F.H. & Toulmond, A. 2002. Evolution of the sulfide-binding function within the globin multigenic family of the deep-sea hydrothermal vent tubeworm *Riftia pachyptila*. *Molecular Biology and Evolution* **19**, 1421–1433.

Bailly, X., Leroy, R., Carney, S., Collin, O., Zal, F., Toulmond, A. & Jollivet, D. 2003. The loss of the hemoglobin H$_2$S-binding function in annelids from sulfide-free habitats reveals molecular adaptation driven by Darwinian positive selection. *Proceedings of the National Academy of Sciences of the United States of America* **100**, 5885–5890.

Bailly, X., Vanin, S., Chabasse, C., Mizuguchi, K. & Vinogradov, S. 2008. A phylogenomic profile of hemerythrins, the nonheme diiron binding respiratory proteins. *BMC Evolutionary Biology* **8**, 244.

Bartolomaeus, T. 1995. Structure and formation of the uncini in *Pectinaria koreni*, *Pectinaria auricoma* (Terebellida) and *Spirorbis spirorbis* (Sabellida): implications for annelid phylogeny and the position of the Pogonophora. *Zoomorphology* **115**, 161–177.

Bartolomaeus, T. 1997. Chaetogenesis in polychaetous Annelida—significance for annelid systematics and the position of the Pogonophora. *Zoology-Analysis of Complex Systems* **100**, 348–364.

Bergquist, D.C., Eckner, J.T., Urcuyo, I.A., Cordes, E.E., Hourdez, S., Macko, S.A. & Fisher, C.R. 2007. Using stable isotopes and quantitative community characteristics to determine a local hydrothermal vent food web. *Marine Ecology Progress Series* **330**, 49–65.

Bergquist, D.C., Urcuyo, I.A. & Fisher, C.R. 2002. Establishment and persistence of seep vestimentiferan aggregations on the upper Louisiana slope of the Gulf of Mexico. *Marine Ecology Progress Series* **241**, 89–98.

Bergquist, D.C., Ward, T., Cordes, E.E., McNelis, T., Howlett, S., Kosoff, R., Hourdez, S., Carney, R. & Fisher, C.R. 2003. Community structure of vestimentiferan-generated habitat islands from Gulf of Mexico cold seeps. *Journal of Experimental Marine Biology and Ecology* **289**, 197–222.

Bergquist, D.C., Williams, F.M. & Fisher, C.R. 2000. Longevity record for deep-sea invertebrate. *Nature* **403**, 499–500.

Beynon, J.D., MacRae, I.J., Huston, S.L., Nelson, D.C., Segel, I.H. & Fisher, A.J. 2001. Crystal structure of ATP sulfurylase from the bacterial symbiont of the hydrothermal vent tubeworm *Riftia pachyptila*. *Biochemistry* **40**, 14509–14517.

Black, M.B. 1991. *Genetic (allozyme) variation in Vestimentifera (Ridgeia spp.) from hydrothermal vents of the Juan de Fuca Ridge (Northeast Pacific Ocean)*. PhD thesis, University of Victoria, Canada.

Black, M.B., Halanych, K.M., Maas, P.A.Y., Hoeh, W.R., Hashimoto, J., Desbruyères, D., Lutz, R.A. & Vrijenhoek, R.C. 1997. Molecular systematics of vestimentiferan tubeworms from hydrothermal vents and cold-water seeps. *Marine Biology* **130**, 141–149.

Boetius, A., Ravenschlag, K., Schubert, C.J., Rickert, D., Widdel, F., Gieseke, A., Amann, R., Jorgensen, B.B., Witte, U. & Pfannkuche, O. 2000. A marine microbial consortium apparently mediating anaerobic oxidation of methane. *Nature* **407**, 623–626.

Boore, J.L. & Brown, W.M. 2000. Mitochondrial genomes of *Galathealinum*, *Helobdella*, and *Platynereis*: sequence and gene arrangement comparisons indicate that Pogonophora is not a phylum and Annelida and Arthropoda are not sister taxa. *Molecular Biology and Evolution* **17**, 87–106.

Bosch, C. & Grassé, P.P. 1984a. Cycle partiel des bactéries chimioautotrophes symbiotiques et leurs rapports avec les bactériocytes chez *Riftia pachyptila* Jones (Pogonophore Vestimentifère). II. L'évolution des bactéries symbiotiques et des bactériocytes. *Comptes Rendus de l'Académie des Sciences, Série III* **299**, 413–419.

Bosch, C. & Grassé, P.P. 1984b. Cycle partiel des bactéries chimioautotrophes symbiotiques et leurs rapports avec les bactériocytes chez *Riftia pachyptila* Jones (Pogonophore Vestimentifère). I. Le trophosome et les bactériocytes. *Comptes Rendus de l'Académie des Sciences, Série III* **299**, 371–376.

Brand, G.L., Horak, R.V., Le Bris, N., Goffredi, S.K., Carney, S.L., Govenar, B. & Yancey, P.H. 2007. Hypotaurine and thiotaurine as indicators of sulfide exposure in bivalves and vestimentiferans from hydrothermal vents and cold seeps. *Marine Ecology* **28**, 208–218.

Bright, M. & Giere, O. 2005. Microbial symbiosis in Annelida. *Symbiosis* **38**, 1–45.

Bright, M., Keckeis, H. & Fisher, C.R. 2000. An autoradiographic examination of carbon fixation, transfer and utilization in the *Riftia pachyptila* symbiosis. *Marine Biology* **136**, 621–632.

Bright, M. & Sorgo, A. 2003. Ultrastructural reinvestigation of the trophosome in adults of *Riftia pachyptila* (Annelida, Siboglinidae). *Invertebrate Biology* **122**, 347–368.

Campbell, K.A. 2006. Hydrocarbon seep and hydrothermal vent paleoenvironments and paleontology: past developments and future research directions. *Palaeogeography Palaeoclimatology Palaeoecology* **232**, 362–407.

Cary, S.C., Felbeck, H. & Holland, N.D. 1989. Observations on the reproductive biology of the hydrothermal vent tube worm *Riftia pachyptila*. *Marine Ecology Progress Series* **52**, 89–94.

Cary, S.C., Warren, W., Anderson, E. & Giovannoni, S.J. 1993. Identification and localization of bacterial endosymbionts in hydrothermal vent taxa with symbiont-specific polymerase chain reaction amplification and *in situ* hybridization techniques. *Molecular Marine Biology and Biotechnology* **2**, 51–62.

Caullery, M. 1914. Sur les Siboglinidae, type nouveau d'invertébrés receuillis par l'expédition du Siboga. *Comptes Rendus de l'Académie des Sciences, Série III* **158**, 2014–2017.

Cavanaugh, C.M., Gardiner, S.L., Jones, M.L., Jannasch, H.W. & Waterbury, J.B. 1981. Prokaryotic cells in the hydrothermal vent tube-worm *Riftia pachyptila* Jones: possible chemoautotrophic symbionts. *Science* **213**, 340–342.

Childress, J.J., Arp, A.J. & Fisher, C.R. 1984. Metabolic and blood characteristics of the hydrothermal vent tube-worm *Riftia pachyptila*. *Marine Biology* **83**, 109–124.

Childress, J.J. & Fisher, C.R. 1992. The biology of hydrothermal vent animals: physiology, biochemistry and autotrophic symbioses. *Oceanography and Marine Biology: An Annual Review* **30**, 337–441.

Childress, J.J., Fisher, C.R., Favuzzi, J.A., Kochevar, R., Sanders, N.K. & Alayse, A.M. 1991. Sulfide-driven autotrophic balance in the bacterial symbiont-containing hydrothermal vent tubeworm, *Riftia pachyptila* Jones. *Biological Bulletin* **180**, 135–153.

Childress, J.J., Lee, R.W., Sanders, N.K., Felbeck, H., Oros, D.R., Toulmond, A., Desbruyères, D., Kennicut II, M.C. & Brooks, J. 1993. Inorganic carbon uptake in hydrothermal vent tubeworms facilitated by high environmental PCO$_2$. *Nature* **362**, 147–149.

Cordes, E.E., Arthur, M.A., Shea, K., Arvidson, R.S. & Fisher, C.R. 2005a. Modeling the mutualistic interactions between tubeworms and microbial consortia. *PLoS Biology* **3**, 497–506.

Cordes, E.E., Arthur, M.A., Shea, K., Arvidson, R.S. & Fisher, C.R. 2005b. Tubeworm may live longer by cycling its sulfur downward. *Plos Biology* **3**, 353.

Cordes, E.E., Bergquist, D.C., Redding, M.L. & Fisher, C.R. 2007a. Patterns of growth in cold-seep vestimenferans including *Seepiophila jonesi*: a second species of long-lived tubeworm. *Marine Ecology* **28**, 160–168.

Cordes, E.E., Bergquist, D.C., Shea, K. & Fisher, C.R. 2003. Hydrogen sulphide demand of long-lived vestimentiferan tube worm aggregations modifies the chemical environment at deep-sea hydrocarbon seeps. *Ecology Letters* **6**, 212–219.

Cordes, E.E., Carney, S.L., Hourdez, S., Carney, R.S., Brooks, J.M. & Fisher, C.R. 2007b. Cold seeps of the deep Gulf of Mexico: community structure and biogeographic comparisons to Atlantic equatorial belt seep communities. *Deep-Sea Research Part I—Oceanographic Research Papers* **54**, 637–653.

Cordes, E.E., Hourdez, S., Predmore, B.L., Redding, M.L. & Fisher, C.R. 2005c. Succession of hydrocarbon seep communities associated with the long-lived foundation species *Lamellibrachia luymesi*. *Marine Ecology Progress Series* **305**, 17–29.

Corliss, J.B., Dymond, J., Gordon, L.I., Edmond, J.M., Herzen, R.P.V., Ballard, R.D., Green, K., Williams, D., Bainbridge, A., Crane, K. & van Andel, T.H. 1979. Submarine thermal springs on the Galapagos rift. *Science* **203**, 1073–1083.

Dando, P.R., Southward, A.J., Southward, E.C., Dixon, D.R., Crawford, A. & Crawford, M. 1992. Shipwrecked tube worms. *Nature* **356**, 667.

Dattagupta, S., Miles, L.L., Barnabei, M.S. & Fisher, C.R. 2006. The hydrocarbon seep tubeworm *Lamellibrachia luymesi* primarily eliminates sulfate and hydrogen ions across its roots to conserve energy and ensure sulfide supply. *Journal of Experimental Biology* **209**, 3795–3805.

Dattagupta, S., Redding, M., Luley, K. & Fisher, C. 2009. Comparison of proton-specific ATPase activities in plume and root tissues of two co-occurring hydrocarbon seep tubeworm species *Lamellibrachia luymesi* and *Seepiophila jonesi*. *Marine Biology* **156**, 779–786.

De Burgh, M.E. 1986. Evidence for a physiological gradient in the vestimentiferan trophosome: size-frequency analysis of bacterial populations and trophosome chemistry. *Canadian Journal of Zoology—Revue Canadienne de Zoologie* **64**, 1095–1103.

De Burgh, M.E., Juniper, S.K. & Singla, C.L. 1989. Bacterial symbiosis in northeast pacific vestimentifera: a TEM study. *Marine Biology* **101**, 97–105.

De Cian, M.-C., Andersen, A.C., Bailly, X. & Lallier, F.H. 2003a. Expression and localization of carbonic anhydrase and ATPases in the symbiotic tubeworm *Riftia pachyptila. Journal of Experimental Biology* **206**, 399–409.

De Cian, M.-C., Andersen, A.C., Toullec, J.-Y., Biegala, I., Caprais, J.-C., Shillito, B. & Lallier, F.H. 2003b. Isolated bacteriocyte cell suspensions from the hydrothermal vent tubeworm *Riftia pachyptila*, a potent tool for cellular physiology in a chemoautotrophic symbiosis. *Marine Biology* **142**, 141–151.

De Cian, M.-C., Bailly, X., Morales, J., Strub, J.M., Van Dorsselaer, A. & Lallier, F.H. 2003c. Characterization of carbonic anhydrases from *Riftia pachyptila*, a symbiotic invertebrate from deep-sea hydrothermal vents. *Proteins—Structure Function and Genetics* **51**, 327–339.

De Cian, M.-C., Regnault, M. & Lallier, F.H. 2000. Nitrogen metabolites and related enzymatic activities in the body fluids and tissues of the hydrothermal vent tubeworm *Riftia pachyptila. Journal of Experimental Biology* **203**, 2907–2920.

Desbruyères, D., Segonzac, M. & Bright, M. 2006. *Handbook of Deep-Sea Hydrothermal Vent Fauna.* Linz-Dornach, Austria: Biologiezentrum der Oberösterreichischen Landesmuseen, 2nd edition.

Di Meo, C.A., Wilbur, A.E., Holben, W.E., Feldman, R.A., Vrijenhoek, R.C. & Cary, S.C. 2000. Genetic variation among endosymbionts of widely distributed vestimentiferan tubeworms. *Applied and Environmental Microbiology* **66**, 651–658.

Distel, D.L. & Felbeck, H. 1988. Pathways of inorganic carbon fixation in the endosymbiont bearing lucinid clam *Lucinoma aequizonata*, Part 1. Purification and characterization of the endosymbiotic bacteria. *Journal of Experimental Zoology* **247**, 1–10.

Dubilier, N., Bergin, C. & Lott, C. 2008. Symbiotic diversity in marine animals: the art of harnessing chemo-synthesis. *Nature Reviews Microbiology* **6**, 725–740.

Elsaied, H., Kimura, H. & Naganuma, T. 2002. Molecular characterization and endosymbiotic localization of the gene encoding D-ribulose 1,5-bisphosphate carboxylase-oxygenase (RuBisCO) form II in the deep-sea vestimentiferan trophosome. *Microbiology* **148**, 1947–1957.

Felbeck, H. 1981. Chemoautotrophic potential of the hydrothermal vent tubeworm, *Riftia pachyptila* Jones (Vestimentifera). *Science* **213**, 336–338.

Felbeck, H. 1985. CO_2 fixation in the hydrothermal vent tube worm *Riftia pachyptila* (Jones). *Physiological Zoology* **58**, 272–281.

Felbeck, H., Arndt, C., Hentschel, U. & Childress, J.J. 2004. Experimental application of vascular and coelomic catheterization to identify vascular transport mechanisms for inorganic carbon in the vent tubeworm, *Riftia pachyptila. Deep-Sea Research Part I—Oceanographic Research Papers* **51**, 401–411.

Felbeck, H. & Jarchow, J. 1998a. Carbon release from purified chemoautotrophic bacterial symbionts of the hydrothermal vent tubeworm *Riftia pachyptila. Physiological Zoology* **71**, 294–302.

Felbeck, H. & Jarchow, J. 1998b. The influence of different incubation media on the carbon transfer from the bacterial symbionts to the hydrothermal vent tube-worm *Riftia pachyptila. Cahiers de Biologie Marine* **39**, 279–282.

Felbeck, H. & Turner, P.J. 1995. CO_2 transport in catheterized hydrothermal vent tubeworms, *Riftia pachyptila* (Vestimentifera). *Journal of Experimental Zoology* **272**, 95–102.

Feldman, R.A., Black, M.B., Cary, C.S., Lutz, R.A. & Vrijenhoek, R.C. 1997. Molecular phylogenetics of bacterial endosymbionts and their vestimentiferan hosts. *Molecular Marine Biology and Biotechnology* **6**, 268–277.

Feldman, R.A., Shank, T.M., Black, M.B., Baco, A.R., Smith, C.R. & Vrijenhoek, R.C. 1998. Vestimentiferan on a whale fall. *Biological Bulletin* **194**, 116–119.

Fisher, C.R., Childress, J.J., Arp, A.J., Brooks, J.M., Distel, D., Favuzzi, J.A., Macko, S.A., Newton, A., Powell, M.A., Somero, G.N. & Soto, T. 1988. Physiology, morphology, and biochemical composition of *Riftia pachyptila* at Rose Garden in 1985. *Deep-Sea Research* **35**, 1745–1758.

Fisher, C.R., Childress, J.J., Macko, S.A. & Brooks, J.M. 1994. Nutritional interactions in Galapagos rift hydrothermal vent communities: inferences from stable carbon and nitrogen isotope analyses. *Marine Ecology Progress Series* **103**, 45–55.

Fisher, C.R., Childress, J.J. & Minnich, E. 1989. Autotrophic carbon fixation by the chemoautotrophic symbionts of *Riftia pachyptila. Biological Bulletin* **177**, 372–385.

Fisher, C.R., Urcuyo, I.A., Simpkins, M.A. & Nix, E. 1997. Life in the slow lane: growth and longevity of cold-seep vestimentiferans. *Marine Ecology—Pubblicazioni della Stazione Zoologica di Napoli I* **18**, 83–94.

Flores, J.F., Fisher, C.R., Carney, S.L., Green, B.N., Freytag, J.K., Schaeffer, S.W. & Royer, W.E. 2005. Sulfide binding is mediated by zinc ions discovered in the crystal structure of a hydrothermal vent tubeworm hemoglobin. *Proceedings of the National Academy of Sciences of the United States of America* **102**, 2713–2718.

Flores, J.F. & Hourdez, S.M. 2006. The zinc-mediated sulfide-binding mechanism of hydrothermal vent tubeworm 400-kDa hemoglobin. *Cahiers de Biologie Marine* **47**, 371–377.

Fornari, D.J., Haymon, R.M., Perfit, M.R., Gregg, T.K.P. & Edwards, M.H. 1998. Axial summit trough of the East Pacific Rise 9°–10°N: Geological characteristics and evolution of the axial zone on fast spreading mid-ocean ridge. *Journal of Geophysical Research* **103**, 9827–9855.

Franzén, A. 1956. On spermiogenesis, morphology of the spermatozoon, and biology of fertilization among invertebrates. *Zoologiska Bidrag fran Uppsala* **31**, 355–482.

Freytag, J.K., Girguis, P.R., Bergquist, D.C., Andras, J.P., Childress, J.J. & Fisher, C.R. 2001. A paradox resolved: sulfide acquisition by roots of seep tubeworms sustains net chemoautotrophy. *Proceedings of the National Academy of Sciences of the United States of America* **98**, 13408–13413.

Fujikura, K., Fujiwara, Y. & Kawato, M. 2006. A new species of *Osedax* (Annelida : Siboglinidae) associated with whale carcasses off Kyushu, Japan. *Zoological Science* **23**, 733–740.

Fujiwara, Y., Kawato, M., Yamamoto, T., Yamanaka, T., Sato Okoshi, W., Noda, C., Tsuchida, S., Komai, T., Cubelio, S.S., Sasakis, T., Jacobsen, K., Kubokawa, K., Fujikura, K., Maruyama, T., Furushima, Y., Okoshi, K., Miyake, H., Miyazaki, M., Nogi, Y., Yatabe, A. & Okutani, T. 2007. Three-year investigations into sperm whale-fall ecosystems in Japan. *Marine Ecology* **28**, 219–232.

Fustec, A., Desbruyères, D. & Juniper, S.K. 1987. Deep-sea hydrothermal vent communities at 13°N on the East Pacific Rise: microdistribution and temporal variations. *Biological Oceanography* **4**, 121–164.

Gaill, F. & Hunt, S. 1986. Tube of deep-sea hydrothermal vent worms *Riftia pachyptila* (Vestimentifera) and *Alvinella pompejana* (annelida). *Marine Ecology Progress Series* **34**, 267–274.

Gaill, F., Shillito, B., Lechaire, J.P., Chanzy, H. & Goffinet, G. 1992. The chitin secreting system from deep sea hydrothermal vent worms. *Biology of the Cell* **76**, 201–204.

Gaill, F., Shillito, B., Menard, F., Goffinet, G. & Childress, J.J. 1997. Rate and process of tube production by the deepsea hydrothermal vent tubeworm *Riftia pachyptila*. *Marine Ecology Progress Series* **148**, 135–143.

Gardiner, S.L. & Hourdez, S. 2003. On the occurrence of the vestimentiferan tube worm *Lamellibrachia luymesi* van der Land and Norrevang, 1975 (Annelida : Pogonophora) in hydrocarbon seep communities in the Gulf of Mexico. *Proceedings of the Biological Society of Washington* **116**, 380–394.

Gardiner, S.L. & Jones, M.L. 1985. Ultrastructure of spermiogenesis in the vestimentiferan tube worm *Riftia pachyptila* (Pogonophora: Obturata). *Transactions of the American Microscopical Society* **104**, 19–44.

Gardiner, S.L. & Jones, M.L. 1993. Vestimentifera. In *Microscopic Anatomy of Invertebrates. Onychophora, Chilopoda and Lesser Protostomata*, F.W. Harrison & M.E. Rice (eds). New York: Wiley-Lyss, 371–460.

Gardiner, S.L. & Jones, M.L. 1994. On the significance of larval and juvenile morphology for suggesting phylogenetic relationships of the Vestimentifera. *American Zoologist* **34**, 513–522.

Gardiner, S.L., McMullin, E. & Fisher, C.R. 2001. *Seepiophila jonesi*, a new genus and species of vestimentiferan tube worm (Annelida : Pogonophora) from hydrocarbon seep communities in the Gulf of Mexico. *Proceedings of the Biological Society of Washington* **114**, 694–707.

Girguis, P.R. & Childress, J.J. 1998. H+ equivalent elimination by the tube-worm *Riftia pachyptila*. *Cahiers de Biologie Marine* **39**, 295–296.

Girguis, P.R. & Childress, J.J. 2006. Metabolite uptake, stoichiometry and chemoautotrophic function of the hydrothermal vent tubeworm *Riftia pachyptila*: responses to environmental variations in substrate concentrations and temperature. *Journal of Experimental Biology* **209**, 3516–3528.

Girguis, P.R., Childress, J.J., Freytag, J.K., Klose, K. & Stuber, R. 2002. Effects of metabolite uptake on proton-equivalent elimination by two species of deep-sea vestimentiferan tubeworm, *Riftia pachyptila* and *Lamellibrachia cf luymesi*: proton elimination is a necessary adaptation to sulfide-oxidizing chemoautotrophic symbionts. *Journal of Experimental Biology* **205**, 3055–3066.

Girguis, P.R., Lee, R.W., Desaulniers, N., Childress, J.J., Pospesel, M., Felbeck, H. & Zal, F. 2000. Fate of nitrate acquired by the tubeworm *Riftia pachyptila*. *Applied and Environmental Microbiology* **66**, 2783–2790.

Glover, A.G., Kallstrom, B., Smith, C.R. & Dahlgren, T.G. 2005. World-wide whale worms? A new species of *Osedax* from the shallow north Atlantic. *Proceedings of the Royal Society B Biological Sciences* **272**, 2587–2592.

Goffredi, S.K. & Childress, J.J. 2001. Activity and inhibitor sensitivity of ATPases in the hydrothermal vent tubeworm *Riftia pachyptila*: a comparative approach. *Marine Biology* **138**, 259–265.

Goffredi, S.K., Childress, J.J., Desaulniers, N.T. & Lallier, F.H. 1997a. Sulfide acquisition by the vent worm *Riftia pachyptila* appears to be via uptake of HS⁻, rather than H₂S. *Journal of Experimental Biology* **200**, 2609–2616.

Goffredi, S.K., Childress, J.J., Desaulniers, N.T., Lee, R.W., Lallier, F.H. & Hammond, D. 1997b. Inorganic carbon acquisition by the hydrothermal vent tubeworm *Riftia pachyptila* depends upon high external PCO_2 and upon proton-equivalent ion transport by the worm. *Journal of Experimental Biology* **200**, 883–896.

Goffredi, S.K., Girguis, P.R., Childress, J.J. & Desaulniers, N.T. 1999. Physiological functioning of carbonic anhydrase in the hydrothermal vent tubeworm *Riftia pachyptila*. *Biological Bulletin* **196**, 257–264.

Goffredi, S.K., Johnson, S.B. & Vrijenhoek, R.C. 2007. Genetic diversity and potential function of microbial symbionts associated with newly discovered species of *Osedax* polychaete worms. *Applied and Environmental Microbiology* **73**, 2314–2323.

Gollner, S., Zekely, J., Govenar, B., Le Bris, N., Nemeschkal, H.L., Fisher, C.R. & Bright, M. 2007. Tubeworm-associated permanent meiobenthic communities from two chemically different hydrothermal vent sites on the East Pacific Rise. *Marine Ecology Progress Series* **337**, 39–49.

Gollner, S., Zekely, J., Van Dover, C.L., Govenar, B., Le Bris, N., Nemeschkal, H.L. & Bright, M. 2006. Benthic copepod communities associated with tubeworm and mussel aggregations on the East Pacific Rise. *Cahiers de Biologie Marine* **47**, 397–402.

Govenar, B., Freeman, M., Bergquist, D.C., Johnson, G.A. & Fisher, C.R. 2004. Composition of a one-year-old *Riftia pachyptila* community following a clearance experiment: insight to succession patterns at deep-sea hydrothermal vents. *Biological Bulletin* **207**, 177–182.

Govenar, B., Le Bris, N., Gollner, S., Glanville, J., Aperghis, A.B., Hourdez, S. & Fisher, C.R. 2005. Epifaunal community structure associated with *Riftia pachyptila* aggregations in chemically different hydrothermal vent habitats. *Marine Ecology Progress Series* **305**, 67–77.

Govenar, B.W., Bergquist, D.C., Urcuyo, I.A., Eckner, J.T. & Fisher, C.R. 2002. Three *Ridgeia piscesae* assemblages from a single Juan de Fuca Ridge sulphide edifice: structurally different and functionally similar. *Cahiers de Biologie Marine* **43**, 247–252.

Grassle, J.F. 1986. The ecology of deep-sea hydrothermal vent communities. *Advances in Marine Biology* **23**, 301–362.

Grieshaber, M.K. & Volkel, S. 1998. Animal adaptations for tolerance and exploitation of poisonous sulfide. *Annual Review of Physiology* **60**, 33–53.

Hahlbeck, E., Pospesel, M.A., Zal, F., Childress, J.J. & Felbeck, H. 2005. Proposed nitrate binding by hemoglobin in *Riftia pachyptila* blood. *Deep-Sea Research Part I—Oceanographic Research Papers* **52**, 1885–1895.

Halanych, K.M. 2005. Molecular phylogeny of siboglinid annelids (a.k.a. pogonophorans): a review. *Hydrobiologia* **535**, 297–307.

Halanych, K.M., Feldman, R.A. & Vrijenhoek, R.C. 2001. Molecular evidence that *Sclerolinum brattstromi* is closely related to vestimentiferans, not to frenulate pogonophorans (Siboglinidae, Annelida). *Biological Bulletin* **201**, 65–75.

Hand, S.C. 1987. Trophosome ultrastructure and the characterization of isolated bacteriocytes from invertebrate-sulfur bacteria symbioses. *Biological Bulletin* **173**, 260–276.

Harmer, T.L., Rotjan, R.D., Nussbaumer, A.D., Bright, M., Ng, A.W., DeChaine, E.G. & Cavanaugh, C.M. 2008. Free-living tube worm endosymbionts found at deep-sea vents. *Applied and Environmental Microbiology* **74**, 3895–3898.

Hashimoto, J., Miura, T., Fujikura, K. & Ossaka, J. 1993. Discovery of vestimentiferan tube-worms in the euphotic zone. *Zoological Science* **10**, 1063–1067.

Haymon, R.M., Fornari, D.J., Edwards, M.H., Carbotte, S., Wright, D. & MacDonald, K.C. 1991. Hydrothermal vent distribution along the East Pacific Rise Crest (9°09′–54′N) and its relationship to magmatic and tectonic processes on fast-spreading mid-ocean ridges. *Earth and Planetary Science Letters* **104**, 513–534.

Haymon, R.M., Fornari, D.J., Von Damm, K.L., Lilley, M.D., Perfit, M.R., Edmond, J.M., Shanks, W.C., III, Lutz, R.A., Grebmeier, J.M., Carbotte, S., Wright, D., McLaughlin, E., Smith, M., Beedle, N. & Olson, E. 1993. Volcanic eruption of the mid-ocean ridge along East Pacific Rise crest at 9°45′–52′N. Direct submersible observations of seafloor phenomena associated with an eruption event in April, 1991. *Earth and Planetary Science Letters* **119**, 85–101.

Heimler, W. 1988. Larvae. In *The Ultrastructure of Polychaeta. Microfauna Marina 4*, W. Westheide & C.O. Hermans (eds). Stuttgart: Gustav Fischer Verlag, 353–371.

Henry, R.P. 1996. Multiple roles of carbonic anhydrase in cellular transport and metabolism. *Annual Review of Physiology* **58**, 523–538.

Hentschel, U. & Felbeck, H. 1993. Nitrate respiration in the hydrothermal vent tubeworm *Riftia pachyptila*. *Nature* **366**, 338–340.

Hentschel, U., Pospesel, M.A. & Felbeck, H. 1998. Evidence for a nitrate uptake mechanism in the hydrothermal vent tube-worm *Riftia pachyptila*. *Cahiers de Biologie Marine* **39**, 301–304.

Hentschel, U., Steinert, M. & Hacker, J. 2000. Common molecular mechanisms of symbiosis and pathogenesis. *Trends in Microbiology* **8**, 226–231.

Hessler, R.R. & Smithey, W.M. 1983. The distribution and community structure of megafauna at the Galápagos Rift hydrothermal vents. In *Hydrothermal Processes at Seafloor Spreading Centers*, P.A. Rona et al. (eds). New York: Plenum Press, 735–770.

Hessler, R.R., Smithey, W.M., Boudrias, M.A., Keller, C.H., Lutz, R.A. & Childress, J.J. 1988. Temporal change in megafauna at the Rose Garden hydrothermal vent (Galapagos Rift; eastern tropical Pacific). *Deep-Sea Research* **35**, 1681–1709.

Hessler, R.R., Smithey, W.M. & Keller, C.H. 1985. Spatial and temporal variation of giant clams, tube worms and mussels at deep-sea hydrothermal vents. *Bulletin of the Biological Society of Washington* **6**, 411–428.

Hilario, A., Young, C.M. & Tyler, P.A. 2005. Sperm storage, internal fertilization, and embryonic dispersal in vent and seep tubeworms (Polychaeta : Siboglinidae : Vestimentifera). *Biological Bulletin* **208**, 20–28.

Hughes, D.J. & Crawford, M. 2006. A new record of the vestimentiferan Lamellibrachia sp. (Polychaeta: Siboglinidae) from a deep shipwreck in the eastern Mediterranean. JMBA2 Biodiversity Records. Available HTTP: http://www.mba.ac.uk/jmba/pdf/5198.pdf (accessed 23 Jan 2010).

Hughes, D.S., Felbeck, H. & Stein, J.L. 1998. Signal transduction and motility genes from the bacterial endosymbionts of *Riftia pachyptila*. *Cahiers de Biologie Marine* **39**, 305–308.

Humes, A.G. 1988. Copepoda from deep-sea hydrothermal vents and cold seeps. *Hydrobiologia* **167/168**, 549–554.

Humes, A.G. & Dojiri, M. 1980. A siphonostome copepod associated with a vestimentiferan from the Galápagos Rift and the East Pacific Rise. *Proceedings of the Biological Society of Washington* **93**, 697–707.

Hunt, H.L., Metaxas, A., Jennings, R.M., Halanych, K.M. & Mullineaux, L.S. 2004. Testing biological control of colonization by vestimentiferan tubeworms at deep-sea hydrothermal vents (East Pacific Rise, 9 degrees 50′N). *Deep-Sea Research Part I—Oceanographic Research Papers* **51**, 225–234.

Hurtado, L.A., Mateos, M., Lutz, R.A. & Vrijenhoek, R.C. 2002. Molecular evidence for multiple species of *Oasisia* (Annelida : Siboglinidae) at eastern Pacific hydrothermal vents. *Cahiers de Biologie Marine* **43**, 377–380.

Ivanenko, V.N. & Defaye, D. 2006. Copepoda. In *Handbook of Hydrothermal Vent Fauna*, D. Desbruyères et al. (eds). Linz-Dornach, Austria: Biologiezentrum der Oberösterreichischen Landesmuseen, 2nd edition, 316–355.

Ivanov, A.V. 1951. On including the genus *Siboglinum* Caullery in the class Pogonophora [In Russian, English summary]. *Doklady Akadami Nauk SSSR* **76**, 739–742.

Jenkins, C.D., Ward, M.E., Turnipseed, M., Osterberg, J. & Van Dover, C.L. 2002. The digestive system of the hydrothermal vent polychaete *Galapagomystides aristata* (Phyllodocidae): evidence for hematophagy? *Invertebrate Biology* **121**, 243–254.

Jennings, R.M. & Halanych, K.M. 2005. Mitochondrial genomes of *Clymenella torquata* (Maldanidae) and *Riftia pachyptila* (Siboglinidae): evidence for conserved gene order in Annelida. *Molecular Biology and Evolution* **22**, 210–222.

Johansson, K.E. 1939. *Lamellisabella zachsi* Uschakow, ein Vertreter eine neuen Tierklasse Pogonophora. *Zoologiska Bidrag fran Uppsala* **18**, 253–268.

Johnson, K.S., Beehler, C.L., Sakamoto-Arnold, C.M. & Childress, J.J. 1986. *In situ* measurements of chemical distributions in a deep-sea hydrothermal vent field. *Science* **231**, 1139–1141.

Johnson, K.S., Childress, J.J. & Beehler, C.L. 1988a. Short-term temperature variability in the Rose Garden hydrothermal vent field: an unstable deep-sea environment. *Deep-Sea Research* **35**, 1711–1721.

Johnson, K.S., Childress, J.J., Hessler, R.R., Sakamoto-Arnold, C.M. & Beehler, C.L. 1988b. Chemical and biological interactions in the Rose Garden hydrothermal vent field, Galapagos spreading center. *Deep-Sea Research* **35**, 1723–1744.

Jollivet, D. 1993. *Distribution et évolution de la faune associée aux sources hydrothermales profondes à 13°N sur la dorsale du Pacifique Oriental: le cas particulier des polychètes Alvinellidae.* PhD thesis, Université de Bretagne Occidentale, Brest, France.

Jones, M.L. 1981. *Riftia pachyptila*, new genus, new species, the vestimentiferan worm from the Galapagos Rift geothermal vents (Pogonophora). *Proceedings of the Biological Society of Washington* **93**, 1295–1313.

Jones, M.L. 1985. On the Vestimentifera, new phylum: six new species, and other taxa, from hydrothermal vents and elsewhere. *Bulletin of the Biological Society of Washington* **6**, 117–158.

Jones, M.L. 1988. The Vestimentifera, their biology, systematic and evolutionary patterns. *Oceanologica Acta* **8**, 69–82.

Jones, M.L. & Gardiner, S.L. 1985. Light and scanning electron microscopic studies of spermatogenesis in the vestimentiferan tube worm *Riftia pachyptila* (Pogonophora : Obturata). *Transactions of the American Microscopical Society* **104**, 1–18.

Jones, M.L. & Gardiner, S.L. 1989. On the early development of the vestimentiferan tube worm *Ridgeia* sp. and observations on the nervous system and trophosome of *Ridgeia* sp. and *Riftia pachyptila*. *Biological Bulletin* **177**, 254–276.

Joye, S.B., Boetius, A., Orcutt, B.N., Montoya, J.P., Schulz, H.N., Erickson, M.J. & Lugo, S.K. 2004. The anaerobic oxidation of methane and sulfate reduction in sediments from Gulf of Mexico cold seeps. *Chemical Geology* **205**, 219–238.

Julian, D., Gaill, F., Wood, E., Arp, A.J. & Fisher, C.R. 1999. Roots as a site of hydrogen sulfide uptake in the hydrocarbon seep vestimentiferan *Lamellibrachia* sp. *Journal of Experimental Biology* **202**, 2245–2257.

Juniper, S.K., Tunnicliffe, V. & Southward, E.C. 1992. Hydrothermal vents in turbidite sediments on a Northeast Pacific spreading center; organisms and substratum at an ocean drilling site. *Canadian Journal of Zoology—Revue Canadienne de Zoologie* **70**, 1792–1809.

Kennicutt, M.C., Brooks, J.M., Bidigare, R.R., Fay, R.R., Wade, T.L. & McDonald, T.J. 1985. Vent-type taxa in a hydrocarbon seep region on the Louisiana slope. *Nature* **317**, 351–353.

Kim, D. & Ohta, S. 2000. TEM observation studies on the chemoautotrophic symbiotic bacteria of invertebrates inhabiting at vents and seeps. *Ocean Research* **22**, 1–13.

Kochevar, R.E. & Childress, J.J. 1996. Carbonic anhydrase in deepsea chemoautotrophic symbioses. *Marine Biology* **125**, 375–383.

Kochevar, R.E., Govind, N.S. & Childress, J.J. 1993. Identification and characterization of two carbonic anhydrases from the hydrothermal vent tubeworm *Riftia pachyptila* Jones. *Molecular Marine Biology and Biotechnology* **2**, 10–19.

Kojima, S., Ohta, S., Yamamoto, T., Miura, T., Fujiwara, Y., Fujikura, K. & Hashimoto, J. 2002. Molecular taxonomy of vestimentiferans of the western Pacific and their phylogenetic relationship to species of the eastern Pacific—II—Families Escarpiidae and Arcovestiidae. *Marine Biology* **141**, 57–64.

Kojima, S., Ohta, S., Yamamoto, T., Miura, T., Fujiwara, Y. & Hashimoto, J. 2001. Molecular taxonomy of vestimentiferans of the western Pacific and their phylogenetic relationship to species of the eastern Pacific. I. Family Lamellibrachiidae. *Marine Biology* **139**, 211–219.

Kojima, S., Ohta, S., Yamamoto, T., Yamaguchi, T., Miura, T., Fujiwara, Y., Fujikura, K. & Hashimoto, J. 2003. Molecular taxonomy of vestimentiferans of the western Pacific, and their phylogenetic relationship to species of the eastern Pacific III. *Alaysia*-like vestimentiferans and relationships among families. *Marine Biology* **142**, 625–635.

Laue, B.E. & Nelson, D.C. 1994. Characterization of the gene encoding the autotrophic ATP sulfurylase from the bacterial endosymbiont of the hydrothermal vent tubeworm *Riftia pachyptila*. *Journal of Bacteriology* **176**, 3723–3729.

Le Bris, N., Govenar, B., Le Gall, C. & Fisher, C.R. 2006a. Variability of physico-chemical conditions in 9°50′N EPR diffuse flow vent habitats. *Marine Chemistry* **98**, 167–182.

Le Bris, N., Rodier, P., Sarradin, P.M. & Le Gall, C. 2006b. Is temperature a good proxy for sulfide in hydrothermal vent habitats? *Cahiers de Biologie Marine* **47**, 465–470.

Le Bris, N., Sarradin, P.M. & Caprais, J.C. 2003. Contrasted sulphide chemistries in the environment of 13 degrees N EPR vent fauna. *Deep-Sea Research Part I—Oceanographic Research Papers* **50**, 737–747.

Lee, R.W. & Childress, J.J. 1994. Assimilation of inorganic nitrogen by marine invertebrates and their chemoautotrophic and methanotrophic symbionts. *Applied and Environmental Microbiology* **60**, 1852–1858.

Lee, R.W., Robinson, J.J. & Cavanaugh, C.M. 1999. Pathways of inorganic nitrogen assimilation in chemoautotrophic bacteria-marine invertebrate symbioses: expression of host and symbiont glutamine synthetase. *Journal of Experimental Biology* **202**, 289–300.

Levin, L.A. 2005. Ecology of cold seep sediments: interactions of fauna with flow, chemistry and microbes. *Oceanography and Marine Biology An Annual Review* **43**, 1–46.

Little, C.T.S., Herrington, R.J., Maslennikov, V.V., Morris, N.J. & Zaykov, V.V. 1997. Silurian hydrothermal-vent community from the southern Urals, Russia. *Nature* **385**, 146–148.

Little, C.T.S. & Vrijenhoek, R.C. 2003. Are hydrothermal vent animals living fossils? *Trends in Ecology and Evolution* **18**, 582–588.

Lopez-Garcia, P., Gaill, F. & Moreira, D. 2002. Wide bacterial diversity associated with tubes of the vent worm *Riftia pachyptila. Environmental Microbiology* **4**, 204–215.

Luther, G.W., Glazer, B.T., Ma, S.F., Trouwborst, R.E., Moore, T.S., Metzger, E., Kraiya, C., Waite, T.J., Druschel, G., Sundby, B., Taillefert, M., Nuzzio, D.B., Shank, T.M., Lewis, B.L. & Brendel, P.J. 2008. Use of voltammetric solid-state (micro)electrodes for studying biogeochemical processes: laboratory measurements to real time measurements with an *in situ* electrochemical analyzer (ISEA). *Marine Chemistry* **108**, 221–235.

Luther, G.W., Rozan, T.F., Taillefert, M., Nuzzio, D.B., Di Meo, C., Shank, T.M., Lutz, R.A. & Cary, S.C. 2001. Chemical speciation drives hydrothermal vent ecology. *Nature* **410**, 813–816.

Lutz, R.A., Shank, T.M., Fornari, D.J., Haymon, R.M., Lilley, M.D., Von Damm, K.L. & Desbruyères, D. 1994. Rapid growth at deep-sea vents. *Nature* **371**, 663–664.

MacAvoy, S.E., Fisher, C.R., Carney, R.S. & Macko, S.A. 2005. Nutritional associations among fauna at hydrocarbon seep communities in the Gulf of Mexico. *Marine Ecology Progress Series* **292**, 51–60.

MacDonald, I.R., Boland, G.S., Baker, J.S., Brooks, J.M., Kennicut II, M.C. & Bidigare, R.R. 1989. Gulf of Mexico hydrocarbon seep communities. II. Spatial distribution of seep organisms and hydrocarbons at Bush Hill. *Marine Biology* **101**, 235–247.

MacDonald, I.R., Guinasso, N.L., Reilly, J.F., Brooks, J.M., Callender, W.R. & Gabrielle, S.G. 1990. Gulf of Mexico hydrocarbon seep communities. VI. Patterns in community structure and habitat. *Geo-Marine Letters* **10**, 244–252.

MacDonald, I.R., Tunnicliffe, V. & Southward, E.C. 2002. Detection of sperm transfer and synchronous fertilization in *Ridgeia piscesae* at Endeavour Segment, Juan de Fuca Ridge. *Cahiers de Biologie Marine* **43**, 395–398.

Madigan, M.T., Martinko, J.M., Dunlap, P.V. & Clark, D.P. 2009. *Brock Biology of Microorganisms*. San Francisco: Pearson Benjamin Cummings, 12th edition.

Malakhov, V.V., Popelyaev, I.S. & Galkin, S.V. 1996a. Microscopic anatomy of *Ridgeia phaeophiale* Jones, 1985 (Pogonophora, Vestimentifera) and the problem of the position of Vestimentifera in the system of animal kingdom. 2. Integument, nerve system, connective tissue, musculature. *Biologiya Morya* **22**, 139–147.

Malakhov, V.V., Popelyaev, I.S. & Galkin, S.V. 1996b. Microscopic anatomy of *Ridgeia phaeophiale* Jones, 1985 (Pogonophora, Vestimentifera) and the problem of the position of Vestimentifera in the system of the Animal Kingdom. III. Rudimentary digestive system, trophosome, and blood vascular system. *Russian Journal of Marine Biology* **22**, 189–198.

Mañé-Garzón, F. & Montero, R. 1985. Sobre una nueve forma de verme tubicola *Lamellibrachia victori* n. sp. (Vestimentifera), proposition de un nueve phylum: Mesoneurophora. *Revista de Biologia del Uruguay* **8**, 1–28.

Markert, S., Arndt, C., Felbeck, H., Becher, D., Sievert, S.M., Hugler, M., Albrecht, D., Robidart, J., Bench, S., Feldman, R.A., Hecker, M. & Schweder, T. 2007. Physiological proteomics of the uncultured endosymbiont of *Riftia pachyptila. Science* **315**, 247–250.

Marotta, R., Melone, G., Bright, M. & Ferraguti, M. 2005. Spermatozoa and sperm aggregates in the vestimentiferan *Lamellibrachia luymesi* compared with those of *Riftia pachyptila* (Polychaeta : Siboglinidae : Vestimentifera). *Biological Bulletin* **209**, 215–226.

Marsh, A.G., Mullineaux, L.S., Young, C.M. & Manahan, D.T. 2001. Larval dispersal potential of the tubeworm *Riftia pachyptila* at deep-sea hydrothermal vents. *Nature* **411**, 77–80.

Martineu, P., Juniper, S.K., Fisher, C.R. & Massoth, G.J. 1997. Sulfide binding in the body fluids of hydrothermal vent alvinellid polychaetes. *Physiological Zoology* **70**, 578–588.

McMullin, E.R., Hourdez, S., Schaeffer, S.W. & Fisher, C.R. 2003. Phylogeny and biogeography of deep sea vestimentiferan tubeworms and their bacterial symbionts. *Symbiosis* **34**, 1–41.

Micheli, F., Peterson, C.H., Mullineaux, L.S., Fisher, C.R., Mills, S.W., Sancho, G., Johnson, G.A. & Lenihan, H.S. 2002. Predation structures communities at deep-sea hydrothermal vents. *Ecological Monographs* **72**, 365–382.

Minic, Z. & Herve, G. 2003. Arginine metabolism in the deep sea tube worm *Riftia pachyptila* and its bacterial endosymbiont. *Journal of Biological Chemistry* **278**, 40527–40533.

Minic, Z. & Herve, G. 2004. Biochemical and enzymological aspects of the symbiosis between the deep-sea tubeworm *Riftia pachyptila* and its bacterial endosymbiont. *European Journal of Biochemistry* **271**, 3093–3102.

Minic, Z., Pastra-Landis, S., Gaill, F. & Herve, G. 2002. Catabolism of pyrimidine nucleotides in the deep-sea tube worm *Riftia pachyptila*. *Journal of Biological Chemistry* **277**, 127–134.

Minic, Z., Simon, V., Penverne, B., Gaill, F. & Herve, G. 2001. Contribution of the bacterial endosymbiont to the biosynthesis of pyrimidine nucleotides in the deep-sea tube worm *Riftia pachyptila*. *Journal of Biological Chemistry* **276**, 23777–23784.

Miura, T. & Kojima, S. 2006. Two new species of vestimentiferan tubeworm (Polychaeta: Siboglinidae a.k.a. Pogonophora) from the Brothers Caldera, Kermadec Arc, South Pacific Ocean. *Species Diversity* **11**, 209–224.

Miura, T., Tsukahara, J. & Hashimoto, J. 1997. *Lamellibrachia satsuma*, a new species of vestimentiferan worms (Annelida: Pogonophora) from a shallow hydrothermal vent in Kagoshima Bay, Japan. *Proceedings of the Biological Society of Washington* **110**, 447–456.

Miyake, H., Tsukahara, J., Hashimoto, J., Uematsu, K. & Maruyama, T. 2006. Rearing and observation methods of vestimentiferan tubeworm and its early development at atmospheric pressure. *Cahiers de Biologie Marine* **47**, 471–475.

Mullineaux, L.S., Fisher, C.R., Peterson, C.H. & Schaeffer, S.W. 2000. Tubeworm succession at hydrothermal vents: use of biogenic cues to reduce habitat selection error? *Oecologia* **123**, 275–284.

Mullineaux, L.S., Mills, S.W. & Goldman, E. 1998. Recruitment variation during a pilot colonization study of hydrothermal vents (9 degrees 50′N, East Pacific Rise). *Deep-Sea Research Part II—Topical Studies in Oceanography* **45**, 441–464.

Mullineaux, L.S., Mills, S.W., Sweetman, A.K., Beaudreau, A.H., Metaxas, A. & Hunt, H.L. 2005. Vertical, lateral and temporal structure in larval distributions at hydrothermal vents. *Marine Ecology Progress Series* **293**, 1–16.

Mullineaux, L.S., Speer, K.G., Thurnherr, A.M., Maltrud, M.E. & Vangriesheim, A. 2002. Implications of cross-axis flow for larval dispersal along mid-ocean ridges. *Cahiers de Biologie Marine* **43**, 281–284.

Naganuma, T., Elsaied, H.E., Hoshii, D. & Kimura, H. 2005. Bacterial endosymbioses of gutless tube-dwelling worms in nonhydrothermal vent habitats. *Marine Biotechnology* **7**, 416–428.

Naganuma, T., Kato, C., Hirayama, H., Moriyama, N., Hashimoto, J. & Horikoshi, K. 1997a. Intracellular ocurrence of ε-Proteobacterial 16S rDNA sequences in the vestimentiferan trophosome. *Journal of Oceanography* **53**, 193–197.

Naganuma, T., Naka, J., Okayama, Y., Minami, A. & Horikoshi, K. 1997b. Morphological diversity of the microbial population in a vestimentiferan tubeworm. *Journal of Marine Biotechnology* **5**, 119–123.

Nakagawa, S. & Takai, K. 2008. Deep-sea vent chemoautotrophs: diversity, biochemistry and ecological significance. *Fems Microbiology Ecology* **65**, 1–14.

Nelson, K. & Fisher, C.R. 2000. Absence of cospeciation in deep-sea vestimentiferan tube worms and their bacterial endosymbionts. *Symbiosis* **28**, 1–15.

Numoto, N., Nakagawa, T., Kita, A., Sasayama, Y., Fukumori, Y. & Miki, K. 2005. Structure of an extracellular giant hemoglobin of the gutless beard worm *Oligobrachia mashikoi*. *Proceedings of the National Academy of Sciences of the United States of America* **102**, 14521–14526.

Nussbaumer, A.D., Fisher, C.R. & Bright, M. 2006. Horizontal endosymbiont transmission in hydrothermal vent tubeworms. *Nature* **441**, 345–348.

Nyholm, S.V., Robidart, J. & Girguis, P.R. 2008. Coupling metabolite flux to transcriptomics: insights into the molecular mechanisms underlying primary productivity by the hydrothermal vent tubeworm *Ridgeia piscesae*. *Biological Bulletin* **214**, 255–265.

Ortega, J.A., Ortega, J.M. & Julian, D. 2008. Hypotaurine and sulfhydryl-containing antioxidants reduce H_2S toxicity in erythrocytes from a marine invertebrate. *Journal of Experimental Biology* **211**, 3816–3825.

Paull, C.K., Hecker, B., Commeau, R., Freeman-Lynde, R.P., Neuman, C., Corso, W.P., Golubic, S., Hokk, J.E., Sikes, E. & Curray, J. 1984. Biological communities at the Florida escarpment resemble hydrothermal vent taxa. *Science* **226**, 965–967.

Peckmann, J., Little, C.T.S., Gill, F. & Reitner, J. 2005. Worm tube fossils from the Hollard Mound hydrocarbon-seep deposit, Middle Devonian, Morocco: palaeozoic seep-related vestimentiferans? *Palaeogeography Palaeoclimatology Palaeoecology* **227**, 242–257.

Pflugfelder, B., Cary, S. & Bright, M. 2009. Dynamics of cell proliferation and apoptosis reflect different life strategies in hydrothermal vent and cold seep vestimentiferan tubeworms. *Cell and Tissue Research* **337**, 149–165.

Pflugfelder, B., Fisher, C.R. & Bright, M. 2005. The color of the trophosome: elemental sulfur distribution in the endosymbionts of *Riftia pachyptila* (Vestimentifera: Siboglinidae). *Marine Biology* **146**, 895–901.

Pleijel, F., Dahlgren, T.G. & Rouse, G.W. 2009. Progress in systematics: from Siboglinidae to Pogonophora and Vestimentifera and back to Siboglinidae. *Comptes Rendus Biologies* **332**, 140–148.

Pospesel, M.A., Hentschel, U. & Felbeck, H. 1998. Determination of nitrate in the blood of the hydrothermal vent tubeworm *Riftia pachyptila* using a bacterial nitrate reduction assay. *Deep-Sea Research Part I—Oceanographic Research Papers* **45**, 2189–2200.

Powell, M.A. & Somero, G.N. 1986. Adaptations to sulfide by hydrothermal vent animals: sites and mechanisms of detoxification and metabolism. *Biological Bulletin* **171**, 274–290.

Pradillon, F., Schmidt, A., Peplies, J. & Dubilier, N. 2007. Species identification of marine invertebrate early stages by whole-larvae *in situ* hybridisation of 18S ribosomal RNA. *Marine Ecology Progress Series* **333**, 103–116.

Pruski, A.M., De Wit, R. & Fiala-Medioni, A. 2001. Carrier of reduced sulfur is a possible role for thiotaurine in symbiotic species from hydrothermal vents with thiotrophic symbionts. *Hydrobiologia* **461**, 9–13.

Pruski, A.M. & Fiala-Medioni, A. 2003. Stimulatory effect of sulphide on thiotaurine synthesis in three hydrothermal-vent species from the East Pacific Rise. *Journal of Experimental Biology* **206**, 2923–2930.

Pruski, A.M., Fiala Medioni, A., Fisher, C.R. & Colomines, J.C. 2000. Composition of free amino acids and related compounds in invertebrates with symbiotic bacteria at hydrocarbon seeps in the Gulf of Mexico. *Marine Biology* **136**, 411–420.

Qian, P.-Y. 1999. Larval settlement of polychaetes. *Hydrobiologia* **402**, 239–253.

Rau, G.H. 1981. Hydrothermal vent clam and tubeworm $^{13}C/^{12}C$: further evidence of nonphotosynthetic food sources. *Science* **213**, 338–340.

Ravaux, J., Chamoy, L. & Shillito, B. 2000. Synthesis and maturation processes in the exoskeleton of the vent worm *Riftia pachyptila*. *Marine Biology* **136**, 505–512.

Ravaux, J., Gay, L., Voss-Foucart, M.F. & Gaill, F. 1998. Tube growth process in the deep-sea hydrothermal vent tube-worm *Riftia pachyptila* (Vestimentifera): synthesis and degradation of chitin. *Cahiers de Biologie Marine* **39**, 99–107.

Renosto, F., Martin, R.L., Borrell, J.L., Nelson, D.C. & Segel, I.H. 1991. ATP sulfurylase from trophosome tissue of *Riftia pachyptila* (hydrothermal vent tube worm). *Archives of Biochemistry and Biophysics* **290**, 66–78.

Roberts, H.H. & Aharon, P. 1994. Hydrocarbon-derived carbonate buildups of the northern Gulf of Mexico continental slope: a review of submersible investigations. *Geo-Marine Letters* **14**, 135–148.

Robidart, J.C., Bench, S.R., Feldman, R.A., Novoradovsky, A., Podell, S.B., Gaasterland, T., Allen, E.E. & Felbeck, H. 2008. Metabolic versatility of the *Riftia pachyptila* endosymbiont revealed through metagenomics. *Environmental Microbiology* **10**, 727–737.

Robinson, J.J. & Cavanaugh, C.M. 1995. Expression of form I and form II Rubisco in chemoautotrophic symbioses: implications for the interpretation of stable carbon isotope values. *Limnology and Oceanography* **40**, 1496–1502.

Robinson, J.J., Scott, K.M., Swanson, S.T., O'Leary, M.H., Horken, K., Tabita, F.R. & Cavanaugh, C.M. 2003. Kinetic isotope effect and characterization of form II RubisCO from the chemoautotrophic endosymbionts of the hydrothermal vent tubeworm *Riftia pachyptila*. *Limnology and Oceanography* **48**, 48–54.

Robinson, J.J., Stein, J.L. & Cavanaugh, C.M. 1998. Cloning and sequencing of a form II ribulose-1,5-biphosphate carboxylase/oxygenase from the bacterial symbiont of the hydrothermal vent tubeworm *Riftia pachyptila*. *Journal of Bacteriology* **180**, 1596–1599.

Rouse, G.W. 1999. Trochophore concepts: ciliary bands and the evolution of larvae in spiralian Metazoa. *Biological Journal of the Linnean Society* **66**, 411–464.

Rouse, G.W. 2001. A cladistic analysis of Siboglinidae Caullery, 1914 (Polychaeta, Annelida): formerly the phyla Pogonophora and Vestimentifera. *Zoological Journal of the Linnean Society* **132**, 55–80.

Rouse, G.W. 2005. Annelid sperm and fertilization biology. *Hydrobiologia* **535/536**, 167–178.

Rouse, G.W. & Fauchald, K. 1997. Cladistics and polychaetes. *Zoologica Scripta* **26**, 139–204.

Rouse, G.W., Goffredi, S.K. & Vrijenhoek, R.C. 2004. *Osedax*: bone-eating marine worms with dwarf males. *Science* **305**, 668–671.

Rouse, G.W. & Jamieson, B.G.M. 1987. An ultrastructural study of the spermatozoon of the polychaetes *Eurythoe complanata* (Amphinomidae), *Clymenella* sp. and *Micromaldane* sp. (Maldanidae), with definition of sperm types in relation to reproductive biology. *Journal of Submicroscopic Cytology* **19**, 573–584.

Rouse, G.W., Worsaae, K., Johnson, S.B., Jones, W.J. & Vrijenhoek, R.C. 2008. Acquisition of dwarf male "harems" by recently settled females of *Osedax roseus* n. sp. (Siboglinidae; Annelida). *Biological Bulletin* **214**, 67–82.

Rousset, V., Rouse, G.W., Siddall, M.E., Tillier, A. & Pleijel, F. 2004. The phylogenetic position of Siboglinidae (Annelida) inferred from 18S rRNA, 28S rRNA and morphological data. *Cladistics* **20**, 518–533.

Ruan, L.W., Bian, X.F., Wang, X., Yan, X.M., Li, F. & Xu, X. 2008. Molecular characteristics of the tubeworm, *Ridgeia piscesae*, from the deep-sea hydrothermal vent. *Extremophiles* **12**, 735–739.

Sanchez, S., Andersen, A.C., Hourdez, S. & Lallier, F.H. 2007a. Identification, sequencing, and localization of a new carbonic anhydrase transcript from the hydrothermal vent tubeworm *Riftia pachyptila*. *FEBS Journal* **274**, 5311–5324.

Sanchez, S., Hourdez, S. & Lallier, F.H. 2007b. Identification of proteins involved in the functioning of *Riftia pachyptila* symbiosis by subtractive suppression hybridization. *BMC Genomics* **8**, article 337.

Sarradin, P.M., Caprais, J.C., Briand, P., Gaill, F., Shillito, B. & Desbruyères, D. 1998. Chemical and thermal description of the environment of the Genesis hydrothermal vent community (13 degrees N, EPR). *Cahiers de Biologie Marine* **39**, 159–167.

Sarrazin, J. & Juniper, S.K. 1999. Biological characteristics of a hydrothermal edifice mosaic community. *Marine Ecology Progress Series* **185**, 1–19.

Sarrazin, J., Juniper, S.K., Massoth, G. & Legendre, P. 1999. Physical and chemical factors influencing species distributions on hydrothermal sulfide edifices of the Juan de Fuca Ridge, northeast Pacific. *Marine Ecology Progress Series* **190**, 89–112.

Sarrazin, J., Levesque, C., Juniper, S.K. & Tivey, M.K. 2002. Mosaic community dynamics on Juan de Fuca Ridge sulphide edifices: substratum, temperature and implications for trophic structure. *Cahiers de Biologie Marine* **43**, 275–279.

Sarrazin, J., Robigou, V., Juniper, S.K. & Delaney, J.R. 1997. Biological and geological dynamics over four years on a high-temperature sulfide structure at the Juan de Fuca Ridge hydrothermal observatory. *Marine Ecology Progress Series* **153**, 5–24.

Schulze, A. 2001a. Comparative anatomy of excretory organs in vestimentiferan tube worms (Pogonophora, Obturata). *Journal of Morphology* **250**, 1–11.

Schulze, A. 2001b. Ultrastructure of opisthosomal chaetae in Vestimentifera (Pogonophora, Obturata) and implications for phylogeny. *Acta Zoologica* **82**, 127–135.

Schulze, A. 2003. Phylogeny of Vestimentifera (Siboglinidae, Annelida) inferred from morphology. *Zoologica Scripta* **32**, 321–342.

Schulze, A. & Halanych, K.M. 2003. Siboglinid evolution shaped by habitat preference and sulfide tolerance. *Hydrobiologia* **496**, 199–205.

Scott, K.M. 2003. A delta^{13}C-based carbon flux model for the hydrothermal vent chemoautotrophic symbiosis *Riftia pachyptila* predicts sizeable CO_2 gradients at the host-symbiont interface. *Environmental Microbiology* **5**, 424–432.

Scott, K.M., Bright, M. & Fisher, C.R. 1998. The burden of independence: inorganic carbon utilization strategies of the sulphur chemoautotrophic hydrothermal vent isolate *Thiomicrospira crunogena* and the symbionts of hydrothermal vent and cold seep vestimentiferans. *Cahiers de Biologie Marine* **39**, 379–381.

Scott, K.M., Bright, M., Macko, S.A. & Fisher, C.R. 1999. Carbon dioxide use by chemoautotrophic endosymbionts of hydrothermal vent vestimentiferans: affinities for carbon dioxide, absence of carboxysomes, and delta ^{13}C values. *Marine Biology* **135**, 25–34.

Scott, K.M. & Fisher, C.R. 1995. Physiological ecology of sulfide metabolism in hydrothermal vent and cold seep vesicomyid clams and vestimentiferan tube worms. *American Zoologist* **35**, 102–111.

Scott, K.M., Fisher, C.R., Vodenichar, J.S., Nix, E.R. & Minnich, E. 1994. Inorganic carbon and temperature requirements for autotrophic carbon fixation by the chemoautotrophic symbionts of the giant hydrothermal vent tube worm, *Riftia pachyptila*. *Physiological Zoology* **67**, 617–638.

Shank, T.M., Fornari, D.J., Von Damm, K.L., Lilley, M.D., Haymon, R.M. & Lutz, R.A. 1998. Temporal and spatial patterns of biological community development at nascent deep-sea hydrothermal vents (9 degree 50′N, East Pacific Rise). *Deep-Sea Research Part II—Topical Studies in Oceanography* **45**, 465–515.

Shillito, B., Lechaire, J.-P. & Gaill, F. 1993. Microvilli-like structures secreting chitin crystallites. *Journal of Structural Biology* **111**, 59–67.

Sibuet, M. & Olu, K. 1998. Biogeography, biodiversity and fluid dependence of deep-sea cold-seep communities at active and passive margins. *Deep-Sea Research Part II—Topical Studies in Oceanography* **45**, 517–567.

Sibuet, M. & Olu-Le Roy, K. 2002. Cold seep communities on continental margins: structure and quantitative distribution relative to geological and fluid venting patterns. In *Ocean Margin Systems*, G. Wefer (ed.). Berlin: Springer-Verlag, 235–251.

Smith, C.R. & Baco, A.R. 2003. Ecology of whale falls at the deep-sea floor. *Oceanography and Marine Biology: An Annual Review* **41**, 311–354.

Sorgo, A., Gaill, F., Lechaire, J.P., Arndt, C. & Bright, M. 2002. Glycogen storage in the *Riftia pachyptila* trophosome: contribution of host and symbionts. *Marine Ecology Progress Series* **231**, 115–120.

Southward, E.C. 1988. Development of the gut and segmentation of newly settled stages of *Ridgeia* (Vestimentifera): implications for relationship between Vestimentifera and Pogonophora. *Journal of the Marine Biological Association of the United Kingdom* **68**, 465–487.

Southward, E.C. 1991. Three new species of Pogonophora, including two vestimentiferans, from hydrothermal sites in the Lau Back-arc Basin (southwest Pacific Ocean). *Journal of Natural History* **25**, 859–882.

Southward, E.C. 1993. Pogonophora. In *Microscopic Anatomy of Invertebrates: Onychophora, Chilopoda, and Lesser Protostomata*, F.W. Harrison & M.E. Rice (eds). New York: Wiley-Liss, 327–369.

Southward, E.C. 1999. Development of Perviata and Vestimentifera (Pogonophora). *Hydrobiologia* **402**, 185–202.

Southward, E.C. 2000. Class Pogonophora. In *Polychaetes and Allies. The Southern Synthesis*, P.L. Beesley et al. (eds). Canberra, Australia: CSIRO, 331–400.

Southward, E.C. & Coates, K.A. 1989. Sperm masses and sperm transfer in a vestimentiferan, *Ridgeia piscesae* Jones, 1985 (Pogonophora : Obturata). *Canadian Journal of Zoology—Revue Canadienne de Zoologie* **67**, 2776–2781.

Southward, E.C. & Galkin, S.V. 1997. A new vestimentiferan (Pogonophora : Obturata) from hydrothermal vent fields in the Manus Back-arc Basin (Bismarck Sea, Papua New Guinea, Southwest Pacific Ocean). *Journal of Natural History* **31**, 43–55.

Southward, E.C., Schulze, A. & Gardiner, S.L. 2005. Pogonophora (Annelida): form and function. *Hydrobiologia* **535**, 227–251.

Southward, E.C., Schulze, A. & Tunnicliffe, V. 2002. Vestimentiferans (Pogonophora) in the Pacific and Indian Oceans: a new genus from Lihir Island (Papua New Guinea) and the Java Trench, with the first report of *Arcovestia ivanovi* from the North Fiji Basin. *Journal of Natural History* **36**, 1179–1197.

Southward, E.C., Tunnicliffe, V. & Black, M. 1995. Revision of the species of *Ridgeia* from northeast Pacific hydrothermal vents, with a redescription of *Ridgeia piscesae* Jones (Pogonophora: Obturata = Vestimentifera). *Canadian Journal of Zoology—Revue Canadienne de Zoologie* **73**, 282–295.

Stewart, F.J., Newton, I.L.G. & Cavanaugh, C.M. 2005. Chemosynthetic endosymbioses: adaptations to oxic-anoxic interfaces. *Trends in Microbiology* **13**, 439–448.

Struck, T., Schult, N., Kusen, T., Hickman, E., Bleidorn, C., McHugh, D. & Halanych, K. 2007. Annelid phylogeny and the status of Sipuncula and Echiura. *BMC Evolutionary Biology* **7**, 57.

Tabita, F., Hanson, T., Li, H., Satagopan, S., Singh, J. & Chan, S. 2007. Function, structure, and evolution of the RubisCO-like proteins and their RubisCO homologs. *Microbiology and Molecular Biology Reviews* **71**, 576–599.

Thiebaut, E., Huther, X., Shillito, B., Jollivet, D. & Gaill, F. 2002. Spatial and temporal variations of recruitment in the tube worm *Riftia pachyptila* on the East Pacific Rise (9 degrees 50′N and 13 degrees N). *Marine Ecology Progress Series* **234**, 147–157.

Thornhill, D.J., Wiley, A.A., Campbell, A.L., Bartol, F.F., Teske, A. & Halanych, K.M. 2008. Endosymbionts of *Siboglinum fiordicum* and the phylogeny of bacterial endosymbionts in Siboglinidae (Annelida). *Biological Bulletin* **214**, 135–144.

Tolstoy, M., Cowen, J.P., Baker, E.T., Fornari, D.J., Rubin, K.H., Shank, T.M., Waldhauser, F., Bohnenstiehl, D.R., Forsyth, D.W., Holmes, R.C., Love, B., Perfit, M.R., Weekly, R.T., Soule, S.A. & Glazer, B. 2006. A sea-floor spreading event captured by seismometers. *Science* **314**, 1920–1922.

Toulmond, A., Lallier, F.H., de Frescheville, J., Childress, J.J., Lee, R., Sanders, N.K. & Desbruyères, D. 1994. Unusual carbon dioxide-combining properties of body fluids in the hydrothermal vent tubeworm *Riftia pachyptila*. *Deep-Sea Research Part I—Oceanographic Research Papers* **41**, 1447–1456.

Tsurumi, M. & Tunnicliffe, V. 2003. Tubeworm-associated communities at hydrothermal vents on the Juan de Fuca Ridge, northeast Pacific. *Deep-Sea Research Part I—Oceanographic Research Papers* **50**, 611–629.

Tunnicliffe, V. 1991. The biology of hydrothermal vents: ecology and evolution. *Oceanography and Marine Biology: An Annual Review* **29**, 319–407.

Tunnicliffe, V., Garrett, J.F. & Johnson, H.P. 1990. Physical and biological factors affecting the behaviour and mortality of hydrothermal vent tube worms (vestimentiferans). *Deep-Sea Research* **37**, 103–125.

Tunnicliffe, V., Embley, R.W., Holden, J.F., Butterfield, D.A., Massoth, G.J. & Juniper, S.K. 1997. Biological colonization of new hydrothermal vents following an eruption on Juan de Fuca Ridge. *Deep-Sea Research Part I—Oceanographic Research Papers* **44**, 1627.

Tunnicliffe, V., McArthur, A.G. & McHugh, D. 1998. A biogeographical perspective of the deep-sea hydrothermal vent fauna. *Advances in Marine Biology* **34**, 353–442.

Tyler, P.A., German, C.R., Ramirez Llodra, E. & Van Dover, C.L. 2002. Understanding the biogeography of chemosynthetic ecosystems. *Oceanologica Acta* **25**, 227–241.

Tyler, P.A. & Young, C.M. 1999. Reproduction and dispersal at vents and cold seeps. *Journal of the Marine Biological Association of the United Kingdom* **79**, 193–208.

Urcuyo, I.A., Bergquist, D.C., MacDonald, I.R., VanHorn, M. & Fisher, C.R. 2007. Growth and longevity of the tubeworm *Ridgeia piscesae* in the variable diffuse flow habitats of the Juan de Fuca Ridge. *Marine Ecology Progress Series* **344**, 143–157.

Urcuyo, I.A., Massoth, G.J., Julian, D. & Fisher, C.R. 2003. Habitat, growth and physiological ecology of a basaltic community of *Ridgeia piscesae* from the Juan de Fuca Ridge. *Deep-Sea Research Part I—Oceanographic Research Papers* **50**, 763–780.

Urcuyo, I.A., Massoth, G.J., MacDonald, I.R. & Fisher, C.R. 1998. *In situ* growth of the vestimentiferan *Ridgeia piscesae* living in highly diffuse flow environments in the main Endeavour Segment of the Juan de Fuca Ridge. *Cahiers de Biologie Marine* **39**, 267–270.

Uschakov, P.V. 1933. Eine neue Form aus der Familie Sabellidae (Polychaeta). *Zoologischer Anzeiger* **104**, 205–208.

van der Land, J. & Nørrevang, A. 1975. The systematic position of *Lamellibrachia* (Annelida, Vestimentifera). *Zeitschrift für Zoologische Systematik und Evolutionsforschung* **1**, 86–101.

van der Land, J. & Nørrevang, A. 1977. Structure and relationships of *Lamellibrachia* (Annelida, Vestimentifera). *Biologiske Skrifter Kongelige Danske Videnskabernes Selskab* **21**, 1–102.

Van Dover, C.L. 1994. *In situ* spawning of hydrothermal vent tubeworms (*Riftia pachyptila*). *Biological Bulletin* **186**, 134–135.

Van Dover, C.L. 2000. *The Ecology of Hydrothermal Vents*. Princeton, NJ: Princeton University Press.

Van Dover, C.L., German, C.R., Speer, K.G., Parson, L.M. & Vrijenhoek, R.C. 2002. Evolution and biogeography of deep-sea vent and seep invertebrates. *Science* **295**, 1253–1257.

Vetter, R.D. & Fry, B. 1998. Sulfur contents and sulfur-isotope compositions of thiotrophic symbioses in bivalve molluscs and vestimentiferan worms. *Marine Biology* **132**, 453–460.

Vrijenhoek, R.C. 2010. Genetics and evolution of deep-sea chemosynthetic bacteria and their invertebrate hosts. *Geobiology* **accepted manuscript**.

Webb, M. 1964. The posterior extremity of *Siboglinum fiordicum* (Pogonophora). *Sarsia* **15**, 33–36.

Webb, M. 1969. *Lamellibrachia barhami*, gen. nov., sp. nov. (Pogonophora), from the North East Pacific. *Bulletin of Marine Science* **19**, 18–47.

Williams, N.A., Dixon, D.R., Southward, E.C. & Holland, P.W.H. 1993. Molecular evolution and diversification of the Vestimentiferan tube worms. *Journal of the Marine Biological Association of the United Kingdom* **73**, 437–452.

Wilmot, D.B. & Vetter, R.D. 1990. The bacterial symbiont from the hydrothermal vent tubeworm *Riftia pachyptila* is a sulfide specialist. *Marine Biology* **106**, 273–283.

Yancey, P.H., Ishikawa, J., Meyer, B., Girguis, P.R. & Lee, R.W. 2009. Thiotaurine and hypotaurine contents in hydrothermal-vent polychaetes without thiotrophic endosymbionts: correlation with sulfide exposure. *Journal of Experimental Zoology Part A—Ecological Genetics and Physiology* **311A**, 439–447.

Yin, M., Palmer, H.R., Fyfe-Johnson, A.L., Bedford, J.J., Smith, R.A. & Yancey, P.H. 2000. Hypotaurine, N-methyltaurine, taurine, and glycine betaine as dominant osmolytes of vestimentiferan tubeworms from hydrothermal vents and cold seeps. *Physiological and Biochemical Zoology* **73**, 629–637.

Young, C.M. 2002. *Atlas of Marine Invertebrate Larvae*. New York: Academic Press.

Young, C.M., Vazquez, E., Metaxas, A. & Tyler, P.A. 1996. Embryology of vestimentiferan tube worms from deep-sea methane/sulphide seeps. *Nature* **381**, 514–516.

Zal, F., Lallier, F.H., Green, B.N., Vinogradov, S.N. & Toulmond, A. 1996a. The multi-hemoglobin system of the hydrothermal vent tube worm *Riftia pachyptila*. 2. Complete polypeptide chain composition investigated by maximum entropy analysis of mass spectra. *Journal of Biological Chemistry* **271**, 8875–8881.

Zal, F., Lallier, F.H., Wall, J.S., Vinogradov, S.N. & Toulmond, A. 1996b. The multi-hemoglobin system of the hydrothermal vent tube worm *Riftia pachyptila*. 1. Reexamination of the number and masses of its constituents. *Journal of Biological Chemistry* **271**, 8869–8874.

Zal, F., Leize, E., Lallier, F.H., Toulmond, A., Van Dorsselaer, A. & Childress, J.J. 1998. S-Sulfohemoglobin and disulfide exchange: the mechanisms of sulfide binding by *Riftia pachyptila* hemoglobins. *Proceedings of the National Academy of Sciences of the United States of America* **95**, 8997–9002.

Zal, F., Suzuki, T., Kawasaki, Y., Childress, J.J., Lallier, F.H. & Toulmond, A. 1997. Primary structure of the common polypeptide chain b from the multi-hemoglobin system of the hydrothermal vent tube worm *Riftia pachyptila*: An insight on the sulfide binding-site. *Proteins—Structure Function and Genetics* **29**, 562–574.

Oceanography and Marine Biology: An Annual Review, 2010, **48**, 267-338
© R. N. Gibson, R. J. A. Atkinson, and J. D. M. Gordon, Editors
Taylor & Francis

HISTORICAL RECONSTRUCTION OF HUMAN-INDUCED CHANGES IN U.S. ESTUARIES

HEIKE K. LOTZE

Biology Department, Dalhousie University, 1355 Oxford Street, Halifax, NS B3H 4J1, Canada
Email: hlotze@dal.ca

Abstract   Estuaries are vital ecosystems that have sustained human and marine life since earliest times. Yet, no other part of the ocean has been so fundamentally shaped by human activities. Understanding the magnitude, drivers and consequences of past changes is essential to determine current trends and realistic management goals. This review provides a detailed account of human-induced changes in Massachusetts, Delaware, Chesapeake, Galveston and San Francisco Bays and Pamlico Sound. Native Americans have lived off these estuaries for millennia, yet left few signs of local resource depletion. European colonisation, commercialisation and industrialisation dramatically depleted and degraded valuable species, habitats and water quality. Exploitation and habitat loss were the main factors depleting 95% of valued species, with 35% being rare and 3% extirpated. Twentieth century conservation efforts enabled 10% of species to recover. Such profound changes in species diversity have altered the structure and functions of estuarine ecosystems as well as their services for human well-being. Thus, undesirable health risks and societal costs have increased over past decades. Protecting and restoring the diversity and vitality of estuaries will enhance their resilience towards current and future disturbances, yet require better governance of these often-neglected ecosystems. Their documented historical richness and essential role for marine life and people may increase the necessary awareness and appreciation.

Introduction

Marine ecosystems have been exploited and influenced by human activities for millennia, especially along the coasts (Jackson et al. 2001, Lotze et al. 2006, Rick & Erlandson 2008). This has resulted in large changes in populations, habitats and water quality over several decades or centuries, yet current management efforts typically rely on the last two to five decades of scientific monitoring data to estimate current states and trends (Lotze & Worm 2009). The total magnitude of change is therefore often obscured by "shifting baselines" (Pauly 1995). For example, from 1970 to 2003 the biomass of Atlantic cod on the Scotian shelf totalled 342,000 t at its peak but was estimated at 1,260,000 t in 1852 (Rosenberg et al. 2005).

Akin to climate change research, knowing past long-term changes in marine populations and ecosystems is essential to understanding present states (Carlton 1998, Roberts 2003, Lotze & Worm 2009). Historical reference points from times prior to strong human influence are needed to derive sound management goals and targets for restoration and recovery. How much has a population or habitat declined from its historical abundance, and to what level could it recover or be restored? If we know historical changes in individual populations, we can further estimate changes in biodiversity, food-web structure and ecosystem functioning (Jackson et al. 2001, Lotze et al. 2005). Finally, understanding the underlying drivers and consequences of change is necessary to derive future projections and set management priorities (Carlton 1998, Clark et al. 2001a).

Recent progress and interdisciplinary collaboration among marine scientists has made it possible to reconstruct historical changes in marine ecosystems over past centuries and millennia (Rick & Erlandson 2008, Starkey et al. 2008, Lotze & Worm 2009). Several studies have aimed at reconstructing historical changes in particular regions, such as the Benguela upwelling system in Africa (Griffiths et al. 2004), the Wadden Sea in Europe (Lotze 2005, 2007, Lotze et al. 2005), the Outer Bay of Fundy in Canada (Lotze & Milewski 2004) and the Gulf of California in Mexico (Sáenz-Arroyo et al. 2005, 2006). Other studies have focused on historical changes in individual species (Rosenberg et al. 2005, McClenachan et al. 2006) or habitats (Orth et al. 2006, Airoldi & Beck 2007). All of these studies provide important insights into the magnitude and range of human-induced changes in marine ecosystems. However, so far few studies have compared historical changes across different systems, such as coral reefs (Pandolfi et al. 2003) or coastal waters (Lotze et al. 2006), highlighting similarities as well as differences in the response to human-induced changes.

In the United States, a detailed, comparative ecological history for marine ecosystems is missing, although such knowledge is needed for current ocean management and conservation (Pew Oceans Commission 2003, U.S. Commission on Ocean Policy 2004). Yet, a wealth of historical information is available for individual species, habitats or other ecosystem components (e.g., Nichols et al. 1986, Kennish 2000, Kemp et al. 2005). Such records have been compiled as part of a synthetic analysis on historical changes in 12 estuaries and coastal seas around the world, but only the summary results have been published (Lotze et al. 2006).

U.S. estuaries have each experienced a similar human history, shaped first by Native American influences and later by European colonisation and expansion. Thus, a quasi-replicated comparison of past ecological changes within a similar historical context can be made. However, each estuary has a unique physical and biological setting and experienced its own sequence of historical events that created estuary-specific histories of change. For example, the gold rush and fur trade were of huge importance in San Francisco Bay (Agriculture and Natural Resources Communication Services [ANR] 2001), the cod fishery and whaling thrived in Massachusetts Bay (Murawski et al. 1999, Claesson 2008), and the oyster fishery and eutrophication are commonly associated with Chesapeake Bay (Kemp et al. 2005). Separating estuary-specific from universal patterns can be of great value in understanding the history, drivers and consequences of human-induced changes in marine ecosystems.

The aim of this review is to provide a comprehensive and detailed history of human-induced ecological changes in selected U.S. estuaries, to compare observed changes across estuaries and to place these changes into a global context. First, the general ecological and societal importance of estuaries is highlighted, including their history of human settlement. It follows a detailed review of historical changes in six large U.S. estuaries on three coasts: Massachusetts Bay, Delaware Bay, Chesapeake Bay, Pamlico Sound, Galveston Bay and San Francisco Bay (Figure 1). Within each estuary, past changes in populations, fisheries and ecosystem characteristics, such as habitat availability and water quality, are described for before and since European colonisation mostly through human activities. Past changes are then compared and summarised across the six case studies to derive a more general picture of change in U.S. estuaries over time. This will include a comparison of the importance of different human drivers and the consequences of historical changes for ecosystem structure, functions and services. Finally, the emerging ecological history of U.S. estuaries is compared with historical changes in other marine ecosystems around the world. Overall, this review compiles essential information on historical changes, drivers and consequences in U.S. estuaries that provides an important basis for current and future ocean management and conservation.

Ecological and social importance of estuaries

Estuaries are among the most important environments in the coastal zone, both biologically (Kennish 2002) and socioeconomically (Limburg 1999), and have been essential in sustaining human and

Figure 1 Map of the United States showing all major and the selected six estuaries chosen as case studies: Massachusetts Bay (Ma), Delaware Bay (De), Chesapeake Bay (Ch), Pamlico Sound (Pa), Galveston Bay (Ga) and San Francisco Bay (SF).

marine life since earliest times. They form protective harbours for human settlement and for spawning, nursing and foraging animals; they support high diversity and primary productivity attracting animal and human consumers; and they link human trading and animal migration routes between rivers and the sea (Limburg 1999). At the land–sea interface, the interactions between humans and marine life have always been interlinked, and a long history of human-induced changes underlies the current and future states of estuarine ecosystems.

Along the U.S. coastline, nearly 900 estuaries cover about 10.9 million ha, which together with coastal lagoons occupy 80–90% of the Atlantic and Gulf of Mexico and 10–20% of the Pacific coasts (Table 1, Kennish 2002). As transition zones between terrestrial, freshwater and marine systems, estuaries create some of the most productive and fertile ecosystems on Earth caused by (1) abundant nutrient supply from land-based sources; (2) efficient nutrient retention and cycling among benthic, wetland and pelagic habitats; (3) maximised autotrophic production by pelagic and benthic algae and plants; and (4) high tidal energy, water circulation and mixing (Kennish 2002). The diverse primary producers provide a rich food supply on which all higher trophic levels depend

Table 1 Characteristics of the selected six U.S. estuaries: surface area, average depth, and estuarine drainage area for each estuary (data from Bricker et al. 1999) and primary productivity (PP) and species richness for fish and marine mammals for large marine ecosystems (LME, with Ma, De, Ch = NE U.S. Shelf, Pa = SE U.S. Shelf, Ga = Gulf of Mexico, SF = California Current, http://www.seaaroundus.org/lme/lme.aspx)

System	Surface (km²)	Depth (m)	Drainage (km²)	PP (mgCm⁻²d⁻¹)	Species richness Fish	Species richness Mammals
Massachusetts Bay (Ma)	478	27.3	3,100	1451	644	42
Delaware Bay (De)	1236	6.3	17,485	1451	644	42
Chesapeake Bay (Ch)	6898	9.4	64,050	1451	644	42
Pamlico Sound (Pa)	4455	4.1	33,102	740	1168	31
Galveston Bay (Ga)	890	6.2	11,502	537	958	31
San Francisco Bay (SF)	829	6.4	17,146	578	804	46

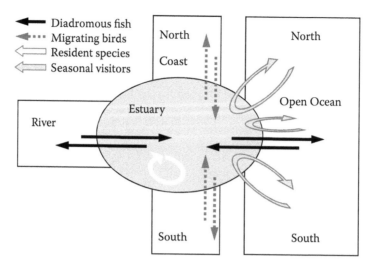

Figure 2 Estuaries are essential habitats for multiple resident species that remain in the estuary year-round (white arrow); seasonal visitors from northern, southern, or other parts of the ocean that come to feed, breed or nurse their young for part of the year (grey arrows); and migratory species that feed and stage in estuaries on their way from rivers to the ocean and vice versa (e.g., diadromous fish, black arrows) and from northern breeding to southern overwintering grounds (e.g., migrating birds, dashed arrows).

(Table 1). Extensive habitats are provided by seagrass and rockweed beds, mangrove forests and salt marshes, as well as oyster reefs, mussel beds, hydrozoan meadows, and other reef- or bank-forming organisms. Together with a mix of soft and hard bottoms, shallow and deep, calm and high-current waters and the surrounding islands, beaches, forests and grasslands, this extraordinary variety of biogeographic habitats provides essential breeding, nursery and foraging grounds and protection from predators (Kennish 2002, Lotze & Milewski 2004).

The rich food supply together with the great habitat diversity provide the special living conditions for a great variety of plants, invertebrates, fish, reptiles, birds and mammals that depend on estuarine food or habitats for their entire or parts of their life cycle (Figure 2). Many resident species live within the estuary's boundaries all their life. Others come to visit seasonally from colder or warmer ocean regions to feed, breed, nurse their young or overwinter. Still others stop over during their migrations from river to sea or northern to southern regions. Together, these species form an estuarine food web that changes dynamically with seasonal variation in environmental parameters and species presence.

Because of their special conditions, estuaries were among the first places settled and have experienced a long history of changing human activities (Limburg 1999, Lotze et al. 2006, Lotze & Glaser 2009). Many of the largest cities around the world have been built around estuaries. Today, it is generally understood that the natural structure and functions of most estuaries have been seriously degraded by human impacts, including pollution with nutrients, chemicals and pathogens; loss and alteration of habitat; overexploitation; freshwater diversion; and introduction of exotic species (Limburg 1999, Kennish 2002). Estuaries are also the regions where future threats from human-induced climate change, especially sea-level rise, will likely have the most severe consequences (Kennish 2002).

History of human settlement and activities in North America

After the melting of the ice sheet that covered large parts of North America during the last ice age, native people settled along the coasts 10–12 thousand yr ago (Tables 2 and 3). Most tribes

Table 2 Cultural periods based on human presence, technology and market conditions

Cultural period	Human presence, technology, market conditions
Prehuman	No human presence; only natural disturbance of ecosystems
Hunter-gatherer	Premarket; low population numbers; seasonal settlements; subsistence exploitation
Agricultural	Premarket; low population numbers; permanent settlements; individual or village-based resource use; subsistence and artisan exploitation
Establishment	European colonisation; establishment of local economy and market; low population numbers; trade between colonies and Europe; mostly subsistence exploitation
Development	Strong growth and expansion of economy, market and trade; rapid rise in population; commercialisation of resource use; development of luxury and fashion markets; industrialisation and technological progress; guns allow mass exploitation of mammals and birds; fishing mostly inshore and seasonal with light gear
Early global (1900–1950)	Global economy and market develop; rise in population; industrialisation and technological progress increase; increasing effort, efficiency and destructiveness of gear; accelerating exploitation, by-catch and habitat destruction, fishing possible in any season but still mostly inshore and coastal
Late global (1950–2000)	Global economy and market; industrial fishing increases after WWII and spreads offshore; multiple unselective and destructive gears enable mass fishing; pollution, eutrophication and other impacts increase; conservation efforts increase

Table 3 Estimates of real time (BC = negative, AD = positive numbers) for the beginning of each cultural period in the six estuaries

Cultural Period	Ma	De	Ch	Pa	Ga	SF
Prehuman	>−8000	>−8000	>−8000	>−8000	>−10000	>−10000
Hunter-gatherer	−8000	−8000	−8000	−8000	−10000	−10000
Agriculture	1400	1200	1200	1200	N/A*	N/A
Colonial establishment	1620	1600	1600	1600	1680	1775
Colonial development	1680	1760	1760	1700	1820	1820
Global market 1	1900	1900	1900	1900	1900	1900
Global market 2	1950	1950	1950	1950	1950	1950

* N/A = not available, meaning this period did not occur.

moved with the seasons between traditional camping, hunting and fishing grounds in pursuit of fish, game, berries and other resources, such as building materials, medicines and ornaments (Bourque 1995, Broughton 2002). Between 1200 and 1400 AD, natives around Massachusetts, Delaware and Chesapeake Bays and Pamlico Sound started to cultivate land, while people around Galveston and San Francisco Bays largely relied on a hunter-gatherer lifestyle until European contact (Table 3). With the arrival of Europeans, native populations in most parts of the continent started to dwindle from direct aggression or disease, while others were forced to give up traditional settlement places and hunting territories (Snow 1980, Sultzman 2000).

European explorers first established settlements between 1600 and 1620 AD in the four estuaries on the Atlantic coast, in 1680 in Galveston and in 1775 in San Francisco Bay (Table 3). Within a matter of years, these settlers transformed the regions culturally, economically and environmentally (Figure 3). In the period of colonial establishment, their subsistence economy mostly served to sustain the colony. People cleared land for settlements and agricultural fields and gained food through fishing, hunting and harvesting of regional resources. This establishment period lasted a few decades to more than a century, depending on the region.

Between the late seventeenth and early nineteenth century, market economies were developed with the commercialisation of fishing, hunting, forestry and shipbuilding (Tables 2 and 3), which

Figure 3 Human settlements through time with an example from San Francisco Bay. (A) Indians in tule boat (by L. Choris, Library of Congress Rare Book and Special Collections Division, Washington, D.C.). (B) San Francisco from the bay, 1847 (by Capt. A.E. Theberge, National Oceanic and Atmospheric Administration/ Department of Commerce). (C) San Francisco, 1862, from Russian Hill (by C.B. Gifford & L. Nagel, http:// www.davidrumsey.com). (D) San Francisco skyline, 2004 (by Capt. A.E. Theberge, National Oceanic and Atmospheric Administration/Department of Commerce).

became the cornerstones of the colonial economies. Over time, other industries were introduced, such as textile and canning industries and saw and pulp mills. Rivers were dammed to gain electricity and wetlands converted to farmland and municipal and industrial construction. The colonial economy, market and trade expanded rapidly, and the human population grew exponentially. Newly developed luxury and fashion markets spurred exploitation of natural resources for non-food purposes, such as furs, feathers and ivory. Industrialisation and technological progress enhanced the capacity for mass exploitation; however, most fishing and hunting still remained inshore and seasonal with light boats and selective gear (Smith 1994).

With the onset of the twentieth century, the globalisation of economy and markets began, and the human population continued to grow exponentially (Tables 2 and 3). Rapid technological progress led to more efficient but also more destructive fishing gears, which accelerated exploitation, by-catch and habitat destruction. With the advent of steam- and motorboats, fishing became possible in any season yet still remained mainly inshore and in coastal regions (Houde & Rutherford 1993, Smith 1994). After WWII, the global economy and market entered a second phase with new dimensions and fewer limits. Industrial fishing strongly increased its effort, efficiency and capacity, thereby spreading from inshore to offshore and towards the deep sea (Pauly et al. 2003, Lotze & Worm 2009). Fishing gear became decreasingly selective and increasingly harmful. Nutrient loading and chemical pollution increased, and aquaculture was developed. Global shipping enhanced the introduction of exotic species, and climate change became a recognised phenomenon. On the other hand, conservation efforts were strongly increased, with the goal of protecting and recovering species that had been driven to low levels (Lotze et al. 2006).

Overall, the general sequence of human settlement, economy and market development was repeated in the different U.S. estuaries. However, the timing of the cultural periods differed, as did the estuary-specific activities, which are reviewed in the next section.

History of ecological changes in selected U.S. estuaries

Human activities have influenced the distribution and abundance of marine species throughout history (Rick & Erlandson 2008, Starkey et al. 2008, Lotze & Worm 2009). This section reviews the detailed historical changes in important species, habitats and water quality in six U.S. estuaries and their adjacent coastal waters (Table 1). Drawing on and updating data collected by Lotze et al. (2006), this covers most economically or ecologically important species for which archaeological, historical, fisheries and ecological records were available. Based on the documented records, a relative abundance index (Table 4) was derived for 22 consistent species groups over time and averaged across six taxonomic groups: mammals, birds, reptiles, fish, invertebrates and vegetation (see Lotze et al. 2006 for details). The summary data are shown for each study system and illustrated by a detailed description of historical changes. Data by species group are compared across U.S. estuaries. Together, the six case studies provide a comprehensive ecological history of U.S. estuaries.

Massachusetts Bay, Gulf of Maine

Massachusetts Bay is the northernmost of the six evaluated estuaries and one of the largest bays on the Atlantic coast. Together with Cape Cod Bay, it is separated from the larger Gulf of Maine by Stellwagen Bank (Figure 4), a very productive area for a variety of groundfish, pelagic fish and whales (Ward 1995). The native population was quite extensive in the region and engaged in hunter-gatherer activities, fishing and shellfish harvesting, with agriculture introduced around 1400 (Table 3). After European contact, the native population was much reduced by conflict and disease, from an estimated 36,700 to 5300 (Snow 1980). European colonisation began in 1620 and expanded after 1680. The backbones of the colonial economy were whaling and fishing, especially for groundfish on the many rich fishing banks in the Gulf of Maine. Stellwagen Bank was well known to fishermen, merchants, explorers and settlers by the seventeenth century (Claesson 2008). For over 400 yr, New England has been identified economically and culturally with the harvest of groundfishes (Murawski et al. 1999), but many other marine and coastal resources were exploited as well, some to extinction. Conservation efforts in the twentieth century enabled some species to recover, and in 1992 Stellwagen Bank was declared a national marine sanctuary (Claesson 2008). The general

Table 4   Judging criteria for relative abundance estimates that integrate quantitative and qualitative records of abundance

Estimate	Quantitative	Qualitative records
Pristine = 100	91–100%	High abundance, no signs of regular human exploitation
Abundant = 90	51–90%	Medium-to-high abundance, regular human exploitation without signs of depletion, species common
Depleted = 50	11–50%	Low-to-medium abundance, strong exploitation with signs of depletion (i.e., at least 50% decline in abundance, catch, CPUE, size or distribution), species depleted
Rare = 10	1–10%	Low-to-very-low abundance, at least 90% decline in abundance, catch, CPUE, size or distribution, collapse of stocks or exploitation, listed on endangered species list, species rare
Extirpated = 0	0%	Locally, regionally or globally extinct, species absent

HEIKE K. LOTZE

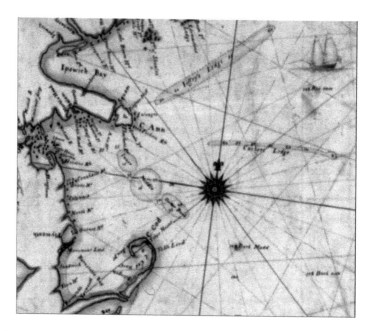

Figure 4 Earliest known nautical chart of Massachusetts Bay and Stellwagen Bank fishing grounds including 'Inner', 'Middle' and 'Outer' Banks (Capt. T. Durrell, *A Large Draught of New England: New Hampshire, York County and Part of Nova Scotia and Accadia*, map, Taunton: U.K. Hydrographic Office, 1734).

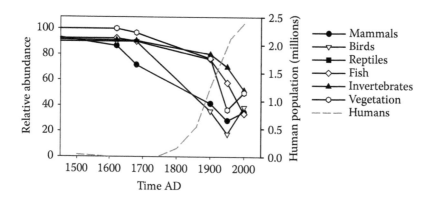

Figure 5 Timeline of changes in relative abundance of six taxonomic groups (left axis) and the human population (right axis) in Massachusetts Bay.

timeline of change in economically and ecologically important species and the human population around Massachusetts Bay is shown in Figure 5.

Marine mammals

Before 1500, an estimated 10,000 northern right whales (*Eubalaena glacialis*) were present in the north-west Atlantic. Between 1530 and 1610, whalers caught about 21,000 right whales in the north-west Atlantic, probably substantially reducing right whale numbers before the time whaling was begun by colonists in the Plymouth area in the early 1600s (Reeves 2001). Stellwagen Bank was famous for its many visiting right whales, but there were also fin (*Balaenoptera physalus*), minke (*B. acutorostrata*) and humpback (*Megaptera novaeangliae*) whales (Ward 1995, Claesson

274

2008). By the late 1600s, right whale populations were severely depleted, but whaling continued from the eastern United States until the early 1900s (Perry et al. 1999). In the 1700s, 29 right whales were killed in Cape Cod Bay, but catches in New England declined, and shore whaling ceased in the 1720s due to the lack of whales (Waring et al. 2001). However, U.S. whalers seeking other whales caught approximately another 250–300 right whales in 1820–1899 (Reeves & Mitchell 1983). The remaining approximately 58–60 right whales that survived were protected by the League of Nations in 1935 (Reeves 2001). By the 1990s, the population had reached about 300 individuals, slowly increasing by about 2.5% per yr (Knowlton et al. 1994).

Another coastal whale that was easy to catch, the Atlantic grey whale (*Eschrichtius robustus*), had become extinct by the late 1800s (Mead & Mitchell 1984). In the nineteenth century, whaling focused on humpback and fin whales along the north-east coast of the United States (Katona et al. 1993), and heavy whaling pressure continued into the early twentieth century. By the 1960s, all great whales were considered rare (Gambell 1999), and commercial whaling ceased with the 1972 Marine Mammal Protection Act (MMPA) and the 1986 moratorium by the International Whaling Commission (IWC) (Perry et al. 1999). Today, all great whales are listed as endangered (U.S. Fish and Wildlife Service [USFWS] 2009). Despite protection, however, recovery of large whales in the North Atlantic is slow, and whales continue to be threatened by collision with ships, entanglement in fishing gear, pollution, habitat degradation and inbreeding depression (Brown & Kraus 1996, Laist et al. 2001).

In the Gulf of Maine, natives hunted harbour seals (*Phoca vitulina*), grey seals (*Halichoerus grypus*) and harbour porpoises (*Phocoena phocoena*) for their meat, skins and oil (Snow 1978, Prescott & Fiorelli 1980). Seals were abundant before European colonisation (Snow 1978) but hunted to low numbers as direct targets and for a bounty paid by the government until the 1960s (Bigelow & Schroeder 1953, Lotze & Milewski 2004). The MMPA ended seal exploitation in 1972, and populations started to rebound (Waring et al. 2001). The harbour porpoise hunt for oil ended in the early twentieth century because of cheaper mineral oil (Prescott & Fiorelli 1980). Still, the harbour porpoise was in decline in the 1960s (Katona et al. 1993, Read 1994) and was considered a candidate species for the endangered species list in the 1990s. The main current threat is by-catch in fisheries (Waring et al. 2001).

A species that became extinct in the course of extensive hunting for its fur was the sea mink (*Mustela macrodon*). This species used to occur in New England and based on archaeological remains had a range from at least Connecticut to the Bay of Fundy (Campbell 1986). It is unclear how abundant the sea mink used to be as it became globally extinct in 1894 (Committee on the Status of Endangered Wildlife in Canada [COSEWIC] 2002).

Birds

Native people throughout North America hunted birds for their meat, eggs and feathers. In archaeological middens near Boston, bone remains of geese, cormorants, ducks and loons were plentiful (Luedtke 1980). During colonial times, geese, swans and many other birds were very abundant: "Some have killed a hundred geese in a week, fifty ducks at a shot, forty teals at another" (Wood 1634). Intense hunting, egg and feather collection and habitat loss reduced many species during the nineteenth century. Swans were considered rare in 1854 (Cronon 1983), and populations of herons, egrets and other birds with beautiful feathers were strongly decimated because of the millinery trade in the late 1800s (Palmer 1962). The now-extinct great auk (*Pinguinus impennis*) used to be common in the north-west Atlantic, including Massachusetts Bay, in the sixteenth and seventeenth centuries (Greenway 1967). Although it occurred in large populations, the great auk was intensely hunted for its meat, oil and feathers to global extinction in 1844 (Kirk 1985). Similarly, the Labrador duck (*Camptorhynchus labradorius*) inhabited the Gulf of Maine but became extinct in 1875, probably as a result of hunting (Chilton 1997).

In the early twentieth century, efforts to increase bird protection started with the introduction of the Migratory Bird Treaty Act in 1918, but many species continued to be affected by ongoing habitat loss and hunting. In the second half of the twentieth century, however, many bird populations started to recover (Palmer 1962). Yet in the 1960s–1970s, pollution from DDT (dichlorodiphenyl-trichloroethane) and other chemicals had strong negative effects on many species by impairing reproduction (Fitzner et al. 1988). For example, only 11 breeding pairs of osprey (*Pandion haliae-tus*) were left in Massachusetts in 1963, but the ban of DDT in the 1970s enabled ospreys to recover to about 350 breeding pairs in 2000 (Massachusetts Division of Fisheries and Wildlife [MDFW] 2009). By the late twentieth century, numbers of many waterfowl and shorebird populations were considered reduced but fairly stable (Veit & Petersen 1993).

Fishes

Archaeological remains show that native people around the Gulf of Maine have harvested a variety of groundfish, pelagic and diadromous fishes over the past millennia (Claesson 2008). For example, remains of large Atlantic cod (*Gadus morhua*) have been very abundant in many prehistoric sites (Steneck 1997, Bourque et al. 2008). With arrival of the Europeans, groundfisheries became the cornerstone of the New England economy, targeting a mixture of bottom-dwelling species (Murawski et al. 1999). Stellwagen Bank supported an abundant and diverse community of Atlantic cod and halibut (*Hippoglossus hippoglossus*) (Figure 6), haddock (*Melanogrammus aeglefinus*), pollock (*Pollachius virens*), skates (Rajidae), yellowtail flounder (*Limanda ferruginea*) and winter flounder (*Pseudopleuronectes americanus*), goosefish (*Lophius americanus*), longhorn sculpin (*Myoxocephalus octodecemspinosus*) and spiny dogfish (*Squalus acanthias*). Common migratory species included herring (*Clupea harengus*), alewive (*Alosa pseudoharengus*), bluefish (*Pomatomus saltatrix*), bluefin tuna (*Thunnus thynnus*), swordfish (*Xiphias gladius*) and mackerel (*Scomber scombrus*) (Ward 1995). Arriving in Massachusetts Bay in 1630, Francis Higginson reported "the abundance of sea fish are almost beyond believing," and "Codfish in these seas are larger than in NewFound Land, six or seven here making a quintal, whereas there they have fifteen to the same weight" (Baird 1873).

Nearshore fish populations significantly deteriorated between 1800 and 1900 (Claesson & Rosenberg 2009). Halibut used to congregate in winter in the gully between Cape Cod and Stell-

Figure 6 Pictures of (A) Atlantic cod in the 1880s (courtesy of R. Steneck) and (B) Atlantic halibut in 1910 caught in the Gulf of Maine (Census of Marine Life, History of Marine Animal Populations Program).

wagen Bank. Around 1820–1825, a demand for halibut developed in the Boston market, and large amounts were caught in Massachusetts Bay. By 1850, halibut had been nearly fished out, and the fishery moved to other parts of the Gulf of Maine and farther (Bigelow & Schroeder 1953). By the 1890s, almost all halibut sold in Gloucester came from Iceland (Murawski et al. 1999). Other large predators, such as swordfish, had been overfished to near extirpation on Stellwagen Bank by the late nineteenth and early twentieth centuries (Claesson & Rosenberg 2009).

In 1906, the first trawler was introduced to American waters in Boston, and by 1930 the trawler fleet had grown to over 300 vessels. Between 1915 and 1940, cod landings remained relatively stable as haddock, redfish (*Sebastes* spp.) and other species were the main targets. Haddock landings in the Gulf of Maine had been relatively low prior to 1900 but increased to over 100,000 t by the 1920s and declined in the early 1930s (Murawski et al. 1999). Still, until the mid-1960s, haddock remained the mainstay of the New England groundfishery. In 1965, landings reached a record high of 154,000 t but rapidly declined afterwards.

In 1960–1965, total groundfish landings increased from 200,000 to 760,000 t, primarily composed of silver hake (*Merluccius bilinearis*), haddock, red hake (*Urophycis chuss*), flounders and cod (Murawski et al. 1999). Yet by 1970, groundfish abundance and landings had severely declined. In 1977, the Magnuson Fishery Conservation and Management Act (Wang & Rosenberg 1997) was implemented and excluded foreign fishing. However, subsequently the U.S. otter-trawl fishing effort essentially doubled. By the mid-1980s, catch rates and abundance had dropped by half, and Georges Bank haddock had collapsed. Because southern New England yellowtail flounder also declined to low levels, the fishery then almost completely relied on cod. Exploitation rates of groundfish reached their highest levels in the early 1990s, and stock biomasses fell in many cases to record lows (Murawski et al. 1999). An emergency closure on Georges Bank has since helped haddock but not cod to increase again (Rosenberg et al. 2006).

In colonial times, diadromous fish such as alewives also occurred "in such multitudes as is almost incredible, pressing up such shallow waters as will scarce permit them to swim" (Wood 1634). Alewives supported a fishery well into the twentieth century, with a peak in 1956 and a decline thereafter (National Marine Fisheries Service [NMFS] 2009). Atlantic salmon (*Salmo salar*) and sturgeons (*Acipenser* spp.) were important subsistence and commercial resources throughout history. Overexploitation, however, together with pollution and habitat loss in rivers resulted in the extirpation of all native salmon runs south of Kennebec River, Maine, in the 1920s (Anderson et al. 1999c) and large declines of Atlantic (*Acipenser oxyrhynchus*) and shortnose (*A. brevirostrum*) sturgeon in the 1800s. Today, only remnant sturgeon populations remain, and the lack of fish passage facilities at dams and poor habitat conditions continue to impair their reestablishment (Friedland 1998).

Invertebrates

Many invertebrates have been used for subsistence by both natives and Europeans (Claesson 2008). Commercial fisheries for clams and oysters in Massachusetts Bay began with European colonisation (Skinner et al. 1995). Banks of the eastern oyster (*Crassostrea virginica*) were "a mile long, and oysters a foot long" (Wood 1634), and "people [were] running over clam banks, huge individual clams as big as a white penny loaf" (Cronon 1983). After more than 200 yr of oyster fishing, mechanical harvesting with tongs started in 1880. The commercial fishery peaked in 1900, strongly declined in the 1920s, and started a series of fishing down the U.S. Atlantic coast (Kirby 2004). Clams were continuously fished over the centuries, but landings increased in the twentieth century, peaking in the 1970s, followed by a decline (Anderson et al. 1999a,b). Landings of lobster (*Homarus americanus*) in the Gulf of Maine have steadily increased since 1940. Although lobster abundance has increased, current fishing mortality is nearly double the overfishing level, and the fishery totally depends on new recruitment (Anderson et al. 1999a).

The decline in traditionally important groundfish and invertebrate fisheries has resulted in the search for new target species, such as sea urchin (*Strongyocentrotus droebachiensis*), for which a fishery was developed in the 1980s. Landings rapidly increased in Massachusetts, reaching the second highest in north-east U.S. nearshore landings with an average yield of 11,400 t. However, catch per unit of effort (CPUE) and landings have steadily declined since 1993 (Anderson et al. 1999b).

Vegetation

Since European colonisation, salt marshes have been important to feed livestock and harvest hay (Chapman 1977), but many have been transformed for agriculture and urban development (Bromberg et al. 2009). Saltmarsh eradication for industrial development and mosquito control began in the midnineteenth century (Russell 1976). By 1940, about 80% of the original salt marshes had been removed or filled in (Chapman 1977). In the 1960s, legislation was developed to protect salt marshes, and restoration began in the 1980s (Chapman 1977, Massachusetts Bays Program [MBP] 2004). Based on historical maps, Massachusetts has lost about 41% of its salt marshes since 1777, with losses in the Boston area as high as 81% (Bromberg & Bertness 2005).

Seagrass beds have also suffered substantial losses, especially in the twentieth century (Wilbur 2009). In the 1930s, an outbreak of wasting disease decimated eelgrass populations by more than 90% throughout the North Atlantic. Eelgrass beds generally recovered in the 1940s, but since the 1950s eelgrass habitat has been drastically reduced by water quality degradation (MBP 2004). Current beds in Massachusetts Bay appear to be stable (Wilbur 2009), but abundance is low (Bricker et al. 1999). Although not located in Massachusetts Bay, quantitative estimates of historical seagrass loss exist for southern Massachusetts. In Buzzards Bay, seagrass covered about 9700 ha in the 1600s. By the 1930s, this was reduced to 81 ha due to the wasting disease, from which it recovered to about 5700 ha in the 1970s (Costa 1988, 2003). In the 1990s, seagrass beds decreased again to about 3200 ha due to eutrophication and disturbance. Thus, overall seagrass loss has been about 67% since the 1600s (Costa 2003). In nearby Waquoit Bay, increased development resulted in an almost complete loss of eelgrass beds by the mid-1990s (Costa 2003).

Water quality

Over time, increased development around Massachusetts Bay has resulted in increasing sewage loads, land run-off and other point and non-point sources of pollution (MBP 2004). A 1990s survey indicated moderate eutrophication levels for Massachusetts Bay and high levels in Boston Harbour (Bricker et al. 1999). High nitrogen inputs have led to increased chlorophyll concentrations, algal blooms and low oxygen concentrations. Over the past 10 yr, some improvements have been made to wastewater treatment, reducing the discharge of sewage solids, organic matter, toxic chemicals and nutrients to Boston Harbour (MBP 2004).

Delaware Bay, Atlantic Coast

In 1609, Henry Hudson entered the Delaware Bay (Figure 7) in search of a trade route to the Far East for a Dutch company (Delaware State History 2009). The Dutch tried to establish a settlement in 1631, which was destroyed by natives. A few years later, Sweden established the first permanent settlement, which was later taken over by England. The native population may have numbered about 20,000 before European contact, but several wars and at least 14 epidemics reduced their numbers to about 4000 by 1700 (Sultzman 2000). The colonial economy flourished under English rule, supported by whaling, sealing, fishing, farming and the lumber industry. By 1760, about 35,000 people lived in the Delaware region, and the population continued to grow rapidly over time (Delaware State History 2009). The general timeline of change is shown in Figure 8.

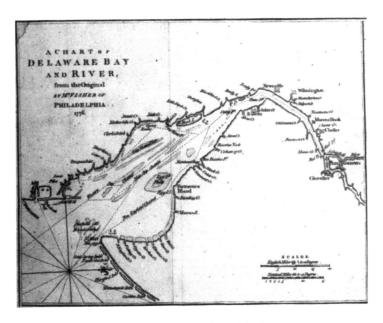

Figure 7 Delaware Bay and River, 1776 (Delaware Public Archives).

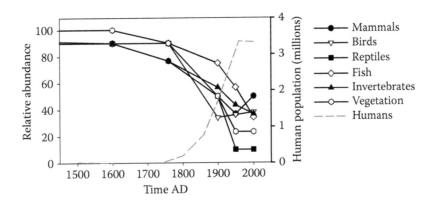

Figure 8 Timeline of changes in relative abundance of six taxonomic groups (left axis) and the human population (right axis) in Delaware Bay.

Marine mammals

In December 1630, the Dutch ship De Walvis sailed from Holland to Delaware Bay with immi-grants, food, cattle and whaling implements. The Dutch planned to open a whale and seal fishery as well as a settlement and plantation for the cultivation of tobacco and grains (Delaware Living History 2009). A persistent shore whaling enterprise was developed in Delaware Bay and along the New Jersey shore (Reeves et al. 1999). The peak of right whaling at the mouth of Delaware Bay was reached in 1680: "Mighty whales roll upon the coast, near the mouth of the bay of the Delaware; eleven caught and worked into oil in one season. We justly hope a considerable profit by whalery, they being so numerous" (William Penn 1683, cited in Watson 1868). Although right whaling was most profitable in the early 1700s, it continued until at least the 1820s. However, a virtual crash of right whaling occurred in 1760, and organised whaling ceased by 1840 (Reeves et al. 1999). Today, fin, humpback and right whales are all listed as endangered in Delaware (USFWS 2009).

Birds

Historical information on birds in Delaware Bay is sparse and largely descriptive. Many species were used by natives and Europeans for their meat, eggs and feathers (Goddard 1978). In the nineteenth century, some uses were commercialised, such as the use of feathers for the millinery industry (Animal Welfare Institute [AWI] 2005). As a consequence, thousands of herons, egrets and other birds with flamboyant plumage were shot for their feathers, while many other species were hunted solely for sport without any regulations in place. Habitat loss, especially the loss and ditching of wetlands, had severe effects on many waterfowl populations (Cole et al. 2001). By the twentieth century, many species of waterfowls, wading birds, sea- and shorebirds were at very low numbers.

Herons were saved from extinction by the federal Lacey Act of 1900, which forbid foreign and interstate trade of wildlife parts, and the Migratory Bird Treaty Act of 1918, which also helped to protect many other species of birds (AWI 2005). However, in the 1940s DDT poisoning resulted in severe declines in ospreys, bald eagles (*Haliaeetus leucocephalus*) and peregrine falcons (*Falco peregrinus*) (Sullivan 1994). After the ban of DDT, ospreys showed improved nesting success in the late 1980s (Clark et al. 2001b). The bald eagle had been listed as endangered in the 1970s but has recently been de-listed (USFWS 2009).

Other birds experienced recent population declines, for example, nine species of wading birds declined from about 12,000 to about 7000 breeding pairs in 1989–2000 (Rattner et al. 2000). Numbers of Canada geese (*Branta canadensis*) have declined since the 1980s, resulting in the closure of the regular hunting season in 1995 along most 'Atlantic flyway' states (Malecki et al. 2001). Some declines have been attributed to the degradation of estuarine habitats (Erwin 1996). Another threat has been the increasing harvest of horseshoe crabs (*Limulus polyphemus*), which provide essential food for up to 1 million migrating birds travelling from South American wintering to Arctic breeding grounds. The timing of their arrival coincides with hundreds of thousands of horseshoe crabs laying their eggs in the sandy beaches, which fuel the long-distance migration of the birds (Walls et al. 2002).

Reptiles

Sea turtles have been harvested for their meat, eggs and shells by natives and Europeans in Delaware Bay (Pearson 1972, Goddard 1978). Delaware Bay provides critical sea turtle habitat, but turtle abundance has declined through hunting, habitat loss and fisheries by-catch. Today, all sea turtles are listed as endangered or threatened in Delaware (USFWS 2009).

Fishes

In a letter in 1683, William Penn described the fishes of Delaware Bay: "Sturgeons play continually in our river. Alloes … and shades are excellent fish. They are so plentiful that six hundred are drawn at a draught. Fish are brought to the door, both fresh and salt" (Watson 1868). The exact time when the fishery for sturgeon reached a considerable extent is unknown, but after 1870 the business expanded rapidly (Cobb 1900). During 1880–1900, Delaware Bay supported the most abundant and commercially important sturgeon population in the United States (Secor & Waldman 1999); "The average catch … is about 1000 fish to a vessel for the month of April" (Collins 1887). In 1890, the abundance of female Atlantic sturgeon, the principal target, was estimated at 180,000, yet "there has been an almost continuous decrease in the number of sturgeon taken" (Cobb 1900). Landings along the U.S. Atlantic coast were about 7 million pounds yr^{-1} (1 pound = 0.4536 kg) just prior to the turn of the century, 90% of which came from Delaware (Figure 9). By 1920, sturgeons

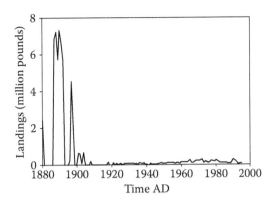

Figure 9 Commercial landings of Atlantic sturgeon for the United States (data from NMFS 2009).

were very rare (Brundage & Meadows 1982). Still, from 1950 to the mid-1990s, landings reached 100,000–250,000 pounds annually along the coast. Overharvesting of sturgeon for flesh and eggs (caviar) continued until the Atlantic States Marine Fisheries Commission (ASMFC) and federal government implemented a coastwide moratorium in the late 1990s (ASMFC 2009). It has been estimated that Atlantic sturgeon stocks are reduced more than 20-fold from their nineteenth century abundances and are now either biologically extinct or extremely depleted throughout their range (Secor & Waldman 1999). There are still about 6000–14,000 shortnose sturgeon in the upper Delaware River (Hastings et al. 1987), which are endangered (USFWS 2009).

American shad (*Alosa sapidissima*) were also very abundant in the Delaware River: "The numbers of shad taken … vary in different seasons. Perhaps it would not be far from the truth to estimate them at 30,000 at each shore fishery. Formerly, when fisheries were fewer, the number far exceeded this amount" (McDonald 1887). Annual landings in 1890–1901 were 11–17 million pounds, several times greater than in any other river system in the United States (Chittenden 1974). Abundance declined rapidly in the early 1900s, and fewer than 0.5 million pounds were landed in 1920. Since then, abundance has remained low (Chittenden 1974). A historically important commercial and recreational fishery also existed for striped bass (*Morone saxatilis*) (Anderson et al. 1999c). Like other diadromous fishes, striped bass has been severely affected by habitat loss and pollution in rivers. Since 1950, commercial catches declined, but recreational fishing increased. In the 1990s, stocks were severely depleted, and protection measures were taken to enhance recruitment (Anderson et al. 1999c).

An important fishery for weakfish (*Cynoscion regalis*) began in the 1800s (Lowerre-Barbieri et al. 1995). "In 1880, while only 10,000 pounds were caught south of Cape Henlopen, 2,608,000 pounds were taken by the fishermen along the shores bordering Delaware Bay" (Collins 1887). In the 1960s, the fishery strongly declined, yet landings and maximum size and age increased during the 1970s (Lowerre-Barbieri et al. 1995). Catches peaked at 1.8 million pounds in 1980, followed by a decline to 24,604 pounds in 2007 (NMFS 2009).

Invertebrates

"Of shell-fish, we have oysters, crabs, coccles, conchs, and mussles; some oysters six inches long, and one sort of coccles as big as the stewing oysters" (William Penn 1683, cited in Ingersoll 1881). European colonists in the 1600s described huge shoals and banks of oysters between Cape Henlopen and Cape May (Miller 1971), and a chart by Peter Lindestrom showed the entire Delaware shore lined with oyster beds (Ford 1997). A commercial oyster fishery began right after European arrival, yet in 1719, a first law was introduced to protect the resource: "That no person … shall rake or gather up any oysters or shells from and off any of the beds … from the tenth day of May to the first day of September" (McCay 1998). Numerous acts followed but were largely ineffective in protecting oysters, which were badly depleted by the late 1700s. In 1820, the use of dredges was banned (McCay 1998), and the first planting of oysters in Maurice River Cove was authorised in 1856 (Hall 1894). The oyster industry declined alarmingly in the mid-1890s but increased again in 1902 (Miller 1971). The first recorded landings in the late 1890s were at 21 million pounds of oyster meats (Ford 1997). Average landings from 1880 to 1931 were 13.9 million pounds but dropped to 6.5 million pounds in 1932–1956. In the late 1950s, the MSX oyster parasite strongly decreased oyster production, and landings were 550,000 pounds in 1963. Resurgence of MSX forced the closure of seed beds in the mid-1980s. A limited oyster fishery opened in 1990, but the Dermo parasite *Haplosporidium nelsoni* caused strong mortality to entire oyster beds (Ford 1997).

A significant hard-shell clam (*Mercenaria mercenaria*) fishery has existed in Delaware Bay since early times (Ford 1997). Large-scale utilisation began in the late 1940s, and landings peaked at 770,000 pounds of meat in 1951, followed by a sharp decline. The dredge fishery ceased to operate in 1966–67 because of clam depletion. The bay's hydrography, tidal exchange and predation by horseshoe crabs were identified as causes for low clam densities. Several surveys in the 1970s confirmed the presence of hard-shell clams in or near oyster beds but at numbers insufficient to support commercial harvesting (Ford 1997).

Horseshoe crabs were commercially harvested for fertiliser and livestock feed from 1870 to the 1960s (Walls et al. 2002). During this period, catch records in Delaware Bay exhibited a progressive decline from more than 4 million to fewer than 100,000 crabs per yr. In the 1950–1960s, the horseshoe crab fishery was minimal, and populations increased at least 13-fold. Since then, a commercial fishery developed to provide bait for catching American eel (*Anguilla rostrata*) and whelk (*Busycon* spp.). The fishery employed trawls, dredges, gill nets and hand gathering and increased dramatically in the 1990s. Horseshoe crabs are also fished for biomedical purposes. Total landings in 2000 were reduced to 40% of 1995–1997 levels, and a marine reserve was implemented at the mouth of Delaware Bay. The commercial harvest is controversial as horseshoe crabs are essential food for migrating birds (Walls et al. 2002).

Vegetation

Today, Delaware Bay has about 160,000 ha of wetland area, which is at least 21–24% less than the original extent (Kennish 2000). In Delaware State, more than half of the wetlands have been converted since the 1780s (U.S. Geological Survey [USGS] 2009a). In the 1930s, ditching of wetlands as a mosquito control programme began. In the 1940s, the common reed (*Phragmites*) invaded coastal marshes comprising one-third of wetland vegetation today (Cole et al. 2001). A 1970s survey estimated that tidal wetlands were reduced from 80,418 to 74,688 acres (1 acre = 0.4047 ha) from 1938 to 1973 (Cole et al. 2001). Compared with other regions, the relatively low loss has been attributed to the lower human population (Sullivan 1994). The Wetland Protection Act of 1970 and the Clean Water Act of 1972 slowed the rate of further wetland loss.

Submerged aquatic vegetation (SAV) is nearly absent from the Delaware Estuary and its coastal bays (Sullivan 1994, Bricker et al. 1999). It is generally assumed that relatively high natural turbidity prevents the establishment of SAV, although some reports from the early twentieth century indicated eelgrass in the backwaters of Cape May County and possibly in Delaware Bay: "It is present to some extent in Cape May Harbor and in back waters of Cape May County, but is very scarce in the upper part of Delaware Bay" (Richards 1929, cited in Sullivan 1994).

Water quality

Delaware Bay water quality began to decline in the early 1800s, with marked increases in pollution and disease since the 1880s (Kennish 2000). In the 1940s–1950s, dissolved oxygen concentrations reached minimum values of 0.1 mg L^{-1} every year. Wastewater treatment plants were built in the 1960s, and water quality began to improve. Minimum oxygen concentrations increased to 3 mg

L^{-1} by the late 1960s and 8 mg L^{-1} by the late 1980s. Nitrogen and phosphorus concentrations decreased 1.33- and 4-fold, respectively, from 1960 to 2000, but loading was still 7500 mmol nitrogen m^{-2} yr^{-1} and 600 mmol phosphorus m^{-2} yr^{-1}. Nitrogen and phosphorus loads in Delaware Bay are about 10 times higher than in Chesapeake and two to four times higher than in San Francisco Bay (Kennish 2000), yet overall eutrophication conditions are low (Bricker et al. 1999).

Chesapeake Bay, Atlantic Coast

Chesapeake Bay is one of the U.S. largest estuaries and probably one of the best studied around the globe (Figure 10). Native Americans have lived in the bay region since the last ice age, beginning agricultural practices around 1200. In 1600, about 25,000–45,000 natives lived in the region (Feest 1978a,b, Ubelaker & Curtin 2001). In 1900, the native population numbered about 2000, compared with 1,120,135 European descendants. European settlement began with the establishment of Jamestown, Virginia, in 1607 (Chesapeake Bay Program [CBP] 2009a), and the colony grew rapidly (Figure 11). Next to agriculture and forestry, the colonial economy was built on an incredible abundance of marine life. Strachey (1612) noted: "Grampus, porpois, seales, stingraies, bretts, mulletts, white salmons, troute, soles, playse, comfish, rockfish, eeles, lampreys, cat-fish, perch of three sorts, shrimps, crefishes, cockles, mishells, and much more like, like needles to name, all good fish", and Burnaby in 1759 indicated: "These waters are stored with incredible quantities of fish, such as sheeps-head, rock-fish, drums, white perch, herrings, oysters, crabs, and several other sorts" (Pearson 1942). The general timeline of historical change is shown in Figure 11.

Marine mammals

Humpback, fin and right whales occur in coastal waters off Virginia and Maryland and occasionally enter Chesapeake Bay (Swingle et al. 1993). Today, these large whales are at low population levels along the U.S. Atlantic coast (Waring et al. 2001) and are listed as endangered (USFWS 2009). Yet, large whales were much more abundant in earlier times. Historical records describe the hunting of large whales in Chesapeake Bay in the 1600s–1700s (Wharton 1957, Pearson 1972). These were probably right or Atlantic grey whales, which were the first whales to be pursued by whalers because

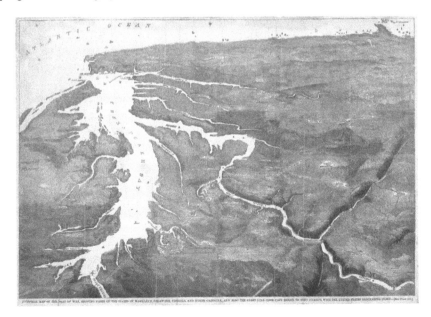

Figure 10 Chesapeake Bay, 1861 (Library of Congress, Geography and Map Division).

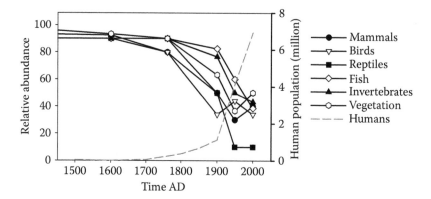

Figure 11 Timeline of changes in relative abundance of six taxonomic groups (left axis) and the human population (right axis) in Chesapeake Bay.

they were coastal, slow swimming, easy to catch and would not sink when dead. In the 1600s and early 1700s, right whales were reduced to very low numbers on the U.S. eastern coast (Reeves et al. 1978), and the Atlantic grey whale was extirpated in the 1800s (Mead & Mitchell 1984).

Historical accounts of Europeans arriving in Chesapeake Bay in the 1500s included reports of dolphins, porpoises, seals and manatees (*Trichechus manatus*) (Carleill 1585, Smith 1910, Bruce 1935, Wharton 1957, Pearson 1972). Coastal bottlenose dolphins (*Tursiops truncatus*) were exploited by a shore-based net fishery from 1797 to 1925, mostly in North Carolina. The historical population numbered 13,748–17,000 animals for the entire coast, while today's stock is listed as depleted under the MMPA (Blaylock 1988, Waring et al. 2001). "A good fish, which is common and found in large numbers is the porpoise," wrote Michel in 1701, and seals were also described as common at that time (Pearson 1942). The repeated sightings of a "sea monster", nicknamed "Chessie", in Chesapeake Bay date to the early twentieth century and possibly include manatee sightings that were not properly identified. The annual migration of manatees from Florida to Chesapeake Bay may have been common in previous centuries (USGS 1996).

Sea turtles

"The waters and especially the tributaries are filled with turtles," wrote Michel in 1701 (Wharton 1957). Sea turtles were found in great numbers (Billings et al. 1986) and have been commercially harvested for their meat, eggs and shells since the early 1600s (Pearson 1972). Years of exploitation have greatly reduced all sea turtle populations, with the leatherback (*Dermochelys coriacea*), Kemp's ridley (*Lepidochelys kempii*) and hawksbill (*Eretmochelys imbricata*) turtles listed as endangered today and the loggerhead (*Caretta caretta*) and green (*Chelonia mydas*) turtles as threatened in Virginia and Maryland (USFWS 2009). Incidental by-catch in commercial shrimp trawling is a continuing source of mortality adversely affecting recovery of some species (Office of Protected Resource [OPR] 2009). Today, 5000–10,000 sea turtles enter Chesapeake Bay each spring and summer, mostly young loggerhead or Kemp's ridley using the Bay as a feeding ground. Leatherback and green turtles also occur, but only one hawksbill turtle has been seen in recent times (Virginia Institute for Marine Science [VIMS] 2009).

Birds

Historically, millions of geese, swans, ducks and other waterfowl overwintered in Chesapeake Bay, supported by profuse seagrass beds, SAV and a rich supply of invertebrates. In the seventeenth century, Alsop reported that waterfowl so blanketed the Bay that "there was such an incessant clattering made with their wings on the water where they rose, and such a noise of those flying higher up,

Table 5 Average number of birds wintering on Susquehanna Flat

Period	Canada geese	Tundra swans	Diving ducks	Dabbling ducks
1959				2500
1959–1962	625	750	2225	1275
1966–1972	2030	770	1660	390
1973–1976	3175	50	275	175

Note: The decline in tundra swans and diving and dabbling ducks was attributed to the decline in submerged aquatic vegetation, while Canada geese switched to feeding on adjacent agricultural fields (data from Bayley et al. 1978).

that it was as if we were all the time surrounded by a whirlwind" (Blankenship 2004). In the 1600s, flock size of ducks was estimated at 7 mi^2 (Bruce 1935). Over time, the deterioration of shallow-water habitats; the loss of wetlands, seagrass beds and SAV; the degradation of water quality; human disturbance and overhunting have greatly reduced waterfowl numbers (Blankenship 2004). By the late nineteenth and early twentieth century, the dwindling number of waterbirds raised concerns. First conservation measures were introduced with the Lacey Act in 1900 and the Migratory Bird Treaty Act in 1918 (Blankenship 2004, AWI 2005). This effectively ended commercial hunting and increased habitat protection, but sport hunting continued. Waterfowl populations never returned to historic levels, and today about 1 million waterfowl spend their winter in Chesapeake Bay, a third of the entire Atlantic flyway waterfowl population (Blankenship 2004).

Since the 1960s, bird numbers have remained relatively constant, but species composition has changed (Table 5; Bayley et al. 1978). Duck populations rely more heavily on aquatic habitats and have decreased nearly 80% since the 1950s, whereas geese populations can feed away from the bay in agricultural fields and have increased (Blankenship 2004). American black ducks (*Anas rubripes*) are particularly intolerant of human disturbance and rely heavily on underwater grasses, which have declined (Krementz et al. 1991, Blankenship 2004). Similarly, the redhead (*Aythya americana*) exclusively feeds on underwater grasses and did not switch to other foods. A few decades ago, about 80,000 wintered here, but only a few thousand remain today (Blankenship 2004). Canvasback (*Aythya valisineria*) was the most abundant diving duck in Chesapeake Bay, but the "large rafts of canvasbacks and other diving ducks that made the flats famous are no longer observed" (Bayley et al. 1978). In the 1930s, about half a million canvasbacks congregated each autumn off Poole's Island in Maryland. They used to feed on wild celery almost exclusively, but the sharp decline in wild celery caused canvasbacks to shift their diet to small clams. Today, only about 50,000 come to the Chesapeake each fall (Blankenship 2004).

Raptors were also abundant around Chesapeake Bay. Historical records indicate that more than 1000 pairs of bald eagles nested around the bay every year in the early 1990s (Reshetiloff 1997). The population declined because of direct killing, habitat loss and decline of prey. In the 1940s, the Bald Eagle Protection Act (AWI 2005) made it illegal to kill, harm, harass or possess bald eagles, alive or dead, including eggs and feathers. However, the increasing use of DDT to control mosquitoes posed a new threat by causing reproductive failure (Reshetiloff 1997). By the 1970s, the number of breeding pairs had dropped to 90, and the bald eagle was listed as endangered. In 1972, DDT was banned in the United States, and nesting success steadily increased. In 1996, there were 179 nesting pairs producing 508 young (Reshetiloff 1997). The bald eagle was reclassified as threatened in 1995 and was removed from the endangered species list in 2007.

Fishes

More than 295 species of fishes are known to occur in the Chesapeake Bay region; 32 of these species are year-round residents (CBP 2009b). Native Americans used a wide variety of fish and shellfish species (Dent 1995). European colonisers first developed subsistence fisheries to supply the new colony and later commercial fisheries to trade with Europe and other regions (Pearson 1942). Important fisheries in Chesapeake Bay included those for sturgeon, shad, menhaden, mackerel and many ground- and flatfish (CBP 2009b).

Atlantic sturgeon was extremely abundant, and hundreds could be seen migrating up the tributaries to spawn every year (Billings et al. 1986). Their average size was more than 3 m in late summer (Bruce 1935), and they were eaten widely by colonists (Smith 1910). "Sturgeon and shad are in such prodigious numbers, that ... some gentlemen in canoes, caught above 600 of the former with hooks, and of the latter above 5000 have been caught at one single haul of the seine," wrote Burnaby in 1759 (Pearson 1942). Sturgeon and caviar were exported to England beginning in the early 1600s (Pearson 1942). Sturgeon was the first primary cash crop of Jamestown, Virginia, and was second only to lobster among important fisheries in the late 1800s (ASMFC 2009). Commercial sturgeon landings in Chesapeake Bay peaked in 1890 at about 250,000 kg (Secor et al. 2000). Afterwards, the mid-Atlantic fishery rapidly declined (ASMFC 2009), and Atlantic sturgeon was thought to be extirpated by the 1920s (Secor et al. 2000). Today, Atlantic sturgeon stocks are extremely depleted and nearly extirpated throughout their range in the United States (Secor & Waldman 1999), and shortnose sturgeon is listed as endangered (USFWS 2009).

American shad have been valued for both their delicious meat and roe. From the mid-1800s to the early 1900s, the American shad fishery was the largest fishery in Chesapeake Bay, with annual catches reaching 17.5 million pounds (Figure 12). By the mid-1930s, annual catches had dropped to less than 4 million pounds and to less than 1 million pounds in the 1980s. Stocks were in such poor condition that a moratorium on taking shad was implemented in Maryland in 1980, while fishing continued in Virginia (Chesapeake Bay Field Office [CBFO] 2009). After catches dropped below 500,000 pounds in 1992, Virginia implemented a moratorium as well. The long decline seems primarily the result of overfishing and habitat degradation in spawning areas (CBFO 2009).

The abundance of menhaden (*Brevoortia tyrannus*) was noted by early European explorers of the mid-Atlantic region (Menhaden Research Council [MRC] 2009). Native Americans showed colonists how to use menhaden as fertiliser, and the colonists soon developed a fishery. Menhaden

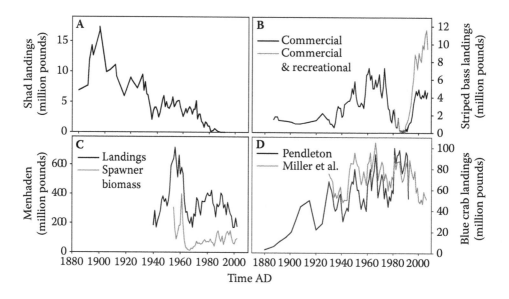

Figure 12 Time series of historical fisheries in Chesapeake Bay for (A) American shad (data from Pendleton 1995), (B) striped bass (data from Pendleton 1995, ASMFC 2008), (C) Atlantic menhaden for the Atlantic coast (data from ASMFC 2004), and (D) blue crabs (data from Pendleton 1995, Miller et al. 2005).

were taken with large seines set from shore, and farmers applied 6000–8000 fish per acre (MRC 2009). The menhaden fishery for reduction had its origins in New England in the early 1800s and spread south after the Civil War. Coal-fired steamers gradually replaced sailing ships in the late 1800s, and diesel and gasoline engines replaced steam engines following WWI (Chesapeake Bay Ecological Foundation [CBEF] 2009). The primary use of menhaden changed from fertiliser to animal feed after WWII, and the industry grew rapidly to peak production of more than 700,000 t in the 1950s (Figure 12). In the 1960s, landings declined to less than 200,000 t, resulting in factory closings and fleet reductions. During the 1970s–1980s, landings improved but declined again during the 1990s (ASMFC 2004). During the same time, the spawning stock biomass of the Atlantic stock declined from 176,000 t in the 1950s–1960s to 32,800 t in 1999 (Figure 12). The population has suffered from poor recruitment in Chesapeake Bay since the 1990s because of not only habitat loss and predation but also disease, toxic algal blooms, parasites and occasional mass mortalities due to low dissolved oxygen levels (CBEF 2009).

Other fish species that have experienced strong declines during the twentieth century include the weakfish, Atlantic croaker (*Micropogonias undulatus*), red drum (*Sciaenops ocellatus*), blue-fish, summer flounder (*Paralichthys dentatus*) and Spanish mackerel (*Scomberomorus maculatus*), among others (Chittenden et al. 1993, CBP 2009b). Yet, not all fish stocks have declined. Striped bass, which has supported an important fishery since colonial times, steeply declined in the 1980s, but strict regulations enabled the species to increase to high abundance levels. Both the commercial and the recreational fishery followed with increasing landings over the past decade (Figure 12; ASMFC 2008).

Invertebrates

When Europeans first visited Chesapeake Bay, they found extensive oyster bars exposed at low tide and in shallow waters (Rothschild et al. 1994). In the Maryland portion of Chesapeake Bay, natural oyster bars covered 116,000 ha in 1907–1912 (Figure 13). Today, these bars are virtually nonexistent (Rothschild et al. 1994). The fishery peaked in 1840–1890 at 600,000 t (Figure 13), accompanied by strong overfishing and destruction of oyster habitat by fishing gears (Héral et al. 1990, Rothschild

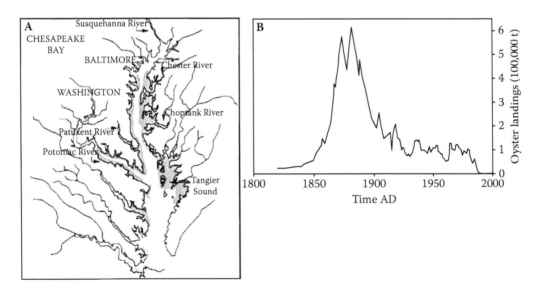

Figure 13 Historical distribution (A) of oyster bars in Chesapeake Bay around 1900 according to Yates 1913 (from Rothschild et al. 1994; reprinted with permission from Inter Research) and oyster landings (B) in Maryland's portion of Chesapeake Bay (data from Rothschild et al. 1994, Wieland 2007).

et al. 1994). Landings strongly decreased in the early 1900s and stabilised at 80,000 t due to failing reseeding plans connected to heavy sedimentation and anoxic summer conditions. In 1981–1988, oyster production further declined due to high mortalities related to disease, predation and management practices. Since 1986, landings have remained under 15,000 t (Figure 13; Rothschild et al. 1994).

In the Maryland portion of Chesapeake Bay, the yield per habitable area, which is proportional to biomass, declined from 550 to 22 g m^{-2} in 1884–1991 (Rothschild et al. 1994). Newell (1988) estimated that the pre-1870 oyster biomass could have potentially filtered the entire water column during the summer in less than 3–6 days, while current oyster stocks need 244–325 days. Thus, the loss in oyster biomass has resulted in a dramatic reduction in filter activity and phytoplankton removal from the water column (Newell 1988).

Another invertebrate that has been harvested since colonial times is the blue crab (*Callinectes sapidus*). A commercial fishery was started in the late nineteenth century, and Maryland and Virginia historically harvested and marketed more blue crabs than any other North American region (Rugolo et al. 1998a). Commercial harvests rapidly increased from 1880 to 1940 and varied between 60 and 100 million pounds annually until the 1990s (Figure 12). However, in recent years baywide harvests fell to around 50 million pounds (Miller et al. 2005). Since 1945, directed effort has increased 5-fold, and CPUE has declined exponentially (Rugolo et al. 1998b). The constant harvest pressure and the loss of SAV habitat, which young crabs require for shelter and food during their development, have contributed to the decline.

Vegetation

At the time of European arrival, forests covered about 95% of the land surrounding Chesapeake Bay, yet forest cover rapidly declined throughout the 1700s and 1800s (Figure 14). By the mid-nineteenth century, deforestation had stabilised, and eventually forest cover slightly increased to 58% of the watershed today (Pendleton 1995, Kemp et al. 2005, CBP 2009c). Chesapeake Bay had also extensive wetlands, which have been partly converted to agricultural, urban, industrial and recreational uses. Since 1780, Maryland has lost about 64% and Virginia 42% of their wetland areas, leaving currently 591,000 acres in Maryland and 1 million acres in Virginia (USGS 2009a). Dominant wetland types include non-tidal forested (60%) and shrub-scrub wetlands (10%) and salt and freshwater marshes (10% each) (Pendleton 1995). Today, wetland conservation has high priority, and any conversion, alteration or development needs to be permitted.

During 1700–1930, SAV consisting of eelgrass (*Zostera marina*) and other macrophytes was a persistent feature of shallow-water habitats in Chesapeake Bay (Orth & Moore 1984). These meadows provided important breeding, nursery and feeding habitats for a variety of invertebrates, fish, birds, sea turtles and manatees. In 1760, land clearing increased sediment loads and water turbidity with possible adverse effects on SAV. Sediment records have revealed declines of *Najas* spp. in the late eighteenth century and eelgrass in the late nineteenth century (Orth & Moore 1984). In the 1930s, extensive losses of eelgrass were caused by the wasting disease, but many eelgrass beds recovered. At six stations in Chesapeake Bay, SAV (mainly *Zostera* and *Ruppia*) increased from 1500 to 2310 ha in 1937–1960 but declined to 570 ha in 1980 (Figure 14; Orth & Moore 1983). Large-scale SAV declines occurred throughout the bay in the 1960s–1970s due to nutrient and sediment loads, turbidity, eutrophication and tropical storms (Bayley et al. 1978, Orth & Moore 1984). Since 1978, aerial surveys have shown an SAV increase from 16,500 to 36,222 ha (Orth et al. 2003), compared with an estimated 83,514 ha that existed in the 1930s–1970s based on historical aerial photographs (Figure 14; Moore et al. 2004).

A long-term estimate of seagrass decline comes from sediment core data (Cooper & Brush 1993), in which the relative abundance of *Cocconeis* (an epiphytic diatom on seagrass leaves) can be used as a proxy for seagrass occurrence. *Cocconeis* abundance was 10% in sediment layers in 550 AD but dropped to 5–7% until 1800 (Figure 14). In the 1800s and early 1900s, values were

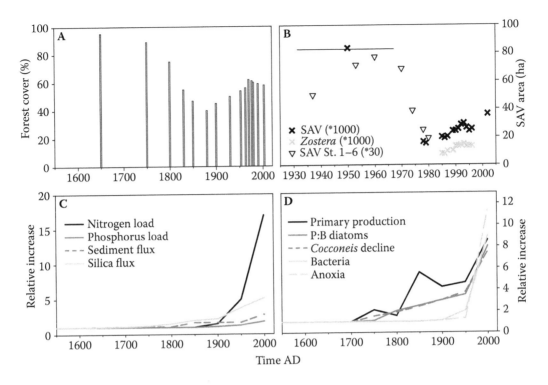

Figure 14 Historical changes in vegetation and water quality in Chesapeake Bay: (A) Forest cover (data from CBP 2009b); (B) submerged aquatic vegetation (SAV) and eelgrass (*Zostera*) in the entire bay (horizontal line = average estimate 1930–1970) and at selected stations (St.1–6) (data from Orth & Moore 1983, Orth et al. 2003, Moore et al. 2004); (C) nutrient and sediment loads and (D) primary production, pelagic:benthic (P:B) diatom ratio, epiphytic diatom *Cocconeis* (relative decline), bacterial matter, and anoxic conditions (data from Cooper & Brush 1993, Zimmerman 2000, Colman & Bratton 2003, and Fisher et al. 2006). See text for details.

1–3%, declining to 0–2% in the 1980s (Cooper & Brush 1993). This represents an 80–100% loss of seagrass at the particular location of the sediment core but may not reflect baywide trends.

Water quality

Since European settlement, Chesapeake Bay's water quality has undergone substantial changes which have been studied in detail using sediment core data (Figure 14). Average sedimentation rate in the estuary increased from 0.06 to 0.19 cm yr^{-1} from before to after European colonisation (Brush 2001), and biogenic silica flow, a measure of nutrient loading, increased more than 5-fold (Colman & Bratton 2003). There are no direct long-term measures of nitrogen and phosphorus loads, but hydro-chemical modeling and land-use yield coefficients suggest that current input rates are 4–20 times higher than under forested conditions 350 yr ago (Fisher et al. 2006). Since the 1960s, fertiliser use increased from 15 to 40 million kg nitrogen yr^{-1} (Kemp et al. 2005). As a consequence, primary production has increased 4- to 12-fold since European contact based on total organic carbon (TOC) and organic matter fluxes (Cooper & Brush 1993, Zimmerman 2000). This was accompanied by an increase in the ratio of planktonic:benthic diatoms, indicating a shift from benthic to pelagic production (Figure 14). Bacteria decompose this increased amount of organic matter and have thus strongly increased, especially since 1950 (Figure 14). Finally, anoxic conditions have increased, as indicated by reactive sulphur, sediment colour and anaerobic bacteria (Zimmerman 2000).

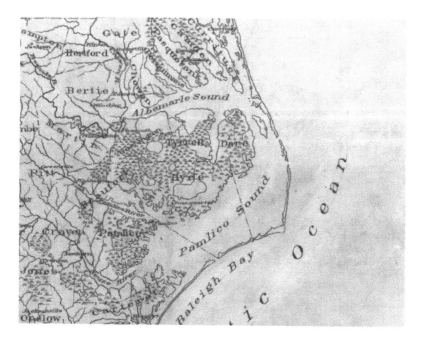

Figure 15   Pamlico Sound, 1893 (Library of Congress, Geography and Map Division).

Pamlico Sound, Atlantic Coast

Similar to the other estuaries, Pamlico Sound (Figure 15) has experienced a long history of human-induced changes. Native people started agriculture about 800 yr ago and Europeans colonised and transformed the land about 400 yr ago. Larson (1970) estimated that there were about 1000 natives in the coastal plains in 1600, who were reduced to fewer than 500 after European contact. Since European colonisation, the human population around the watershed increased to more than 215,000 in 1990 (Forstall 1996). Over the last two to three centuries, overexploitation, habitat loss and pollution have resulted in the depletion of many marine species that have been of economic or ecological importance (Figure 16). Of the 44 marine mammals, birds, reptiles, fishes, invertebrates and plants

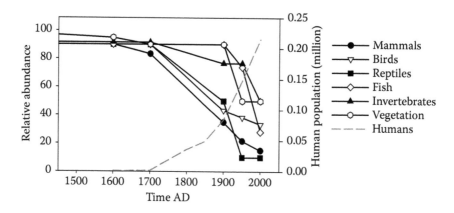

Figure 16   Timeline of changes in relative abundance of six taxonomic groups (left axis) and the human population (right axis) in Pamlico Sound.

reviewed, 24 have become depleted (>50% decline from former abundance), 19 rare (>90% decline), and 1 extirpated (100% decline) by the year 2000 (Lotze et al. 2006).

Marine mammals

Native Americans hunted whales all along the coast southward of Canada; catch records only occur for the Tequesta Indians in Florida, who hunted whales from the Keys to Cape Canaveral in the 1500s (Larson 1970). In the 1660s, Europeans began hunting whales along the coast off Pamlico Sound, especially the Atlantic grey whale, which was probably the only whale that entered the sound (Larson 1970, Mead & Mitchell 1984). With the youngest grey whale skeleton carbon dated to 1675, extinction must have occurred later, probably in the 1800s (Mead & Mitchell 1984). Whales were described as numerous along the Carolina coast by early explorers (Lawson 1712, Brickell 1737), but shore-based whaling reduced their numbers rapidly. In the early 1800s, coastal whaling strongly declined, and whalers moved offshore to target sperm whales (*Physeter macrocephalus*) (Reeves & Mitchell 1988). Today, fin, right, humpback and sperm whales are all listed as endangered in North Carolina (USFWS 2009).

In the early 1700s, "the bottle-nose are found everywhere in Pamlico Sound" (Brickell 1737), and the colonists developed a shore-based net fishery. Mitchell (1975) estimated that the coastal population had at least 13,748 bottlenose dolphins in the 1800s, but the fishery strongly depleted the population and ended in 1925 (Mead 1975). The population was at high levels again prior to an anomalous mortality event in 1987–88, which reduced the numbers by as much as 53% (Northeast Fisheries Science Center [NEFSC] 1997). In the 1990s, the western North Atlantic coastal stock was listed as depleted under the MMPA, with a minimum population estimate of 2482 dolphins (NEFSC 1997). Surveys counted 2114–2544 dolphins in 2002 along the coast (NEFSC 2005) and 950–1050 in the estuaries of North Carolina (Read et al. 2003). Annual by-catch or other fishery-related mortality and serious injury exceed 10%, which is considered to be significant (NEFSC 2005).

Another marine mammal hunted by natives and Europeans was the West Indian manatee (Larson 1970, Hartman 1979, Catesby 1996). In 1720, they were common in North Carolina sounds (Hartman 1979). Over time, however, numbers were reduced historically by exploitation and more recently by decline of seagrass (its main food and habitat), boat accidents and other threats. In 1967, the manatee was added to the endangered species list, and only few have been spotted in North Carolina in the last 20 yr (USFWS 2008).

Birds

Archaeological records suggest that there were many thousands of brown pelicans (*Pelecanus occidentalis*) in Pamlico Sound that were used by natives (Larson 1970). European colonisers in the early 1700s also described thousands and thousands of pelicans (Catesby 1996). Over time, hunting and more recent DDT poisoning in the 1960s lead to their listing as endangered in the 1970s (Larson 1970). Since then, the population has increased and is now considered stable with approximately 4000 nesting pairs. Major current threats include human disturbances at nesting sites, loss of nesting habitat, fish kills and fishing lines (National Audubon Society [NAS] 1997a).

Pamlico Sound was and still is a haven for waterbirds. Geese, swans, ducks, herons, egrets and many more waterfowl and wading birds occurred in great abundance in the past, although they were hunted throughout history. Brickell (1737) noted about wild geese: "They are plenty here all the Winter, come and go with the Swans ...; vast numbers are shot every year". He also described great numbers of many kinds of ducks and mallard, and that the eggs of the black ducks were taken (Brickell 1737). In the late nineteenth century, geese, swans and other waterfowl were heavily hunted by European settlers, strongly reducing their numbers by the early twentieth century (Serie et al. 2002). In the 1930s, the loss of seagrass contributed to further declines in waterfowl directly feeding on seagrass or associated invertebrates (Mallin et al. 2000). With the recovery of seagrass, some species have increased in population size since the 1950s (Serie et al. 2002), yet habitat degradation and loss continue to be a major threat (Golder 2004).

Birds with beautiful feathers were heavily hunted in the late nineteenth century for the millinery industry. Terns and gulls were highly prized for their wings and skins; great egrets (*Ardea alba*) and snowy egrets (*Egretta thula*) were sought for their aigrettes, and these species were shot by the tens of thousands (NAS 1997b). Those nesting in colonies were easy targets for market hunters: "They don't much like to leave their young. I have often shot at these 'strikers' [terns] so fast that I had to put my gun overboard to cool the barrels" (NAS 1997b). This relentless shooting took a tremendous toll on populations, and several species of herons, egrets and terns were pushed nearly to extinction (NAS 1997b, Golder 2004). The least tern (*Sterna antillarum*) once numbered in the thousands but became a rare sight in the early 1900s (NAS 1997b). The extinction of many species was prevented by the Lacey Act in 1900 (AWI 2005) and the Migratory Bird Act of 1918 (AWI 2005). Since then, the main threats to these species are habitat degradation and loss in breeding, migratory stopover and wintering areas (Golder 2004). Surveys in the 1970s–1980s revealed alarming declines in nesting sites of colonial waterbirds due to chronic human disturbances, lack of suitable nesting habitats and erosion. This led to the establishment of the North Carolina Coastal Islands sanctuary system in 1989 and to intensified efforts to protect coastal birds and their habitat (NAS 1997b). Today, waterfowl number about 10,000 and shorebirds about 70,000 on Cape Hatteras (Golder 2004). Many gulls show stable or increasing population trends, while several terns, herons and egrets are declining and listed or proposed for listing on the endangered species list (NAS 1997a).

Reptiles

American alligators (*Alligator mississippiensis*) and sea turtles were hunted by natives throughout history (Larson 1970, Catesby 1996). Because Pamlico Sound is the northern extent of the range of the alligators, they may have never been very plentiful in this area. However, in the early 1700s, European explorers described them as frequent along the sides of the rivers (Lawson 1712). "Alligators are common and reach 18 feet," reported Brickell (1737). In the early 1900s, populations were severely affected by commercial hunting for their belly skin (Brandt 1991, Britton 2009). Commercial hunting pressure was particularly strong in Louisiana and Florida, and by the 1920s, alligator numbers were severely depleted throughout their range. In the 1940s, alligators were only occasionally spotted in North Carolina. Hunting was prohibited in the 1960s, but illegal poaching continued into the 1970s. Because of the risk of extinction, additional changes were introduced in the law to control the movement of hides, which helped the species survive. Since the 1980s, populations have improved considerably throughout their range and are now only considered to be threatened in a few areas by habitat degradation (Brandt 1991, Britton 2009). Today, the entire population is estimated at 1 million, and the species is listed under the Convention on International Trade in Endangered Species (CITES) (Appendix II) and as low risk and least concern by the International Union for Conservation of Nature (IUCN; Britton 2009).

European explorers encountered lots of sea turtles in the North Carolina Sounds, which were hunted first by natives and later by Europeans (Catesby 1996). In the nineteenth century, commercial harvesting increased, and by 1880, a reduction in sea turtle numbers was noticed (Epperly et al. 1995). However, the inshore waters of North Carolina harboured enough green turtles to support a commercial fishery before exploitation ceased in the early 1900s (Epperly et al. 1995). In the 1970s, high mortality occurred through by-catch in the shrimp fishery, and sea turtles were listed as endangered (USFWS 2009). In the 1990s, the waters of Pamlico Sound continued to provide important developmental habitats for loggerhead, green, and Kemp's ridley, while leatherback and hawksbill turtles were infrequently found (Epperly et al. 1995). Today, all sea turtles are listed as endangered or threatened in North Carolina (USFWS 2009).

Fishes

Native Americans caught a large variety of fishes in Pamlico Sound, including large hammerhead (*Sphyrna* spp.) and small sharks (Larson 1970). European explorers described "Shovel-nose sharks [hammerheads] four to six thousand pounds," and "dogfish weigh normally 20 lbs or more, and are commonly caught when fishing for mackerel" (Brickell 1737). There is not much historical information on sharks in Pamlico Sound. In recent decades, the two most abundant species in North Carolina were sandbar (*Carcharhinus plumbeus*) and Atlantic sharpnose (*Rhizoprionodon terraenovae*) sharks, and sandbar and blacktip (*Carcharhinus limbatus*) sharks are the main targets of the North Carolina shark fishery (North Carolina Division of Marine Fisheries [NCDMF] 2008). The status of large sharks in general is of concern, and the sandbar shark is considered overfished. Although harvest restrictions have been in place since 1993 and a closure to commercial harvest occurred in 1997–2006, there is no conclusive evidence to suggest that stocks as a whole are recovering for these slow-growing, late-maturing animals (NCDMF 2008). Since the 1980s, there was also an increase in fishery yields for small sharks, especially spiny dogfish. The population declined during the 1990s and was considered overfished in 2003. Although the stock was considered recovering in 2008, the continued decrease in female abundance, imbalance in sex ratio and low recruitment are concerns for the current stock status (NCDMF 2008).

"The Drum-fish, whereof there are two sorts, ... the Red and the Black. ... There are greater numbers of them to be met with in Carolina, than any other sort of Fish" (Brickell 1737). But by the twentieth century, a decline in the fishery for drums occurred due to high fishing pressure and loss of reef habitat (Lenihan et al. 2001). The red drum was listed as overfished but is now recovering, with the adult stock protected from exploitation (NCDMF 2008). Similarly, there have been declines in other reef- or bottom-dwelling fish, including the gag grouper (*Mycteroperca microlepis*), weakfish, southern flounder (*Paralichthys lethostigma*) and summer flounder, mainly caused by overfishing and loss of reef habitat (Lenihan et al. 2001). Today, these species are considered as depleted or of concern. Among all 73 species of reef fish in the region, 17 are considered overfished (NCDMF 2008).

Other fish species that have been valued and fished at least since colonial times are striped mullets (*Mugil cephalus*), sturgeon and shad, all of which were described as plentiful in the 1700s (Brickell 1737, Larson 1970). Today, their abundance is much reduced. Since the 1950s, striped mullets produced consistently large annual landings of more than 2 million pounds, which ranked the species among the top seven finfish fisheries in North Carolina (NCDMF 2004). However, rapid surges in roe value in the late 1980s, followed by rising commercial fishing effort and landings through the mid-1990s, caused concern for the stock. Because commercial exploitation targets pre-spawned, roe-carrying adults, this directly reduces the yearly reproductive output. Recreational exploitation of juveniles for bait is also of concern (NCDMF 2004). In 2008, striped mullet was considered viable (NCDMF 2008).

In the 1700s, Lawson (1712) described the abundance of sturgeon: "In May, they run upwards the heads of rivers, where you see hundreds of them in one day", and Brickell (1737) noted "The Sturgeon is the first of these whereof we have great plenty. ... The Indians kill great Numbers of them with their Fish-gigs and Nets". The Europeans also fished Atlantic sturgeon over the centuries, and a heavy fishery for roe operated in the twentieth century, leading to strong declines in the population. In North Carolina, landings have been low since the 1960s, and in 1991 the NC Marine Fisheries Commission made it illegal to possess sturgeon in North Carolina (NCDMF 2004). Currently, the stock is considered depleted along the entire Atlantic coast (NCDMF 2008). American shad was another highly abundant and highly valued species. At one time, North Carolina produced more American shad than any other state, mostly in the Neuse River (Smith 1907). Shad catches have plummeted from more than 8 million pounds in 1896 to an average of 252,469 pounds in 1998–2007, and the stock is considered of concern (Smith 1907, NCDMF 2008).

Invertebrates

Many invertebrate species, including oysters, clams, scallops and crabs, were used by natives and Europeans around Pamlico Sound. Among the crustaceans, blue crabs were abundant in Pamlico Sound and easily trapped (Catesby 1996). Thus, a small-scale fishery was developed in the early 1700s (Lawson 1712). In the twentieth century, commercial and recreational harvesting became very popular, and in 1996 over 65.5 million pounds of blue crabs were harvested commercially. After these record landings, catches were strongly reduced in 2000–2002, which has caused increased concern for the health of the resource (NCDMF 2004). In 2007, the fishery yielded the lowest landings since 1998, and the stock is currently considered of concern (NCDMF 2008).

Among molluscs, the bay scallop (*Argopecten irradians*), hard clam, and oysters were much used by natives, and fisheries were developed by European colonists, first on a small scale but expanding over the centuries (Larson 1970, Catesby 1996). Declines in bay scallops were noticed in the 1980s due to reduced water quality, toxic red tides, loss of seagrass habitat and heavy fishing pressure (Peterson et al. 1996). Another recent factor negatively affecting the bay scallop population is increased predation by cownose rays (*Rhinoptera bonasus*), which have increased due to declines in their predators, large sharks (Myers et al. 2007). Thus, in the early 2000s, landings decreased essentially to nothing; the harvest was closed in 2007 due to limited availability of scallops, and the stock is now considered depleted (Myers et al. 2007, NCDMF 2008). Hard clam declines in the 1980s were also related to disturbance of seagrass beds, the impact of mechanical harvesting and high fishing pressure (Peterson et al. 1987). Currently, data are insufficient to evaluate the status of this stock (NCDMF 2008).

In the early 1900s, Kellogg (1910) noted, "The history of the oyster industry in Pamlico Sound is a record of the usual series of events. Natural beds were discovered, dredging became excessive, the beds were soon impoverished, many of them being completely destroyed". He continued, "The ruin of a large natural source of wealth was begun. All this occurred much more rapidly than in Chesapeake Bay". In the 1700s, oysters were found in every creek and gut of saltwater (Lawson 1712). In several places, there were such quantities of large oyster banks that they were very troublesome to vessels (Brickell 1737). However, as the fishery took its course, North Carolina passed "An Act to Prevent the Destruction of Oysters" in the 1820s (Campbell 1998). Mechanical harvesting began in the 1880s, and Pamlico Sound gained commercial importance as an oyster-producing region in 1889, when the scarcity of oysters in Chesapeake Bay let the industry move to North Carolina (Grave 1905). The fishery peaked in 1900, followed by strong declines caused by overfishing, habitat destruction and disease (Kellogg 1910, Lenihan & Peterson 1998). Although the stock may have improved slightly in recent years, it is considered of concern (NCDMF 2008).

Vegetation

Historical accounts from the late 1800s indicate that the bays and waterways near the mainland once had extensive beds of seagrass, while today seagrass is limited to the landward side of the barrier islands (Mallin et al. 2000). A progressive decline of seagrass beds started in the early 1900s (Stanley 1992a). In the 1930s, wasting disease caused extensive seagrass losses. In most regions, seagrass recovered, yet in the 1970s, strong die-offs occurred in Pamlico Sound that were even more severe than those in the 1930s (Stanley 1992a). Long-term estimates of seagrass decline come from the relative abundance of *Cocconeis* in sediment cores. In Pamlico Sound, *Cocconeis* abundance was 13% in 1540, varied between 7 and 9% during 1700–1900, then declined to less than 7% in the early 1900s, less than 2% in the 1980s, and 0.4% in 2000, an overall 97% decline in seagrass at the location where the sediment core has been taken (Figure 17; Cooper 2000).

Before European colonisation, North Carolina had about 11 million acres of wetlands, of which only 5.7 million still exist today. About one-third of the wetland conversion, mostly to managed forests and agriculture, has occurred since the 1950s (USGS 2009a). Today, the coastal plains contain

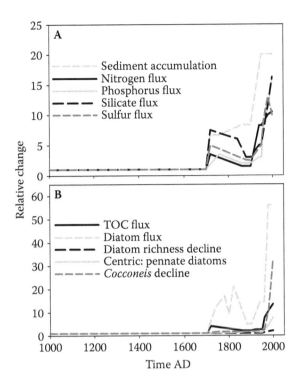

Figure 17 Historical changes in water quality in Pamlico Sound as measured in sediment cores: (A) relative increase in sediment accumulation and nutrient fluxes and (B) fluxes of total organic carbon (TOC) and diatoms, increase in the centric:pennate diatom ratio and declines in diatom species richness and the abundance of *Cocconeis*, an epiphytic diatom on seagrass. (Data from core M4, Cooper 2000, Cooper et al. 2004.)

about 95% of North Carolina's wetlands. Since 1850, the amount of cropland has increased 3.5-fold, and today 20% of the basin area consists of agricultural land, 60% is forested and 2% urbanised (Stanley 1992a). Over the last three decades, the production of swine has tripled and the area of fertilised cropland almost doubled (Cooper et al. 2004).

Water quality

Changes in land use patterns and increased point and non-point nutrient loading have induced multiple changes in water quality, with greatest changes occurring during the last 50–60 yr (Cooper et al. 2004). Sediment core data from the Pamlico Estuary (core M4, Cooper 2000, Cooper et al. 2004) indicated an overall increase in sediment accumulation (20-fold) and the fluxes of nitrogen (10-fold), phosphorus (16-fold), TOC (13.5-fold), silicate (6.25-fold), sulphur (10-fold) and diatoms (56-fold) from baseline levels in 1000–1700 (Figure 17). Moreover, the ratio of planktonic:benthic diatoms increased 7.75-fold, indicating a shift from benthic to planktonic primary production (Cooper et al. 2004).

Phosphorus loading was greatly enhanced by phosphate mining, which began in 1964 and accounts for half of the total phosphorus loading (Copeland & Hobbie 1972, Stanley 1992a). Nitrogen loading from point sources in summer months accounts for up to 60–70% and atmospheric nitrogen deposition for 15–32% of total nitrogen load (Steel 1991, Paerl et al. 2002). High surface sediment concentrations of arsenic, chromium, copper, nickel and lead are found in the Neuse Estuary, possibly associated with industrial and military operations, while high cadmium and silver levels most likely result from phosphate mining discharges (Cooper et al. 2004). In 1960, hypoxia was first reported in the Pamlico Estuary (Hobbie et al. 1975). Since then, records of hypoxic and anoxic waters were mostly of short duration but have resulted in many fish kills (Cooper et al. 2004). Nuisance and toxic algal blooms are reported periodically (Bricker et al. 1999).

Galveston Bay, Gulf of Mexico

Galveston Bay (Figure 18) has been the focus of many human activities throughout history. About 14,000 yr ago, Paleo-Indians hunted mammoth, mastodon and bison around the bay, leaving traces of their activities in shell middens from about 8000 BC (Galveston Bay National Estuary Program [GBNEP] 1994, Lester & Gonzalez 2002). Native Americans continued to live around and use the bay for food, and Spanish and French explorers began visiting in the sixteenth to eighteenth centuries. European colonists encountered a rich abundance of wildlife to serve subsistence needs. The seemingly endless flocks of ducks, geese and swans and a bounty of fish and shellfish established a viewpoint that the New World's wildlife resources were inexhaustible (LaRoe et al. 1995). The native population probably never exceeded 10,000 (Markowitz 1995), yet the European colony grew rapidly, especially after the establishment of the first towns in 1820. The first settlers were mainly farmers, cotton planters and merchants, but marine resources were also exploited (Figure 19, GBNEP 1994). Over time, the watershed has been transformed for urban development, agriculture, and petroleum and petrochemical production, while the estuary itself has been used for fisheries, transportation, oil and gas production and recreation (GBNEP 1994). Today, Galveston Bay is surrounded by the eighth largest metropolitan area in the United States and has experienced heavy industrial pollution. Half of the wastewater of Texas gets discharged into the estuary; more than 50% of all chemical products manufactured in the United States are produced here, and 17% of the oil produced in the Gulf of Mexico is refined here.

Marine mammals

Natives of Galveston Bay used to hunt harbour porpoise, of which they offered pieces to the Europeans (Cox 1906). Stevenson (1893) reported: "Porpoises are numerous on the Texas coast, and large schools of them are often seen in the bays as well as outside along the coasts." He also noted: "It is reported, however, that they have never yet been taken for commercial purposes." Today, the most common marine mammal in Galveston Bay is the bottlenose dolphin, yet it is listed

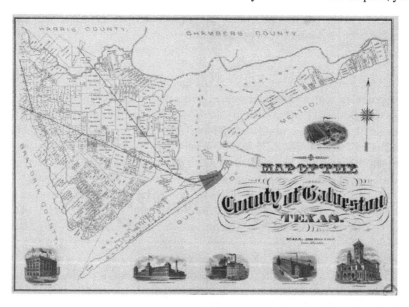

Figure 18   Map of the county of Galveston, Texas, 1902 (Library of Congress, Geography and Map Division).

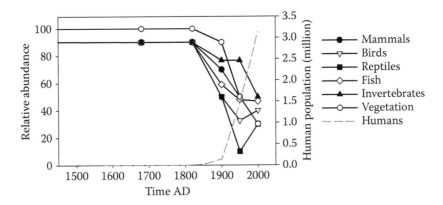

Figure 19 Timeline of changes in relative abundance of six taxonomic groups (left axis) and the human population (right axis) in Galveston Bay.

as depleted under MMPA (Swartz et al. 1999, Lester & Gonzalez 2002). Dolphins in Galveston, East, and Trinity Bays, considered the Gulf of Mexico estuarine and sound stock, number 107–152 (Waring et al. 2001). Many factors have led to the decline in bottlenose dolphin. During 1972–1989, 490 bottlenose dolphins from the estuarine and sound stock were removed to oceanaria for research and public display (Waring et al. 2001). In the 1990s, mortality events caused high numbers of dolphin deaths, suggesting these stocks are stressed. Some dolphins had pesticide concentrations at levels of possible toxicological concern. Finally, about 3% of stranded bottlenose dolphins showed evidence of human interactions as the cause of death, including gear entanglement, mutilation and gunshot wounds (Waring et al. 2001).

Of the large whales, several species occur in the Gulf of Mexico, and some may occasionally enter Galveston Bay. Right whales are rare along the Texas coast, but minke and Bryde's (*Balaenoptera edeni*) whales are found (Waring et al. 2001). The latter may represent a resident stock, but consisting only of about 35 animals. Sperm whales occur during all seasons in the northern Gulf of Mexico, but generally in deeper waters, with an estimated abundance of about 530 animals. A commercial fishery for sperm whales operated in the Gulf of Mexico during the late 1700s to the early 1900s, but the exact number of whales taken is not known (Waring et al. 2001). The only pinniped that might have occurred in the Galveston Bay area is the Caribbean monk seal (*Monachus tropicalis*), which did live in the Gulf of Mexico, Florida and the West Indies but became extinct in 1950 (Vermeij 1993, McClenachan & Cooper 2008). Some sightings, although unconfirmed, have been found along the coast of the western Gulf of Mexico from before the 1800s and around 1900 (McClenachan & Cooper 2008).

Birds

Pelican Island was a noted nesting site for brown pelicans in 1820 (Lester & Gonzalez 2002). In 1918, about 5000 pelicans nested on the Texas coast, but relentless killing by fishermen and pesticide pollution in the 1960s strongly reduced their numbers and nesting success. In 1967–1974, fewer than 10 pairs bred each year along the Texas coast, and the species was listed as endangered. The elimination of certain pesticides together with legal protection has enabled brown pelicans to re-establish their nesting colonies in Galveston Bay since 1993. The population grew to more than 800 breeding pairs in 2000 (Figure 20; Lester & Gonzalez 2002).

Early European colonists found huge flocks of ducks, geese and swans in Galveston Bay. Waterfowl hunting has a rich tradition in Texas and along the Gulf of Mexico coast. In 1998–1999, there were 58,177 hunting licenses sold in the five counties surrounding Galveston Bay. Duck breeding population estimates compiled by the USFWS for the years 1955–2001 indicate declines in

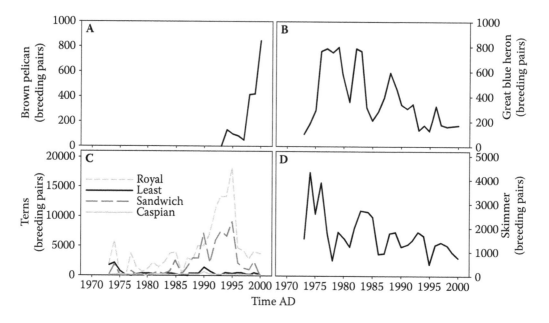

Figure 20 Changes in the number of breeding pairs of (A) brown pelican, (B) great blue heron, (C) terns, and (D) skimmer in Galveston Bay. (Data from Lester & Gonzalez 2002.)

breeding populations of mallard (*Anas platyrhynchos*), American wigeon (*A. americana*), northern pintail (*A. acuta*) and scaup (*Aythya affinis*) over the 46-yr period. However, increasing trends are seen for gadwall (*Anas strepera*), blue-winged teal (*A. discors*), green-winged teal (*A. carolinensis*), northern shoveler (*A. clypeata*), redhead (*Aythya americana*) and canvasback (*A. valisineria*). Some geese populations have also increased since 1955 (Lester & Gonzalez 2002).

Most wading birds, seabirds, shorebirds and raptors have historically been reduced by hunting, habitat loss, pollution and disturbance. Protective laws in the early twentieth century enabled many populations to increase again, yet others continue to be impacted, especially by habitat loss and disturbance of nesting sites. During 1973–1998, six colonial waterbird species exhibited increasing trends: the white ibis (*Eudocimus albus*), brown pelican, neotropic cormorant (*Phalacrocorax brasilianus*), and the sandwich tern (*Thalasseus sandvicensis*), gull-billed tern (*Gelochelidon nilotica*) and royal tern (*Thalasseus maximus*) (Figure 20). Ten species experienced stable trends, and four species declined: the great blue heron (*Ardea herodias*), roseate spoonbill (*Platalea ajaja*), black skimmer (*Rynchops niger*) and least tern (*Sternula antillarum*) (Figure 20; McFarlane 2001). In 2000, the three most commonly sighted colonial waterbirds that utilised Galveston Bay to feed and nest were the laughing gull (*Leucophaeus atricilla*), cattle egret (*Bubulcus ibis*) and royal tern (Lester & Gonzalez 2002).

Reptiles

The French explorer René Robert Cavelier, Sieur de La Salle, explored the Gulf of Mexico in the seventeenth century, noting: "There are also many alligators in the rivers, some of them of a frightful magnitude and bulk. ... I have shot many of them dead" (Cox 1906). Alligators living in the fresh and brackish waters and wetlands around the bay have been hunted throughout history, but especially in the early 1900s when their belly skin was highly prized for its leather. By the mid-twentieth century, alligators were very rare and became listed as an endangered species. Since then, regulated hunting and trade of alligator products have enabled the species to increase in number. Currently, the greatest threat to alligator populations around Galveston Bay is that posed by encroaching development (Lester & Gonzalez 2002).

René Robert Cavelier also commented on turtles: "We had plenty of land and sea tortoises, whose eggs served to season our sauces" (Cox 1906). Turtles were among other seafood species that were being sold at the Galveston dock during colonial times (Lester & Gonzalez 2002). In the late nineteenth century, Stevenson (1893) reported: "Large green turtle *(Chelonia mydas)* occur more or less abundantly all along the Texas coast", but he also described the increasing pressure on the population: "Green turtle are gradually becoming less abundant on the coast of Texas, yet on account of the increasing demand for them, the annual catch is probably increasing." A brackish water turtle, the diamondback terrapin *(Malaclemys terrapin)*, was also highly valued as a delicacy and exploited heavily in the 1800s (LaRoe et al. 1995). After strong declines over the last 150 yr, terrapin populations have recovered somewhat in recent decades but are now threatened by habitat loss and degradation (Lester & Gonzalez 2002). Today, the most common turtle in nearshore waters is Kemp's ridley. In 1940–1990, the Galveston Bay region was among the three most important areas for this species (Manzella & Williams 1992). A population decline in recent years has been linked to declines in blue crabs, their primary prey (Lester & Gonzalez 2002). All sea turtles are listed as endangered or threatened in Texas today (USFWS 2009).

Fishes

Early European colonisers made use of a wide variety of seafood, including redfish, flounder, mullet, skate and many other fishes (Lester & Gonzalez 2002). The fishing industry for the whole state of Texas was small, with only 291 fishermen listed as full time in 1880. Most fishing was done inside Galveston Bay, and the most popular market fish were redfish, sea trout *(Cynoscion* spp.), mullets, croakers and sheepshead *(Archosargus probatocephalus)* (Table 6; Stevenson 1893, GBNEP 1994). A marked decline in fish populations was noted before 1900, and in 1907, the total fish catch from Galveston Bay was 185,119 pounds. This was not enough to supply local markets, and breeding season closures were instituted in most of the bay to protect nursery grounds. Fish hatcheries were also

Table 6 Commercial finfish and shellfish landings (1000 pounds) from Galveston Bay in the years 1890 and 1989)

Species common name	1890	1989
Fish		
Red drum	404.2	0.0
Black drum	4.0	21.8
Flounder	46.0	14.6
Mullet	39.3	108.0
Sheepshead	17.0	16.2
Striped bass	5.0	0.0
Trout	427.4	0.0
Other fish	542.9	60.5
Total fish	1485.8	221.1
Shellfish		
Oyster	1647.1	705.5
Crabs	162.5	2149.5
Shrimp	138.0	4056.1
Terrapins (turtles)	2.4	0.0
Other shellfish	0.0	13.4
Total shellfish	1950.0	6924.5
Total fish and shellfish	3435.8	7145.6

Source: Data from Green et al. 1992.

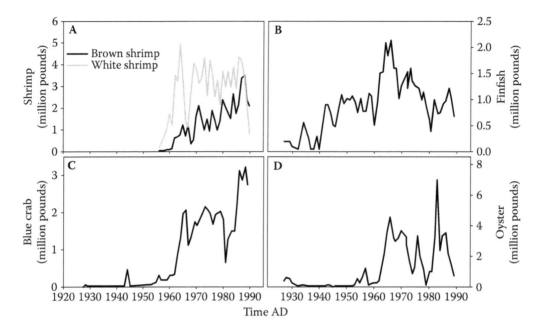

Figure 21 Commercial landings for (A) shrimp, (B) finfish, (C) blue crab, and (D) oyster in Galveston Bay. (Data from Green et al. 1992.)

proposed to overcome the shortage. However, controversy between fishermen and the commissioner over the closures prevented restrictive laws from being passed until 1929, when all littoral waters were closed permanently to drag seines, which were replaced by gill nets (Tucker 1929).

Significant increases in commercial landings occurred in the 1950s (Figure 21) when the pet food industry began harvesting fish in the northern Gulf of Mexico (Massley & Slater 1999). Commercial landings of drums (red drum and black drum, *Pogonias cromis*) and croaker peaked at more than 32,000 t in 1956, more than 20,000 t above that of 1953 (Massley & Slater 1999). The strong increase in fishing pressure resulted in an overall decline in finfish landings from Galveston Bay in the 1960s and 1970s (Figure 21). Yet, commercial landings of red drum from offshore stocks increased rapidly in the 1980s, when public demand suddenly grew for a new seafood preparation called blackened redfish (Massley & Slater 1999). In the 1970s, however, red drum abundance strongly declined and stayed low throughout the 1980s. Thus, in 1981 the Texas State Legislature banned the sale of red drum and spotted seatrout (*Cynoscion nebulosus*) to stem the decline of these two species (Stanley 1992b).

Another species that has dramatically declined over the last 150 yr is the striped bass. In addition to fishing, this species has been severely affected by habitat alteration in the form of dams, bay closures and shell dredging (Lester & Gonzalez 2002). From the late nineteenth to the late twentieth century, commercial landings of finfish have declined 7-fold, while landings of shellfish, especially shrimp and crab, increased almost 4-fold (Table 6). Commercial finfish catches account for only 14% of overall catches from Galveston Bay; most of these catches are recreational (Lester & Gonzalez 2002).

Invertebrates

Around 1840, a young Irishman described the shrimp fishery in Galveston Bay. Fishermen would use seines and catch the shrimp as they migrated out of the bay at the inlets at each end of Galveston Island. The quantity of one seine haul was 70 pails of shrimp plus many pails of fish, with shrimp

6 to 7 in. in size (Lester & Gonzalez 2002). The shrimp harvest was done by hand until after 1900. In addition to large migrating shrimp, small shrimp were harvested with cast nets and haul seines in the marshes. These were sun-dried or pickled in salt for export (Iversen et al. 1993). After 1920, shrimp fishing became a vital commercial fishing industry in Galveston Bay and the Gulf of Mexico. Catches of brown (*Penaeus aztecus*) and white (*P. setiferus*) shrimp increased significantly from the late 1950s to 1990, with recent years showing a slight decrease from these maximum values (Figure 21; Nance & Harper 1999). The shrimp grow up in estuaries like Galveston Bay and migrate out into the gulf as they mature.

Blue crabs have been a major source of seafood throughout history in Galveston Bay. Only in the mid-twentieth century, however, did commercial landings strongly increase (Figure 21). After a strong decline in the early 1980s, landings have sharply risen again to an all-time high of more than 3 million pounds in 1986 (Stanley 1992b). In 1998, more than 2.6 million pounds of blue crab were harvested in Galveston Bay. However, scientific surveys indicated that blue crab populations have declined in abundance over the past three decades. Recent declines in adult size classes also indicate overharvesting from high fishing pressure (Lester & Gonzalez 2002).

Oysters and clams have been important for people living around Galveston Bay for thousands of years. Shell middens containing *Rangia cuneata* clams and oysters occur along the Texas coast from Corpus Christi Bay to the Louisiana border (Aten 1983). The oyster industry in Galveston Bay was limited to local trade until after the 1870s, when cold shipping and processing industries were developed. In 1885, the oyster industry employed 50 boats and 500 men working in the bay. Galveston shipped about 25,000 oysters daily to markets within the state, but the city itself was said to consume 25,000–30,000 oysters daily during the season (Lester & Gonzalez 2002). Stevenson (1893) reported that, "Galveston Bay has a greater area of natural oyster beds than any other bay in Texas, but the reefs are not so plentifully supplied with oysters as some others in the State. This is to some extent due to overfishing". In 1895, the state tried to control the heavy fishing pressure on natural oyster reefs and established the Oyster and Fish Commission.

In the twentieth century, commercial shell dredging started and operated until the 1960s. This destructive practice greatly diminished oyster reefs. Ward (1993) estimated that 135,000 acre-feet (1 acre-foot = 1233.5 m^3) of shell were removed by shell dredgers, not oyster fishermen, between 1910 and 1969. Prior to the 1950s, tongers harvested 20–60% of the oysters, but tongers were replaced almost entirely by dredgers by the 1960s (Hofstetter 1982, 1983). In 1970, Carter (1970) wrote: "Fifty years ago perhaps nearly a fifth of the bay bottom was covered by exposed oyster shell, much of it lying in extensive semifossilized shell reefs". Old maps of the Texas coast indicate that there were much more extensive oyster reefs along the shoreline than seen today in Galveston Bay. One study reported a significant increase in the extent of oyster reefs over the last 20 yr, but this has not yet replaced the large amounts of shell that were removed (Lester & Gonzalez 2002). Despite the decline in reef area, harvesting of oysters for food strongly increased after the 1960s (Figure 21). As oyster production in many other estuaries such as Chesapeake plummeted, Galveston Bay became an important supplier for cities throughout the United States. Yet, landings strongly declined in the late 1970s and in the 1980s. Between 1994 and 1998, the annual oyster harvest from Galveston Bay averaged close to 4 million pounds (Lester & Gonzalez 2002).

Vegetation

Texas has lost about 50% of its original wetlands and 35% of its coastal marshes as a result of agricultural conversion, overgrazing, urbanisation, construction of navigation canals, water table declines and other causes (Texas Environmental Profiles [TEP] 2009, USGS 2009a). In Galveston Bay, coastal marshes declined from 165,500 to 130,400 acres between 1950 and 1989. In the 1990s, Pulich & Hinson (1996) estimated 120,132 acres of coastal marshes based on aerial photographs. In recent years, about 4500 acres of marsh habitat have been restored, protected or created (Lester & Gonzalez 2002).

Most SAV once present in Galveston Bay has been lost since the late 1950s (Pulich & White 1991). By 1989, the distribution of shoal grass (*Halodule wrightii*) and widgeon grass (*Ruppia maritima*) decreased from 2500 to 800 acres (White et al. 1993). In 1995, there were 280 acres in Christmas Bay plus some amount of widgeon grass in Trinity Bay. The overall decline in SAV was about 80% of the 1950s extent (White et al. 1993, Lester & Gonzalez 2002). The most significant losses occurred along the margins of western Galveston Bay, caused by subsidence and Hurricane Carla in 1970. In West Bay, nearly 2200 acres of seagrasses have been completely lost due to industrial, residential and commercial development; wastewater discharges; chemical spills and increased turbidity from boat traffic and dredging (Pulich & White 1991).

Water quality

Historically, Galveston Bay had good water quality because of its shallow, well-mixed, and well-aerated conditions (Lester & Gonzalez 2002). Galveston Bay undergoes a total water exchange more than four times a year due to freshwater inflow and tidal action. However, land-based industrial and municipal activities, especially in the western, most urbanised part of Galveston Bay, strongly reduced water quality in the twentieth century. Prior to the mid-1970s, portions of the Houston Ship Channel were among the 10 most polluted water bodies in the United States (Lester & Gonzalez 2002). Nitrogen loading increased almost 4-fold from the early 1900s to the 1970s (Figure 22), causing eutrophication and oxygen depletion in the tributaries and bays. Oxygen levels were really low in the 1960s, with regular hypoxia (<2 mg O_2 L^{-1}) and frequent anoxia (0 mg O_2 L^{-1}) in bottom waters (Kennish 2000). The biological oxygen demand (BOD) loading increased 20-fold from 1920 to the late 1960s (Figure 22). In 1971, stringent discharge goals were established for industrial and municipal point sources, and wastewater treatment facilities were upgraded and expanded (Lester & Gonzalez 2002). Nitrogen and BOD loading dropped by half, and oxygen concentrations started to improve in the 1970s (Figure 22, Kennish 2000). Over the last three decades, concentrations of ammonia and phosphorus have decreased baywide, and chlorophyll-a has declined to less than 25% of its 1975 value (Lester & Gonzalez 2002). Chemical and heavy metal concentrations have declined as well. Some of the most polluted parts of Galveston Bay, such as the Houston Ship Channel, have experienced considerable improvements in water quality and the return of some species of fish (GBNEP 1994).

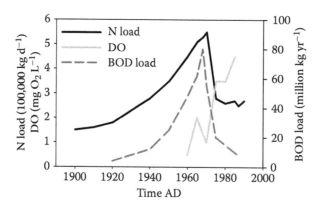

Figure 22 Trends in water quality in Galveston Bay: nitrogen (N) loads, dissolved oxygen concentrations (DOs) and biological oxygen demand (BOD) loading. (Data from Kennish 2000.)

San Francisco Bay, Pacific Coast

San Francisco Bay (Figure 23) has been inhabited by Native Californians for at least 10,000 yr (Ainsworth 2002). With declining sea levels about 6000 yr ago, the San Francisco, San Pablo and Suisun Bays emerged, and thousands of acres of intertidal mudflats and salt and brackish marshes developed. Living conditions were good; the large number and size of shell middens suggest that the native population was fairly large. They utilised a wide variety of terrestrial and marine resources, including shellfish, fish, birds and marine mammals, and constructed a cultivated landscape through controlled burning, pruning, weeding, seeding and tillage over many centuries. From about 2600 to 700 yr ago, the native population grew significantly. At the same time, indications of resource depression appeared in the archaeological record, suggesting that some valuable resources were overexploited (Ainsworth 2002, Broughton 2002). Spanish explorers arriving in 1769–1775 described a very dense native population with many villages around the bay (Ainsworth 2002). They also described many whales, dolphins, sea otters, grizzly bears and thousands of pelicans in the bay (Galvin 1971). Over the following decades and centuries, San Francisco Bay was rapidly and substantially transformed (see Figure 3). Europeans brought the fur trade, whaling and the gold rush, among other enterprises. The general timeline of human population growth and changes in marine fauna and flora is depicted in Figure 24.

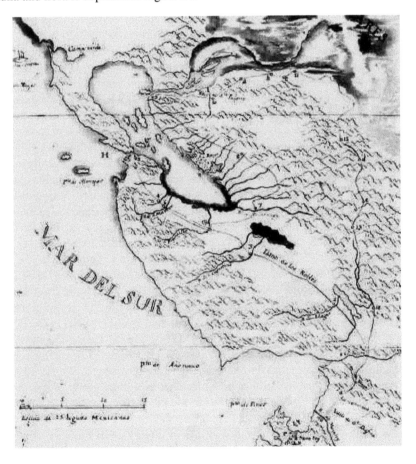

Figure 23 Font's map of the entrance to San Francisco Bay, 1776 (The John Carter Brown Library at Brown University).

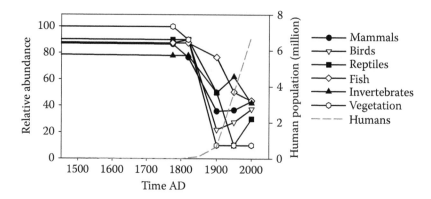

Figure 24 Timeline of changes in relative abundance of six taxonomic groups (left axis) and the human population (right axis) in San Francisco Bay.

Marine mammals

Native Californians hunted a large variety of marine mammals whose remains have been found abundantly in shell middens around San Francisco Bay (Ainsworth 2002, Broughton 2002). From 2600 to 700 yr ago, remains of large-bodied pinnipeds such as Steller's sea lion (*Eumetopias jubatus*), California sea lion (*Zalophus californianus*) and northern fur seal (*Callorhinus ursinus*) decreased relative to smaller harbour seal (*Phoca vitulina*) and sea otter (*Enhydra lutris*), likely caused by intense hunting and overexploitation (Broughton 1997). Relative abundance of sea otter first increased but later decreased, indicating intensified use followed by resource depletion. This was accompanied by an increase in the ratio of adult to juvenile otter remains, suggesting that breeding colonies or microhabitats favoured by females with young were abandoned as hunting pressure increased (Broughton 2002). There is little evidence that native people systematically hunted whales prior to European influence, but stranded whales were scavenged. Frequent remains of killer whales (*Orcinus orca*) in shell middens suggest that they were common in the bay (Margolin 1978).

With the arrival of Europeans, resource use changed. The Spanish trade in sea otter pelts began in 1786 and was the most important industry in coastal California for several decades (ANR 2001). San Francisco Bay was filled with sea otters (Galvin 1971, Margolin 1978), and they were killed by the thousands (Skinner 1962). The Russians hunted them from canoes under the guns of the Spanish fort (Eldredge 1912). In 1812, per week 700–800 sea otters were taken (Huff 1957), and in the 1820s, about 2000 otters were removed every year in the bay and up the coast (von Chamisso 1822, Ogden 1941). As William Heath Davis described in the 1820s: "Otters were then numerous in the bay and their skins plentiful. Murphy hunted them and sold their pelts to the Boston traders for from $40 to $60 each" (San Francisco News Letter [SFNL] 1925). Until 1830, the Russians took 5000–10,000 sea otters per year along the California coast (Eldredge 1912), and the sea otters grew scarce and eventually became extinct in San Francisco Bay (Olofson 2000). By 1900, sea otters were widely regarded as extinct in California, but fortunately a small group of less than 50 to 100 survivors were discovered in 1914 near Point Sur in Monterey County (ANR 2001). It took many decades for the population to increase again in numbers and spread north and south from Point Sur. Until the 1990s, there were no documented sightings of otters in San Francisco Bay, but several sightings were confirmed since then (Olofson 2000). In the early 1980s, population growth ceased, and concerns were raised regarding incidental entanglement and drowning in fishing gear, with annual mortality rates of 80–100 animals (ANR 2001).

Fur seals were also highly sought by fur traders. They were abundant along the coast and on the islands but were rapidly and drastically reduced. In 1810–1811 alone, the Russian ship ALBATROSS took 73,402 fur seals, and fur seals became scarce by 1840 (Eldredge 1912). It took until 1911, however, for fur seals and sea otters to be protected by the International Fur Seal Treaty (ANR 2001). The Guadalupe fur seal (*Arctocephalus townsendi*) was believed to be extinct until 1926 but is making a very gradual recovery. Today, the eastern North Pacific fur seal stock is listed as depleted under the MMPA, but numbers are increasing (ANR 2001).

In the 1860s–1870s, many pinnipeds, including harbour seal, northern elephant seal (*Mirounga angustirostris*) and sea lions, were killed for their oil or body parts, and many females were captured for display. Pinnipeds were hunted commercially until 1938, when California law implemented complete protection. Nevertheless, sport and commercial fishermen were allowed to kill sea lions and harbour seals that were interfering with fishing operations (ANR 2001). Since the passage of the MMPA in 1972, the number of California sea lions has increased and seasonally occur in Central Bay (Schoenherr 1992, Olofson 2000). However, harbour seals have not increased in San Francisco Bay since the early 1970s, despite steadily growing numbers along the California coast (International Marine Mammal Project [IMMP] 2009). It has been suggested that shoreline development has reduced many haul-out areas and breeding beaches used in the past and thus the number of animals that could survive in the region. In 1991, harbour seal numbers severely dropped at the primary pupping site in southern San Francisco Bay (IMMP 2009).

In 1822, the first whalers anchored in San Francisco Bay (Huff 1957). The early shore-based whaling industry in California primarily caught Pacific grey and humpback whales within 10 mi of the coastline. Occasionally, right, blue and fin whales were also caught, which were highly prized due to the greater oil content of their blubber (ANR 2001). In 1860, about 1000 barrels of whale oil were processed per day in Sausalito (Huff 1957). The completion of the transcontinental railroad in 1869 made it possible to ship whale products over land and facilitate trade. Yet, whale numbers started to decline. In 1895, only 40 barrels were processed per day in Sausalito, and whaling for profit came to an end (Huff 1957). Nevertheless, modern whaling vessels continued to catch some grey and many humpback whales in California waters (ANR 2001).

Around 1900, there were about 15,000 humpback whales in the North Pacific Ocean, which were reduced to dangerously low levels. In 1966, the IWC established a harvest moratorium, and the humpback was listed as endangered. Since then, the population has recovered to about 8000 individuals, with the California feeding population counting about 1000 animals and growing at about 8% per yr (ANR 2001). With protection, the Pacific grey whale also increased from about 1000 animals in the early 1900s to 18,000–29,000 animals in recent census counts (Rugh et al. 2005, Christensen 2006). Estimates of prewhaling population size for grey whales based on catch data range between 19,480 and 35,430 and suggest that the current population may have recovered. However, alternative preexploitation estimates are higher, including 70,000 from population models (Wade 2002) and 96,000 based on genetic diversity (Alter et al. 2007).

Birds

Archaeological evidence suggests that native people around San Francisco Bay abundantly used and depleted some highly valued large geese and cormorants (Broughton 2002). Double-crested (*Phalacrocorax auritus*) and Brandt's (*P. penicillatus*) cormorants were among the most abundant bird taxa in shell middens. While remains of geese consistently declined from 2600 to 700 yr ago, remains of cormorants increased over the first 800 yr and strongly declined thereafter. An increasing ratio of adult to young cormorants indicates the loss or abandonment of local breeding colonies (Broughton 2002). There also used to occur a flightless duck, *Chenodytes lawi*, in California and Oregon, which became extinct in prehistoric times (Vermeij 1993).

The first European observers reported that migrating birds blackened the sky over the estuary (Skinner 1962): "Thousands of pelicans took to the air from a rocky island as we passed by entering the bay" (Galvin 1971). Canvasbacks were historically very abundant, and in 1776 Jose Canizares mapped areas in the northern estuary as "forests of the red duck" (Olofson 2000). Yet, marine birds suffered relentless exploitation from European settlers, especially at the Farallon and other islands during and after the gold rush from 1850 to 1900 (ANR 2001). Hunters delivered millions of waterbirds, shorebirds and their eggs to the tables of a growing population and egret plumes to hat makers (Skinner 1962). Prior to 1880, the snowy egret was considered locally common (Grinnell & Miller 1944), but hunting for its feathers was devastating and nearly wiped this species out. By the early 1900s, the snowy egret was thought to be extinct in California, yet some rare stragglers were noted at two locations in the bay in the 1920s (Grinnell & Wythe 1927, Olofson 2000).

Common murres (*Uria aalge*) were heavily exploited for their eggs. There were no regulations, and the murre population declined by an order of magnitude by the 1900s; only a few thousand individuals were left in the 1930s. The murre population did not recover for several decades and even now is far below numbers of the 1800s. Today, the breeding population numbers 363,200 with stable or increasing trends (ANR 2001). Shorebird populations such as plovers (Charadriinae), dowitchers (*Limnodromus* spp.) and godwits (*Limosa* spp.) were markedly reduced in the late twentieth century due to market hunting and destruction of breeding habitat (Olofson 2000).

With the Migratory Bird Treaty Act in 1918, many bird populations started to recover (Olofson 2000), but habitat loss and DDT pollution continued to threaten many species (Skinner 1962). The San Francisco Bay region is 1 of 34 waterfowl habitats of major concern in the North American Waterfowl Management Plan. Also, half the migratory birds along the Pacific flyway, about 500,000 birds, use the bay's wetlands for wintering each year (The Bay Institute [TBI] 1998, Olofson 2000). Since the 1950s, however, waterfowl surveys indicated a 25% decrease in abundance, with some species suffering much larger declines. For example, wintering northern pintails declined from 200,000 to fewer than 20,000 since the 1950s, and canvasbacks declined by 50% since the 1970s. Other species experienced more favourable population trends, including snow geese, egrets, cormorants, pelicans and bald eagles (Olofson 2000).

Fishes

Requiem sharks (Carcharhinidae), bat ray (*Myliobatis californica*), sturgeon and salmon were very abundant among fish remains in shell middens around San Francisco Bay (Broughton 1997). From 2600 to 700 BP, relative abundance and size of white sturgeon (*Acipenser transmontanus*) declined significantly, indicating that native people had a substantial impact on this species (Broughton 1997, 2002). White sturgeon represented the highest-value fish available, reaching up to 6.1 m long, weighing 816 kg, and with a high-fat content (Broughton 1997). Between 1860 and 1901, Europeans developed a commercial fishery for white sturgeon, stimulated by growing demand for smoked sturgeon and caviar on the eastern coast of America. The fishery concentrated in San Francisco Bay. Green sturgeon (*A. medirostris*) was also taken but was of minor importance (ANR 2001). The commercial catch peaked at 1.65 million pounds in 1887 but declined to 0.3 and 0.2 million pounds in 1895 and 1901, respectively, when the commercial fishery was closed. However, sport fishing for sturgeon increased dramatically in 1964, and 2258 sturgeon were landed in 1967 (ANR 2001). In 1990, angling regulations reduced harvest to less than 50% of 1980s levels. Adult abundance of white sturgeon fluctuated in 1967–1998, with a high of 142,000 individuals in 1997 (ANR 2001).

Chinook (*Oncorhynchus tshawytscha*) and Coho (*O. kisutch*) salmon were also highly valued and abundant. A few hundred years ago, one visitor reported salmon runs so dense that the rivers looked like silver "pavements" (Skinner 1962). Salmon fisheries existed long before European arrival, with estimated harvests exceeding 8.5 million pounds annually (ANR 2001). A small commercial river fishery began in the early 1800s (Smith & Kato 1979) and a large-scale fishery with the gold rush (ANR 2001, Francis et al. 2001). After 1849, the gold rush caused high siltation, destroying important river and bay habitat (Skinner 1962). In 1860, about 3220 t of salmon were caught, but populations were in decline (Smith & Kato 1979). In 1880, there were 20 canneries operating in the Sacramento-San Joaquin river system. Fishing was intense, with peak landings of 12 million pounds in 1882 (ANR 2001). Shortly after, the fishery collapsed, and dramatic population declines were linked to pollution and habitat degradation combined with high fishing pressure. The last cannery closed in 1919, and one by one all rivers were closed to commercial fishing (ANR 2001). Both species are on the endangered species list today (USFWS 2009).

California halibut (*Paralichthys californicus*) was the first groundfish targeted by commercial fisheries in the late nineteenth century (Francis et al. 2001). The highest recorded catch was 4.7 million pounds in 1919, followed by a decline to 950,000 pounds in 1932. Average catch has remained at about 910,000 pounds since then (ANR 2001). Other groundfish were lightly exploited until the 1960s (Rogers & Builder 1999), but catches rapidly increased thereafter (Francis et al. 2001). In the 1980s, rockfish (*Sebastes* spp.) stocks declined, and overall groundfish landings decreased by 60% in the 1990s. The current status of many rockfish and lingcod (*Ophiodon elongatus*) off the western coast is poor, and the fishery was closed in 2000 (ANR 2001).

In 1936–1944, a shark fishery for vitamin oil boomed, with more than 24 million pounds landed, mainly soupfin sharks (*Galeorhinus galeus*). The fishery collapsed in the mid-1940s from overexploitation and development of synthetic vitamins. Yet, because of the strong decimation of soupfin sharks, particularly in nursery areas in San Francisco and Tomales Bays, the population never fully recovered (Olofson 2000). In the mid-1970s, interest in shark fishing renewed, this time for their meat for human consumption (Olofson 2000). The commercial fishery peaked off California in the mid-1980s, with local resource depletion and large catches of immature sharks. Since then, regulations have reduced total fishing effort and catches, and at least one species, the common thresher (*Alopias vulpinus*), may be recovering (Anderson et al. 1999b). In contrast to large sharks, skate and ray landings increased about 10-fold in the 1990s, but CPUE decreased, and concerns of overfishing have been raised (Olofson 2000).

A pelagic fishery for Pacific herring (*Clupea pallasi*) began in the early 1800s and fluctuated with markets over time (Skinner 1962). The fishery off California peaked at 3600 t in 1916–1919, 4500 t in 1947–53, and more than 10,000 t in 1982 (Jacobson et al. 1999). Since 1965, there was also a lucrative herring roe-on-kelp fishery in San Francisco and Tomales Bays. San Francisco Bay has the largest spawning population of herring and supplies over 90% of the U.S. herring catch (Jacobson et al. 1999). Other coastal pelagics were much more abundant 100 yr ago. The biomass of pelagic predators such as hake and mackerel declined by 75% from 1900 to 1950 and another 50% from 1950 to 2000 (Francis et al. 2001). Sardine (*Sardinops sagax*) abundance fluctuated with climate conditions throughout the last 2000 yr (Francis et al. 2001). In the nineteenth century, abundance was low, but a legendary California sardine fishery began in the early 1900s and supported the largest fishery in the Western Hemisphere in the 1930s–1940s (Jacobson et al. 1999). However, the fishery began to decline from north to south in the 1920s, strongly declined after WWII, and finally collapsed in the late 1950s. Sardine biomass remained negligibly low for about 40 yr but has increased since 1986 (Jacobson et al. 1999).

Invertebrates

Shellfish were a staple food for native people around San Francisco Bay and are abundant in shell middens (Ainsworth 2002). Archaeological evidence suggests that large molluscs such as California oyster (*Ostrea conchaphila*) and bay mussel (*Mytilus edulis*) decreased significantly over time relative to smaller species (Broughton 2002). In the 1840s, a small fishery for native oyster served the San Francisco market (Nichols et al. 1986). In 1867, native oysters died after an earthquake and heated bottom waters (Ingersoll 1881). The American oyster (*Crassostrea virginica*) was introduced in 1869 and displaced the remaining native population (Shaw 1997). Maps from the early 1900s depict extensive deposits of native oyster shells across the bottom of the bay, indicating the former abundance of this species (Packard 1918). These shells were dredged for various purposes in the early twentieth century (Galtsoff 1930). Today, only remnant populations of native oysters exist in the bay.

California Indians fished abalones extensively in coastal areas and on the Channel Islands (ANR 2001). During the 1850s, Chinese Americans started a fishery targeting intertidal green (*Haliotis fulgens*) and black (*H. cracherodii*) abalones, with peak landings of 4.1 million pounds of meat and shell in 1879 (ANR 2001). Landings crashed in the 1890s (Anderson et al. 1999a), and shallow waters were closed to commercial harvest in 1900 (ANR 2001). An increase in landings of red abalone (*H. rufescens*) began in 1916, with a peak of 3.9 million pounds in 1935, followed by a decline to 164,000 pounds in 1942 (ANR 2001). Catches averaged 2.1 million pounds in 1931–1967 but have strongly declined since the mid-1960s. The harvest of black abalone was closed in 1993, and white abalone (*H. sorenseni*) has been proposed as an endangered species (Anderson et al. 1999a). By 1990, landings of red abalones declined to 17% of the 1931–1967 average, and all abalone species are considered depleted (ANR 2001).

Dungeness crabs (*Cancer magister*) were harvested by natives and taken commercially from San Francisco Bay in 1848 (ANR 2001). Before 1944, the fishery centred in the San Francisco area with average annual landings of 2.6 million pounds. While the bay fishery declined in the early 1900s (Wright & Phillips 1988), commercial catches increased to 10 million pounds (Galvin 1971). The fishery was relatively stable until 1956 but declined to 710,000 pounds in the early 1960s and remained seriously depressed up to 1985. Recently, catches increased again to about 2 million pounds; however, fishing intensity is extreme, and in most years 80–90% of all available legal-sized male crabs are captured (ANR 2001).

A commercial bay shrimp (*Crangon franciscorum*) fishery began in San Francisco Bay in the early 1860s. By 1871, Chinese immigrants established fishing camps along the shores and exported large quantities of dried shrimp meal to China. At the height of the fishery in the 1890s, more than 5 million pounds were landed each year (ANR 2001). Concerns were raised regarding damage of bottom habitat and by-catch of other species, and seasonal closure and prohibition of Chinese shrimp nets were implemented in 1911 but modified in 1915 (ANR 2001). Shortly after, commercial shrimp harvesters introduced beam trawl nets, and landings increased, peaking at 3.4 million pounds in 1935. Afterwards, landings steadily declined to 1500 pounds in the early 1960s, and no shrimp were landed in 1964 (ANR 2001). Since 1965, the bay shrimp fishery mostly supplies live bait for sport fishing, with landings of 75,000–150,000 pounds (ANR 2001).

Archaeological evidence shows that sea urchins have been fished by coastal American Indians for centuries, whereas a commercial fishery only developed in the last 30 yr for red sea urchin (*Strongylocentrotus franciscanus*) (ANR 2001). The fishery began in southern California in 1971 and expanded to northern California in the late 1970s–1980s (Anderson et al. 1999a). The northern California fishery rapidly grew to 30 million pounds in 1988 but declined to less than 5 million pounds in the late 1990s. Still, the sea urchin fishery has been one of California's most valuable fisheries for more than a decade (ANR 2001).

Vegetation

Data on the historic areal extent of eelgrass within San Francisco Bay are limited, although it is believed that extensive eelgrass meadows occurred in the past. The gold rush and hydraulic mining beginning in the 1850s caused large amounts of siltation and sediment loading that covered the central portion of the bay, creating mudflats that were once seagrass beds (Nichols et al. 1986). In the 1950s, there were about 480 acres of seagrass beds, and some recovery was noticed, but increasing turbidity caused new seagrass declines in the 1980s (Olofson 2000). By 1989, eelgrass beds were limited to relatively small patches of 316 acres. Overall, at least one-third of historical eelgrass beds have been lost to fill and development (ANR 2001).

Before 1850, San Francisco Bay sustained about 1400 km^2 of freshwater wetlands and 800 km^2 of salt marshes (Figure 25). From 1850 to 1970, intense sedimentation due to the gold rush, wetland conversion and filling reduced the amount of undiked marshes to only 125 km^2, a 95% loss of crucial habitat (Atwater et al. 1979, Nichols et al. 1986).

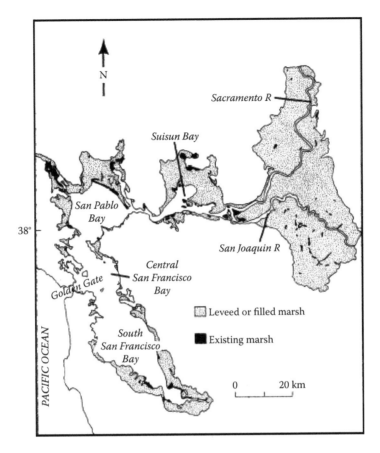

Figure 25 Distribution of undiked tidal marshes around San Francisco Bay before 1850 (grey) and in the 1980s (black). (From Nichols et al. 1986. *Science* **231**. Reprinted with permission from AAAS and F.H. Nichols.)

Water quality

San Francisco Bay is located at the mouth of the Sacramento-San Joaquin river system, which drains about 40% of California's surface areas (Nichols et al. 1986). In the course of European settlement and the gold rush, sediment and nutrient loads rapidly increased, while natural filters such as marshes, wetlands and oysters were diminished. From 1852 to 1914, sediment transport to San Francisco Bay increased from 1.5 to 14 million m^3 yr^{-1}, and large deposits of these sediments still reside in San Francisco Bay today (van Geen & Luoma 1999). As the city of San Francisco grew, so did municipal and industrial discharges. From 1950 to 1980, nitrate concentrations in the San Joaquin River mouth increased about 7.5-fold (Nichols et al. 1986, Cloern 2001). By the 1990s, about 3.2 billion L of wastewater entered the bay every day plus 4500 to 36,000 t of toxic pollutants (Monroe and Kelly, 1992).

Although discharges have continually increased, inputs of at least some contaminants have declined, especially after 1970, when advanced waste treatment was introduced and some chemicals such as DDT and PCBs were banned. Also, several heavy industries have closed since 1970, and mining activities ceased, reducing pollutant and heavy metal loads. Thus, ammonia concentrations declined 10-fold in the 1970s (Figure 26; San Francisco Bay Institute [SFBI] 2003), and phytoplankton concentrations declined from an annual average of 10.27 µg chlorophyll L^{-1} in 1976–1980 to 2.38 µg L^{-1} in 1997–2001 (Figure 26; SFBI 2004). In 1986, the invasive clam *Potamocorbula amurensis* was introduced to San Francisco Bay and rapidly increased in abundance (Figure 26).

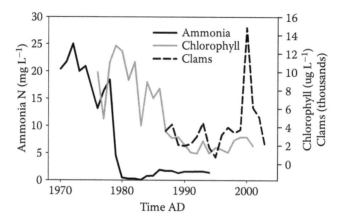

Figure 26 Recent trends in water quality in San Francisco Bay: ammonia concentrations in wastewater effluents, chlorophyll-a concentrations in Suisun Bay and abundance of the invasive clam *Potamocorbula* in Suisun and San Pablo Bays. (Data from SFBI 2003, 2004.)

This filter-feeder probably contributes to keeping phytoplankton at low levels compared with other estuaries (SFBI 2004). Oxygen concentrations have been relatively stable from 1972 to 1978 and 1993 to 2001 and remained mostly at or above the water quality standard of 5 mg L^{-1}, and harmful algal blooms occur infrequently (SFBI 2003).

Comparison of historical changes across U.S. estuaries

Magnitude of ecosystem degradation and species declines

Despite their wide geographic distribution and unique regional histories, the six estuaries showed remarkably similar trajectories of species decline, habitat loss and water quality degradation over time. Averaged across taxonomic groups, overall ecosystem degradation was relatively small during the hunter-gatherer, agriculture and colonial establishment periods (Table 2) but accelerated in the colonial development and global market I periods before slowing and stabilising in the global market II period (Figure 27). The low levels of ecosystem deterioration documented during native and early European occupation suggests that humans had limited impact on marine resources when exploitation was primarily for subsistence purposes and the human population was small. Yet, significant local resource depletion by hunter-gatherers has been documented in San Francisco Bay, where the native population was quite dense (Broughton 2002). Although no such signs were found in the other five estuaries, increasing archaeological research may uncover more rather than less ancient human impacts (Rick & Erlandson 2008).

During the periods of colonial development and global market I, ecosystem changes accelerated (Figure 27). This was a period of rapid human population growth accompanied by increasing demand for natural resources. Exploitation for food, oil, furs, feathers and luxury items was intensified and commercialised, and technological progress allowed more efficient but also less-selective and more destructive harvesting. Many marine mammal, bird and reptile populations were already depleted by 1900 and driven to even lower levels by 1950, while many fish, plant and oyster populations declined continually during this time (Figure 28). At the end of the development period in 1900, San Francisco Bay was by far the most degraded estuary, probably due to the strong impact of the gold rush and associated human activities (Figure 27).

In the global market II period, essentially 1950–2000, overall degradation trends slowed in most regions with enhanced management and conservation efforts (Figure 27). Several species groups

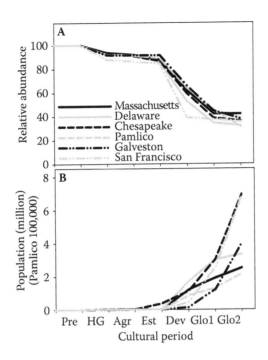

Figure 27 Ecosystem change across cultural periods in the six study systems: (A) average relative decline across six taxonomic groups and (B) human population growth. (Data adapted from Lotze et al. 2006.)

showed signs of stabilisation or recovery, including large whales, all bird groups, reptiles and plants (Figure 28). Yet, all fish groups continued to decline, and all invertebrates except oysters showed accelerated losses. This mirrors the enhanced focus on invertebrate fisheries after the decline of traditional finfish fisheries, a phenomenon known as 'fishing down the food web' (Pauly et al. 1998). At the end of the twentieth century, the average relative abundance of taxonomic groups in the six estuaries was 31–43% of the prehuman state (Figure 27), with Massachusetts Bay the least degraded. Declines of species groups ranged from 91% of pre-exploitation levels for large whales to 36% for pinnipeds and otters (Figure 28) and were similar in all estuaries, as indicated by the relatively small error bars. These general trajectories suggest that degradation of estuarine resources and ecosystems was driven by human history rather than natural change.

Species depletions, extinctions and recoveries

Native Americans and early European visitors encountered a rich abundance of marine animals in the estuaries they populated across the United States. By the end of the twentieth century, however, an average of 95% of all recorded species in the estuaries were depleted (reduced to <50% of former abundance; Table 4), with 35% rare (<10%), 3% extirpated or extinct (0%), and only 10% of species recovering (from extirpated or rare to >10%) (Lotze et al. 2006). Patterns were relatively similar across the six study systems (Figure 29). Depleted and rare species belonged to all taxonomic groups, while extirpated species were dominated by fish and recovering species by birds. Pamlico Sound had the most depleted or rare species, while Massachusetts Bay had the most extirpated species (7%). This masks trends in some systems, though, where species had been extirpated by the late 1800s or early 1900s but re-established themselves or were reintroduced later. The list of currently endangered or threatened marine animals in each estuary (by state) ranges from 11 in Delaware Bay to 27 in San Francisco Bay and includes species from all taxonomic groups (Table 7). Despite

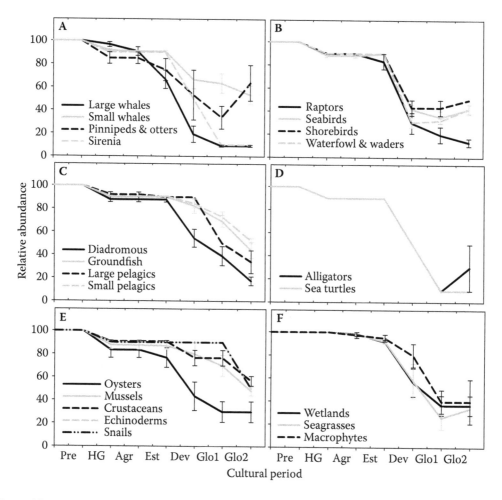

Figure 28 Change in relative abundance of different species groups (average ± SE of six estuaries) within (A) mammals, (B) birds, (C) fish, (D) reptiles, (E) invertebrates, and (F) vegetation. (Data adapted from Lotze et al. 2006.)

the high number of endangered species, San Francisco Bay has experienced the highest percentage (20%) of recovering species (Figure 29). Although affected by an intense history, including the gold rush, the fur trade and general human impacts from a huge population, today's California has a good record with conservation successes.

Habitat alteration and loss

Coastal wetlands, seagrass meadows and oyster reefs are important breeding, nursery, foraging and staging habitats for numerous marine and coastal animals (Beck et al. 2001). Moreover, they are important natural filters and buffer zones between land and sea because they retain and cycle nutrients, sediments and organic matter (Costanza et al. 1997). Today, habitat conservation and restoration are high priorities in many estuaries and coastal states of the United States (USGS 2009a). However, large extents of original habitat have been lost since European colonisation. In the six estuaries, an average of 55% of historical wetland area has been lost or transformed, with the highest loss of 94% occurring in San Francisco Bay and the lowest in Galveston and Delaware Bays (Figure 30). Pamlico Sound had by far the highest extent of past and present wetland area.

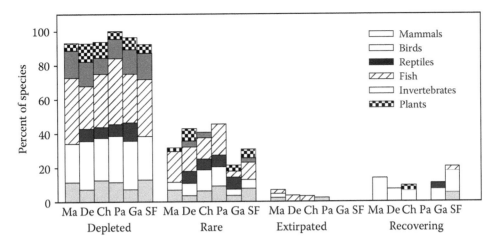

Figure 29 Percent of recorded species by taxonomic group that were depleted (<50%), rare (<10%), extinct or extirpated (0%), and recovering (from <10% to >10%) at the end of the twentieth century in the six study systems.

According to the USGS, wetland loss by state since 1780 ranged from 28% in Massachusetts to 91% in California (USGS 2009a).

Historically, most U.S. estuaries had extensive seagrass beds and other SAV that provided habitat for many species and food for waterfowl, invertebrates, green turtles and manatees. A 1990s survey (Bricker et al. 1999) estimated current SAV status as low (10–25% cover) to very low (0–10% cover) for most systems and non-existent for Delaware Bay (Figure 31). Estimates of past SAV area are limited and mostly confined to the twentieth century. In Chesapeake Bay, SAV extent was reduced by 56% in 1930–1970 (Moore et al. 2004). In Galveston Bay, seagrass area has decreased by 86% and in San Francisco Bay by 34% since the 1950s (White et al. 1993, ANR 2001). Yet, seagrass beds were affected much earlier in many estuaries. Sediment core data from Pamlico Sound indicate a decline in the seagrass epiphyte *Cocconeis* from 13% in 1540 to 0.4% in 2000, an overall decline of 97% (Cooper 2000). Similarly, in Chesapeake Bay *Cocconeis* declined by 80–100% since 550 AD (Cooper & Brush 1993). Combining these different estimates (Figure 31) gives an average seagrass or SAV loss of 68–75% in the estuaries studied.

Oyster banks or reefs also provided important habitat in most estuaries, yet the pressure and destructive practices of the oyster fishery strongly decimated the abundance and health of most oyster reefs (Kirby 2004). Although the long-term trend of the oyster fishery can be well reconstructed from catch data (Figure 28), there is little quantitative information on the former extent of oyster reef area. Historical reports and maps, however, indicate that oyster reefs were extensive in the bays and along the shorelines in most of the studied estuaries. In Chesapeake Bay, one of the more well-studied estuaries in the country, it is known that about 111,600 ha of natural oyster bar habitat originally existed on the Maryland side. This declined by more than 50% from 1907 to 1982, with localised losses of up to 95% (Rothschild et al. 1994).

Water quality

Comparable data on the history of changes in water quality for U.S. estuaries are very limited. However, a survey in the 1990s (Bricker et al. 1999) indicated that all six estuaries were experiencing low-to-moderate signs of eutrophication, which were most pronounced in Chesapeake Bay and lowest in Delaware Bay (Figure 32). A comparison of primary and secondary symptoms as well as the main drivers and trends of eutrophication in the main bays and subsystems are shown in

Table 7 Endangered (E) and threatened (T) marine species listed for each state related to the six study systems

Common name	Scientific name	MA	DE	VI	MD	NC	TX	CA
Mammals								
West Indian manatee	*Trichechus manatus*						E	
Southern sea otter	*Enhydra lutris nereis*							T
Guadalupe fur seal	*Arctocephalus townsendi*							T
Steller sea lion	*Eumetopias jubatus*							T/E
Killer whale	*Orcinus orca*							E
Blue whale	*Balaenoptera musculus*	E						E
Finback whale	*Balaenoptera physalus*	E	E	E	E	E	E	E
Humpback whale	*Megaptera novaeangliae*	E	E	E	E	E	E	E
Right whale	*Balaena glacialis*	E	E	E	E	E		
Sei whale	*Balaenoptera borealis*	E						E
Sperm whale	*Physeter catodon*					E		E
Birds								
Short-tailed albatross	*Phoebastria albatrus*							E
Whooping crane	*Grus Americana*						E	
Eskimo curlew	*Numenius borealis*	E	E	E	E	E	E	E
Marbled murrelet	*Brachyramphus marmoratus*							T
Brown pelican	*Pelecanus occidentalis*					E	E	E
Piping plover	*Charadrius melodus*	T	T	T	T	T	T	
Western snowy plover	*Charadrius alexandrinus nivosus*							T
California least tern	*Sterna antillarum browni*							E
Least tern	*Sterna antillarum*						E	
Roseate tern	*Sterna dougallii dougallii*	E		E		E		
Reptiles								
Green sea turtle	*Chelonia mydas*		T	T	T	T	T	T
Hawksbill turtle	*Eretmochelys imbricata*	E	E	E	E	E	E	
Kemp's ridley turtle	*Lepidochelys kempii*	E	E	E	E	E	E	
Leatherback turtle	*Dermochelys coriacea*	E	E	E	E	E	E	E
Loggerhead turtle	*Caretta caretta*	T	T	T	T	T	T	T
Olive ridley turtle	*Lepidochelys olivacea*							T
Fishes								
Tidewater goby	*Eucyclogobius newberryi*							E
Chinook salmon	*Oncorhynchus tshawytscha*							T/E
Coho salmon	*Oncorhynchus kisutch*							T/E
Smalltooth sawfish	*Pristis pectinata*					E	E	
Smelt, delta	*Hypomesus transpacificus*							T
Steelhead	*Oncorhynchus mykiss*							T/E
Unarmoured threespine stickleback	*Gasterosteus aculeatus williamsoni*							E
Shortnose sturgeon	*Acipenser brevirostrum*	E	E	E	E	E		
Invertebrates								
White abalone	*Haliotis sorenseni*							E
Shasta crayfish	*Pacifastacus fortis*							E
Total		13	11	12	11	15	14	27

Source: Data from USFWS 2009.

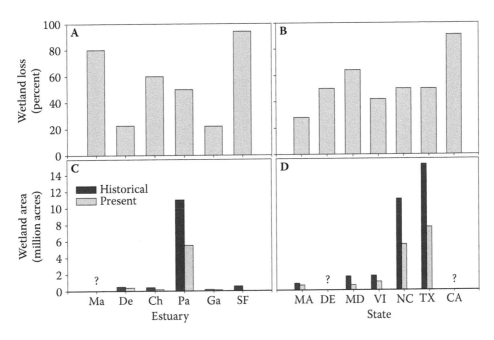

Figure 30 Estimated percent loss of wetland area by (A) estuary and (B) state and historical and present extent of wetland area by (C) estuary and (D) state. (Estuary data from Chapman 1977, Nichols et al. 1986, White et al. 1993, Kennish 2000, USGS 2009a.)

Table 8. All systems showed increased chlorophyll concentrations, indicating enhanced pelagic production, while macroalgae and epiphytes seem only problematic in Chesapeake and Massachusetts Bays. Secondary symptoms, including low oxygen levels, SAV loss or harmful algal blooms, are also found in all systems to varying degrees. Human impacts, especially nitrogen inputs, have been consistently identified as the main driver of eutrophication, but the susceptibility of each system is also an important factor. At the time of the survey, projected trends toward 2020 indicated improvement only in Boston Harbour in Massachusetts Bay, while suggested trends for other systems were stable or worsening (Bricker et al. 1999). Where long-term sediment core or hydrographical data were available, trajectories indicated that degradation of water quality started with European colonisation (Figure 32; Lotze et al. 2006). Initially, land clearing mobilised sediments and nutrients, which enhanced primary production in the estuary. These trends stabilised until the late 1800s, when discharges began to increase with growing municipal and industrial activities. After WWII, artificial fertilisers further enhanced nutrient loads, and eutrophication became a well-known phenomenon with increasing occurrence of algal blooms, SAV loss and oxygen depletion (Cloern 2001).

Species invasions

Another change humans brought to estuarine systems was the introduction of exotic species. This was sometimes intentionally done (e.g., by importing exotic aquaculture species), but often invaders arrived via ship hulls, ballast water or associated with aquaculture species. This includes a variety of species, from viruses to fish and mammals, but the bulk of known estuarine and marine invaders consists of invertebrates, plants and algae (Cohen & Carlton 1998, Carlton 2003, Fofonoff et al. 2003). Time series of recorded species invasions indicated an increase in numbers after European colonisation and further acceleration in the late nineteenth century (Figure 33), likely driven by increasing global navigation and commercial exchange, but also increased awareness and recording (Ruiz et al. 1997). The overall rate of invasions increased about 5-fold from an average of 2.6 per

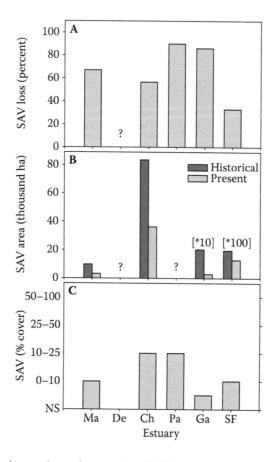

Figure 31 Changes in submerged aquatic vegetation (SAV) in each estuary: (A) percent loss of SAV area (with data from Buzzards Bay for Ma and estimates based on *Cocconeis* decline for Pamlico Sound); (B) historical and present SAV area (data for Galveston multiplied by 10 and San Francisco Bay by 100) (see case studies for the various data sources); and (C) today's SAV status estimated by Bricker et al. 1999.

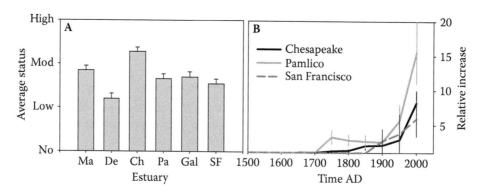

Figure 32 Average current eutrophication signs (A) in the six estuaries (based on data by Bricker et al. 1999) and timeline (B) for water quality degradation based on sediment core and hydrographic data (mean ± SE across parameters; see case studies for various data sources).

Table 8 Eutrophication signs in the six estuaries and subsystems ranking from no (—), low, moderate (mod) to high

		Primary symptoms			Secondary symptoms			Influencing factors			Trend 2020
	Overall	Chl.a	Macroalgae	Epiphytes	Low DO	SAV loss	HAB	Human	Susc.	N input	
Massachusetts Bay	Mod	Mod	—	High	—	Low	Mod	Mod	Low	High	▼
Boston Harbor	High	Mod	High	Mod	Mod	Low	—	High	Mod	High	▲
Delaware Bay	Low	Mod	—	—	Low	—	—	High	Mod	High	—
Chesapeake Bay	High	High	High	—	High	High	—	High	High	Mod	▼
Patuxent River	High	High	?	—	Mod	High	High	High	High	Mod	▼
Potomac River	High	High	?	Low	High	Mod	Mod	High	High	Mod	▼
Rappahannock River	High	High	—	—	Mod	—	—	High	High	Mod	▼
Choptank River	Mod	Mod	High	Mod	Low	High	Mod	High	High	Mod	▼
Tangier/Pocomoke Sound	High	High	High	High	Low	High	Low	High	High	High	▼
Pamlico Sound	?	Mod	—	—	Low	Low	Mod	Mod	?	Mod	—
Pamlico/Pungo Rivers	High	High	—	—	Mod	Mod	Mod	High	High	Mod	—
Neuse River	High	High	—	—	High	Low	High	High	High	High	—
Albermale Sound	Low	Low	—	—	Low	Low	Low	Mod	Mod	Mod	▼
Galveston Bay	High	Mod	—	—	Low	High	Low	High	Mod	Mod	▼
San Francisco Bay	High	High	—	—	Low	High	—	High	Mod	Mod	▼
Central/Suisun/San Pablo	Mod	Mod	—	—	Low	Mod	High	Low	Low	Mod	—

Source: Data from Bricker et al. 1999.

Note: Primary symptoms include chlorophyll levels (Chl.), macrogalae and epiphyte abundance; secondary symptoms include low dissolved oxygen (DO) levels, submerged aquatic vegetation (SAV) loss, and harmful algal blooms (HAB); influencing factors include human impacts, the system's susceptibility (Susc.) and nitrogen (N) inputs; projected trends toward 2020 are worsening ▼, improving ▲, or stable —.

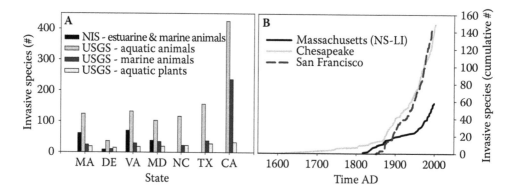

Figure 33 Recorded non-indigenous species (NIS) (A) by state, with VA and MD for Chesapeake Bay (NIS data from Ray 2005 for four NE U.S. states, USGS data from USGS 2009b), and (B) over time, with data for Massachusetts referring to Nova Scotia–Long Island Sound (data from Cohen & Carlton 1998, Carlton 2003, and Fofonoff et al. 2003).

decade in the first 50 yr of available time series to 13.2 in the last 50 yr, with higher increases in Chesapeake and San Francisco Bays compared with the area from Nova Scotia to Long Island Sound, which includes Massachusetts Bay (Cohen & Carlton 1998, Carlton 2003, Fofonoff et al. 2003). Today, different agencies and organisations are recording invasive species in the United States by state. In the northeastern United States, Ray (2005) reported between 8 (Delaware) and 70 (Virginia, Chesapeake Bay) invasive estuarine and marine animals. Data by the USGS (2009b) indicated between 36 (Delaware) and 425 (California) aquatic (freshwater and marine) invasive species (Figure 33). Delaware Bay had consistently the lowest and San Francisco Bay the highest number of recorded invasive species.

Causes and consequences of historical changes

Importance of human impacts

The six case studies and the summary reveal that humans have had a variety of impacts on estuarine and coastal species and ecosystems. However, exploitation clearly stood out as the major driver of species declines and extirpations, followed by habitat loss (Figure 34). Pollution, general human disturbance, disease and eutrophication were less-prominent drivers of decline. This is consistent with studies summarising marine species extinctions, depletions and their threats (Dulvy et al. 2003, Kappel 2005). Importantly, cumulative human impacts, mostly the combined effects of exploitation and habitat loss and in some cases pollution, were important in 58% of the recorded extinctions and 41% of depletions (Lotze et al. 2006). The records also suggest that reversing these human impacts enabled 10% of recorded species to recover (Figure 34). Reducing or banning exploitation and protecting habitat were the main drivers for recovery, but controlling pollution and disease also helped in some cases. Notably, in 77% of recoveries it was the reversal of multiple human impacts that worked, in most cases the combination of reduced exploitation and habitat protection. This indicates that cumulative human impacts can pose severe threats to species survival, and that basic needs, including shelter, food, habitat and health, need to be met to promote species recovery. Therefore, management of multiple human impacts needs to be integrated to maintain biodiversity and ecosystem function.

Yet, current management also needs to take new human impacts into account. Among past drivers of depletion and extinction, invasive species and climate change did not play a major role. However, these factors become more prevalent in coastal waters worldwide (Ruiz et al. 1997, Scavia

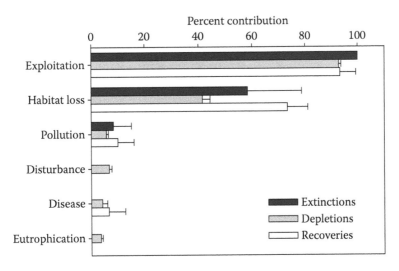

Figure 34 Importance of different human impacts (average ± SE) in causing or contributing to species extinctions and depletions, and the reduction or ban of those has an impact on enabling recoveries. (Data adapted from Lotze et al. 2006.)

et al. 2002, Harley et al. 2006), causing additional threats and possible interactions with existing human impacts. Climate warming may enhance the effects of eutrophication (Lotze & Worm 2002), disease risk (Harvell et al. 2002) and invasions (Stachowicz et al. 2002). On the other hand, management aimed at reducing overexploitation, protecting habitat and controlling pollution may decrease the importance of these threats to estuarine species in the future.

Species richness and diversity

The number of species extinctions and invasions changes local species richness and diversity. In many estuaries and coastal waters, invaders by far outnumber extirpated species (Lotze et al. 2005, 2006, Byrnes et al. 2007), thereby enhancing actual species richness. However, many species have been depleted by more than 90% compared with historical abundance and may be considered rare or ecologically extinct, which would strongly reduce the functioning species diversity in those ecosystems (Byrnes et al. 2007, Jackson 2008). Also, the species groups most strongly affected by depletions and extinctions do not correspond to those affected by invasions. While declines have occurred predominantly among large mammals, birds, reptiles and fishes, invasions mostly consist of smaller invertebrates, plants and microscopic algae, protozoans, viruses and bacteria (Lotze et al. 2006). This mismatch creates a strong shift in species composition and diversity of estuarine and coastal systems where large, long-lived, slow-growing and late-maturing species have been replaced by small, fast-growing and high-turnover species (Byrnes et al. 2007). Given past and current trends, this shift in diversity is likely increasing in the future although partly dampened by recoveries.

Ecosystem structure, functions and services

The recorded changes in diversity have likely profoundly altered the structure and function of estuarine ecosystems (Hooper et al. 2005), but the question is how? Not only have all taxonomic groups experienced depletions in the estuaries examined, but also all important functional groups, including large and small carnivores, herbivores, suspension-feeders and submerged vegetation and wetlands (Figure 35). This means that estuarine ecosystems have lost a large amount of top-down control, while increasing nutrient loads have enhanced bottom-up control. Changing the strength of these

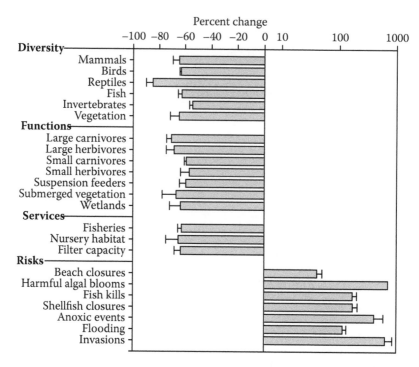

Figure 35 Summary of historical ecosystem changes in U.S. estuaries (average ± SE). Diversity has been changed by declines in relative abundance in all taxonomic groups. This has also affected relative abundance of different functional groups and the services the estuarine ecosystems perform. The loss of diversity, functions and services has contributed to the increase in a range of risks to environmental and human health. (Data for diversity and functions adapted from Lotze et al. 2006, for services and risks from Worm et al. 2006.)

basic forces likely has severe consequences on species abundance, food web interactions, ecosystem productivity and functioning (Worm et al. 2002, Worm & Duffy 2003). The loss of predation may have released prey or competitors, while recovering top predators may suffer from depletion of their prey (Dayton et al. 2002). The loss of essential habitat and water quality degradation may have altered the carrying capacity for many species. An important open question is how—if 'everything' was more abundant in the past—could former ecosystems have sustained that abundance? Relative to current ecosystem structure, alternative past scenarios may involve very slow growth rates and turnover of large animals or more efficient transfer of primary production to higher trophic levels (Steele & Schumacher 2000). In both cases, the composition and dynamics of estuarine ecosystems have profoundly changed, with potentially severe consequences for the maintenance of their diversity, productivity and stability.

Changes in ecosystem structure and functioning translate into changes in ecosystem services that benefit people, including food, clean water, fibres, shelter and aesthetic and cultural services (Daily 1997, Millennium Ecosystem Assessment [MEA] 2005). Over the past 50 yr, human activities have changed ecosystem structure and function more rapidly and extensively than in any other period in history (Vitousek et al. 1997, MEA 2005). This has degraded approximately 60% of global ecosystem services, with consequences on all components of human well-being (MEA 2005). In estuaries and coastal waters, the decline in taxonomic and functional groups has affected the provision of fisheries and seafood, spawning and nursery habitats to sustain marine resources and filter capacity to maintain good water quality (Figure 35). The losses of these functions and services have brought rising health risks and costs to society, such as increasing beach closures, harmful algal blooms, fish kills, shellfish closures, oxygen depletion, flooding and invasions (Worm et al. 2006).

To reverse these undesirable effects and ensure the continued provision of important ecosystem services, it is essential to strengthen the resilience of both the ecosystem and the social or governance system. While the first can be achieved by promoting biodiversity and limiting human impacts, the latter rises with societal abilities to learn and adapt (Lotze & Glaser 2009).

National and global comparisons

Other U.S. estuaries

The six reviewed estuaries are among the largest and possibly most impacted estuaries of the United States, some of them surrounded by major cities and industries. However, other U.S. estuaries have experienced similar histories of human-induced changes with potentially similar consequences on the status of species, habitats and water quality. Historically, more than 50% of all U.S. fishery yields came from estuaries or estuarine-dependent species (Houde & Rutherford 1993). Yet today, in 22 of the 28 estuaries designated as National Estuary Program sites in the United States, declines of fish and wildlife populations are considered to be high- or medium-priority problems (Kennish 2002). This is not just driven by past and present exploitation pressure. A survey of the status and impacts of eutrophication across 138 U.S. estuaries (Bricker et al. 1999) indicated that 50% of systems experience impaired resource use in commercial and recreational fisheries, shellfish harvesting, fish consumption, swimming, boating and tourism. Overall, 44 estuaries were identified as highly and 40 as moderately eutrophic, with projected trends to worsen in 86 estuaries (Bricker et al. 1999). The main management targets of concern to curb eutrophication in the 138 estuaries included agriculture, wastewater treatment and urban run-off. Past and present habitat loss is also of concern in many U.S. estuaries. At the time of European settlement, there were about 221 million acres of wetlands within the area currently covered by the political boundaries of the United States (Dahl & Allord 1999). By the mid-1980s, only 103 million acres remained. Six of the 50 U.S. states have lost 85% or more of their original wetlands, and 22 lost more than 50%. The estuaries selected for this review lost 22–94% of their wetlands.

Global context

This review indicated very similar histories of change in the selected estuaries, likely because they experienced the same historical context of first Native American and later European occupation and finally becoming part of the United States. In a comparative analysis of 12 estuaries and coastal seas around the world, these six U.S. systems showed a higher magnitude of degradation compared with Canadian but less degradation than European systems (Lotze et al. 2006). Although the Canadian Bay of Fundy and Gulf of St. Lawrence experienced the same North American historical context, human population density and impact is generally less intense in these regions, with smaller cities and fewer industrial activities along the coast. However, the European systems experienced a much longer and more severe history of human activities compared with the United States. Resource depletion, wetland loss and water quality changes in the northern Adriatic Sea can be traced to Roman times 2000 yr ago (Lotze et al. 2006). In the Wadden and Baltic Seas, strong human influence started in medieval times 1000 yr ago. Although past changes in resources, habitats and water quality followed similar trajectories in all systems, they played out more rapidly in North America and Australia in the course of European colonisation in the past 150–300 yr (Lotze et al. 2006). Detailed historical reconstructions of ecological changes in the Bay of Fundy (Lotze & Milewski 2004) and the Wadden Sea (Lotze 2005, 2007, Lotze et al. 2005) mirror the changes described for the U.S. estuaries in this review.

Human impacts have not only affected estuarine environments in the past. Evidence is increasing that many coastal ecosystems, including the Benguela upwelling system (Griffiths et al. 2004), coral reefs (Pandolfi et al. 2003) and offshore islands (Craig et al. 2008), have experienced similar histories of ecological change. For example, across 14 coral reefs in the Atlantic, Red Sea and Australia, very similar trajectories of change emerged, with the rapid depletion of large carnivores and herbivores in the course of European colonisation, followed by declines in smaller animals and architectural species (Pandolfi et al. 2003). The reefs of today were 30–80% degraded from pristine levels. Together, these studies revealed a much longer history of human-induced changes in estuarine and coastal ecosystems than previously assumed. On the continental shelves, human-induced changes generally occurred only in the past 100–200 yr and moved from there into the open ocean about 50 yr ago and into the deep sea 10–20 yr ago. This spatial expansion of resource use was accompanied by a temporal acceleration of induced changes in population depletion (Lotze 2007, Lotze & Worm 2009).

An increasing body of literature in marine historical ecology is also revealing stronger declines in valuable species groups than previously assumed. A review across 256 estimates of past population changes from 95 studies indicated an average decline of 89% from historical abundance levels. Recent recovery in about 15% of the species reduced this estimate to an 84% average decline today (Lotze & Worm 2009). These rates of decline are comparable to those of the species considered rare (>90% decline) in the U.S estuaries of this study. A review of more recent changes in 232 fish stocks indicated an average 83% decline from maximum breeding population size over 10–73 yr (Hutchings & Reynolds 2004). Similarly, an analysis of exploited marine mammal populations suggested declines of 76% across all species and 81% for the great whales since the beginning of exploitation (Christensen 2006). Finally, a meta-analysis of cod stocks in the North Atlantic indicated an average decline of 96% from carrying capacity (Myers & Worm 2005). All these figures indicate that many populations have been reduced to very low levels that correspond to threatened or endangered status following criteria by the IUCN, and many species have been listed as threatened or endangered (e.g., Table 7).

In addition to species depletion, habitat loss is a global phenomenon. A review of historical habitat losses around Europe suggests that more than 50% of coastal wetlands and seagrass beds have been lost or transformed, with losses of more than 80% in many regions (Airoldi & Beck 2007). Native oyster reefs were ecologically extinct by the 1950s along most European coastlines and in many estuaries and bays well before that. Today, less than 15% of the European coastline is considered in 'good' condition (Airoldi & Beck 2007). In tropical systems, about 20% of coral reefs have been lost and 20% degraded over the past 50 yr, and 35% of mangrove area has been lost over the past two decades (MEA 2005).

Outlook

Scientific progress in marine historical ecology

Over the past decade, scientists from a variety of disciplines, including palaeontology, archaeology, history, fisheries science and ecology have made tremendous progress in marine historical ecology, tracing human impacts on marine animal populations back hundreds and sometimes thousands of years (Rick & Erlandson 2008, Starkey et al. 2008, Lotze & Worm 2009). Palaeontological records have identified natural long-term, often climate-driven changes in population abundance that place recent anthropogenic changes into context. Archaeological studies suggest significant human impacts on local marine resource abundance, distribution and size long before commercial and industrial exploitation started. Furthermore, historical records imply that many highly valued species were severely depleted before the midtwentieth century, often reaching their low point decades to more than a century ago. Thus, marine historical ecology expands our usual ecological data horizon of a few years or decades way into the past, thereby revealing dramatic changes before general scientific inquiry began. It also shows that wherever we study marine ecosystems today, we should be aware of their 'unnatural' state because of the long history of change they have undergone prior to our assessment (Carlton 1998, Jackson et al. 2001, Roberts 2003).

Knowledge gaps and future research

Many studies in marine historical research have focused on single species, and few studies have aimed at combining results for various species or functional groups (such as Pandolfi et al. 2003, Lotze et al. 2006). Even past trajectories of multiple ecosystem components do not answer the question of how past food webs and ecosystems have functioned. There are many unanswered questions regarding how trophic flows, system biomass and productivity have changed ecosystem performance over time. The best way to explore this may involve ecosystem modeling tools (Pitcher et al. 2002, Coll et al. 2006). Also, historical research has revealed the importance of exploitation and habitat loss in driving population declines over the past centuries, while climate variability played a major role in population fluctuations over much longer periods. However, climate change may become critical for predicting future trends in populations and ocean ecosystems (Brander 2007), and a renewed focus on the consequences of past climate fluctuations might help project future ecosystem changes (Rose 2004, Enghoff et al. 2007).

Using historical reference points for management and conservation

Around the world, many fisheries and coastal ecosystems are in trouble, and although many management agencies are aiming at halting or reversing past trends of depletion and degradation, there are some major factors that perpetuate the current crisis. First, the historical context is generally overlooked, thereby failing to acknowledge the total magnitude of population declines and ecosystem change. If management relies on the past 20–50 yr of scientific monitoring data, the necessary reference points are obscured by shifting baselines (Pauly 1995, Sáenz-Arroyo et al. 2005, Pinnegar & Engelhard 2007). Second, the ecosystem context is often ignored, yet it is not single species or stocks that have changed, but their predators, prey and habitat as well, thereby changing the overall ecosystem structure and functioning (Steele & Schumacher 2000). This may prevent recovery, as, for example, witnessed in the non-recovery of North Atlantic cod. Finally, management continues to focus on single species and single human impacts, which are easier to assess and manage. Although recognition of an ecosystem approach to management is increasing, there is little understanding of how to actually do this. The Pew Oceans Commission (2003) and U.S. Commission on Ocean Policy (2004) outlined goals for an ecosystem-based management approach that include the restoration of historical levels of native biodiversity and a vision for healthy, productive, resilient marine ecosystems that provide stable fisheries, abundant wildlife, clean beaches, vibrant coastal communities and healthy seafood. But, what are historical levels of native biodiversity? And, to which levels should ecosystems be restored? The largest gap in this approach is the lack of historical baselines as basic reference points to determine sound management and conservation goals (Carlton 1998, Roberts 2003). The present review may fill some of these gaps for U.S. estuaries.

Recovery, restoration and resilience

The environmental or ecological history of an ecosystem can provide perspective on the possible diversity, abundance and health of an ecosystem that is otherwise missing. Although we may never be able to restore an ecosystem to 'pristine' levels, the past provides a vision for conservation and recovery (Carlton 1998, Clark et al. 2001a). Many populations and ecosystems are quite robust and can recover from perturbations, although they may not necessarily bounce back to their original abundance or state (Palumbi et al. 2008). In general, it is most important to keep all the parts, which means maintaining the biodiversity of the system. Higher diversity has been linked to greater stability, productivity and recovery potential both in small-scale experiments and large-scale ecosystem comparisons (Worm et al. 2006).

Fortunately, despite several extinctions, most species and functional groups still persist in U.S. estuaries today, albeit in greatly reduced numbers. Thus, the potential for recovery remains, and where human efforts have been focused on protection and conservation, recovery has been possible, albeit often with significant lag times. Long-lived species may need several decades to rebound to previous levels of abundance (Caddy & Agnew 2004, Lotze & Milewski 2004), and life-history characteristics and the magnitude of depletion are important determinants for recovery. Across 232 fish populations, only 12% recovered 15 yr after a collapse, mostly clupeids, while 40% showed no recovery (Hutchings & Reynolds 2004). A review of historical population changes found that 15% of 256 examples, mostly marine mammals, have experienced some recovery in past decades, increasing on average from 13% to 39% of former abundance levels (Lotze & Worm 2009).

Although several species, habitats and water quality parameters have shown recovery or improvement in the presented U.S. estuaries, it is unclear whether this contributed to an overall recovery of the ecosystem. But, what would that look like? An example from the Thames estuary in the United Kingdom shows that improving oxygen levels led to the return of about 110 fish species that were absent during times of heavy pollution (Cloern 2001). A 10-fold reduction in nitrogen loads in Tampa Bay in Florida led to reduced phytoplankton levels, increased water clarity, decreased blue-green algae and, finally, after 10 yr, the return of seagrass (Cloern 2001). Studies of marine protected areas show that species richness and productivity increased within compared with outside protected zones (Worm et al. 2006). These examples indicate that at least partial ecosystem recovery can be possible, even in heavily degraded systems. Today, pollution controls, habitat protection and restoration and no-exploitation or no-go zones are increasingly part of coastal management plans. Thus, the major drivers of past changes are increasingly addressed and may eventually lead to enhanced recovery in the future.

Governance and a sustainable future

A major challenge is to reverse ecosystem degradation while meeting increasing societal demands for the services these systems provide. Yet, protecting and restoring marine biodiversity are key to maintaining essential ecosystem services (Worm et al. 2006). This concept is fairly new, however, and lacking broad public awareness and acknowledgement. Balancing exploitation and protection, degradation and restoration needs good and strong governance, but economic, social and political incentives often favour over- rather than underutilisation.

In the United States, the Magnuson-Stevens Act requires recovery plans for overexploited fish stocks. By 2004, half of the 74 fish stocks requiring recovery showed some signs of increase, but only 3 reached their recovery target (Rosenberg et al. 2006). The latter were stocks for which fishing pressure had been significantly reduced toward or below the target fishing mortality. Of the other stocks, 54 were considered overfished, and 34 continued to experience overfishing despite having a recovery plan. Thus, a recovery plan does not help in itself; it also needs to be implemented, which falls into the realm of governance and political will rather than scientific analysis and advice. Unfortunately, a survey of fisheries management effectiveness around the world indicates that only 7% of coastal states have rigorous scientific assessments, 1.4% also have a participatory and transparent conversion of scientific advice into policy, and 0.95% also have mechanisms that ensure compliance with regulations, while none was also free of excess fishing capacity or subsidies (Mora et al. 2009). Yet, in systems where management measures have been taken seriously and implemented, even heavily exploited large-scale ecosystems such as the California Current or small-scale regions such as Kenya's coastal waters can achieve recovery of important fish and fisheries (Worm et al. 2009).

One coastal ecosystem that was heavily degraded but has undergone remarkable recovery over the past decades is Monterey Bay in California (Palumbi 2009). Comparable to the ecological histories experienced in the U.S. estuaries of this review, Monterey Bay went through periods of whaling, fur and feather trade, heavy exploitation of fish, birds and abalone, and intense pollution at the height of the canning industry in early to mid-1900s. After the collapse of sardine stocks, the last major fishery, it was the vision, perseverance and governance led by one person, Julia Platt, that turned the tide for Monterey Bay. Over the past decades, extensive marine and coastal reserves, a commitment to conservation education and nature tourism and a balance between resource use and protection has enabled the bay's ecosystem to recover and once again teem with life (Palumbi 2009). This hopeful example is proof of the ability of marine ecosystems to recover and of human communities to make it possible.

Many estuaries and coastal bays around the world have undergone similar histories of depletion and degradation as Monterey Bay, yet recovery signs and prospects seem so far more limited. This review provides a detailed account of such ecological histories in selected U.S. estuaries. On the one hand, the reconstruction of former species and habitat abundance and ecosystem health may serve as historical reference points for setting sound future management and conservation targets. On the other hand, the rediscovery of the past offers a vision of the former richness and importance of estuaries that may inspire better governance for estuarine diversity and productivity supporting animal and human life.

Acknowledgements

I would like to thank all my colleagues who helped in compiling and interpreting data from different disciplines and discussing analytical approaches and results, including H. Lenihan, B. Bourque, R. Bradbury, R. Cooke, M. Kay, S. Kidwell, M. Kirby, C. Peterson, J. Jackson and K. Selkoe. Many thanks also to B. Worm and C. Muir for valuable support and comments on this work. Financial support was provided by the Lenfest Ocean Program, the Census of Marine Life and the National Center for Ecological Analysis and Synthesis.

References

Agriculture and Natural Resources Communication Services (ANR). 2001. California's living marine resources: a status report. Agriculture and Natural Resources Communication Services, ANR Publication SG01–11. Oakland, California: University of California.

Ainsworth, D. 2002. San Francisco Bay: a 5,000-year perspective on the human transformation of the bay. *University of California Berkeley Campus News*. Online. Available HTTP: http://www.berkeley.edu/news/media/releases/2002/06/13_sfbay.html (accessed May 31, 2009).

Airoldi, L. & Beck, M.W. 2007. Loss, status, and trends for coastal marine habitats of Europe. *Oceanography and Marine Biology An Annual Review* **45**, 345–405.

Alter, S.E., Rynes, E. & Palumbi, S.R. 2007. DNA evidence for historic population size and past ecosystem impacts of gray whales. *Proceedings of the National Academy of Sciences of the United States of America* **104**, 15162–15167.

Anderson, E.D., Cadrin, S.X., Hendrickson, L.C., Idoine, J.S., Lai, H.L. & Weinberg, J.R. 1999a. Unit 4—Northeast Invertebrate Fisheries. In Our living oceans. Report on the status of U.S. living marine resources, 1999. Silver Spring, MD: National Oceanic and Atmospheric Administration (NOAA). Tech. Memo. NMFS-F/SPO-41.

Anderson, E.D., Creaser, T., MacKenzie, C.L., Jr., Bennett, J., Woodby, D.A., Low, L.L., Smith, S.E. & Hamm,D.C. 1999b. Unit 21—Nearshore fisheries. In Our living oceans. Report on the status of U.S. living marine resources, 1999. Silver Spring, MD: NOAA. Tech. Memo. NMFS-F/SPO-41.

Anderson, E.D., Kocik, J.F. & Shepherd, G.R. 1999c. Unit 3—Atlantic anadromous fisheries. In Our living oceans.report on the status of U.S. living marine resources, 1999. Silver Spring, MD: NOAA. Tech. Memo. NMFS-F/SPO-41.

Animal Welfare Institute (AWI). 2005. *Endangered Species Handbook*. Washington, D.C: AWI. Online. Available HTTP: http://www.endangeredspecieshandbook.org/legislation.php (accessed 10 May 2009).

Aten, L.E. 1983. *Indians of the Upper Texas Coast*. New York: Academic Press.

Atlantic States Marine Fisheries Commission (ASMFC). 2004. Atlantic menhaden stock assessment report for peer review. Washington, D.C.: ASMFC. Stock Assessment Report No. 04–01 (Suppl.).

Atlantic States Marine Fisheries Commission (ASMFC). 2008. Atlantic striped bass species profile. Washington, D.C.: ASFMC. Online. Available HTTP: http://www.asmfc.org/speciesDocuments/stripedBass/profiles/speciesprofile.pdf (accessed 5 May 2009).

Atlantic States Marine Fisheries Commission (ASMFC). 2009. Species Profile: Atlantic sturgeon. Washington, D.C.: ASFMC. Online. Available HTTP: http://www.asmfc.org/speciesDocuments/sturgeon/sturgeonProfile.pdf (accessed 10 May 2009).

Atwater, B.F., Conrad, S.G., Dowden, J.N., Hedel, C.W., MacDonald, R.L. & Savage, W. 1979. History, landforms, and vegetation of the estuary's tidal marshes. In *San Francisco Bay: The Urbanized Estuary*, T.J. Conomos (ed.). San Francisco, Calfornia: Pacific Division American Association for the Advancement of Science, 347–386.

Baird, S. F. 1873. *Part I: Report on the Condition of the Sea Fisheries of the South Coast of New England in 1871 and 1872*. Washington, D.C.: USGPO.

Bayley, S., Stotts, V.D., Springer, P.F. & Steenis, J. 1978. Changes in submerged aquatic populations at the head of Chesapeake Bay, 1958–1975. *Estuaries* **1**, 73–84.

Beck, M.W., Heck, J., Kenneth, L., Able, K.W., Childers, D.L., Eggleston, D.B., Gillanders, B.M., Halpern, B., Hays, C.G., Hoshino, K., Minello, T.J., Orth, R.J., Sheridan, P.F. & Weinstein, M.P. 2001. The identification, conservation, and management of estuarine and marine nurseries for fish and invertebrates. *BioScience* **51**, 633–641.

Bigelow, H. B. & Schroeder, W.C. 1953. Fishes of the Gulf of Maine. *Fishery Bulletin of the Fish and Wildlife Service* **74** (53), 577 p.

Billings, W.M., Selby, J.E. & Tate T.W. 1986. *Colonial Virginia, a History*. White Plains, New York: KTO Press.

Blankenship, K. 2004. Chesapeake Bay Gateways Network: Ducks! Ducks! Geese and swans! Alliance for the Chesapeake Bay, *Bay Journal* **14**(9). Online. Available HTTP: http://www.bayjournal.com/article.cfm?article=2445 (accessed 12 January 2010).

Blaylock, R.A. 1988. Distribution and abundance of the bottlenose dolphin, *Tursiops truncatus* (Montagu, 1821), in Virginia. *Fishery Bulletin* **86**, 797–805.

Bourque, B.J. 1995. *Diversity and Complexity in Prehistoric Maritime Societies: A Gulf of Maine Perspective*. New York: Plenum Press.

Bourque, B.J., Johnson, B.J. & Steneck, R.S. 2008. Possible prehistoric fishing effects on coastal marine food webs in the Gulf of Maine. In *Human Impacts on Ancient Marine Ecosystems: A Global Perspective*, T.C. Rick & J.M. Erlandson (eds). Berkeley, California: University of California Press.

Brander, K.M. 2007. Global fish production and climate change. *Proceedings of the National Academy of Sciences of the United States of America* **104**, 19709–19714.

Brandt, L.A. 1991. Long-term changes in a population of *Alligator mississippiensis* in South Carolina. *Journal of Herpetology* **25**, 419–424.

Brickell, J. 1737. *The Natural History of North Carolina with an Account of the Trade, Manners, and Customs of Christians and Indian Inhabitants*. Dublin, Ireland: Carson.

Bricker, S.B., Clement, C.G., Pirhalla, D.E., Orlando, S.P. & Farrow, D.R.G. 1999. *National Estuarine Eutrophication Assessment: Effects of Nutrient Enrichment in the Nation's Estuaries*. Silver Spring, Maryland: NOAA, National Ocean Service, Special Projects Office and the National Center for Coastal Ocean Science.

Britton, A. 2009. *Alligator mississippiensis*. Crocodilians Natural History and Conservation. Gainesville, FL: Florida Natural History Museum. Online. Available HTTP: http://www.flmnh.ufl.edu/natsci/herpetology/brittoncrocs/csp_amis.htm (accessed 18 May 2009).

Bromberg, K.D. & Bertness, M.D. 2005. Reconstructing New England salt marsh losses using historical maps. *Estuaries* **28**, 823–832.

Bromberg, K.D., Silliman, B.R., & Bertness, M. D. 2009. Centuries of human-driven change in salt marsh ecosystems. *Annual Reviews of Marine Science* **1**, 117–141.

Broughton, J.M. 1997. Widening diet breath, declining foraging efficiency, and prehistoric harvest pressure: ichthyofaunal evidence from the Emeryville Shellmound, California. *Antiquity* **71**, 845–862.

Broughton, J.M. 2002. Prey spatial structure and behaviour affect archaeological tests of optimal foraging models: examples from the Emeryville Shellmound vertebrate fauna. *World Archaeology* **34**, 60–83.

Brown, M.W. & Kraus, S.D. 1996. North Atlantic right whales. *Environment Canada Atlantic Region, Occasional Report* **8**, 92–94.

Bruce, P.A. 1935. *Economic History of Virginia in the Seventeenth Century*. New York: Peter Smith.

Brundage, HM & Meadows, R.E. 1982. Occurrence of the endangered shortnose sturgeon, *Acipenser breviro-strum*, in the Delaware River Estuary. *Estuaries* **5**, 203–208.

Brush, G.S. 2001. Natural and anthropogenic changes in Chesapeake Bay during the last 1000 years. *Human and Ecological Risk Assessment* **7**, 1283–1296.

Byrnes, J.E., Reynolds, P.L. & Stachowicz, J.J. 2007. Invasions and extinctions reshape coastal food webs. *PloS One* **2**(3), e295.

Caddy, J.F. & Agnew, D.J. 2004. An overview of recent global experience with recovery plans for depleted marine resources and suggested guidelines for recovery planning. *Reviews in Fish Biology and Fisheries* **14**, 43–112.

Campbell, R., Jr. 1998. *Performance Audit of the Department of Environment and Natural Resources, Division of Marine Fisheries*. Raleigh, North Carolina: Office of the State Auditor, State of North Carolina.

Campbell, R.R. 1986. *Status Report on the Sea Mink,* Mustela macrodon, *in Canada*. Ottawa, Canada: Committee on the Status of Endangered Wildlife in Canada.

Carleill, C. 1585. Chesapeake Bay. In *New American World: A Documentary History of North America to 1612: Volume 3*, D.B. Quinn (ed.). (1979). New York: Arno Press.

Carlton, J.T. 1998. Apostrophe to the ocean. *Conservation Biology* **12**, 1165–1167.

Carlton, J.T. 2003. *A Checklist of the Introduced and Cryptogenic Marine and Estuarine Organisms from Nova Scotia to Long Island Sound*. Mystic, Connecticut: Maritime Studies Program, Williams College—Mystic Seaport, 2nd edition.

Carter, L.J. 1970. Galveston Bay: test case of an estuary in crisis. *Science* **167**, 1102–1108.

Catesby, M. 1996. The *Natural History of Carolina, Florida and the Bahama Islands (1682–1749)*. London: Alecto Historical Editions.

Chapman, V.J. 1977. *Wet Coastal Ecosystems*. Amsterdam: Elsevier.

Chesapeake Bay Ecological Foundation (CBEF). 2009. *Atlantic menhaden*. Easton, MD: CBEF. Online. Available HTTP: http://www.chesbay.org/forageFish/menhaden.asp (accessed 12 May 2009).

Chesapeake Bay Field Office (CBFO). 2009. *American shad*. Annapolis, MD: U.S. Fish and Wildlife Service. Online. Available HTTP: http://www.fws.gov/chesapeakebay/SHAD.HTM (accessed 12 May 2009).

Chesapeake Bay Program (CBP). 2009a. *Bay history*. Online. Available HTTP: http://www.chesapeakebay.net/bayhistory.aspx?menuitem=14591 (accessed 12 May 2009).

Chesapeake Bay Program (CBP). 2009b. *Fish*. Online. Available HTTP: http://www.chesapeakebay.net/fish1.htm (accessed 12 May 2009).

Chesapeake Bay Program (CBP). 2009c. *Bay watershed forest cover*. Online. Available HTTP: http://www.chesapeakebay.net/status_watershedforests.aspx?menuitem=26067 (accessed 12 May 2009).

Chilton, G. 1997. Labrador duck (*Camptorhynchus labradorius*). In *The Birds of North America Online*, A. Poole (ed.). Ithaca, New York: Cornell Lab of Ornithology. Online. Available HTTP: http://bna.birds.cornell.edu/bna/species/307 (accessed 2 May 2009).

Chittenden, M.E., Jr. 1974. Trends in the abundance of American shad, *Alosa sapidissima*, in the Delaware River Basin. *Chesapeake Science* **15**, 96–103.

Chittenden, M.E., Jr., Barbieri, L.R., & Jones, C.M. 1993. Fluctuations in abundance of Spanish mackerel in Chesapeake Bay and the mid-Atlantic region. *North American Journal of Fisheries Management* **13**, 450–458.

Christensen, L. B. 2006. Marine mammal populations: reconstructing historical abundances at the global scale. *UBC Fisheries Centre Research Reports* **14**. Vancouver, BC: University of British Columbia.

Claesson, S.H. 2008. *Sustainable development of maritime cultural heritage in the Gulf of Maine*. PhD thesis, University of New Hampshire, Durham, New Hampshire, United States.

Claesson, S.H. & Rosenberg, A.A. 2009. Stellwagen Bank marine historical ecology. Final report. Durham, New Hampshire: Gulf of Maine Cod Project, University of New Hampshire.

Clark, J.S., Carpenter, S.R., Barber, M., Collins, S., Dobson, A., Foley, J.A., Lodge, D.M., Pascual, M., Pielke, R., Jr., Pizer, W., Pringle, C., Reid, W.V., Rose, K.A., Sala, O., Schlesinger, W.H., Wall, D.H. & Wear, D. 2001a. Ecological forecasts: an emerging imperative. *Science* **293**, 657–660.

Clark, K.E., Stansley, W. & Niles, L.J. 2001b. Changes in contaminant levels in New Jersey osprey eggs and prey, 1989 to 1998. *Archives of Environmental Contamination and Toxicology* **40**, 277–284.

Cloern, J.E. 2001. Our evolving conceptual model of the coastal eutrophication problem. *Marine Ecology Progress Series* **210**, 223–253.

Cobb, J.N. 1900. The sturgeon fishery of Delaware River and Bay. *Reports of the United States Commission of Fish* **25**, 369–380.

Cohen, A.N. & Carlton, J.T. 1998. Accelerating invasion rate in a highly invaded estuary. *Science* **279**, 555–558.

Cole, K.B., Carter, D.B., Finkbeiner, M. & Seaman, R. 2001. Quantification of the spatial extent of sub-merged aquatic vegetation in Delaware's inland bays using remote sensing. In *Proceedings of the 2nd Biennial Coastal GeoTools Conference*. Charleston, South Carolina: U.S. National Oceanic and Atmospheric Administration.

Coll, M., Palomera, I., Tudela, S. & Sarda, F. 2006. Trophic flows, ecosystem structure and fishing impacts in the South Catalan Sea, Northwestern Mediterranean. *Journal of Marine Systems* **59**, 63–96.

Collins, J.W. 1887. The shore fisheries of southern Delaware. In *The Fisheries and Fishery Industries of the United States*, G.B. Goode (ed.). **1**(5), 527–541. Washington, D.C.: Bureau of Fisheries.

Colman, S.M. & Bratton, J.F. 2003. Anthropogenically induced changes in sediment and biogenic silica fluxes in Chesapeake Bay. *Geology* **31**, 71–74.

Committee on the Status of Endangered Wildlife in Canada (COSEWIC). 2002. *Canadian Species at Risk, May 2002*. Ottawa, Canada: Committee on the Status of Endangered Wildlife in Canada.

Cooper, S.R. 2000. The history of water quality in North Carolina estuarine waters as documented in the strati-graphic record. University of North Carolina Water Resources Research Institute Report **327**. Raleigh, North Carolina: University of North Carolina.

Cooper, S.R. & Brush, G.S. 1993. A 2,500-year history of anoxia and eutrophication in Chesapeake Bay. *Estuaries* **16**, 617–626.

Cooper, S.R., McGlothlin, S.K., Madritch, M. & Jones, D.L. 2004. Paleoecological evidence of human impacts on the Neuse and Pamlico Estuaries of North Carolina, USA. *Estuaries* **27**, 617–633.

Copeland, B.J. & Hobbie, J.E. 1972. Phosphorus and eutrophication in the Pamlico River estuary, N.C. 1966–1969—a summary. *Report 1972–65*. Raleigh, North Carolina: University of North Carolina Water Resources Research Institute.

Costa, J.E. 1988. *Distribution, production, and historical change in abundance of eelgrass (Zostera marina) in southeast Massachusetts*. PhD thesis, Boston University, Boston, Massachusetts, United States.

Costa, J.E. 2003. Eelgrass in Buzzards Bay. East Wareham, MA: Buzzards Bay National Estuary Program. Online. Available HTTP: http://www.buzzardsbay.org/eelgrass.htm (accessed 3 May 2009).

Costanza, R., d'Arge, R., de Groot, R., Farber, S., Grasso, M., Hannon, B., Limburg, K., Naeem, S., O'Neill, R.V., Paruelo, J., Raskin, R.G., Sutton, P. & van den Belt, M. 1997. The value of the world's ecosystem services and natural capital. *Nature* **387**, 253–260.

Cox, I.J. 1906. *The Journeys of René Robert Cavelier Sieur de LaSalle*. New York: Allerton Book.

Craig, P., Green, A. & Tuilagi, F. 2008. Subsistence harvest of coral reef resources in the outer islands of American Samoa: modern, historic and prehistoric catches. *Fisheries Research* **89**, 230–240.

Cronon, W. 1983. *Changes in the Land*. New York: Farrar, Straus and Giroux.

Dahl, T.E. & Allord, G.J. 1999. History of wetlands in the conterminous United States. National Water Summary on Wetland Resources, United States Geological Survey Water Supply Paper 2425. Reston, VA: USGS.

Daily, G.C. 1997. *Nature's Services. Societal Dependence on Natural Ecosystems*. Washington, D.C.: Island Press.

Dayton, P.K., Thrush, S.F. & Coleman, F.C. 2002. *Ecological Effects of Fishing in Marine Ecosystems of the United States*. Arlington, Virginia: Pew Oceans Commission.

Delaware living history. 2009. Wilmington, DE: Delaware Living. Online. Available HTTP: http://www.delawareliving.com/history.html (accessed 10 May 2009).

Delaware state history. 2009. Riverton, UT: Things to Do. Online. Available HTTP: http://www3.thingstodo.com/states/DE/history.htm (accessed 10 May 2009).

Dent, R.J. 1995. *Chesapeake Prehistory: Old Traditions, New Directions*. New York: Plenum Press.

Dulvy, N.K., Sadovy, Y. & Reynolds, J.D. 2003. Extinction vulnerability in marine populations. *Fish and Fisheries* **4**, 25–64.

Eldredge, Z.S. 1912. *The Beginnings of San Francisco. From the Expedition of Anza, 1774, to the City Charter of April 15, 1850*. New York: Rankin.

Enghoff, I.B., MacKenzie, B.R. & Nielsen, E.E. 2007. The Danish fish fauna during the warm Atlantic period (ca. 7000–3900 BC): forerunner of future changes? *Fisheries Research* **87**, 167–180.

Epperly, S.P., Braun, J. & Veishlow, A. 1995. Sea-turtles in North Carolina waters. *Conservation Biology* **9**, 384–394.

Erwin, R.M. 1996. Dependence of waterbirds and shorebirds on shallow-water habitats in the mid-Atlantic coastal region: an ecological profile and management recommendations. *Marine and Estuarine Shallow Water Science and Management* **19**(2A), 213–219.

Feest, C.F. 1978a. Nanticoke and neighboring tribes. In *Handbook of North American Indians*, W.C. Sturtevant (ed.). Washington, D.C.: Smithsonian Institution, 240–252.

Feest, C.F. 1978b. Virginia Algonquins. In *Handbook of North American Indians*, W.C. Sturtevant (ed.). Washington, D.C.: Smithsonian Institution, 271–281.

Fisher, T.R., Hagy, J.D., Boynton, W.R. & Williams, M.R. 2006. Cultural eutrophication in the Choptank and Patuxent estuaries of Chesapeake Bay. *Limnology and Oceanography* **51**, 435–447.

Fitzner, R.E., Blus, L.J., Henny, C.J. & Carlile, D.W. 1988. Organochlorine residues in Great Blue Herons from the northwestern United States. *Colonial Waterbirds* **11**, 293–300.

Fofonoff, P.W., Ruiz, G.M., Steves, B., Hines, A.H. & Carlton J.T. 2003. *National Exotic Marine and Estuarine Species Information System*. Edgewater, MD: Smithsonian Environmental Research Center. Online. HTTP: http://invasions.si.edu/nemesis/ (accessed 10 May 2009).

Ford, S.E. 1997. History and present status of molluscan shellfisheries from Barnegat Bay to Delaware Bay. Silver Spring, MD: NOAA. NOAA Technical Report NMFS **127**, 119–140.

Forstall, R.L. 1996. *Population of States and Counties of the United States: 1790 to 1990*. Washington, D.C.: U.S. Department of Commerce, Bureau of the Census.

Francis, R.C., Field, J., Holmgren, D. & Strom, A. 2001. Historical approaches to the northern California current ecosystem. In *The Exploited Seas: New Directions for Marine Environmental History*, P. Holm et al. (eds). St. John's, Newfoundland, Canada: International Maritime Economic History Association and Census of Marine Life, 123–139.

Friedland K. 1998. Atlantic and shortnose sturgeon. In *Status of the Fishery Resources of the Northeastern United States*, S.H. Clark (ed.). Woods Hole, MA: Resource Evaluation and Assessment Division, Northeast Fisheries Science Center.

Galtsoff, P.S. 1930. Oyster industry of the Pacific coast of the United States. In *Report of the United States Commissioner of Fisheries for the Fiscal Year 1929*. Washington, D.C.: Bureau of Fisheries, 367–400.

Galveston Bay National Estuary Program (GBNEP). 1994. Chapter four: The human role, past and present. In *The State of the Bay: A Characterization of the Galveston Bay Ecosystem*. Publication GBNEP-44. Webster, Texas: Galveston Bay Estuary Program, 39–66.

Galvin, J. 1971. *The First Spanish Entry into San Francisco Bay, 1775*. San Francisco: Howell Books.

Gambell, R. 1999. The International Whaling Commission and the contemporary whaling debate. In *Conservation and Management of Marine Mammals*, J.R. Twiss Jr. & R.R. Reeves (eds). Washington, D.C.: Smithsonian Institution Press, 179–198.

Goddard, I. 1978. Delaware. In *Handbook of North American Indians*, W. Sturtevant (ed.). Washington, D.C.: Smithsonian Institution, 583–587.

Golder, W. 2004. *Important Bird Areas of North Carolina*. Chapel Hill, NC: Audubon Society of North Carolina.

Grave, C. 1905. Investigations for the promotion of the oyster industry of North Carolina. In *Report of the U.S. Commissioner for Fish and Fisheries 1903*. Washington, D.C.: Bureau of Fisheries, 247–341.

Green, A., Osborn, M., Chai, P., Lin, J., Loeffler, C., Morgan, A., Rubec, P., Spanyers, S., Walton, A., Slack, R.D., Gawlik, D., Harpole, D., Thomas, J., Buskey, E., Schmidt, K., Zimmerman, R., Harper, D., Hinkley, D. & Sager, T. 1992. *Status and Trends of Selected Living Resources in the Galveston Bay System*. Publication GBNEP-19. Webster, Texas: Galveston Bay National Estuary Program.

Greenway, J.C., Jr. 1967. *Extinct and Vanishing Birds of the World*. New York: Dover.

Griffiths, C.L., van Sittert, L., Best, P.B., Brown, A.C., Clark, B.M., Cook, P.A., Crawford, R.J.M., David, J.H.M., Davies, B.R., Griffiths, M.H., Hutchings, K., Jerardino, A., Kruger, N., Lamberth, S., Leslie, R., Melville-Smith, R., Tarr, R. & van der Lingen, C.D. 2004. Impacts of human activities on marine animal life in the Benguela: a historical overview. *Oceanography and Marine Biology An Annual Review* **42**, 303–392.

Grinnell, J. & Miller, A.H. 1944. The distribution of the birds of California. Pacific Coast Avifauna **27**. Berkeley, California: Cooper Ornithological Club.

Grinnell, J. & Wythe, M.W. 1927. Directory to the bird life of the San Francisco Bay region. Pacific Coast Avifauna **18**. Berkeley, California: Cooper Ornithological Club.

Hall, A. 1894. Oyster industry of New Jersey. In *Report of the U.S. Commissioner of Fish and Fisheries for 1892*. Washington, D.C.: Bureau of Fisheries, 463–528.

Harley, C.D.G., Hughes, A.R., Hultgren, K.M., Miner, B.G., Sorte, C.J.B., Thornber, C.S., Rodriguez, L.F., Tomanek, L. & Williams, S.L. 2006. The impacts of climate change in coastal marine systems. *Ecology Letters* **9**, 228–241.

Hartman, D.S. 1979. West Indian Manatee. In *Threatened, Rare and Endangered Biota of Florida, Vol. 1, Mammals*, J.N. Layane (ed.). Gainesville, Florida: University Presses of Florida, 27–39.

Harvell, C.D., Mitchell, C.E., Ward, J.R., Altizer, S., Dobson, A.P., Ostfeld, R.S. & Samuel, M.D. 2002. Climate warming and disease risks for terrestrial and marine biota. *Science* **296**, 2158–2162.

Hastings, W.N., O'Herron, J.C., Schick, K. & Lazzari, M.A. 1987. Occurrence and distribution of shortnose sturgeon, *Acipenser brevirostrum*, in the upper tidal Delaware River. *Estuaries* **10**, 337–341.

Héral, M., Rothschild, B.J. & Goulletquer, P. 1990. *Decline of Oyster Production in the Maryland Portion of the Chesapeake Bay: Causes and Perspectives*. Copenhagen: International Council for the Exploration of the Sea (ICES).

Hobbie, J.E., Copeland, J.B. & Harrison, W.G. 1975. Sources and fates of nutrients in the Pamlico River estuary, North Carolina. In *Estuarine Research, Volume 1: Chemistry, Biology and the Estuarine System*, L.E. Cronin (ed.). New York: Academic Press, 287–302.

Hofstetter, R.P. 1982. Gulf Coast region: status of the Texas oyster industry. In *Proceedings of the North American Oyster Workshop*, K.K. Chew (ed.). Baton Rouge, Louisiana: Louisiana State University, 96–100.

Hofstetter, R.P. 1983. Oyster population trends in Galveston Bay 1973–1978. Texas Parks and Wildlife Department Management Data Series **51**. Austin, Texas: Coastal Fisheries Branch.

Hooper, D.U., Chapin, F.S., Ewel, J.J., Hector, A., Inchausti, P., Lavorel, S., Lawton, J.H., Lodge, D.M., Loreau, M., Naeem, S., Schmid, B., Setälä, H., Symstad, A.J., Vandermeer, J. & Wardle, D.A. 2005. Effects of biodiversity on ecosystem functioning: a consensus of current knowledge. *Ecological Monographs* **75**, 3–35.

Houde, E.D. & Rutherford, E.S. 1993. Recent trends in estuarine fisheries: predictions of fish production and yield. *Estuaries* **16**, 161–176.

Huff, B. 1957. *El Puerto de los Balleneros: Annals of the Sausalito Whaling Anchorage*. Los Angeles: Dawson.

Hutchings, J.A. & Reynolds, J.D. 2004. Marine fish population collapses: consequences for recovery and extinction risk. *BioScience* **54**, 297–309.

Ingersoll, E. 1881. The oyster-industry. *The History and Present Condition of the Fishery Industries, Tenth Census of the United States*. Washington, D.C.: U.S. Department of the Interior, 251 pp.

International Marine Mammal Project (IMMP). 2009. *San Francisco Bay Seal Project*. San Francisco, California: Earth Island Institute. Available HTTP: http://www.earthisland.org/immp/seal.html (accessed 31 May 2009).

Iversen, E.S., Higman, A. & Higman, J.B. 1993. *Shrimp Capture and Culture Fisheries of the United States*. New York: Wiley.

Jackson, J.B.C. 2008. Ecological extinction and evolution in the brave new ocean. *Proceedings of the National Academy of Sciences of the United States of America* **105**, 11458–11465.

Jackson, J.B.C., Kirby, M.X., Berger, W.H., Bjorndal, K.A., Botsford, L.W., Bourque, B.J., Bradbury, R.H., Cooke, R., Erlandson, J., Estes, J.A., Hughes, T.P., Kidwell, S., Lange, C.B., Lenihan, H.S., Pandolfi, J.M., Peterson, C.H., Steneck, R.S., Tegner, M.J. & Warner, R.R. 2001. Historical overfishing and the recent collapse of coastal ecosystems. *Science* **293**, 629–638.

Jacobson, L.D., Funk, F.C. & Goiney, B.J. 1999. Unit 14—Pacific Coast and Alaska Pelagic Fisheries. In Our living oceans. Report on the status of U.S. living marine resources, 1999. U.S. Department of Commerce. Silver Spring, MD: NOAA. NOAA Technical Memorandum NMFS-F/SPO-41.

Kappel, C.V. 2005. Losing pieces of the puzzle: threats to marine, estuarine, and diadromous species. *Frontiers in Ecology and the Environment* **3**, 275–282.

Katona, S.K., Rough, V., & Richardson, D.T. 1993. *A Field Guide to Whales, Porpoises, and Seals from Cape Cod to Newfoundland*. Washington, D.C.: Smithsonian Institute Press.

Kellogg, J.L. 1910. *Shell-fish Industries*. New York: Holt.

Kemp, W., Boynton, W., Adolf, J., Boesch, D., Boicourt, W., Brush, G., Cornwell, J., Fisher, T., Glibert, P., Hagy, J., Harding, L., Houde, E., Kimmel, D., Miller, W., Newell, R., Roman, M., Smith, E. & Stevenson, J. 2005. Eutrophication of Chesapeake Bay: historical trends and ecological interactions. *Marine Ecology Progress Series* **303**, 1–29.

Kennish, M.J. (ed.). 2000. *Estuary Restoration and Maintenance: The National Estuary Program*. New York: CRC Press.

Kennish, M.J. 2002. Environmental threats and environmental futures of estuaries. *Environmental Conservation* **29**, 78–107.

Kirby, M.X. 2004. Fishing down the coast: historical expansion and collapse of oyster fisheries along continental margins. *Proceedings of the National Academy of Sciences of the United States of America* **101**, 13096–13099.

Kirk, D. 1985. *Status Report on the Great Auk,* Pinguinis impennis. Ottawa, Canada: Committee of the Status of Endangered Wildlife in Canada.

Knowlton, A.R., Kraus, S.D. & Kenney, R.D. 1994. Reproduction in North Atlantic right whales (*Eubalaena glacialis*). *Canadian Journal of Zoology* **72**, 1297–1305.

Krementz, D.G., Stotts, V.D., Stotts, D.B., Hines, J.E. & Funderburk, S.L. 1991. Historical changes in laying date, clutch size, and nest success of American black ducks. *Journal of Wildlife Management* **55**, 462–466.

Laist, D.W., Knowlton, A.R., Mead, J.G., Collet, A.S. & Podesta, M. 2001. Collisions between ships and whales. *Marine Mammal Science* **17**, 35–75.

LaRoe, E.T., et al. (eds). 1995. *Our Living Resources: A Report to the Nation on the Distribution, Abundance and Health of U.S. Plants and Animals, and Ecosystems.* Washington, D.C.: U.S. Department of the Interior, National Biological Service.

Larson, L.H., Jr. 1970. *Aboriginal subsistence technology on the southeastern coastal plain during the late prehistoric period.* PhD thesis, University of Michigan, Ann Arbor, Michigan.

Lawson, J.D. 1712. *A New Voyage to Carolina. Accounts from 1700–1701 Journey.* Chapel Hill, North Carolina: University of North Carolina Press (1984).

Lenihan, H.S. & Peterson, C.H. 1998. How habitat degradation through fishery disturbance enhances impacts of hypoxia on oyster reefs. *Ecological Applications* **8**, 128–140.

Lenihan, H.S., Peterson, C.H., Byers, J.E., Grabowski, J.H., Thayer, G.W. & Colby, D.R. 2001. Cascading of habitat degradation: oyster reefs invaded by refugee fishes escaping stress. *Ecological Applications* **11**, 748–764.

Lester, J. & Gonzalez, L. 2002. *The State of the Bay—A Characterization of the Galveston Bay Ecosystem.* Webster, Texas: Galveston Bay Estuary Program, 2nd edition.

Limburg, K.E. 1999. Estuaries, ecology, and economic decisions: an example of perceptual barriers and challenges to understanding. *Ecological Economics* **30**, 185–188.

Lotze, H.K. 2005. Radical changes in the Wadden Sea fauna and flora over the last 2000 years. *Helgoland Marine Research* **59**, 71–83.

Lotze, H.K. 2007. Rise and fall of fishing and marine resource use in the Wadden Sea, southern North Sea. *Fisheries Research* **87**, 208–218.

Lotze, H.K. & Glaser, M. 2009. Ecosystem services of semi-enclosed marine systems. In *Watersheds, Bays and Bounded Seas*, E.R. Urban Jr. et al. (eds). Washington, D.C.: Island Press, 227–249.

Lotze, H.K., Lenihan, H.S., Bourque, B.J., Bradbury, R.H., Cooke, R.G., Kay, M.C., Kidwell, S.M., Kirby, M.X., Peterson, C.H. & Jackson, J.B.C. 2006. Depletion, degradation, and recovery potential of estuaries and coastal seas. *Science* **312**, 1806–1809.

Lotze, H.K. & Milewski, I. 2004. Two centuries of multiple human impacts and successive changes in a North Atlantic food web. *Ecological Applications* **14**, 1428–1447.

Lotze, H.K., Reise, K., Worm, B., van Beusekom, J.E.E., Busch, M., Ehlers, A., Heinrich, D., Hoffmann, R.C., Holm, P., Jensen, C., Knottnerus, O.S., Langhanki, N., Prummel, W. & Wolff, W.J. 2005. Human transformations of the Wadden Sea ecosystem through time: a synthesis. *Helgoland Marine Research* **59**, 84–95.

Lotze, H. K. & Worm, B. 2002. Complex interactions of climatic and ecological controls on macroalgal recruitment. *Limnology and Oceanography* **47**, 1734–1741.

Lotze, H.K. & Worm, B. 2009. Historical baselines for large marine animals. *Trends in Ecology & Evolution* **24**, 254–262.

Lowerre-Barbieri, S.K., Chittenden, M.E. & Barbieri, L.R. 1995. Age and growth of weakfish, *Cynoscion regalis*, in the Chesapeake Bay region with a discussion of historical changes in maximum size. *Fishery Bulletin* **93**, 643–656.

Luedtke, B.E. 1980. The Calf Island Site and the late Prehistoric period in Boston Harbor. *Man in the Northeast* **20**, 25–76.

Malecki, R.A., Batt, B.D.J. & Sheaffer, S.E. 2001. Spatial and temporal distribution of Atlantic population Canada geese. *Journal of Wildlife Management* **65**, 242–247.

Mallin, M.A., Burkholder, J.M., Cahoon, L.B. & Posey, M.H. 2000. North and South Carolina Coasts. *Marine Pollution Bulletin* **41**, 56–75.

Manzella, S.A. & Williams, J.A. 1992. The distribution of Kemp's Ridley sea turtles *(Lepidochelys kempi)* along the Texas Coast: an atlas. NOAA Technical Reports NMFS 110. Galveston, Texas: National Marine Fisheries Service.

Margolin, M. 1978. *The Ohlone Way*. Berkeley, California: Heyday Books.

Markowitz, H. 1995. *American Indians*. Pasadena, California: Salem Press.

Massachusetts Bays Program (MBP). 2004. *State of the Bays Report*. Boston: Massachusetts Bays Program.

Massachusetts Division of Fisheries and Wildlife (MDFW). 2009. *Species conservation*. Westborough, MA: MDFW. Online. HTTP: http://www.mass.gov/dfwele/dfw/nhesp/conservation/conservation_home.htm (accessed 9 May 2009).

Massley, H. & Slater, B. 1999. Unit 9—Southeast drum and croaker fisheries. In Our living oceans. Report on the status of U.S. living marine resources, 1999. Silver Spring, MD: NOAA. NOAA Tech. Memo. NMFS-F/SPO-41. U.S. Department of Commerce.

McCay, B.J. 1998. *Oyster Wars and the Public Trust*. Tucson, Arizona: University of Arizona Press.

McClenachan, L. & Cooper, A. B. 2008. Extinction rate, historical population structure and ecological role of the Caribbean monk seal. *Proceedings of the Royal Society London Series B* **275**, 1351–1358.

McClenachan, L., Jackson, J.B.C. & Newman, M.J. 2006. Conservation implications of historic sea turtle nesting beach loss. *Frontiers in Ecology and the Environment* **4**, 290–296.

McDonald, M. 1887. The fisheries of the Delaware River. In *The Fisheries and Fishery Industries of the United States*, G.B. Goode (ed.). Washington, D.C.: Bureau of Fisheries. **1(5)**, 654–657.

McFarlane, R.W. 2001. Population trends of colonial waterbirds in Galveston Bay, Texas. In *Proceedings of the Fifth Galveston Bay Estuary Program State of the Bay Symposium*, C.L.P. Palmer et al. (eds). Galveston Bay Estuary Program Publication GBEP T-5 (CTF 09/01). Webster, Texas.

Mead, J.G. 1975. Preliminary report on the former net fisheries for *Tursiops truncatus* in the western North Atlantic. *Journal of the Fisheries Research Board of Canada* **32**, 1155–1162.

Mead, J.G. & Mitchell, E.D. 1984. Atlantic gray whales. In *The Gray Whale* Eschrichtius robustus, M.L. Jones et al. (eds). Orlando, Florida: Academic Press, 33–53.

Menhaden Research Council (MRC). 2009. *History of the menhaden fishery*. Washington, D.C.: MRC. Online. Available HTTP: http://www.menhaden.org/history.htm (accessed 12 May 2009).

Millennium Ecosystem Assessment (MEA). 2005. *Ecosystems and Human Well-Being: Synthesis*. Washington, D.C.: Island Press.

Miller, M.E., 1971. The Delaware oyster industry. *Delaware History* **14**, 238–254.

Miller, T.J., Martell, S.J.D., Bunnell, D.B., Davis, G., Fegleu, L., Sharov, A., Bonzek, C., Hewitt, D., Hoenig, J. & Lipicius, R.N. 2005. Stock assessment of blue crab in Chesapeake Bay 2005. Final report. Technical Report Series No. TS-487-05 of the University of Maryland Center for Environmental Science, Cambridge, MD.

Mitchell, E. 1975. Porpoise, dolphin and small whale fisheries of the world. Status and problems. *International Union for Conservation of Natural Resources Monograph* **3**, 1–129.

Monroe, M.W. & Kelly, J. 1992. *State of the Estuary: A Report on Conditions and Problems in the San Francisco Bay Sacramento–San Joaquin Delta Estuary*. Oakland, California: San Francisco Bay Estuary Project.

Moore, K.A., Wilcox, D.J., Anderson, B., Parham, T.A. & Naylor, M.D. 2004. *Historical analysis of SAV in the Potomac River and analysis of bay-wide historic SAV to establish a new acreage goal*. Report prepared for the Chesapeake Bay Program (CB983627–01).

Mora, C., Myers, R.A., Coll, M., Libralato, S., Pitcher, T.J., Sumaila, R.U., Zeller, D., Watson, R., Gaston, K.J. & Worm, B. 2009. Management effectiveness of the world's marine fisheries. *PLOS Biology*, **7** (6): e1000131.

Murawski, S.A., Brown, R.W., Cadrin, S.X., Mayo, R.K., O'Brien, L., Overholtz, W.J. & Sosebee, K.A. 1999. New England groundfish. In *Our living oceans. Report on the status of U.S. living marine resources*. Silver Spring, MD: NOAA. Tech. Memo. NMFS-F/SPO-41.

Myers, R.A., Baum, J.K., Shepherd, T.D., Powers, S.P. & Peterson, C.H. 2007. Cascading effects of the loss of apex predatory sharks from a coastal ocean. *Science* **315**, 1846–1850.

Myers, R.A. & Worm, B. 2005. Extinction, survival or recovery of large predatory fishes. *Philosophical Transactions of the Royal Society London B: Biological Sciences* **360**, 13–20.

Nance, J.M. & Harper, D. 1999. Unit 11—Southeast and Caribbean invertebrate fisheries. In *Our living oceans. Report on the status of U.S. living marine resources, 1999*. Silver Spring, MD: NOAA. Tech. Memo. NMFS-F/SPO-41. U.S. Department of Commerce.

National Audubon Society (NAS). 1997a. *Colonial Waterbirds of North Carolina—Waterbirds Index*. National Audubon Society, North Carolina Sanctuaries. Wilmington, NC: NAS. Online. Available HTTP: http://www.audubon.org/chapter/nc/nc/waterbirds_nest.html (accessed 19 May 2009).

National Audubon Society (NAS). 1997b. *The Colonial Waterbirds of North Carolina: Making It on a Wing and a Prayer*. National Audubon Society, North Carolina Sanctuaries. Wilmington, NC: NAS. Online. Available HTTP: http://www.audubon.org/chapter/nc/nc/nc_waterbirds.html (accessed 19 May 2009).

National Marine Fisheries Service (NMFS). 2009. *Fisheries Statistics and Economics*. Online. Available HTTP: http://www.st.nmfs.gov/st1/commercial/landings/annual_landings.html (accessed 10 May 2009).

Newell, R.I.E. (1988). Ecological changes in Chesapeake Bay: are they the result of overharvesting the American oyster, *Crassostrea virginica*? In *Understanding the Estuary. Advances in Chesapeake Bay Research*. Boca Raton, Florida: Chesapeake Bay Research Consortium, CRC Publication **129**, 536–546.

Nichols, F.H., Cloern, J.E., Luoma, S.N. & Peterson, D.H. 1986. The modification of an estuary. *Science* **231**, 567–573.

North Carolina Division of Marine Fisheries (NCDMF). 2004. *Stock Status of Important Coastal Fisheries in North Carolina, 2004*. Morehead City, NC: North Carolina Department of Environment and Natural Resources, Division of Marine Fisheries.

North Carolina Division of Marine Fisheries (NCDMF). 2008. *Stock Status of Important Coastal Fisheries in North Carolina, 2008*. Morehead City, NC: North Carolina Department of Environment and Natural Resources, Division of Marine Fisheries.

Northeast Fisheries Science Center (NEFSC). 1997. *Bottlenose dolphin (*Tursiops truncatus*): Western North Atlantic Coastal Stock*. NEFSC Publication, Woods Hole, MA.

Northeast Fisheries Science Center (NEFSC). 2005. *Bottlenose dolphin (*Tursiops truncatus*): Western North Atlantic Coastal Morphotype Stocks*. NEFSC Publication, Woods Hole, MA.

Office of Protected Resource (OPR). 2009. *Marine Turtles*. NOAA Fisheries Service. Silver Spring, MD: NOAA. Online. Available HTTP: http://www.nmfs.noaa.gov/prot_res/PR3/Turtles/turtles.html (accessed 12 May 2009).

Ogden, A. 1941. *The California Sea Otter Trade 1784–1848*. University of California Publications in History **26**. Berkeley, CA: University of California Press.

Olofson, P.R. (ed.). 2000. *Baylands Ecosystem Species and Community Profiles: Life Histories and Environmental Requirements of Key Plants, Fish and Wildlife*. Prepared by the San Francisco Bay Area Wetlands Ecosystem Goals Project. Oakland, California: San Francisco Bay Regional Water Quality Control Board.

Orth, R.J., Carruthers, T.J.B., Dennison, W.C., Duarte, C.M., Fourqurean, J.W., Heck, K.L., Jr., Hughes, A.R., Kendrick, G.A., Kenworthy, W.J., Olyarnik, S., Short, F.T., Waycott, M. & Williams, S.L. 2006. A global crisis for seagrass ecosystems. *BioScience* **56**, 987–996.

Orth, R.J. & Moore, K.A. 1983. An unprecedented decline in submerged aquatic vegetation (Chesapeake Bay). *Science* **22**, 51–53.

Orth, R.J. & Moore, K.A. 1984. Distribution and abundance of submerged aquatic vegetation in Chesapeake Bay: an historical perspective. *Estuaries* **7**, 531–540.

Orth, R.J., Nowak, J.F. & Wilcox, D.J. 2003. Distribution and abundance of submerged aquatic vegetation in the Chesapeake Bay and tributaries and the coastal bays—2002. VIMS Special Scientific Report 139. Gloucester Point, VA: Virginia Institute of Marine Science.

Packard, E.L. 1918. *The molluscan fauna from San Francisco Bay*. University of California Publications in Zoology, **14**. Berkeley, CA: University of California Press.

Paerl, H.W., Dennis, R.L. & Whitall, D.R. 2002. Atmospheric deposition of nitrogen: implications for nutrient overenrichment of coastal waters. *Estuaries* **25**, 677–693.

Palmer, R.S. 1962. *Handbook of North American Birds, Vol. 1: Loons through Flamingos*. New Haven, Connecticut: Yale University Press.

Palumbi, S.R. 2010. *Monterey Bay Reborn*. Washington, D.C.: Island Press.

Palumbi, S.R., McLeod, K.L. & Grünbaum, D. 2008. Ecosystems in action: lessons from marine ecology about recovery, resistance and reversibility. *Bioscience* **58**, 33–41.

Pandolfi, J.M., Bradbury, R.H., Sala, E., Hughes, T.P., Bjorndal, K.A., Cooke, R.G., McArdle, D., McClenachan, L., Newman, M.J.H., Paredes, G., Warner, R.R. & Jackson, J.B.C. 2003. Global trajectories of the long-term decline of coral reef ecosystems. *Science* **301**, 955–958.

Pauly, D. 1995. Anecdotes and the shifting baseline syndrome of fisheries. *Trends in Ecology and Evolution* **10**, 430.

Pauly, D., Alder, J., Bennett, E., Christensen, V., Tyedmers, P. & Watson, R. 2003. The future for fisheries. *Science* **302**, 1359–1361.

Pauly, D., Christensen, V., Dalsgaard, J., Froese, R. & Torres, F., Jr. 1998. Fishing down marine food webs. *Science* **279**, 860–863.

Pearson, J.C. (ed.). 1972. *The Fish and Fisheries of Colonial North America Part III. The Middle Atlantic States*. Washington, D.C.: US Fish and Wildlife Service.

Pearson, J.C. 1942. The fish and fisheries of colonial Virginia. *William and Mary College Quarterly Historical Magazine* **22**, 213–220.

Pendleton, E. 1995. Natural resources in the Chesapeake Bay Watershed. In *Our Living Resources: A Report to the Nation on the Distribution, Abundance, and Health of U.S. Plants, Animals, and Ecosystems*, E.T. LaRoe et al. (eds). Washington, D.C.: U.S. Department of the Interior, National Biological Service, 263–267.

Perry, S.L., DeMaster, D.P. & Silber, G.K. 1999. The great whales: history and status of six species listed as endangered under the U.S. Endangered Species Act of 1973. *Marine Fisheries Review Special Issue* **61**, 1–74.

Peterson, C.H., Summerson, H.C. & Fegley, S.R. 1987. Ecological consequences of mechanical harvesting of clams. *Fishery Bulletin* **85**, 281–298.

Peterson, C.H., Summerson, H.C. & Luettich, R. 1996. Response of bay scallops to spawner transplants: a test of recruitment limitation. *Marine Ecology Progress Series* **132**, 93–107.

Pew Oceans Commission. 2003. *America's Living Oceans: Charting a Course for Sea Change*. Arlington, Virginia: Pew Oceans Commission.

Pinnegar, J.K. & Engelhard, G.H. 2007. The "shifting baseline" phenomenon: a global perspective. *Reviews in Fish Biology and Fisheries* **18**, 1–16.

Pitcher, T. J., Heymans, J. J. S. & Vasconcellos, M. 2002. Ecosystem models of Newfoundland for the time periods 1995, 1985, 1900 and 1450. *Fisheries Centre Research Report* **10**(5), 76 pp.

Prescott, J.H., Fiorelli, P.M. 1980. *Review of the Harbor Porpoise* (Phocoena phocoena) *in the U.S. North Atlantic*. Final Report to U.S. Marine Mammal Commission, MMC-78/08. Washington, D.C.: Marine Mammal Commission.

Pulich, W.M., Jr. & Hinson, J. 1996. *Coastal Studies Technical Report No. 1: Development of Geographic Information System Data Sets on Coastal Wetlands and Land Cover*. Austin, Texas: Texas Parks and Wildlife Department.

Pulich, W.M., Jr. & White, W.A. 1991. Decline of submerged vegetation in the Galveston Bay system: chronology and relationship to physical processes. *Journal of Coastal Research* **4**, 1125–1138.

Rattner, B.A., Hoffman, D.J., Melancon, M.J., Olsen, G.H., Schmidt, S.R. & Parsons, K.C. 2000. Organochlorine and metal contaminant exposure and effects in hatching black-crowned night herons (*Nycticorax nycticorax*) in Delaware Bay. *Archives of Environmental Contamination and Toxicology* **39**, 38–45.

Ray, G.L. 2005. Invasive estuarine and marine animals of the North Atlantic. ANSRP Technical Notes Collection (ERDC/TN ANSRP-05-1), Vicksburg, MS: U.S. Army Engineer Research and Development Center.

Read, A.J. 1994. Interactions between cetaceans and gillnet and trap fisheries in the Northwest Atlantic. *Report of the International Whaling Commission Special Issue* **15**, 133–147.

Read, A.J., Foster, B., Urian, K., Wilson, B. & Waples, D. 2003. Abundance of bottlenose dolphins in the bays, sounds, and estuaries of North Carolina. *Marine Mammal Science* **19**, 59–73.

Reeves, R.R. 2001. Overview of catch history, historic abundance and distribution of right whales in the western North Atlantic and in Cintra Bay, West Africa. *Journal of Cetacean Research and Management Special Issue* **2**, 187–192.

Reeves, R.R., Breiwick, J.M. & Mitchell, E.D. 1999. History of whaling and estimated kill of right whales, *Balaena glacialis*, in the Northeastern United States, 1620–1924. *Marine Fisheries Review* **61**, 1–36.

Reeves, R.R., Mead, J.G. & Katona, S. 1978. The right whale, *Eubalaena glacialis*, in the Western North Atlantic. *Report to the International Whaling Commission* **28**, 303–312.

Reeves, R.R. & Mitchell, E. 1983. Yankee whaling for right whales in the North Atlantic Ocean. *Whalewatcher* **17**, 3–8.

Reeves, R.R. & Mitchell, E. 1988. History of whaling in and near North Carolina. NOAA Technical Report NMFS 65. Rockville, Maryland: U.S. Department of Commerce.

Reshetiloff, K. 1997. Bay's bald eagles making comeback; threats still loom. Alliance for the Chesapeake Bay, *Bay Journal* 7(5). Online. Available HTTP: http://www.bayjournal.com/article.cfm?article=1958 (accessed 12 January 2010).

Rick, T.C. & Erlandson, J.M. 2008. *Human Impacts on Ancient Marine Ecosystems: A Global Perspective.* Berkeley, California: University of California Press.

Roberts, C.M. 2003. Our shifting perspectives on the oceans. *Oryx* **37**, 166–177.

Rogers, J.B. & Builder, T.L. 1999. Unit 15—Pacific coast groundfish fisheries. In *Our living oceans. Report on the status of U.S. living marine resources, 1999.* U.S. Department of Commerce, Silver Spring, MD: NOAA. Technical Memorandum NMFS-F/SPO-41.

Rose, G.A. 2004. Reconciling overfishing and climate change with stock dynamics of Atlantic cod (*Gadus morhua*) over 500 years. *Canadian Journal of Fisheries and Aquatic Sciences* **61**, 1553–1557.

Rosenberg, A.A., Bolster, W.J., Alexander, K.E., Leavenworth, W.B., Cooper, A.B. & MacKenzie, B.R. 2005. The history of ocean resources: modeling cod biomass using historical records. *Frontiers in Ecology and the Environment* **3**, 84–90.

Rosenberg, A.A., Swasey, J.H. & Bowman, M. 2006. Rebuilding US fisheries: progress and problems. *Frontiers in Ecology and the Environment* **4**, 303–308.

Rothschild, B.J., Ault, J.S., Goulletquer, P. & Héral, M. 1994. Decline of the Chesapeake Bay oyster population: a century of habitat destruction and overfishing. *Marine Ecology Progress Series* **111**, 29–39.

Rugh, D., Hobbs, R.C., Lerczak, J.A. & Breiwick, J.M. 2005. Estimates of abundance of the eastern North Pacific stock of gray whales (*Eschrichtius robustus*) 1997–2002. *Journal of Cetacean Research and Management* **7**, 1–12.

Rugolo, L.J., Knotts, K.S., Lange, A.M. 1998a. Historical profile of the Chesapeake Bay blue crab (*Callinectes sapidus* Rathbun) fishery. *Journal of Shellfish Research* **17**, 383–394.

Rugolo, L.J., Knotts, K.S., Lange A.M., & Crecco, V.A. 1998b. Stock assessment of Chesapeake Bay blue crab (*Callinectes sapidus* Rathbun). *Journal of Shellfish Research* **17**, 493–518.

Ruiz, G.M., Carlton, J.T., Grosholz, E.D. & Hines, A.H. 1997. Global invasions of marine and estuarine habitats by non-indigenous species: mechanisms, extent, and consequences. *American Zoologist* **37**, 621–632.

Russell, H.S. 1976. *A Long, Deep Furrow: Three Centuries of Farming in New England.* Hanover, New Hampshire: University Press of New England.

Sáenz-Arroyo, A., Roberts, C.M., Torre, J., Cariño-Olvera, M. & Enríquez-Andrade, R.R. 2005. Rapidly shifting environmental baselines among fishers of the Gulf of California. *Proceedings of the Royal Society London Series B* **272**, 1957–1962.

Sáenz-Arroyo, A., Roberts, C.M., Torre, J., Cariño-Olvera, M. & Hawkins, J.P. 2006. The value of evidence about past abundance: marine fauna of the Gulf of California through the eyes of 16th to 19th century travellers. *Fish and Fisheries* **7**, 128–146.

San Francisco Bay Institute (SFBI). 2003. San Francisco Bay water quality index. *The Bay Institute Ecological Scorecard*, October 17, 1–21.

San Francisco Bay Institute (SFBI). 2004. San Francisco Bay (Suisun Bay) food web index. *The Bay Institute Ecological Scorecard*, April 6, 1–31.

San Francisco News Letter (SFNL). 1925. *From the 1820's to the Gold Rush.* Museum of the City of San Francisco, 2002. Online. Available HTTP: http://www.sfmuseum.org/hist1/early.html (accessed 31 May 2009).

Scavia, D., Field, J.C., Boesch, D.F., Buddemeier, R.W., Burkett, V., Cayan, D.R., Fogarty, M., Harwell, M.A., Howarth, R.W., Mason, C., Reed, D.J., Royer, T.C., Sallenger, A.H. & Titus, J.G. 2002. Climate change impacts on U.S. coastal and marine ecosystems. *Estuaries* **25**, 149–164.

Schoenherr, A.A. 1992. *A Natural History of California.* Berkeley, California: University of California Press.

Secor, D.H., Niklitschek, E.J., Stevenson, J.T., Gunderson, T.E., Minkkinen, S.P., Richardson, B., Florence, B., Mangold, M., Skjeveland, J., & Henderson-Arzapalo, A. 2000. Dispersal and growth of yearling Atlantic sturgeon, *Acipenser oxyrinchus*, released into Chesapeake Bay. *Fishery Bulletin* **98**, 800–810.

Secor, D.H. & Waldman, J.R. 1999. Historical abundance of Delaware Bay Atlantic sturgeon and potential rate of recovery. *American Fisheries Society Symposium* **23**, 203–216.

Serie, J.R., Luszcz, D. & Rafrovich, R.V. 2002. Population trends, productivity, and harvest of eastern population tundra swans. *Waterbirds* **25**, Supplement 1, 32–36.

Shaw, W.N. 1997. The shellfish industry of California—past, present and future. In *The History, Present Condition, and Future of the Molluscan Fisheries of North and Central America and Europe*, C.L. MacKenzie et al. (eds). Seattle, Washington: NOAA. Technical Report NMFS 128:57–74.

Skinner, J., Durand, B., Skinner, T., Snow-Cotter, S. & Babb-Brott, D. 1995. Shellfish bottom and off-bottom culture. *Massachusetts Aquaculture White Paper.* Boston: Massachusetts Office of Coastal Zone Management.

Skinner, J. E. 1962. An historical review of the fish and wildlife resources of the San Francisco Bay Area. California Department of Fish and Game Special Projects Branch Report 1. Sacramento, CA: California Dept. of Fish and Game.

Smith, H.M. 1907. *The Fishes of North Carolina, Volume II.* Raleigh, North Carolina: North Carolina Geological Survey.

Smith, J. 1910. *Travels and Works of Captain John Smith: President of Virginia and Admiral of New England, 1580–1631,* E. Arbor (ed.). Edinburgh: Grant.

Smith, S.E. & Kato, S. 1979. The fisheries of San Francisco Bay: past present and future. In *San Francisco Bay: The Urbanized Estuary,* T.J Conomos (ed.). San Francisco, California: Pacific Division, American Association for the Advancements of Science, 445–468.

Smith, T.D. 1994. *Scaling Fisheries: The Science of Measuring the Effects of Fishing 1855–1955.* Cambridge, U.K.: Cambridge University Press.

Snow, D.R. 1978. Eastern Abenki. In *Handbook of North American Indians, Vol. 15,* B.G. Trigger (ed.). Washington, D.C.: Smithsonian Institution, 137–145.

Snow, D.R. 1980. *The Archaeology of New England.* New York: Academic Press.

Stachowicz, J.J., Terwin, J.R., Whitlatch, R.B. & Osman, R.W. 2002. Linking climate change and biological invasions: ocean warming facilitates nonindigenous species invasions. *Proceedings of the National Academy of Sciences of the United States of America* **99**, 15497–15500.

Stanley, D.W. 1992a. Historical trends: water quality and fisheries, Albemarle-Pamlico Sound, with emphasis on the Pamlico River Estuary. North Carolina Sea Grant College Program Publication UNC-SG-92-04. Greenville, North Carolina: Institute for Coastal Marine Resources, East Carolina University.

Stanley, D.W. 1992b. Historical trends: water quality and fisheries, Galveston Bay. University of North Carolina Sea Grant College Program Publication UNC-SG-92-03. Greenville, North Carolina: Institute for Coastal and Marine Resources, East Carolina University.

Starkey, D.J., Holm, P. & Barnard, M. 2008. *Oceans Past: Management Insights from the History of Marine Animal Populations.* London: Earthscan.

Steel, J. 1991. Albemarle-Pamlico estuarine system: technical analysis of status and trends. Albemarle-Pamlico Estuarine Study Report 91-01. Raleigh, North Carolina: Environmental Protection Agency National Estuary Program.

Steele, J.H. & Schumacher, M. 2000. Ecosystem structure before fishing. *Fisheries Research* **44**, 201–205.

Steneck, R.S. 1997. Fisheries-induced biological changes to the structure and function of the Gulf of Maine ecosystem. In *Proceedings of the Gulf of Maine Ecosystems Dynamics Symposium and Workshop, RARGOM Report 91–1.* G.T. Wallace & E.F. Braasch (eds). Hanover, New Hampshire: Regional Association for Research in Gulf of Maine, 151–165.

Stevenson, C.H. 1893. Report on the coast fisheries of Texas. In *Report U.S. Commissioner for Fish and Fisheries 1889–1891,* Washington, D.C.: Bureau of Fisheries, 373–420.

Strachey, W. 1612. *The Historie of Travaile into Virginia Britannia.* London: Hakluyt Society (1849).

Sullivan, J.K. 1994. Habitat status and trends in the Delaware estuary. *Coastal Management* **22**, 49–79.

Sultzman, L. 2000. *Delaware history.* Online. Available HTTP: http://www.tolatsga.org/dela.html (accessed 20 May 2009).

Swartz, S., Palka, D.L., Waring, G.T. & Clapham, P.J. 1999. Unit 24—Marine mammals of the Atlantic region and the Gulf of Mexico. In *Our living oceans. Report on the status of U.S. living marine resources, 1999.* Silver Spring, MD: NOAA. Tech. Memo. NMFS-F/SPO-41. U.S. Department of Commerce.

Swingle, W.M., Barco, S.G., Pitchford, T.D., McLellan, W.A. & Pabst, D.A. 1993. Appearance of juvenile humpback whales feeding in the nearshore waters of Virginia. *Marine Mammal Science* **9**, 309–315.

Texas Environmental Profiles (TEP). 2009. *Wetlands: Critical Habitats*. Texas Environmental Profiles, Environmental Defense and Texas Center for Policy Studies, Austin, TX. Online. Available HTTP: http://www.texasep.org/html/wld/wld_5wet.html (accessed 21 May 2009).

The Bay Institute (TBI). 1998. *From the Sierra to the Sea: The Ecological History of the San Francisco Bay-Delta Watershed*. San Rafael, California: The Bay Institute of San Francisco.

Tucker, W.J. 1929. *Review of Texas Wildlife and Conservation*. Austin, Texas: Game, Fish and Oyster Commission.

Ubelaker, D.H. & Curtin, P.D. 2001. Human biology of populations in the Chesapeake watershed. In *Discovering the Chesapeake: The History of an Ecosystem*, P.D. Curtin et al. (eds). Baltimore: Johns Hopkins University Press, 127–148.

U.S. Commission on Ocean Policy. 2004. *An Ocean Blueprint for the 21st Century*. Washington, D.C.: U.S. Commission on Ocean Policy.

U.S. Fish and Wildlife Service (USFWS). 2009. *Endangered Species Program*. Washington, D.C.: U.S. Fish and Wildlife Service. Online. Available HTTP: http://www.fws.gov/Endangered/wildlife.html (accessed 15 May 2009).

U.S. Fish and Wildlife Service (USFWS). 2008. *West Indian Manatee (*Trichechus manatus*) Factsheet*. Washington, D.C.: U.S. Fish and Wildlife Service. Online. Available HTTP: http://www.fws.gov/endangered/factsheets/manatee.pdf (accessed 13 May 2009).

U.S. Geological Survey (USGS). 2009a. National water summary on wetland resources. United States Geological Survey Water Supply Paper **2425**. Washington, D.C.: U.S. Geological Survey.

U.S. Geological Survey (USGS). 2009b. *NAS—Nonindigenous Aquatic Species*. Washington, D.C.: U.S. Geological Survey. Online. Available HTTP: http://nas.er.usgs.gov/queries/default.asp (accessed 20 May 2009).

U.S. Geological Survey (USGS). 1996. News release: Update on Chessie the manatee. July 17. Washington, D.C.: U.S. Department of the Interior.

van Geen, A. & Luoma, A. N. 1999. The impact of human activities on sediments of San Francisco Bay, California: an overview. *Marine Chemistry* **64**, 1–6.

Veit, R.R. & Petersen, W.R. 1993. *Birds of Massachusetts*. Lincoln, Massachusetts: Massachusetts Audubon Society.

Vermeij, G.J. 1993. Biogeography of recently extinct marine species: implications for conservation. *Conservation Biology* **7**, 391–397.

Virginia Institute for Marine Science (VIMS). 2009. *Sea Turtle Stranding Program*. Department of Fisheries Science, Gloucester Point, VA: VIMS. Online. Available HTTP: http://www.fisheries.vims.edu/turtle-tracking/stsp.html (accessed 12 May 2009).

Vitousek, P.M., Mooney, H.A., Lubchenco, J. & Melillo, J.M. 1997. Human domination of Earth's ecosystems. *Science* **277**, 494–499.

von Chamisso, A. 1822. *A Sojourn at San Francisco Bay 1816*. San Francisco, CA: The Book Club of California 1936.

Wade, P.R. 2002. A Bayesian stock assessment of the eastern Pacific gray whale using abundance and harvest data from 1967–1996. *Journal of Cetacean Research and Management* **4**, 85–98.

Walls, E.A., Berkson, J. & Smith, S.A. 2002. The horseshoe crab, *Limulus polyphemus*: 200 million years of existence, 100 years of study. *Reviews in Fisheries Science* **10**, 39–73.

Wang, S.D.H. & Rosenberg, A.A. 1997. U.S. New England groundfish management under the Magnuson–Stevens Fishery Conservation and Management Act. *Marine Resource Economics* **12**, 361–366.

Ward, G.H. 1993. Dredge and fill activities in Galveston Bay. Galveston Bay National Estuary Program Publication GBNEP-28. Webster, Texas.

Ward, N. 1995. *Stellwagen Bank: A Guide to the Whales, Seabirds, and Marine Life of the Stellwagen Bank National Marine Sanctuary*. Camden, Maine: Downeast Books.

Waring, G.T., Quintal, J.M. & Swartz, S.L. 2001. U.S. Atlantic and Gulf of Mexico marine mammal stock assessments 2001. NOAA Technical Memorandum NMFS-NE-168. Woods Hole, Massachusetts: NOAA.

Watson, J.F. 1868. *Annals of Philadelphia and Pennsylvania, in the Olden Time*. Philadelphia: Lippincott.

Wharton, J. 1957. *The Bounty of the Chesapeake: Fishing in Colonial Virginia*. Williamsburg, VA: Virginia 350th Anniversary Celebration Corp.

White, W.A., Tremblay, T.A., Wermund, E.G., Jr. & Handley, L.R. 1993. Trends and status of wetland and aquatic habitats in the Galveston Bay system, Texas. Galveston Bay National Estuary Program, Publ. GBNEP-31. Webster, Texas.

Wieland, R. 2007. *Managing Oyster Harvests in Maryland's Chesapeake Bay*. Annapolis, MD: NOAA Chesapeake Bay Office, Non-native Oyster Research Program.

Wilbur, A.R. 2009. *Spotlight on Eelgrass: A Species and Habitat at Risk*. Massachusetts Office of Coastal Zone Management, Boston, MA. Online. Available HTTP: http://www.mass.gov/czm/coastlines/2004–2005/habitat/e_grass.htm (accessed 2 May 2009).

Wood, W. 1634. *New England's Prospect*. Printed as Wood's *New England's Prospect* (1865). Boston: Wilson and Son, Publications of the Prince Society.

Worm, B., Barbier, E.B., Beaumont, N., Duffy, J.E., Folke, C., Halpern, B.S., Jackson, J.B.C., Lotze, H.K., Micheli, F., Palumbi, S.R., Sala, E., Selkoe, K.A., Stachowicz, J.J. & Watson, R. 2006. Impacts of biodiversity loss on ocean ecosystem services. *Science* **314**, 787–790.

Worm, B. & Duffy, J.E. 2003. Biodiversity, productivity and stability in real food webs. *Trends in Ecology and Evolution* **18**, 628–632.

Worm, B., Hilborn, R., Baum, J.K., Branch, T.A., Collie, J.S., Costello, C., Fogarty, M.J., Fulton, E.A., Hutchings, J.A., Jennings, S., Jensen, O.P., Lotze, H.K., Mace, P.M., McClanahan, T.R., Minto, C., Palumbi, C.R., Parma, A.M., Ricard, D., Rosenberg, A.A., Watson, R. & Zeller, D. 2009. Rebuilding global fisheries. *Science*, **325**, 578–585.

Worm, B., Lotze, H.K., Hillebrand, H. & Sommer, U. 2002. Consumer versus resource control of species diversity and ecosystem functioning. *Nature* **417**, 848–851.

Wright, D.A. & Phillips, D.J.H. 1988. Chesapeake and San Francisco Bays: a study in contrasts and parallels. *Marine Pollution Bulletin* **19**, 405–413.

Zimmerman, A.R. 2000. *Organic matter composition of sediments and the history of eutrophication and anoxia in the mesohaline Chesapeake Bay*. Ph.D. dissertation, Virginia Institute of Marine Sciences, College of William and Mary, Gloucester Point, Virginia.

AUTHOR INDEX

References to complete articles are given in **bold type**, references to sections of article are in *italics* and references to pages are given in regular type.

A

Abbiati, M., 200
   See Corriero, G., 202
   See Costantini, F., 203
   See Santangelo, G., 209
Abbott, I.A., 34
Abdelmajid, D., 200
Abernathy, K. *See* Parrish, F.A., 208
Able, K.W. *See* Beck, M.W., 35, 138, 326
Abrahamsson, M. *See* Gislén, A., 204
Adaime, R.R. *See* Schaeffer-Novelli, Y., 154
Adam, P. *See* Mitchell, M.L., 150
Adams, A.J. *See* Dalgren, C.P., 141
Adams, J.B., 137
Adkins, B.E. *See* Breen, P.A., 35
Adkins, J.F., 200
   See Thresher, R., 210
Adolf, J. *See* Kemp, W., 330
Afzal Rafii, Z. *See* Dodd, R.S., 142
Agegian, C. *See* Foster, M.S., 37
Agnew, D.J. *See* Caddy, J.F., 327
Agriculture and Natural Resources Communication
   Services (ANR), 325
Aharon, P. *See* Roberts, H.H., 261
Ahn, O., 34
Ainsworth, D., 325
Airoldi, L., 325
Alayse, A.M. *See* Childress, J.J., 253
Alberic, P., 251
Alberte, R.S. *See* Koehl, M.A.R., 38
Albrecht, D. *See* Markert, S., 259
Albright, L.J., 137
Alder, J. *See* Pauly, D., 334
Alderslade, P. *See* Fabricius, K., 204
Alexander, K.E. *See* Rosenberg, A.A., 335
Alfaro, A.C., 137
Allain, V. *See* Hall-Spencer, J., 205
Allaway, W.B. *See* Goulter, P.F.E., 144
Allaway, W.G. *See* Clarke, P.J., 140
   See Curran, M., 141
   See Hovenden, M.J., 146
Allee, W.C., 200
Allemand, D. *See* Cvejic, J., 203
Allen, E.E. *See* Robidart, J.C., 261
Allen, H. *See* Tracey, D.M., 210
Allen, J.A. *See* Krauss, K.W., 147
Allen, J.R.L., 137
Allison, M.A., 137
Allord, G.J. *See* Dahl, T.E., 328
Ally, J.R. *See* Gotshall, D.W., 37
Alongi, D.M., 137

Altenberger, A. *See* Swales, A., 156
Alter, S.E., 325
Altizer, S. *See* Harvell, C.D., 330
Amann, R. *See* Boetius, A., 252
Amend, M. *See* Fox, D., 37
Amos, C.L. Green, M.O., 145
Amsler, C.D., 34
   See Brzezinski, M.A., 35
   See Reed, D.C., 40
Andaloro, F., 200
Andersen, A.C., 251
   See De Cian, M.-C., 254
   See Sanchez, S., 262
Andersen, R.A. *See* Coyer, J.A., 36
Andersen, T.J., 137
Anderson, B. *See* Moore, K.A., 332
Anderson, C.J. *See* Mitsch, W.J., 150
Anderson, E. *See* Cary, S.C., 253
Anderson, E.D., 325
Anderson, S.H. *See* Deng, Y., 142
Anderson, T.W. *See* Steele, M.A., 40
Andras, J.P. *See* Freytag, J.K., 255
Andrews, A.H., 200
   See Love M.S., 207
Andrews, H.L., 34
Anger, K. *See* Duke, N.C., 142
Angiolillo, M., 200
Animal Welfare Institute (AWI), 325
Anthony, E.J. *See* Gratiot, N., 145
Antrim, L.D., 35
Aperghis, A.B. *See* Govenar, B., 256
Arai, M.N. *See* Cairns, S.D., 202
Aranguren, M. *See* Tsounis, G., 210
Araújo da Silva, C., 137
Areki, F. *See* Gilman, E., 144
Arena, P., 200
Arkema, K.K. *See* Gutiérrez, J.L., 205
Armenteros, M., 137
Armonies, W., 137
Arnaud-Haond, S., 137
Arndt, C., 251
   See Felbeck, H., 254
   See Markert, S., 259
   See Sorgo, A., 263
Arnott, S.A. *See* Conover, D.O., 202
Arntz, W.E., 200
   See Tsounis, G., 210
Arp, A.J., 251
   See Childress, J.J., 253
   See Fisher, C.R., 254
   See Julian, D., 258
Arthur, M.A. *See* Cordes, E.E., 253

C

H

M

SYSTEMATIC INDEX

References to complete articles are given in **bold** type, references to sections of article are in *italics* and references to pages are given in regular type.

SUBJECT INDEX

References to complete articles are given in **bold** type, references to sections of article are in *italics* and references to pages are given in regular type.

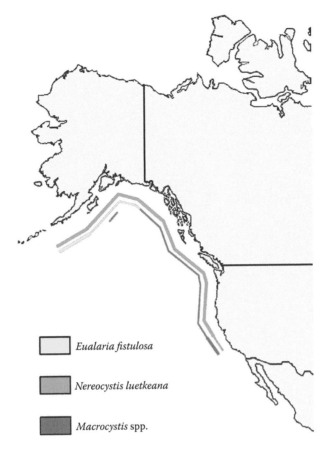

Colour Figure 3 (Springer et al.) Geographic distribution of *Nereocystis luetkeana* indicating areas of co-occurrence with two other surface canopy-forming kelps: giant kelp *Macrocystis* spp. and *Eualaria* (formally *Alaria*). Distributional patterns based on personal communications with M. Foster, M. Graham, B. Konar, and S. Lindstrom. Line width proportional to levels of relative abundance across the range of the species.

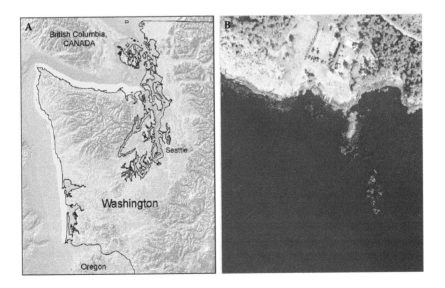

Colour Figure 5 (Springer et al.) (A) Study area for long-term monitoring of canopy-forming kelp in Washington State conducted by the Department of Natural Resources. (B) Colour-infrared imagery collected by areal surveys. Floating kelp canopies appear as red areas on the dark water surface. Photo interpretation is used to classify red floating kelp as canopy area. Bed area is delineated by grouping classified kelp canopies with a distance threshold of 25 m. (Courtesy of Helen D. Berry, Nearshore Habitat Program, Washington State Department of Natural Resources.)

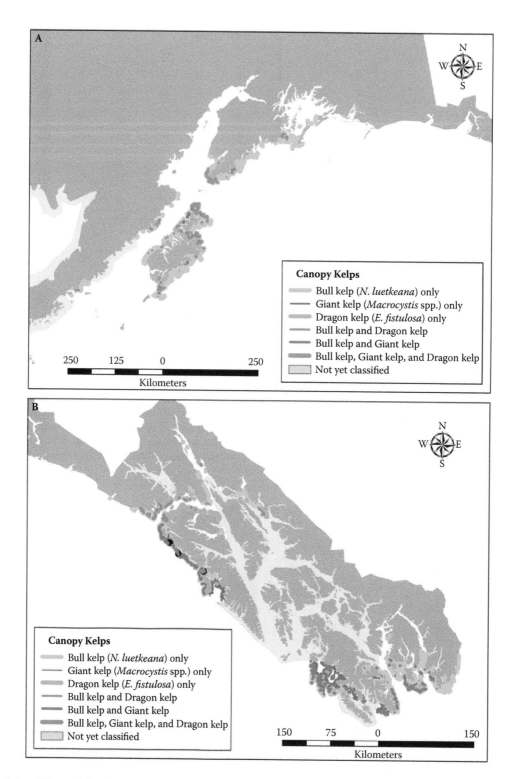

Colour Figure 6 (Springer et al.) Canopy kelp distribution of *Nereocystis luetkeana*, *Macrocystis* spp., and *Eualaria fistulosa* in (A) Gulf of Alaska and (B) south-east Alaska. (Courtesy Alaska ShoreZone. Program materials available at http://alaskafisheries.noaa.gov/habitat/shorezone/szintro.htm)

Colour Figure 1 (Tsounis et al.)   Unusually large red coral colony of nearly 20 cm (above) and an average sized colony of 3 cm height (below). (Courtesy of Georgios Tsounis (above) and Sergio Rossi (below).)

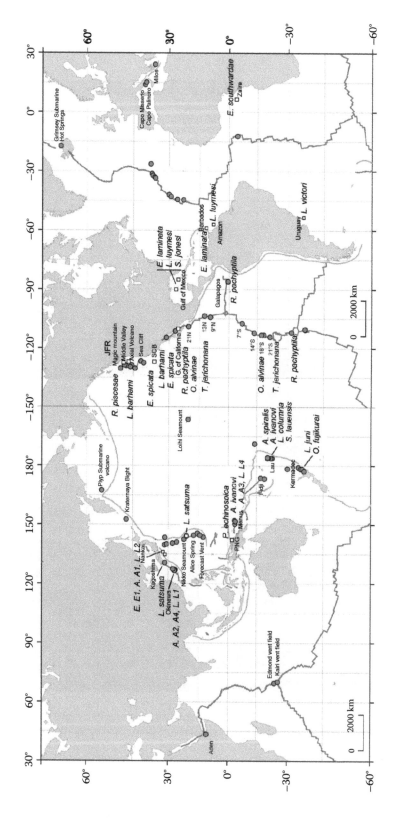

Colour Figure 3 (Bright & Lallier) Geographic localisation of hydrothermal vent and cold seep sites explored to date (2008) with indication of identified Vestimentifera species. Refer to Table 1 for complete species names and references. (Map modified from Desbruyères et al. 2006.)

Colour Figure 4 (Bright & Lallier)   The two contrasted ecological settings of Vestimentifera are illustrated by the most-studied species: a bush of *Riftia pachyptila* from the EPR at 12°50′N (© Ifremer-Hope 1999) and one of *Lamellibrachia luymesi* from the Gulf of Mexico (© C.R. Fisher). Diagrams on the right show (top) how vent species such as *Riftia pachyptila*, fixed to the hard rock substratum, get both oxygen and sulphide through their branchial plume from the mixed fluid and deliver them through circulation to the internally located but environmentally acquired bacteria (black triangle) and (bottom) in seep species such as *Lamellibrachia luymesi,* sulphide is acquired from the sediment through the tapering, buried tube and trunk 'roots'. BR = branchial plume; TR = trophosome.

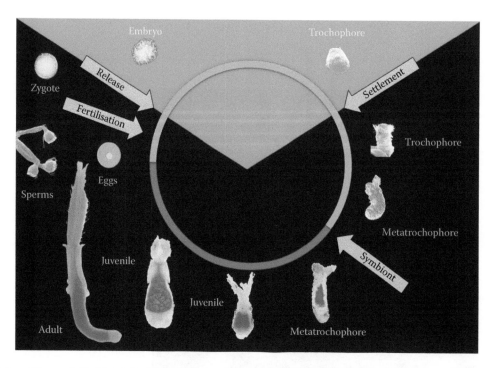

Colour Figure 5 (Bright & Lallier) Schematic life cycle of Vestimentifera. Adults with separate sexes produce sperm and eggs. Sperm bundles are released and taken up by females, in which internal fertilisation takes place, and aposymbiotic zygotes are released into the water column and disperse. Embryonic and larval development into a trochophore takes place in the pelagic environment. On settlement and further larval development into a metatrochophore, metamorphosis and uptake of symbionts from the environment are initiated. The symbiotic metatrochophore with a trophosome develops into a small juvenile in which the trophosome is present as a one-lobule stage. On growth, the trophosome expands into a multilobule stage, and animals become mature. Green = aposymbiotic life stages; red = symbiotic stages (photographs of sperm bundles from Marotta et al. 2005; zygote, embryo, and pelagic trochophore from Marsh et al. 2001; sessile trochophore from Gardiner & Jones 1994).

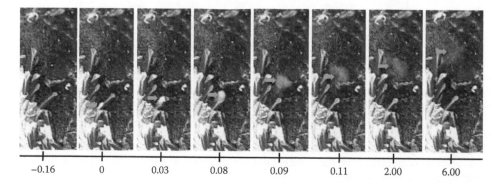

Colour Figure 6 (Bright & Lallier) Video sequence of the expulsion of a white cloud from one individual of *Riftia pachyptila* at the East Pacific Rise 9°50′N region, location Tica, in December 2003, Alvin dive 3948. Red arrows mark the location of the released product. Time is given in seconds as ± from the time of release (0).

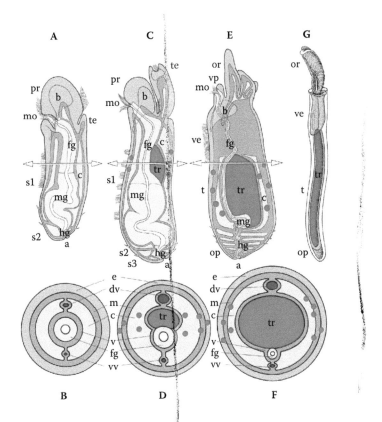

Colour Figure 7 (Bright & Lallier) Schematic drawing of life-history stages: (A) aposymbiotic metatrochophore with (B) corresponding cross section; (C) symbiotic metatrochophore with symbionts invading and (D) corresponding cross section; (E) symbiotic juvenile with trophosome in one-lobule stage and (F) corresponding cross section; (G) adult with trophosome in multilobule stage. Pink, symbiont-housing trophosome or symbionts (tr); blue, digestive system, mouth opening (mo), ventral process (vp), foregut (fg), midgut (mg), hindgut (hg) and anus (a); purple, blood vascular system with dorsal blood vessel (dv) and ventral blood vessel (vv). Body regions of larvae: pr = prototroch; s1–3 = chaetigers 1 to 3; and te = palps. Body regions of juvenile and adult: or = obturacular region; ve = vestimentum; t = trunk; op = opisthosome. Tissues: e = epidermis; m = muscles; c = coelom; v = visceral mesoderm. (From Nussbaumer et al. 2006.)

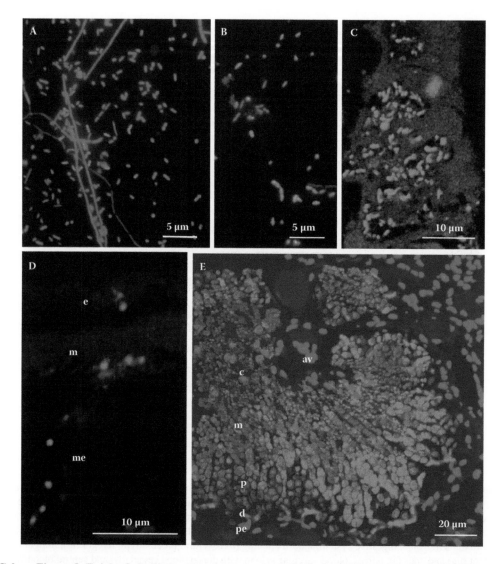

Colour Figure 8 (Bright & Lallier) Fluorescent *in situ* hybridisation with symbiont-specific (pink) and eubacterial (blue) probes. (A) Free-living bacterial community containing symbionts (pink) colonising on glass slide during 1-yr deployment (courtesy of A.D. Nussbaumer). (B) Free-living bacterial community containing symbionts (pink) on and in developing tube of metatrochophore (courtesy of A.D. Nussbaumer). (C) Symbionts (pink) in developing trophosome in metatrochophore. (D) Symbionts (pink) in epidermis (e), muscles (m), and undifferentiated mesoblastem (me). (E) Juvenile one-lobule stage with symbionts (pink) and host nuclei (blue): av = axial blood vessel; distinct zonation of central (c), median (m), peripheral (p) and few degrading bacteriocytes (d), peritoneum (pe) (C, D and E from Nussbaumer et al. 2006).